Enhancing Extraction Processes in the Food Industry

Contemporary Food Engineering

Series Editor

Professor Da-Wen Sun, Director

Food Refrigeration & Computerized Food Technology
National University of Ireland, Dublin
(University College Dublin)
Dublin, Ireland
http://www.ucd.ie/sun/

Contemporary Food
Engineering Series
Da-Wen Sun, Series Editor

Enhancing Extraction Processes in the Food Industry

Edited by
Nikolai Lebovka
Eugene Vorobiev
Farid Chemat

CRC Press
Taylor & Francis Group
Boca Raton   London   New York

CRC Press is an imprint of the
Taylor & Francis Group, an **informa** business

CRC Press
Taylor & Francis Group
6000 Broken Sound Parkway NW, Suite 300
Boca Raton, FL 33487-2742

First issued in paperback 2016

© 2012 by Taylor & Francis Group, LLC
CRC Press is an imprint of Taylor & Francis Group, an Informa business

No claim to original U.S. Government works

ISBN 13: 978-1-138-19933-0 (pbk)
ISBN 13: 978-1-4398-4593-6 (hbk)

Library of Congress Cataloging-in-Publication Data

Enhancing extraction processes in the food industry / editors, Nikolai Lebovka, Eugene Vorobiev, and Farid Chemat.
    p. cm. -- (Contemporary food engineering)
    Includes bibliographical references and index.
    ISBN 978-1-4398-4593-6 (hardback)
    1. Extraction (Chemistry) 2. Plant extracts. 3. Food industry and trade. I. Lebovka, N. I. (Nikolai Ivanovich) II. Vorobiev, Eugène. III. Chémat, Farid.

TP156.E8E65 2012
664--dc23   2011022667

Visit the Taylor & Francis Web site at
http://www.taylorandfrancis.com

and the CRC Press Web site at
http://www.crcpress.com

Contents

List of Figures

List of Tables

Series Preface

Food engineering is the multidisciplinary field of applied physical sciences combined with the knowledge of product properties. Food engineers provide the technological knowledge transfer essential to the cost-effective production and commercialization of food products and services. In particular, food engineers develop and design processes and equipment in order to convert raw agricultural materials and ingredients into safe, convenient, and nutritious consumer food products. However, food engineering topics are continuously undergoing changes to meet diverse consumer demands, and the subject is being rapidly developed to reflect market needs.

In the development of food engineering, one of the many challenges is to employ modern tools and knowledge, such as computational materials science and nanotechnology, to develop new products and processes. Simultaneously, improving quality, safety, and security remains a critical issue in the study of food engineering. New packaging materials and techniques are being developed to provide more protection to foods, and novel preservation technologies are emerging to enhance food security and defense. Additionally, process control and automation regularly appear among the top priorities identified in food engineering. Advanced monitoring and control systems are developed to facilitate automation and flexible food manufacturing. Furthermore, energy savings and minimization of environmental problems continue to be important issues in food engineering, and significant progress is being made in waste management, efficient utilization of energy, and reduction of effluents and emissions in food production.

The *Contemporary Food Engineering* book series, which consists of edited books, attempts to address some of the recent developments in food engineering. Advances in classical unit operations in engineering related to food manufacturing are covered as well as such topics as progress in the transport and storage of liquid and solid foods; heating, chilling, and freezing of foods; mass transfer in foods; chemical and biochemical aspects of food engineering and the use of kinetic analysis; dehydration, thermal processing, nonthermal processing, extrusion, liquid food concentration, membrane processes, and applications of membranes in food processing; shelf-life, electronic indicators in inventory management, and sustainable technologies in food processing; and packaging, cleaning, and sanitation. These books are aimed at professional food scientists, academics researching food engineering problems, and graduate-level students.

The editors of these books are leading engineers and scientists from all parts of the world. All of them were asked to present their books in such a manner as to address the market needs and pinpoint the cutting-edge technologies in food engineering. Furthermore, all contributions are written by internationally renowned experts who have both academic and professional credentials. All authors have attempted to provide critical, comprehensive, and readily accessible information on the art and science of a relevant topic in each chapter, with reference lists for further information. Therefore, each book can serve as an essential reference source to students and researchers in universities and research institutions.

Da-Wen Sun
Series Editor

Preface

Extraction has been used probably since the discovery of fire. Egyptians and Phoenicians, Jews and Arabs, Indians and Chinese, Greeks and Romans, and even Mayans and Aztecs all utilized innovative extraction and distillation for processing of perfumes or food. Nowadays, we cannot find a production line in the food industry that does not use extraction processes (e.g., maceration, solvent extraction, steam distillation or hydrodistillation, cold pressing, squeezing, etc.). With the increasing energy costs and the drive to reduce carbon dioxide emissions, food industries are under a challenge to find new technologies in order to reduce energy consumption, to meet legal requirements on emissions, product/process safety and control, and for cost reduction and increased quality as well as functionality. For example, existing extraction technologies have considerable technological and scientific bottlenecks to overcome, often requiring up to 50% of investments in a new plant and more than 70% of total process energy used in food industries. These shortcomings have led to the consideration of the use of enhanced extraction techniques, which typically require less solvent and energy, such as microwave extraction, supercritical fluid extraction, ultrasound extraction, flash distillation, and controlled pressure drop process.

Although there are a number of books that explain the innovative unit operations in food technology and describe how to conduct conventional extraction, there are few books that focus on understanding the actual instruments used in innovative and enhanced extraction. This book was prepared by a team of chemists, biochemists, chemical engineers, physicians, and food technologists who have extensive personal experience in the research of innovative extraction techniques at the laboratory and industrial scales. The book provides valuable information about the newly developed processes and methods for extraction.

The book comprises a preface, a contributors list, and 16 chapters, which take the reader through accessible descriptions of enhanced extraction techniques and their applications in food laboratory and industry. The book is addressed primarily to science graduate students, chemists, and biochemists in industry and food quality control, as well as researchers and persons who participate in continuing education and research systems.

Nikolai I. Lebovka
Eugene Vorobiev
Farid Chemat

MATLAB® is a registered trademark of The MathWorks, Inc. For product information, please contact:

The MathWorks, Inc.
3 Apple Hill Drive
Natick, MA 01760-2098 USA
Tel: 508 647 7000
Fax: 508-647-7001
E-mail: info@mathworks.com
Web: www.mathworks.com

Acknowledgments

We thank all the authors who have collaborated in the writing of this book. Particular thanks are due to the Editor-in-Chief of the book series *Contemporary Food Engineering*, Member of the Royal Irish Academy, Professor Da-Wen Sun for his kind advice and help during the preparation of this book.

Series Editor

Born in southern China, Professor Da-Wen Sun is a world authority in food engineering research and education; he is a member of the Royal Irish Academy, which is the highest academic honor. His main research activities include cooling, drying, and refrigeration processes and systems; quality and safety of food products; bioprocess simulation and optimization; and computer vision technology. His innovative studies on vacuum cooling of cooked meats, pizza quality inspection by computer vision and edible films for shelf-life extension of fruits and vegetables have especially been widely reported in national and international media. The results of his work have especially been published in over 500 papers, including about 250 peer-reviewed journal papers. He has also edited 12 authoritative books. According to Thomson Scientific's Essential Science Indicator[SM], updated as of July 1, 2010, based on data derived over a period of 10 years and 4 months (January 1, 2000, to April 30, 2010) from the ISI Web of Science, a total of 2,554 scientists are among the top 1% of the most frequently cited scientists in the category of Agricultural Sciences, and professor Sun tops the list with his ranking of 31.

Sun received his BSc honors (first class), his MSc in mechanical engineering, and his PhD in chemical engineering in China before working in various universities in Europe. He became the first Chinese national to be permanently employed in an Irish university when he was appointed as college lecturer at the National University of Ireland, Dublin (University College Dublin), in 1995, and was then continuously promoted in the shortest possible time to senior lecturer, associate professor, and full professor. He is currently the professor of food and biosystems engineering and the director of the Food Refrigeration and Computerized Food Technology Research Group at the University College Dublin (UCD).

Sun has contributed significantly to the field of food engineering as a leading educator in this field. He has trained many PhD students who have made their own contributions to the industry and academia. He has also regularly given lectures on advances in food engineering in international academic institutions and delivered keynote speeches at international conferences. As a recognized authority in food engineering, he has been conferred adjunct/visiting/consulting professorships from over 10 top universities in China, including Zhejiang University, Shanghai Jiaotong University, Harbin Institute of Technology, China Agricultural University, South China University of Technology, and Jiangnan University. In recognition of his significant contributions to food engineering worldwide and for his outstanding leadership in this field, the International Commission of Agricultural and Biosystems Engineering (CIGR) awarded him the CIGR Merit Award in 2000 and again in 2006. The Institution of Mechanical Engineers (IMechE) based in the United Kingdom named him Food Engineer of the Year in 2004. In 2008, he was awarded the CIGR Recognition Award in honor of his distinguished achievements in the top 1% of agricultural engineering scientists in the world. In 2007, he was presented with the AFST(I) Fellow Award by the Association of Food Scientists and Technologists (India), and in 2010 he was presented with the CIGR Fellow Award. The title of Fellow is the highest honor in CIGR and is conferred upon individuals who have made sustained, outstanding contributions worldwide.

Sun is a Fellow of the Institution of Agricultural Engineers and a Fellow of Engineers Ireland (the Institution of Engineers of Ireland). He has received numerous awards for teaching and research excellence, including the President's Research Fellowship and the President's Research Award of University College Dublin on two occasions. He is the editor-in-chief of *Food and Bioprocess Technology—An International Journal* (Springer) (2010 Impact Factor = 3.576, ranked at the 4th position among 126 food science and technology journals); the former editor of *Journal of Food Engineering* (Elsevier); and an editorial board member for *Journal of Food Engineering* (Elsevier), *Journal of Food Process Engineering* (Blackwell), *Sensing and Instrumentation for Food Quality and Safety* (Springer), and *Czech Journal of Food Sciences*. He is a chartered engineer.

On May 28, 2010, he was awarded membership of the Royal Irish Academy (RIA), which is the highest honor that can be attained by scholars and scientists working in Ireland, and at the 51st CIGR General Assembly held during the CIGR World Congress in Quebec City, Canada on June 13–17, 2010, he was elected incoming president of CIGR, and will become CIGR President in 2013–2014—the term of his CIGR presidency is six years, two years each for serving as incoming president, president, and past president.

Editors

Nikolai I. Lebovka was born in Kiev, Ukraine, in 1954. He received his PhD in molecular physics from Taras Shevchenko National University of Kyiv (1986) and Dr. Habil in physics of colloids from the Biocolloid Chemistry Institute, Ukraine (1995). He is currently head of the Physical Chemistry Department of the Biocolloid Chemistry Institute and professor of physics at Taras Shevchenko National University of Kiev. He studies electric field–induced effects in biological and food materials and is also active in the fields of colloids and biocolloids, theory and applications of nanocomposites, computation physics, and theory and practice of percolation phenomena. He has published more than 230 papers in peer-reviewed journals and several chapters in books, and was a member of the organizing committee of several international conferences.

Eugene Vorobiev is a full professor at the Chemical Engineering Department and head of Laboratory for Agro-Industrial Technologies at the Université de Technologie de Compiègne (UTC), France. He received his PhD in Food Engineering (1980, Ukraine) and his Dr. Habil in Chemical Engineering (1997, France). His main research interests are focused on mass transfer phenomena, theory and practice of solid–liquid separation, and innovative food technologies (especially electrotechnologies). He has published more than 200 peer-reviewed papers and is the author of 18 patents. He is a member of the editorial board of several journals (*Separation and Purification Technology*, *Food Engineering Reviews*, *Filtration*) and president of the Scientific Council of IFTS ("Institut de la Filtration et des Techniques Séparatives"). He was awarded the Gold Medal of the Filtration Society (2001) and is a Laureate of the Price for the innovative technique for the environment (Ademe, 2008). He acted as a chairman of several international conferences.

Farid Chemat is a full professor of chemistry and director of the Laboratory for Green Extraction Techniques of Natural Products (GREEN) at the Université d'Avignon et des Pays de Vaucluse, France. Born in Blida (1968), he received his PhD (1994) in innovative process engineering from the Institut National Polytechnique de Toulouse. His main research interests are focused on innovative and sustainable extraction techniques (especially microwave, ultrasound, and green solvents) for food, pharmaceutical, and cosmetic applications. His research activities are documented by more than 100 scientific peer-reviewed papers and 6 patents. He is coordinator of a new group named "France Eco-Extraction," which deals with international dissemination of research and education on green extraction technologies.

Contributors

Karim Salim Allaf
Department of Process Engineering
University of La Rochelle
La Rochelle, France

Tamara Sabrine Vicenta Allaf
Department of Process Engineering
ABCAR-DIC Process
La Rochelle, France

Abdellah Arhaliass
GEPEA-UMR CNRS
University of Nantes
Saint-Nazaire, France

Ramesh Y. Avula
Research and Development
Cherry Central Inc.
Traverse City, Michigan

Baya Berka
Department of Process Engineering
University of La Rochelle
La Rochelle, France

Régis Baron
BRM Department
IFREMER
Nantes, France

Colette Besombes
Department of Process Engineering
University of La Rochelle
La Rochelle, France

Nadia Boussetta
Chemical Engineering Department
University of Technology of Compiègne
Compiègne, France

Kerry Alan Campbell
Department of Chemical and Biological
  Engineering
Iowa State University
Ames, Iowa

Farid Chemat
UMR, INRA-UAPV
Université d'Avignon et des Pays du
  Vaucluse
Avignon, France

Giancarlo Cravotto
Dipartimento di Scienza e Tecnologia
  del Farmaco
Università di Torino
Torino, Italy

Zbigniew J. Dolatowski
Department of Meat Technology and
  Food Quality
University of Life Sciences in Lublin
Lublin, Poland

André B. de Haan
Department of Chemical Engineering
  and Chemistry
Eindhoven University of Technology
Eindhoven, the Netherlands

H. Umesh Hebbar
Department of Food Engineering
Central Food Technological Research
  Institute
Mysore, India

A. B. Hemavathi
Department of Food Engineering
Central Food Technological Research
  Institute
Mysore, India

Elena Ibañez
Bioactivity and Food Analysis
    Department
Food Research Institute (CIAL-CSIC)
Madrid, Spain

Amin Ismail
Department of Nutrition and Dietetics
Universiti Putra Malaysia
Selangor, Malaysia

Yue Ming Jiang
South China Botanical Garden
Chinese Academy of Sciences
Guangzhou, People's Republic of China

Lawrence Johnson
Department of Food Science and
    Human Nutrition
Iowa State University
Ames, Iowa

Stephanie Jung
Department of Food Science and
    Human Nutrition
Iowa State University
Ames, Iowa

Raymond Kaas
BRM Department
IFREMER
Nantes, France

Magdalena Kristiawan
Department of Process Engineering
University of La Rochelle
La Rochelle, France

M. C. Lakshmi
Department of Food Engineering
Central Food Technological Research
    Institute
Mysore, India

Jean-Louis Lanoisellé
Chemical Engineering Department
University of Technology of Compiègne
Compiègne, France

Nikolai I. Lebovka
Department of Physical Chemistry of
    Disperse Minerals
National Academy of Sciences of
    Ukraine
Kiev, Ukraine

Jack Legrand
GEPEA-UMR CNRS
University of Nantes
Saint-Nazaire, France

**Juliana Maria Leite Nóbrega de
Moura**
Department of Food Science
Iowa State University
Ames, Iowa

Philip J. Lloyd
Energy Institute
Cape Peninsula University of
    Technology
Cape Town, South Africa

Igor O. Lomovsky
Department of Solid State Chemistry
Siberian Branch of the Russian
    Academy of Science
Novosibirsk, Russia

Oleg I. Lomovsky
Department of Solid State Chemistry
Siberian Branch of the Russian
    Academy of Science
Novosibirsk, Russia

María Dolores Luque de Castro
Department of Analytical Chemistry
University of Córdoba
Córdoba, Spain

M. C. Madhusudhan
Department of Food Engineering
Central Food Technological Research
    Institute
Mysore, India

Krishna Murthy Nagendra Prasad
Department of Nutrition and Dietetics
Universiti Putra Malaysia
Selangor, Malaysia

Feliciano Priego-Capote
Department of Analytical Chemistry
University of Córdoba
Córdoba, Spain

Karumanchi S. M. S. Raghavarao
Department of Food Engineering
Central Food Technological Research
    Institute
Mysore, India

Thierry Reess
Department of Electrical Engineering
University of Pau
Pau, France

John Shi
Food Research Center
Agriculture and Agri-Food Canada
Guelph, Ontario, Canada

Rakesh K. Singh
Department of Food Science and
    Technology
University of Georgia
Athens, Georgia

Vaclav Sobolik
Department of Process Engineering
University of La Rochelle
La Rochelle, France

Dariusz M. Stasiak
Department of Meat Technology and
    Food Quality
University of Life Sciences in Lublin
Lublin, Poland

Charlotta Turner
Department of Chemistry
Lund University
Lund, Sweden

Jessy van Wyk
Department of Food Technology
Cape Peninsula University of
    Technology
Bellville, South Africa

Peggy Vauchel
ProBioGEM Laboratory
Lille 1 University
Villeneuve d'Ascq, France

Eugene Vorobiev
Département de Génie des Procédés
    Industriels
Université de Technologie de
    Compiègne
Compiègne, France

Paul Willems
Faculty of Science and Technology
University of Twente
Enschede, the Netherlands

Main Abbreviations

AC	alternative current
AEP	aqueous extraction processing
AMF	anhydrous milk fat
AOT	sodium bis(2-ethyl-1-hexyl) sulfosuccinate
AP	affinity partitioning
ARMES	affinity-based reverse micellar extraction and separation
ATPE	aqueous two-phase extraction
ATPS	aqueous two-phase system
BAS	biologically active substance
BDBAC	N-benzyl-N-dodecyl-N-bis(2-hydroxy ethyl) ammonium chloride
BE	back extraction
BHT	butylated Hydroxytoluene
BO	butter oil
BPR	back pressure regulator
BSA	bovine serum albumin
CAEP	cellulase/pectinase-assisted aqueous extraction processing
CAN-BD	carbon dioxide assisted nebulization with a bubble dryer
CCD	countercurrent distribution
CDAB	cetyldimethylammonium bromide
CE	capillary electrophoresis
CE	conventional extraction
CED	cohesion energy density
CEL	conventional assisted extract of longan
CE-MS	capillary electrophoresis coupled to mass spectrometry
CLA	conjugated linoleic
CMC	critical micellar concentration
CTAB	cetyltrimethyl ammonium bromide
DC	direct current
DDGS	dried distiller's grains with solubles
DH	degrees of protein hydrolysis
DHA	docosahexaenoic acid
DIC	instant controlled pressure drop technology (from the French, détente instantanée contrôlée)
DMSO	dimethyl sulfoxide
DSP	downstream processing
DTAB	dodecyltrimethylammonium bromide
DTDPA	di(tridecyl) phosphoric acid

EAEP	enzyme-assisted aqueous extraction processing
ECD	electrochemical detection
EDTA	ethylenediaminetetraacetic acid
EE	extraction efficiency
EHS	environmental, health, and safety
EMASE	focused microwave-assisted Soxhlet extraction
EPA	eicosapentaenoic acid
FE	forward extraction
FFA	free fatty acid
FID	flame ionization detector
FMASE	focused microwave assisted Soxhlet extraction
GAME	gas-assisted mechanical expression
GC	gas chromatography
HD	hydrodistillation
HLB	hydrophilic–lipophilic balance
HLW	hot liquid water
HPE	high pressure–assisted extraction
HPEL	high pressure–assisted extract of longan
HPH	high-pressure homogenization
HPLC	high-performance liquid chromatography
HRP	horseradish peroxidase
IEP	isoelectric precipitate
IRR	internal rate of return
LCA	life cycle assessment
LDH	lactate dehydrogenase
LFP	litchi fruit pericarp
LLE	liquid–liquid extraction
LPO	lactoperoxidase
MAE	microwave-assisted extraction
MAEE	microwave-assisted enzymatic extraction
MAHD	microwave-assisted hydrodistillation
MAP	microwave-assisted process
MHG	microwave hydrodiffusion and gravity
MIR	middle infrared spectrometry
MIS	microwave integrated Soxhlet
MMLLE	microporous membrane liquid–liquid extraction
MRM	mixed reverse micellar
MS	mass spectrometry
MTBE	methyl tert-butyl ether
MW	microwave
NaDEHP	sodium di-2-ethyl hexyl phosphate

NP	nonylphenol polyethoxylate
PAED	pulsed arc electrohydraulic discharges
PAEP	protease-assisted aqueous extraction processing
PCBs	polychlorinated biphenyls
PCED	pulsed corona electrohydraulic discharges
PED	pulsed electrical discharges
PEF	pulsed electric field
PEG	polyethylene glycol
PFE	pressurized fluid extraction
PHWE	pressurized hot water extraction
PLE	pressurized liquid extraction
PTFE	polytetrafluoroethylene
RI	retention index
RM	reverse micelle
RME	reverse micellar extraction
RNA	ribonucleic acid
RPLC–GC	reversed-phase liquid chromatography–GC
RSM	response surface methodology
RTD	residence time distribution
SBM	soybean meal
SC	supercritical
SCE	supercritical carbon dioxide extraction
SCFE	supercritical fluid extraction (SCFE)
SD	steam distillation
SDS	sodium dodecyl sulfate
SFE	supercritical fluid extraction
SFME	solvent-free microwave extraction or hydrodistillation
SHP	soy hull peroxidase
SIDMS	speciated isotope dilution mass spectrometry
SLR	solids-to-liquid ratio
SME	specific mechanical energy
SPC	soy protein concentrate
SPE	solid phase extraction
SPI	soy protein isolate
SPME	solid phase microextraction
TMA-PEG	trimethylamino-PEG
TOMAC	trioctylmethyl ammonium chloride
TPG	total pressure gradient
TSMF	two-stage membrane filtration
UAE	ultrasound assisted extraction
UAED	ultrasonic-assisted enzymatic digestion

UAEE	ultrasonic-assisted enzymatic extraction
UF	ultrafiltration
UHPE	ultrahigh pressure–assisted extractions
UMAE	combination of UAE and MAE
US	ultrasound
WEPO	water extraction with particle formation on-line

1 Introduction to Extraction in Food Processing

Philip J. Lloyd and Jessy van Wyk

CONTENTS

1.1   WHAT THIS CHAPTER IS ABOUT

This chapter is strictly introductory. It aims to provide an overview of solvent extraction technology in general, so that the reader can place in context the detailed topics in subsequent chapters. It contains essentially no new information, so the reader will look in vain for detailed references to most of the issues discussed. Much can be found in standard chemical engineering texts. Texts such as Rydberg et al. (2004) or the earlier Lo et al. (1983) handbook provide much depth about the technology, but nothing about its application in food processing. Schügerl's (1994) monograph has some very relevant material, although its focus is definitely on biotechnology rather than food technology. A recent encyclopedic review of food technology (Campbell-Platt 2009) devotes a scant two pages to the topic of solvent extraction.

1.2   WHAT IS MEANT BY EXTRACTION

One of the oldest recorded methods of separation is solvent extraction, which dates back to the Palaeolithic age (Herrero et al. 2010). In food processing, extraction is defined as the transfer of one or more components of a biological feed from its source material into a fluid phase, followed by separation of the fluid phase and recovery of the component(s) from the fluid. The feed is usually of plant origin, but the principles of extraction remain the same if the material is animal or piscine in origin.

Extraction is a process that is growing in importance. It is generally more energy efficient than competitive processes such as expression—the pressing of biological feed materials to liberate fluids. For example, sugar is extracted from sugar beets with hot water, which yields a sucrose stream free of contaminants and of higher concentration (typically 15% sugar) than can be achieved by expression. Solvent extraction can be made selective for specific components of the feed. For instance, supercritical carbon dioxide ($SC-CO_2$) will selectively dissolve caffeine from coffee beans to yield decaffeinated coffee. The extracted caffeine can then be recovered for sale as a pharmaceutical. Extraction can recover thermally labile components that would be degraded by heating, such as gelatin from collagen. Table 1.1 gives some examples of typical extraction processes employed industrially.

The intent of this chapter is to give an overview of the broad principles underlying extraction, to provide a basis for understanding the rationale behind some of the technical advances described in later chapters.

TABLE 1.1

Some Examples of Industrial Extraction Processes

Solvent	Feed	Product	Component
Water	Apple pulp	Apple juice	–
	Malted barley	Brewing worts	Sugars, grain solutes
	Kelp	Carrageenan	–
	Manioc	Cassava	Cyanogenetic glycosides
	Citrus press residues	Citrus molasses	–
	Papaya latex	Papain	Papain
	Rosemary leaves	Rosemary essential oil	Rosemary essential oil
	Citrus peel	Citrus essential oils	Citrus essential oils
Acidic water	Collagen	Gelatin	Gelatin
	Citrus peel	Pectin	Pectin
	Hog stomach	Pepsin	Pepsin
Alkaline water	Defatted soy flour	Soya protein	–
Aqueous ethanol	Red beets	Betalains	Betalains
	Animal pancreas	Insulin	Insulin
	Spices	Spice extracts	–
	Vanilla beans	Vanilla essence	–
Methylene chloride	Green coffee beans	Decaffeinated coffee	Caffeine
Supercritical CO_2	Green coffee beans	Decaffeinated coffee	Caffeine
	Hops	Hops extract (resin)	Hops essential oils (myrcene, humulene, caryophyllene, and farnesene), alpha and beta acids
	Ginger rhizomes	Ginger extract	Gingerols
	Pomegranate seeds	Pomegranate seed oil	Pomegranate seed oil
	Vanilla beans	Vanilla essence	
	Spices (turmeric, nutmeg, mace, cardamom, etc.)	Spice extracts	
	Egg yolk	Decholesterolized egg yolks	Cholesterol
	Wheat germ	Wheat germ oils rich in tocopherols	–
Hexane	Soybeans	Soybean oil	–
Methyl ethyl ketone	Spices	Spice oleoresins	–
Tributyl phosphate	Phosphoric acid	Food-grade phosphoric acid	–

Source: Schwarztberg, H.G., in *Handbook of Separation Process Technology (Chapter 10)*, R.W. Rousseau (Ed.), New York: Wiley-Interscience. ISBN: 0471 89558X, 1987.

1.3   PHYSICAL PRINCIPLES OF EXTRACTION

1.3.1   NOMENCLATURE OF EXTRACTION

A component that it is desired to be removed from the feed through extraction is called the "solute." The phase that is mixed with the feed to remove the solute is the "solvent." After the solvent has been mixed with the feed and the solute has transferred from the feed phase into the solvent phase, the solvent phase is called the "extract" and the feed phase is now called the "raffinate." It must be stressed that, in food processing, the feed is usually solid, semisolid, or gel-like, whereas much of the science of extraction is based on liquid feeds. However, there are very close parallels provided allowance is made for the impact of the nature of the feed on mass transfer properties, as further discussed in Section 1.3.3. Indeed, much of the rest of this text is concerned with means of improving the rate of mass transfer so that the science derived from liquid feeds may be better applied to the processing of food products.

1.3.2   SOLUBILITY

When a feed containing a solute is contacted with a solvent in which the solute is reasonably soluble, then the solute will distribute itself between the feed and the solvent until there is equilibrium between the feed and the solvent phases. When this occurs, the chemical potential of the solute in each phase is the same. The chemical potential is made up of two terms—the concentration of the solute and its activity in the phase concerned. However, in processing foods, it is rarely possible to measure the activity of the solute in the feed; thus, the primary concern is with the solubility of the solute in the solvent. Figure 1.1, for instance, shows the solubility of caffeine in SC-CO_2 and in SC-CO_2–ethanol mixtures. The solubility increases with pressure and with the addition of ethanol to the solvent, but decreases with temperature.

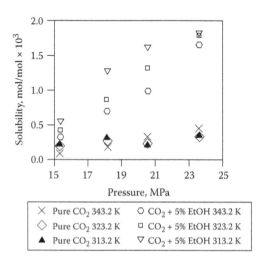

FIGURE 1.1   Solubility of caffeine in SC-CO_2 and CO_2–ethanol. (From Kopcak, U. and Mohamed, R.S., *J Supercrit Fluids* 34, 209, 2005. With permission.)

Determination of solubility is covered in many standard texts. Today, there are powerful packages for estimating chemical properties, which have a wide range of experimental data and strong theoretical bases to permit almost any solubility to be estimated. A recent paper (Kumhom et al. 2010) gives an introduction to these methods, for biochemical solids dissolved in mixtures such as the SC-CO_2–ethanol mixtures illustrated in Figure 1.1. A typical software package is Aspen Properties, part of the AspenTech suite (AspenTech 2010), while the U.S. Environmental Protection Agency's SPARC suite is a free resource (U.S. EPA 2010). Because these methods now find widespread use, there is little purpose in an extended discussion of solubility in a text such as this.

It is, however, useful to understand how solubility can be represented graphically. Figure 1.2 shows the triangular coordinates used to represent solubility. This is an equilateral triangle, with the concentrations of each component along each side as shown.

The composition at point M is then 50% solute, 20% solvent, and 30% feed. A useful feature of such diagrams is that mixtures can be represented by straight lines, and the result of mixing two streams of different composition is determined by simple geometry. Consider two streams of masses D and E and with compositions given by points D and E. Then the result of mixing these two streams will give a composition that lies on the straight line DE and has a composition given by point F such that $E/D = DF/FE$. This feature will be used in much of the discussion that follows.

The equilibrium between the various phases can readily be shown on triangular coordinates. Figure 1.3 illustrates this.

The line WXY represents the boundary between a single-phase and a two-phase region. Above the curve, the presence of the solute allows the feed and solvent phases to dissolve in each other, which is of course undesirable from the point of view of separation. The difference between the single-phase and the two-phase region is critically important in extraction because it is the existence of a second phase that

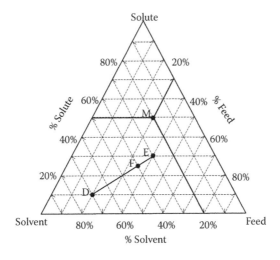

FIGURE 1.2 Use of triangular coordinates.

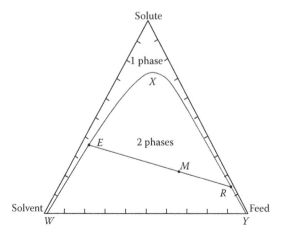

FIGURE 1.3 Solubility representation in a ternary diagram.

permits extraction to take place. The challenge in selecting a solvent is often to find one that will give as large a two-phase region as possible. Ideally, the feed and solvent will be essentially immiscible, and the presence of the solute will not change this immiscibility. However, in many biological systems, it is difficult to reach this ideal state. It then becomes necessary to allow for the effect of a single-phase region and operate so that its influence is minimized.

Because the side of the triangle between the solvent and feed corners represents the line of zero solute, then point W represents the solubility of the feed in the solvent and point Y represents the solubility of the solvent in the feed. Line WXY represents the boundary of the two-phase region.

Consider mixing some solvent with a feed, such that the average composition of the mixture is given by point M in the diagram. Then, provided M is within the two-phase region, the mixture will split and give two phases whose compositions will be given by points E and R on line WXY. These represent the concentrations in the extract and raffinate, respectively. Line ER is called a "tie line," and it joins the composition of two phases at equilibrium. There is a whole series of such tie lines, depending on the starting conditions. EMR is a straight line, as indicated previously, and the mass ratio of solvent to feed is given by the ratio of the lengths EM/MR.

These diagrams refer to conditions at a constant temperature, and the equilibrium line WXY is often referred to as an "isotherm" for this reason.

1.3.3 Mass Transfer

During the process of extraction, one or more compounds ("solutes") transfer from the biological feed material into the solvent. The physical process underlying the transfer is that the concentration* of the solute in the solvent is less than its

---

* Strictly speaking, concentration is an approximation to the "chemical potential" of the solute in the feed or solvent.

concentration in the feed, so that the solute diffuses from the feed into the solvent. However, the diffusion process is hindered by a number of phenomena.

First, there will be some interface between the feed and the solvent. The feed may also be liquid, but the interface between two liquids will slow the diffusion. Consider two liquid phases fully mixed in the bulk of the phases. Thus, the concentration in the bulk of each phase will be the same everywhere in that phase, except close to the interface where diffusion is occurring. This is illustrated in Figure 1.4.

On the feed side, there is a concentration difference between the bulk concentration and the concentration at the feed side of the interface, and it is in this non–fully mixed zone where diffusion takes place. The width of the diffusion zone may be less than a millimeter. Similarly, on the solvent side, there is a concentration difference between the bulk concentration and the concentration at the solvent side of the interface, and diffusion occurs across this narrow layer. At the interface itself, there is a drop in concentration that reflects the difference in chemical potential on either side of the interface.

Thus, three physical layers resist the transfer of mass from the feed to the solvent:

- The diffusion layer on the feed side
- The resistance to transfer of the interface itself
- The diffusion layer on the solvent side

In the case of food processing, the last of these is of little significance. The solvent is generally a liquid of relatively low viscosity, which means that it can relatively simply be fully mixed and the thickness of the boundary diffusion layer can be reduced to a minimum. The challenge in applying extraction to food processing is to minimize the first two of the resistances—that in the feed phase and that at the interface.

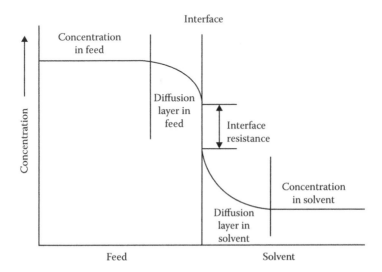

FIGURE 1.4 Transfer of solute between two liquid phases.

In food processing, the feed phase is generally not a low-viscosity fluid. It may be semifluid or gel-like. It may be semifluid contained within cellular structures. It may be quite solid. Whatever its state, it will resist mass transfer to a far greater extent than a low-viscosity fluid. Much of the rest of this book is devoted to ways and means of increasing the rate of mass transfer from real-life feeds, and doing so in such a way as not to change the properties of the desired solute and not to enhance the extraction of additional solutes that might detract from the properties of the desired solute.

Similarly, in food processing, the interface is generally not the simple interface between two low-viscosity fluids. It may, as we have seen above, be the interface between a solvent and a semisolid or even a solid. It may be attractive to surface-active substances present in the feed, which thicken the interface and thus increase its resistance. It may be a cell wall designed by nature specifically to resist the release of the desired solute. It may even be a solid and even, in some cases, a crystalline solid. Methods for coping with all these forms of resistance to mass transfer are described in later chapters.

The reason for being concerned about mass transfer is that the slower the rate of mass transfer, the longer the feed and solvent must be in contact. This means, other things being equal, that the longer the two are in contact, the larger must be the equipment in which the contact takes place—and larger equipment is inherently more expensive than smaller equipment. Also, the longer the two are in contact, the greater the chance of the solvent dissolving other solutes from the feed, or for the desired solute to be degraded by temperature or exposure to the atmosphere—or even by reaction with the solvent.

1.3.4 DIFFUSION

The rate of diffusion of a single species in a single fluid can be described by Fick's law

$$J_A = -cD_{AB} \frac{dx_A}{dz} \tag{1.1}$$

where J_A is the rate of diffusion (mol m^{-2} s^{-1}), c is the concentration (mol m^{-3}), D_{AB} is the diffusion coefficient of solute A in solvent B (m^2 s^{-1}), x_A is the mole fraction of A in B, and z is the direction of diffusion.

What this makes clear is that the greater the area over which diffusion can take place, then the greater will be the rate of mass transfer. Thus, when extracting from a liquid feed, the solvent and the feed are agitated together to increase the surface area between the two to maximize the rate of transfer. Similarly, when extracting solute from a solid feed, the solid should be reduced in size as far as possible to maximize the area through which mass transfer can take place (Pronyk and Mazza 2009).

There are, of course, some constraints on this maximization of surface area to enhance the rate of extraction. With two liquids, for instance, it is possible to mix them so intimately that one emulsifies in the other, so that separation of the solvent after extraction becomes difficult. With a solid feed, the very process of size reduction may be so energy intensive that labile substances are altered. While as large an area as possible can enhance the rate of extraction of a desired solute, it may also make the rate of extraction of a less-desired solute sufficiently high that the desired solute is unacceptably contaminated. Nevertheless, the general rule holds good—a large surface area will speed the rate of extraction.

1.3.5 CHOICE OF SOLVENT

The solvent should, naturally, be capable of dissolving the desired solute. It is useful if the solubility of the solute in the solvent is high because this will reduce the quantity of solvent needed to extract a given quantity of solute. However, this is not an essential requirement. Other factors guide the choice of solvent. For instance, in many cases, it is desirable to ensure maximum selectivity—that is, that the solvent dissolves the desired solute preferentially to other potentially soluble materials present in the feed. Water is generally nonselective, but in some cases, as Table 1.1 illustrates, even it can be sufficiently selective given the right feed and solute.

An important requirement is that the solvent should be reasonably stable and that it should not react chemically with the solute in such a way as to adversely affect the properties of the solute. There are classes of solvents that are inherently acidic or basic, and they can be used to extract anionic or cationic solutes, respectively. Contact with an acidic or basic solution will then recover the solute and regenerate an acidic or basic solvent. In these cases, there is a chemical reaction between the solvent and the solute, but it is employed beneficially and does not adversely affect the properties of the solute.

A further requirement is that the solute should be reasonably readily recovered from the extract (i.e., the solution after the extraction process) in those cases where the desired product is the solute. As Figure 1.1 indicates, merely reducing the pressure suffices to lower the solubility of caffeine in SC-CO_2. If the extract is saturated at high pressure, then the pressure can be reduced, caffeine will crystallize from the solution, and the depleted solvent can then be repressurized for reuse. In some cases the solute is not thermally labile, and the solvent can be recovered and separated from the solute by distillation. A further possibility is that the recovery of the solvent can take place through extractive distillation. For example, in the recovery of acetic acid from dilute aqueous solutions with methyl *tert*-butyl ether (MTBE), the extract contains both water and acetic acid; however, when the solvent is distilled, an azeotrope of water and MTBE is formed and the solute (acetic acid) is recovered as anhydrous glacial acetic acid (de Klerk 2008). The azeotrope is then cooled, when it separates into an aqueous and an MTBE layer, and MTBE can be recycled directly.

A further consideration in the choice of solvent is that it should readily be separated from the raffinate (i.e., the feed material following extraction). Generally density differences suffice to bring about a high degree of separation, although in some cases the difference is so small that centrifuges must be employed. Centrifuges are also employed when the raffinate is a pulp that tends to entrain the extract. It may be necessary to wash such pulps with fresh solvent to recover trapped extract, particularly if a high yield of the solute is required.

The solvent will dissolve to some extent in the raffinate. One seeks to minimize the quantity lost in this way because it represents an economic loss, but possibly more importantly because it may contaminate the raffinate. The choice of a solvent with a very low solubility in the raffinate will naturally assist, but other considerations may force the choice of a solvent that has a significant solubility over one with a very low solubility. Removal of dissolved solvent from the raffinate may rely on the vapor pressure of the solvent being sufficiently lower than that of the raffinate that vapor scrubbing or distillation can remove it. Alternatively, if the raffinate is reasonably liquid, residual solvent may be stripped by adsorption on a solid such as activated carbon or clay. In extreme cases, it may be necessary to employ a second solvent that has a very low solubility in the raffinate to remove the solvent that was first used to remove the solute.

1.4 ENGINEERING CONSIDERATIONS

1.4.1 BATCH OPERATIONS

In the simplest extraction step, a feed is mixed with a solvent for a period sufficient for the solute to reach equilibrium between the two phases. The two phases are then separated, and the solute is recovered from the solvent. The question is how much of the solute will be extracted.

A simple mass balance with feed F, solvent S, extract E, and raffinate R, and c_i being the concentration of the solute in the ith stream, gives

$$F + S = E + R = M \tag{1.2}$$

$$Fc_f + Sc_s = Ec_e^* + Rc_r^* \tag{1.3}$$

where c^* represents the concentration at equilibrium. This is illustrated in Figure 1.5.

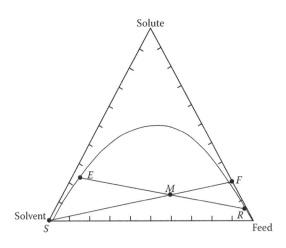

FIGURE 1.5 Extraction in a single stage.

The position of M on line SMF is found from the relationship

$$\frac{F}{S} = \frac{\overline{MS}}{\overline{FM}} \tag{1.4}$$

where overstrike indicates the length of the section of the line. The equilibrium relationship, shown as the tie line EMR in Figure 1.5, links c_e^* and c_r^* such that

$$c_e^*/c_r^* = D_T \tag{1.5}$$

at any one temperature T. For many systems, the distribution coefficient, D_T, is close to constant at low concentrations, but will vary with temperature.

If the solvent is recycled free of solute ($c_s = 0$), the quantity of solute extracted will be given by Ec_e^* and the fraction extracted by Ec_e^*/Fc_f. Because the solvent is usually chosen to be reasonably insoluble in the feed, the volume of extract will be close to the volume of solvent fed (i.e., $E \approx S$), so that the fraction extracted will depend strongly on the volume ratio of feed to solvent, S/F. However, increasing the volume ratio will reduce the concentration of the solute in the extract, which will in turn increase the cost of removing the solute from the solvent. Therefore, there is an economic balance to be struck that sets a limit on the maximum quantity of solvent that can be employed and the maximum achievable recovery of the solute.

1.4.2 DIFFERENTIAL BATCH OPERATIONS

It is possible to maximize the recovery of a solute by repeated extraction with fresh solvent. A practical way of achieving this is shown in Figure 1.6.

The feed and solvent are mixed together, then overflow to a settler and allowed to separate. The raffinate is returned to the mixer and the extract passes to a solvent recovery stage where the product is removed and the recovered solvent is recycled

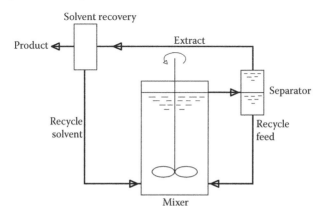

FIGURE 1.6   Differential extraction circuit.

to the mixer. This type of process is employed where a highly valued solute is at low concentration in the feed. It is widely used in the essential oils industry for extracting trace aromas or flavorants from plant material. Particular care must be taken to avoid trace contaminants in the fresh solvent, as there is a risk that these will also be concentrated in the product and in turn contaminate it unacceptably.

1.4.3 COUNTERCURRENT OPERATIONS

1.4.3.1 Batch Countercurrent

In the process shown in Figure 1.6, the quantity of solvent that must be separated from an increasingly dilute extract becomes very large. An alternative, which has the advantage of reducing the quantity of solvent for a given duty, is to operate in such a way that the fresh feed is extracted with solvent containing nearly the maximum quantity of solute. The raffinate from that stage is then extracted with solvent containing even less solute, and the extract from that stage then forms the feed to the first stage. This can be done batch-wise, as illustrated in Figure 1.7.

In Figure 1.7, there are four tanks, the first of which is empty in the first stage of operations. Solvent enters the fourth tank, overflows to the third tank, which

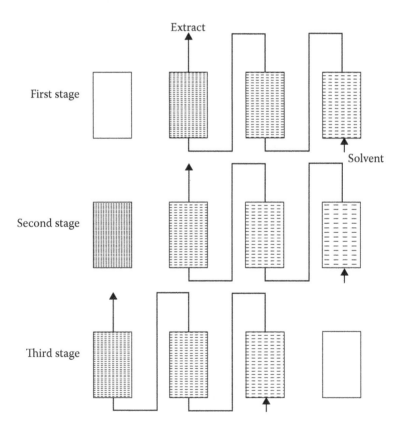

FIGURE 1.7 Batch countercurrent operation.

overflows in turn through the second. The extract comprises the solution leaving the second tank. The feed material is stripped of solute progressively from the second to the fourth tank.

In the second stage, the first tank is filled with fresh feed, while the material in the other tanks is progressively depleted. Eventually the material in the fourth tank loses all its solute, and the operation enters its third stage.

In the third stage of operations, the solvent enters the third tank and the extract leaves the first tank, while the raffinate is emptied from the fourth. The system is then effectively back at its starting condition, and the process continues.

The type of batch operation is frequently used where the feed material does not flow readily. It is somewhat expensive, partly because of the need to operate a large number of valves in a strict sequence, which involves control and maintenance challenges, and partly because the emptying of solids from the tank once the extraction is complete is not necessarily straightforward. Nevertheless, this type of extraction is quite widely applied in the industry.

1.4.3.2 Mixer–Settlers

Where the feed is reasonably fluid, then continuous countercurrent extraction is preferable to batch operation. In one variant, the feed and solvent are physically mixed, and then allowed to settle, usually under gravity. The settled extract phase then passes to the next mixer upstream while the settled raffinate passes to next mixer downstream. Figure 1.8 illustrates this for three stages of mixer–settler.

The two flows are separated at the end of the settler by a simple weir arrangement. Figure 1.9 shows this.

The lighter extract phase overflows a weir at the end of the settler and is withdrawn. The heavier raffinate phase flows out at the bottom of the settler into a chamber, from the top of which it overflows. The height of the raffinate weir determines the depth of each phase in the settler, because a simple pressure balance in the settler and the raffinate chamber gives:

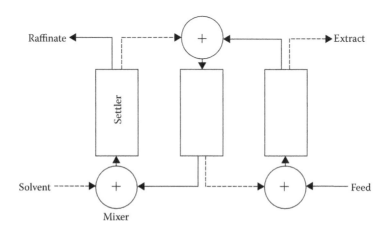

FIGURE 1.8   Continuous countercurrent mixer–settler.

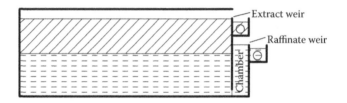

FIGURE 1.9   Elevation view of a settler, showing weir arrangement to separate phases.

$h_{rs}\rho_r + h_{es}\rho_e$ = pressure at base of settler = $h_{rc}\rho_r$ = pressure at base of chamber

where h is the height, ρ is the density of the phase, subscript r refers to raffinate, subscript e refers to extract, subscript s refers to settler, and subscript c refers to the chamber.

It is often found that any solids in the feed tend to collect at the interface between the extract and the raffinate in the settler, where they may interfere with the efficiency of separation. For this reason many settlers also have an arrangement by which material can be withdrawn from the region of the interface and treated separately for the recovery of the extract or raffinate.

The phases move from one mixer–settler unit to the next either by pumping or by gravity. Some designs incorporate a pump function in the mixer, which minimizes the number of pumps required. Each unit may be placed on a different level, so that one phase may gravitate while the other is pumped between stages.

Mixers require careful design. They have to disperse one phase in the other, and to do so without creating such fine droplets that the phases will not separate readily in the settler. Care must be taken to keep surface-active agents out of the system, as they will lower the interfacial tension between the phases and thus cause a fine dispersion that will not settle. Many biological systems contain natural surfactants; thus, it is often necessary in developing extraction systems to pilot them carefully to ensure that substances that affect the interfacial tension are not present—or, if they are present, to design the mixer to minimize the influence of the surfactant.

It is possible to employ centrifugal forces to hasten settling, and centrifuges are occasionally used instead of gravity settlers. There are designs of centrifugal mixer–settlers where several stages are packed within a single centrifugal unit. However, these tend to foul if there are any solids in the feed, and thus have not found widespread use in food processing.

1.4.3.3   Columns

It is possible to contact two liquid phases countercurrently in a column, as Figure 1.10 shows.

In the simplest design, one phase may be dispersed in the other by spraying through nozzles. The droplets rise or fall (as the case may be) through a counterflowing stream of the other phase. In Figure 1.10, the solvent phase is continuous, and droplets of the feed fall through a rising stream of solvent before forming a pool at the bottom of the column, from where the raffinate is removed. At the top of the column, the extract merely overflows.

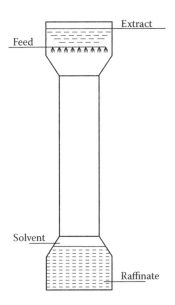

FIGURE 1.10   Extraction column operated with the solvent phase continuous.

This simple arrangement is not very efficient for a variety of reasons. One is that the surface area of the droplets is not very large. Another is that the falling droplets drag solvent molecules with them, and this means that solvent containing a lot of solute is mixed with lean solvent lower in the column. This reduces the concentration difference on which mass transfer depends, and thus reduces the efficiency.

For these reasons most columns contain a packing, the function of which is to increase the surface area of the phase that preferentially wets the packing. Because there is a larger area, the velocity of the dispersed phase is reduced and consequently there is less backmixing. There is a wide range of proprietary packing that may be employed. Some have less tendency than others to collect suspended solids, and are therefore preferred in food processing applications.

1.4.3.4   Extent of Extraction

It is obviously necessary to be able to estimate how many stages of countercurrent extraction are necessary to achieve a desired degree of extraction. It may be necessary to remove essentially all of a particular solute from the feed. Alternatively, there may be little point in recovering the last traces of a solute if the value of those last traces is too low to justify the expense of building additional extraction stages and operating them. Whichever the case, methods for estimating the number of stages required to perform a given extraction are essential.

With the equilibrium information available, it is obviously possible to solve the various mass balance equations numerically. However, for more than two or three stages, solving the resultant set of equations becomes tedious even in environments such as MATLAB®, and, if anything, it is self-defeating, because it is not possible to achieve 100% efficiency. Thus, a simple estimate suffices in the majority of cases.

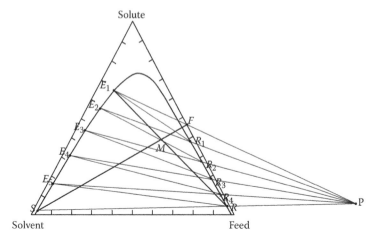

FIGURE 1.11 Graphical estimation of number of countercurrent stages.

For this reason graphical methods still find use. Figure 1.11 gives an example.

The feed concentration F and the solvent composition S should be known. Then assume a raffinate concentration R and an extract composition E_1. The extent of extraction obviously follows from $1 - (R/F)$. Construct SR and E_1F and extend until they meet at P, the "operating point," because

$$F + S = E_1 + R, \; F - E_1 = R - S = P \tag{1.6}$$

Then there is a tie line from E_1 to R_1, the raffinate concentration in equilibrium with E_1. A mass balance over the first and second stages gives

$$F + E_2 = E_1 + R, \; F - E_1 = R - E_2 = P \tag{1.7}$$

Therefore, a construction from P through R_1 will give E_2, which in turn will give a tie line to R_2, and a construction from P through R_2 will give E_3. Continuing in this way, eventually the graphical R_n will be at or below R. In the example given in Figure 1.10, $R_6 \approx R$; thus, six stages are needed to reduce the concentration of the solute from F to R. The ratio of solvent to feed is given as was the case in Equation 1.4 by the position of M.

There is a special case when there is no mutual solubility between the extract and the raffinate. In this case, the equilibrium shown in Figure 1.5 is much simplified and can be shown on rectilinear coordinates with, typically, the raffinate concentration of the solute on the ordinate and the extract concentration of the solute on the abscissa. This is illustrated in Figure 1.12.

As before, the feed concentration F and the solvent composition S should be known, and a raffinate concentration R and an extract composition E_1 is assumed. Then, with reference to Figure 1.12, the equilibrium between the two phases is given by the curve shown. Construct an operating line AB where point A has the coordinates $[F, E_1]$ and point B the coordinates $[R,S]$. Mass balance considerations show

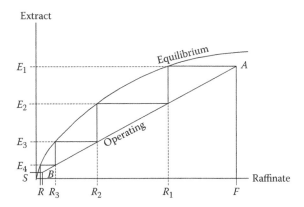

FIGURE 1.12 Graphical estimation of number of countercurrent stages where extract and raffinate are mutually insoluble (McCabe–Thiele diagram).

that the slope of this line is the same as the phase ratio (i.e., the ratio of volumetric flow of extract to the volumetric flow of raffinate). This is readily shown from mass balance considerations and, of course, only holds true provided the raffinate and extract are mutually insoluble.

Then if equilibrium is attained in each stage, the feed at concentration F will be in equilibrium with the final extract at a concentration E_1; thus, the raffinate concentration will fall from F to R_1. In the next stage R_1 will be extracted to equilibrium with E_2, and the raffinate concentration will fall further to R_2. Continuing in steps in this way, there will come a point where the raffinate concentration is below the desired final concentration R. In the example given in Figure 1.12, four stages suffice to reduce the concentration to the desired level. The graphical representation of countercurrent extraction in this manner is known as the McCabe–Thiele diagram after the chemical engineers who originally derived it.

Note that if the flow of feed is reduced—that is, the phase ratio is increased—then the slope of the operating line will be increased. It is not possible to reduce the feed flow indefinitely—at some point the operating line will cross the equilibrium line, and point A will lie above the equilibrium. It is no longer possible to construct the diagram, and E_1 must be reduced until point A is again below the equilibrium line at the new phase ratio.

1.5 TREATMENT OF THE EXTRACT AND RAFFINATE

1.5.1 EXTRACT

The extract will need to be treated to recover both the solvent and the solute. A range of separation technologies is available for this purpose. If the solute is not thermally labile, then separation may be effected by distillation, with recovery of the solvent in a condenser. Vacuum distillation can often be employed if the solute is thermally labile, although this may give rise to considerable cost because of the need to run the condenser at low temperatures to recover the solvent. It may also be possible to recover the solute by reducing the temperature of the extract and thus crystallizing the solute and settling, filtering, or centrifuging to remove the solvent from the solute crystals.

Chemical methods may also be employed. For instance, soy protein is extracted by alkaline water at approximately pH 9. Acidifying the water to about pH 4 precipitates the protein as a curd, which is centrifuged to remove the remaining water.

The solute can also be back-extracted from the extract. For instance, quinine is extracted from the bark of the cinchona tree into warm mineral oil and the solute is stripped from the extract using sulfuric acid–acidified water. The acidic water containing the quinine solute is filtered to remove insoluble material, and the quinine is recovered by adding alkali when the sulfate salt precipitates.

Similar principles can be applied to many extracts. The methods to be employed in any particular case will require testing before application, but in the majority of cases a simple and cost-effective method can be found for recovering solute and solvent separately. It is noteworthy that complete removal of the solute from the extract is not essential. The solvent may contain some of the solute and still be recycled to extract more material.

1.5.2 RAFFINATE

The principal task in raffinate treatment is usually the complete removal of the last traces of solvent. In food processing, the raffinate often contains significant quantities of solids, which adds complications. However, considerable separation may be possible by agitating the pulp with water, when the solvent will coalesce above the aqueous layer and can be removed that way. Oily solvents can be removed by washing with a much lighter, volatile hydrocarbon such as hexane, from which the residual solvent can be recovered by distillation, while any hexane left in the raffinate is removed by heating.

A wider range of separation processes is available if the raffinate is liquid and contains minimal solids. In that case residual solvent may be removed by adsorption on activated carbon or even clay. Centrifuging will take advantage of the density difference to remove droplets that are too fine to settle under gravity. In some cases, fine air bubbles have been used to scavenge traces of solvent. The surface of the bubbles is hydrophobic, which therefore attracts the solvent, and can be removed from the surface as a scum.

1.6   NEW TECHNOLOGIES

1.6.1 GENERAL

Increasing energy costs and the global imperative to reduce the carbon footprint has sparked the development of a number of new separation techniques for the chemical, pharmaceutical, and food industries (Bousbia et al. 2009a). Currently, many extraction processes in the food industry involve the use of organic solvents. However, these solvents not only present as atmospheric pollutants but also remain in the raffinate, as well as in the extracts, detracting from their purity (Reverchon 2003; Temelli 2009). While water, not organic solvents, is most often used in the extraction of essential oils (Berka-Zougali et al. 2010), water is also becoming a scarce commodity, sufficient to engender an interest in processes that conserve this precious solvent. Hence, to satisfy the growing demand for product purity, nonpolluting and energy-efficient processes, with the added advantage of using less solvent, alternative processes to the aforementioned solvent extraction methods are being sought (Reverchon 2003; Bousbia et al. 2009a). At present, supercritical fluid extraction (SFE) is the most extensively used alternative to solvent extraction with many commercially produced SFE compounds already available (Reverchon 2003; Brunner 2005; Bousbia et al. 2009a). More recent extraction techniques developed in the quest to create commercially viable, efficient, energy-saving, safe, compact, and sustainable extraction processes include extraction assisted by pulsed electric field (Loginova et al. 2011), solvent-free microwave extraction (SFME) (Bayramoglu et al. 2008), instant controlled pressure drop technology (DIC; from the French, *Détente Instantanée Contrôlée*) (Berka-Zougali et al. 2010), microwave hydrodiffusion and gravity (MHG) (Bousbia et al. 2009a), ultrasound assisted extraction, subcritical water extraction (Bousbia et al. 2009a), high pressure–assisted extraction (Jun 2009), aqueous two-phase extraction (Chethana et al. 2007), and enzyme-assisted aqueous extraction (Niranjan and Hanmoungjai 2004). Some examples and features of these processes will be discussed in the following sections.

1.6.1.1 Supercritical Fluid Extraction

Owing to its low cost and ready availability at high purity, the mostly commonly used supercritical fluid solvent in food applications is carbon dioxide (CO_2). CO_2 is not flammable, with a moderate critical temperature (31°C) and pressure (7.4 MPa), ensuring its safety in handling, while its easy removal from the extract to sound physiological levels brokered its "generally regarded as safe" status—that is, it can be used in food processing without declaration. Moreover, when recycled during the process—for example, after recovery by reducing the pressure during caffeine extraction as mentioned in an earlier section—it does not contribute to the carbon footprint. Some examples of everyday products where SC-CO_2 extraction is used are decaffeinated coffee and tea, flavor-enhanced orange juice, dealcoholized wine and beer, defatted meat and French fries, beer brewed with CO_2 hop extracts, rice parboiled using CO_2, spice extracts, and vitamin E– and β-carotene–enriched natural products (Davarnejad et al. 2008; Sahena et al. 2009; Temelli 2009; Herrero et al. 2010). Until recently, SFE was considered too costly for producing low-value, high-volume commodity oils, and only viable if applied to high-value, low-volume specialty oils (Brunner 2005). However, increasingly stringent environmental regulations, particularly regarding the use of hexane, has led to significant progress in optimization of design and operation of large-scale supercritical oil extraction plants, such that their cost structure is now comparable with that of conventional plants (Pronyk and Mazza 2009; Temelli 2009). However, to realize the full potential of SFE in terms of oil extraction, it should be extended to include oil refining, as well as further extraction of valuable biomass components (e.g., proteins and carbohydrates). Hence, SC-CO_2 extraction, combined with subcritical and supercritical water extraction, may result in biorefineries that are "green" in the true sense of the word (Temelli 2009).

1.6.1.2 Pulsed Electric Field-Assisted Extraction

Pulsed electric field (PEF)-assisted extraction has been shown to enhance solid–liquid extraction processes in the food industry. PEF had been successfully employed to extract anthocyanins and phenolics in red wine must (Puértolas et al. 2010); sucrose, proteins, and inulin from chicory (Loginova et al. 2010); betanine from beetroot (López et al. 2009), sugar beet, apple, and carrot juice; and maize germ and olive oils (Toepfl et al. 2006). The high extract yield, lower operating temperatures (preventing thermal degradation), high product quality and purity, high process efficiency, shorter extraction times, and lower energy cost of PEF extraction methods are key advantages of this technique, heralding its potential for energy-efficient and environmentally friendly food processing.

1.6.1.3 Microwave-Assisted Extraction

Essential oils, the single most widely extracted commodity, are extracted from herbs and spices and other botanicals for flavors, fragrances, and antimicrobial applications (Anon 2004; Bakkali et al. 2008; Shrinivas 2008). Conventional extraction techniques, namely steam distillation, hydrodistillation, and extraction with lipophilic solvents, have been successfully replaced by SC-CO$_2$ (Bakkali et al. 2008; Bayramoglu et al. 2008), with the attendant cost savings in terms of energy (lower process time and temperature), as well as superior product quality (Atti-Santos et al. 2005). The most recent developments in this field are the use of microwave-assisted hydrodistillation (MAHD) (Wang et al. 2010), SFME (Bayramoglu et al. 2008), DIC, and MHG. Compared with hydrodistillation, MAHD, SFME, and MHG not only produce superior quality essential oils but also result in higher yields and significant savings in process time (Luchessi et al. 2007; Bayramoglu et al. 2008; Bousbia et al. 2009a, 2009b; Wang et al. 2010); in addition, energy savings and lower water consumption are achieved (Bousbia et al. 2009a). The savings in energy and solvent is due to the absence of distillation or solvent extraction, those being the unit operations responsible for high energy and solvent consumption (Bousbia et al. 2009a). MHG, in particular, shows great promise for industrial-scale operations. With both energy consumption and carbon emission being 6% that of hydrodistillation, as well as its very short process time, nonrequirement for water or other solvent nor for postprocess waste water treatment, and higher-purity final product, this process ticks all the boxes for a green, effective alternative to conventional solvent extraction techniques (Bousbia et al. 2009a).

1.6.1.4 Ultrasound-Assisted Extraction

Ultrasound-assisted extraction (UAE) is another emerging technology with industrial-scale processing equipment designs available. Applications to food processes include extraction of vanillin, almond oils, herbal extracts, soy protein (with enhanced removal of flatulence-causing soluble sugars), polyphenols, and caffeine from green tea. UAE also offers a process that reduces the dependence on solvents such as hexane, with improved economics and environmental benefits (mainly due to an increased yield of extracted components), increased extraction rate, reduced extraction time, and higher process throughput (Vilkhu et al. 2008; Jadhav et al. 2009; Karki et al. 2010). A more recent modification of UAE, namely ultrasound-assisted dynamic extraction, was used to extract chickpea oil. This process entails circulation of the solvent while the sample and solvent are subjected to ultrasound, and results in further reductions in solvent consumption and extraction time and, therefore, environmental impact (Lou et al. 2010).

1.6.1.5 Subcritical Water Extraction

Subcritical water, also known as pressurized low-polarity water, or pressurized hot water extraction (PHWE) (Chapter 8), utilizes hot water (100–374°C) under pressure (1000–6000 kPa) to replace organic solvents (Herrero et al. 2006). PHWE delivers higher extraction yields from solid samples than conventional solvents. This technique was used effectively to extract peppermint oil, carotenoids from microalgae, carnosic acid and aroma compounds from rosemary, quercetin from onion skins, and rice bran oil through simultaneous lipase inactivation, to name a few. Compared with conventional extraction methods (i.e., solid–liquid extraction, hydrodistillation, and organic solvents), PHWE offers several advantages, namely shorter extraction times, higher-quality extracts, a less costly extracting solvent, and an environmentally friendly process (Herrero et al. 2006; Pourali et al. 2009; Pronyk and Mazza 2009; Ko et al. 2011).

1.6.1.6 High Pressure–Assisted Extraction

High pressure–assisted extraction has also gained ground as an environmentally friendly alternative to solvent extraction. Advantages include shorter extraction times, higher yields, extract purity, and lower energy consumption (Jun 2009). A variation on this, known as DIC, uses high-pressure steam to extract essential oils, followed by rapid transfer and cooling to a vacuum chamber. The rapid condensation in the vacuum tank produces a microemulsion of water and essential oils (Berka-Zougali et al. 2010). Further information on this process is given in Chapter 9.

1.6.1.7 Aqueous Two-Phase Extraction

Aqueous two-phase extraction was successfully employed for purification and concentration of betalains (a natural colorant from beetroot), resulting in a simpler and therefore more environmentally friendly process (Chethana et al. 2007).

1.6.1.8 Enzyme-Assisted Aqueous Extraction

This process has been applied to extract oils from various oil seeds and some fruits. The advantages of the method reside in the fact that processing occurs at relatively low temperatures and use water as a solvent, ensuring superior product quality and making the method safe and environmentally friendly. The presence of food-grade enzymes improves the oil yield (Niranjan and Hanmoungjai 2004).

1.6.2 Impact of Refining

Another aspect of vegetable oil processing is that the steps after solvent extraction, namely solvent recovery and refining, also have high-energy demands, consume large quantities of water and other chemical reagents, and produce significant quantities of effluent. The above concerns had been addressed by employing membrane technology, particularly ultrafiltration and nanofiltration. The technology has been applied successfully at the pilot scale for solvent recovery from soybean and cotton seed oil and shows great promise with regard to deacidification and degumming, obviating the use of sodium hydroxide. It had been estimated that using membrane technology for solvent recovery rather than heating could effect a savings of 2.1 $\times$ 10^{12} kJ year^{-1} in the United States alone, while having considerably less noxious effect on the environment (Coutinho et al. 2009).

1.6.3 COMBINED METHODS AND SAMPLE EXTRACTION

As will be seen in later chapters, combinations of the above processes have also been researched (Chapter 6), while many of the techniques are also used as alternatives to solvent extraction as applied to food analytical techniques—for example, microwave-assisted extraction (Chapter 3).

1.7 CONCLUSION

In this short introduction, it had not been possible to do more than outline some of the principles underlying solvent extraction technology. However, the reader should now be equipped to appreciate the many advances reported in the remaining chapters of this book. We need advances, because extraction can make a significant contribution to the safe and environmentally friendly processing of food. Extraction has not realized its full potential partly because of the difficulty of ensuring efficient transfer of the solutes from the food to the solvent, and perhaps also because the equipment available to deploy the developing technologies has not yet reached the market.

ACKNOWLEDGMENTS

The authors are grateful to their respective departments at the Cape Peninsula University of Technology for permitting them time to prepare this introductory chapter.

REFERENCES

Anonymous. 2004. Aroma chemicals derived from essential oils. http://www.nedlac.org.za/media/5906/essential.pdf (accessed November 2010).
AspenTech. 2010. http://www.aspentech.com/products/aspen-properties.cfm (accessed October 2010).
Atti-Santos, A.M., Rossato, M., Serafini, L.A., Cassel, E., Moyna, P. 2005. Extraction of essential oils from lime (*Citrus latifolia* Tanaka) by hydro-distillation and supercritical carbon dioxide. *Braz Arch Biol Technol* 48: 155–160.
Bakkali, F., Averbeck, S., Averbeck, D., Idaomar, M. 2008. Biological effects of essential oils—A review. *Food Chem Toxicol* 46: 446–475.
Bayramoglu, B., Sahin, S., Sumnu, G. 2008. Solvent-free microwave extraction of essential oils from oregano. *J Food Eng* 88: 535–540.
Berka-Zougali, B., Hassani, A., Besombes, C., Allaf, K. 2010. Extraction of essential oils from Algerian myrtle leaves using instant pressure drop technology. *J Chromatogr A* 1217: 6134–6142.

Bousbia, N., Vian, M.A., Ferhat, M.A., Meklati, B.Y., Chemat, F. 2009a. A new process for extraction of essential oil from citrus peels: Microwave hydrodiffusion and gravity. *J Food Eng* 90: 409–413.

Bousbia, N., Vian, M.A., Ferhat, M.A., Petitcolas, E., Meklati, B.Y., Chemat, F. 2009b. Comparison of two isolation methods for essential oil from rosemary leaves: Hydrodistillation and microwave hydrodiffusion and gravity. *Food Chem* 114: 355–362.

Brunner, G. 2005. Supercritical fluids: Technology and application to food processing. *J Food Eng* 67: 21–33.

Campbell-Platt, G. (Ed.). 2009. *Food Science and Technology*. Oxford: Blackwell-Wiley. ISBN: 978-0-632-06421-2.

Chethana, S., Nayak, C.A., Raghavarao, K.S.M.S. 2007. Aqueous two phase extraction for purification and concentration of betalains. *J Food Eng* 81: 679–687.

Coutinho, C.M., Chiu, M.C., Basso, R.C., Ribeiro, A.P.B., Gonçalves, L.A.G., Viotto, L.A. 2009. State of the art application of membrane technology to vegetable oils: A review. *Food Res Int* 42: 536–550.

Davarnejad, R., Kassim, A.Z., Sata, S.A. 2008. Supercritical fluid extraction of β-carotene from crude palm oil using CO_2. *J Food Eng* 89: 472–478.

de Klerk, A. 2008. Fischer-Tropsch Refining. PhD dissertation. South Africa: Department of Chemical Engineering, University of Pretoria.

Herrero, M., Cifuentes, A., Ibáñez, E. 2006. Sub- and supercritical fluid extraction of functional ingredients from different sources: Plants, food by-products, algae and microalgae. A review. *Food Chem* 98: 136–148.

Herrero, M., Mendiola, J.A., Cifuentes, A., Ibáñez, E. 2010. Supercritical fluid extraction: Recent advances and applications. *J Chromatogr A* 1217: 2495–2511.

Jadhav, D., Rekha, B.N., Gogate, P.R., Rathod, V.K. 2009. Extraction of vanillin from vanilla pods: A comparison study of conventional Soxhlet and ultrasound assisted extraction. *J Food Eng* 93: 421–426.

Jun, X. 2009. Caffeine extraction from green tea leaves assisted by high pressure processing. *J Food Eng* 94: 105–109.

Karki, B., Lamsal, B.P., Jung S., van Leeuwen, J., Pometto, A.L., Grewell, D., Khanal, S.K. 2010. Enhancing protein and sugar release from defatted soy flakes using ultrasound technology. *J Food Eng* 96: 270–278.

Ko, M., Cheigh, C., Cho, S., Chung, M. 2011. Subcritical water extraction of flavonol quercetin from onion skin. *J Food Eng* 102: 327–333.

Kopcak, U., Mohamed, R.S. 2005. Caffeine solubility in supercritical carbon dioxide/co-solvent mixtures. *J Supercrit Fluids* 34: 209–214.

Kumhom, T., Douglas, P.L., Douglas, S., Pongamphai, S., Teppaitoon, W. 2010. Prediction of solubilities of solid biomolecules in modified supercritical fluids using group contribution methods and equations of state. *Ind Eng Chem Res* 49: 2433–2441.

Lo, T.C., Baird, M.H., Hanson, C. (Eds.). 1983. *Handbook of Solvent Extraction*. New York: Wiley & Sons. ISBN: 0-471-04164-5.

Loginova, K.V., Shynkaryk, M.V., Lebovka, N.I., Vorobiev E. 2010. Acceleration of soluble matter extraction from chicory with pulsed electric fields. *J Food Eng* 96: 374–379.

Loginova, K.V., Vorobiev, E., Bals, O., Lebovka, N.I. 2011. Pilot study of countercurrent cold and mil heat extraction of sugar from sugar beets, assisted by pulsed electric fields. *J Food Eng* 102: 340–347.

López, N., Puértolas, E., Condón, S., Raso, J., Álvarez, I. 2009. Enhancement of the extraction of betanine from red beetroot by pulsed electric fields. *J Food Eng* 90: 60–66.

Lou, Z., Wang, H., Zhang, M., Wang, Z. 2010. Improved extraction of oil from chickpea under ultrasound in a dynamic system. *J Food Eng* 98: 13–18.

Luchessi, M.E., Smadja, J., Bradshaw, S., Louw, W., Chemat, F. 2007. Solvent free microwave extraction of *Elletaria cardamomum* L.: A multivariate study of a new technique for the extraction of essential oil. *J Food Eng* 79: 1079–1086.

Niranjan, K., Hanmoungjai, P. 2004. Enzyme-aided aqueous extraction, in *Nutritionally Enhanced Edible Oil Processing*, N.T. Dunford and H.B. Dunford (Eds.). Champaign, Illinois: AOCS Publishing, eBook ISBN: 978-1-4398-2227-2, DOI: 10.1201/9781439822272.ch5.

Pourali, O., Asghari, F.S., Yoshida, H. 2009. Simultaneous rice bran oil stabilization and extraction using sub-critical water medium. *J Food Eng* 95: 510–516.

Pronyk, C., Mazza, G. 2009. Design and scale-up of pressurized fluid extractors for food bio-products *J Food Eng* 95: 215–226.

Puértolas, E., López, N., Saldaña, G., Álvarez, I., Raso, J. 2010. Evaluation of phenolic extraction during fermentation of red grapes treated by a continuous pulsed electric fields process at pilot-scale. *J Food Eng* 98: 120–125.

Reverchon, E. 2003. Supercritical fluid extraction, in *Encyclopedia of Food Sciences and Nutrition*. 2nd Ed. B. Cabelloro, P. Finglas, and L. Trugo (Eds.), pp. 5680–5687. Elsevier Science Ltd. (Online). ISBN: 978-0-12-227055-0.

Rydberg, J., Cox, M., Musikas, C., Choppin, G.R. (Eds.). 2004. *Solvent Extraction. Principles and Practice*. 2nd Edn. New York: Marcel Dekker. ISBN: 0-8247-5063-2.

Sahena, F., Zaidul, I.S.M., Jinap, S., Karim, A.A., Abbas, K.A., Norulaini, N.A.N., Omar, A.K.M. 2009. Application of supercritical CO_2 in lipid extraction—A review. *J Food Eng* 95: 240–253.

Schügerl, K. 1994. *Solvent Extraction in Biotechnology*. Springer-Verlag: Berlin ISBN 3540576940.

Schwarztberg, H.G. 1987. Leaching – organic materials, in *Handbook of Separation Process Technology (Chapter 10)*. R.W. Rousseau (Ed.). New York: Wiley-Interscience. ISBN: 0471 89558X.

Shrinivas, P.K. 2008. Market research data on essential oils and absolutes in the fragrance and flavour industries. http://goarticles.com/cgi-bin/showa.cgi?C=1175757 (accessed November 2010).

Temelli, F. 2009. Perspectives on supercritical fluid processing of fats and oils. *J Supercrit Fluids* 47: 583–590.

Toepfl, S., Mathys, A., Heinz, V., Knorr, D. 2006. Potential of high hydrostatic pressure and pulsed electric fields for energy efficient and environmentally friendly food processing. *Food Rev Int* 22: 405–423.

U.S. EPA. 2010. SPARC suite. http://www.epa.gov/athens/research/projects/sparc/ (accessed October 2010).

Vilkhu, K., Mawson, R., Simons, L., Bates, D. 2008. Applications and opportunities for ultrasound assisted extraction in the food industry—A review. *Innov Food Sci Emerg Technol* 9: 161–169.

Wang, H., Liu, Y., Wei, S., Yan, Z., Lu, K. 2010. Comparison of microwave-assisted and conventional hydrodistillation in the extraction of essential oils from mango (*Mangifera indica* L.) flowers. *Molecules* 15: 7715–7723.

2 Pulse Electric Field-Assisted Extraction

Eugene Vorobiev and Nikolai I. Lebovka

CONTENTS

2.1 INTRODUCTION

Pulsed electric fields (PEFs) are now attracting strong interest in food engineering research. This minimally invasive method allows avoidance of undesirable changes in a biological material, which are typical for other techniques such as thermal, chemical, and enzymatic ones (Knorr et al. 2001; Vorobiev et al. 2005; Toepfl et al. 2005; Vorobiev and Lebovka 2006, 2008, 2010; Toepfl and Knorr 2006; Raso and Heinz 2006; Toepfl et al. 2007a,b; Ravishankar 2008; Donsì et al. 2010; Lebovka and Vorobiev 2010; Sack et al. 2010; Toepfl and Heinz 2010). A supplementary advantage of PEF treatment for food applications is its potential to kill microorganisms (Barbosa-Cánovas et al. 1998, 2000; Barbosa-Cánovas and Cano 2004; Altunakar et al. 2007; Vega-Mercado et al. 2007; Tewari and Juneja 2007).

Many useful examples of PEF application for enhancing pressing, drying, extraction, and diffusion in the processing of materials of biological origin have already been demonstrated (Figure 2.1). PEF-assisted techniques display unusual synergetic

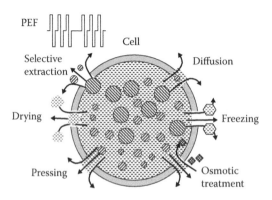

FIGURE 2.1   The PEF-assisted technique.

effects and present the possibility of "cold diffusion," "cold drying," and improved osmotic and freezing treatment.

The PEF technology is not simple in application, and has a long history. The main historical landmarks in the field are summarized in Table 2.1. The most important of them are the discovery of bioelectricity by Luigi Galvani in 1791 and the discovery of electroporation in the 1960s–1970s (see, e.g., Weaver and Chizmadzhev 1996; Pavlin et al. 2008; Pakhomov et al. 2010; Saulis 2010). Many efforts were also aimed at industrial implementations of alternative current (AC), direct current (DC), and PEF treatments. They started at the beginning of the past century by application of the said methods for microbial killing, canning, ohmic heating, and others (Stone 1909; Beattie 1914; Anderson and Finkelshtein 1919; Prescott 1927; Fettermann 1928; Getchell 1935). Different electrical apparatus for treatment of fluid foods were patented (Jones 1897; Anglim 1923; Ball 1937). Later on, Flaumenbaum (1949) and Zagorulko (1958) reported applications of DC and AC electric fields for treatment of prunes, apples, grapes, and sugar beets. They demonstrated acceleration of extraction by electrical breakage of cellular membranes. This phenomenon was called electroplasmolysis.

TABLE 2.1
Main Historical Landmarks in the Progress of PEF Applications

Development	Authors and Data	Comment
Discovery of bioelectricity	Luigi Galvani (1791)	
Microbial killing	Prochownick and Spaeth (1890)	DC and AC
Different practical applications of DC and AC for microbial killing and ohmic heating	Different authors (1900–1940); for a review, see, e.g., de Alwis and Fryer (1990)	DC and AC
Increase in juice yield from fruits	Flaumenbaum (1949)	(220 V, 50 Hz)
Electroplasmolysis, extraction of juice from sugar beets	Zagorulko (1958)	DC and AC
Electrophysiological model of biotissues, derivation of transmembrane potential	Schwan (1957), Foster and Schwan (1989)	
Disintegration of biomaterials, killing of bacteria	Doevenspeck (1961), Sale and Hamilton (1967)	PEF
Reversible electrical breakdown of biomembranes, discovery of electroporation	Stampfli (1958), Neumann and Rosenheck (1972)	PEF
Earlier industrial applications (canning and wine production; treatment of apples, sugar beets, etc.)	Different authors (1965–1980); for a review, see Rogov and Gorbatov (1974)	DC and AC
Current applications of PEF for microbial killing and disintegration of plant tissues	Different authors (1965–1980); for a review, see Vorobiev and Lebovka (2010), Donsì et al. (2010), Lebovka and Vorobiev (2010), Sack et al. (2010), Barbosa-Cánovas and Cano (2004), Jaeger and Knorr (2010)	PEF

The most important steps were made in the 1960s–1970s when the first applications of PEF were reported (Doevenspeck 1961; Sale and Hamilton 1967), the first industrial AC setups were implemented (Flaumenbaum 1968; Kogan 1968; Matov and Reshetko 1968; Rogov and Gorbatov 1974, 1988), and the concept of membrane electroporation was theoretically worked out (Weaver and Chizmadzhev 1996; Pavlin et al. 2008; Pakhomov et al. 2010; Saulis 2010).

Starting from the early 1990s, many new practical PEF-assisted techniques have been tested, and their usability for microbial killing, food preservation, and acceleration of drying, pressing, diffusion, and selective extraction has been demonstrated (Gulyi et al. 1994; Knorr et al. 1994; Toepfl et al. 2007a; Ravishankar et al. 2008; Vorobiev and Lebovka 2008). Since then, new types of higher-voltage PEF generators, new designs of treatment chambers, and new pilot schemes have been developed (Barbosa-Cánovas et al. 1998; Vorobiev and Lebovka 2008).

This chapter reviews the current state of the art in food engineering, existing fundamental knowledge on the mechanism of PEF-induced effects in biomaterials, impact of PEF on functional food ingredients, recent experiments in the field, practical applications of PEF and their examples for different food materials, and perspectives on the industrial applications of PEF-assisted extraction techniques.

2.2   PEF-INDUCED EFFECTS

2.2.1   Basics of Electroporation

The impact of PEF on biomaterials is reflected by the loss of membrane barrier functions. A membrane envelope around the cell restricts the exchange of inter- and intracellular media. The application of an electric field induces the formation of pores inside the membrane and increases its permeability. Traditionally this phenomenon is called "electroporation" or "electropermeabilization" (Weaver and Chizmadzhev 1996; Pakhomov et al. 2010).

2.2.1.1   Transmembrane Potential

The degree of electroporation depends on the potential difference across a membrane, or the transmembrane potential, u_m. Electroporation requires some threshold value of u_m, typically 0.5–1.5 V. Depending on treatment conditions, the value of u_m, and PEF exposure time (t_{PEF}), a temporary (reversible) or irreversible loss of barrier function may occur. It is assumed that electroporation involves membrane charging, membrane polarization (charging time $t_c > 1$ μs), expansion of pore radii, and aggregation of pores (during the first 100 μs). On turning off the electric field, pore resealing and memory effects (lasting from seconds to hours) may be observed (Teissié et al. 2005; Pavlin et al. 2008). A number of theoretical models have been proposed for the description of the electroporation of membranes at the micro level. These theories considered different mechanisms, such as electromechanical, electrohydrodynamic, electroosmotic, and the development of viscoelastic instabilities. However, the mechanism of membrane electroporation is not yet fully understood, and there are a lot of discrepancies between theoretical and experimental results (Weaver and Chizmadzhev 1996; Pakhomov et al. 2010).

For a spherical cell in the external field, the induced transmembrane potential u_m is a function of the cell radius R, field strength E, and position of the observation point on the surface of a membrane (Schwan 1957):

$$u_m = 1.5REe \cos \theta(1 - \exp(-t/\tau_C)) \tag{2.1}$$

Here θ is the angle between the external field E and radius vector R, e is the electroporation factor that is dependent on geometry and electrophysical properties of cells, and τ_C (≈ 1–10 μs) is the time constant reflecting the process of charging the membrane capacity C (Figure 2.2) (Pavlin et al. 2008). Note that for anisotropic cells, the value of u_m is a function not only of electric field intensity and cell size but also of the cell shape and orientation.

2.2.1.2 Effects of Cell Size and Electrical Conductivity Contrast

The value of u_m is directly proportional to the cell radius R, while the drop of potential is highest at the cell poles and decreases toward zero at $\theta = \pm\pi/2$. Thus, larger cells become damaged before smaller ones, and the probability of damage is at maximum at the cell poles.

Typically the width of membrane d (≈ 5 nm) is very small as compared with the cell radius R ($R \approx 50$ μm for plant cells and $R < 10$ μm for microbial cells). The electric field strength inside the membranes can be estimated as $E_m = u_m/d \approx ER/d \sim 10^4E$. The experimentally estimated threshold value E_t required for a noticeable electroporation is of the order of 100 V/cm for plant cells (Vorobiev and Lebovka 2006) and 10 kV/cm for small microbial cells ($R \approx 1$–10 μm) (Barbosa-Cánovas et al. 1998). In practice, the degree of electropermeabilization also depends on the properties of materials and details of the pulse protocol (Vorobiev and Lebovka 2006). A considerable damage to plant tissues can be observed at $E = 500$–1000 V/cm and treatment

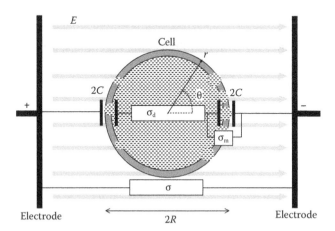

FIGURE 2.2 Electrophysical schema of a cell. Here R is the radius of the cell; d is the membrane width; θ is the angle between the external field E and radius vector r at the surface of membrane; C is the membrane capacitance; and σ_m, σ, and σ_d are the electrical conductivities of the membrane, extracellular medium, and cytoplasm, respectively.

time within 10^{-4}–10^{-1} s. For microbial killing, higher field strengths ($E = 20$–50 kV/cm) and shorter treatment times (10^{-5}–10^{-4} s) are required.

The general expression for electroporation factor e is rather complex (Kotnik et al. 1998)

$$e = (3d/R)\sigma_d\sigma/[(\sigma_m + 2\sigma)(\sigma_m + 0.5\sigma_d) - (1 - 3d/R)(\sigma - \sigma_m)(\sigma_d - \sigma_m)] \quad (2.2)$$

where d is the membrane width (≈ 5 nm) and σ_m, σ, and σ_d are the electrical conductivities of the membrane, external medium, and cytoplasm, respectively.

At $\sigma_m \ll \sigma$, $\sigma_m \ll \sigma_d$, Equation 2.2 reduces to $e \approx 1$. This approximation works well for suspensions of small biological cells, where the typical electrical conductivity values are $\sigma_m = 3 \times 10^{-7}$ S/m and $\sigma_d = 0.3$ S/m (Pavlin et al. 2008). Figure 2.1 shows that at $\sigma_m = 3 \times 10^{-7}$ S/m, the value of e is an increasing function of σ/σ_d, which approaches 1 at $\sigma/\sigma_d > 0.2$.

The σ_m value of a plant cell is unknown; however, it can be estimated from the conductivities of intact ($\sigma = \sigma_i$ when all membranes are intact) and completely damaged ($\sigma \approx \sigma_d$ when all membranes are disrupted) plant tissues. For the serial model of cell packing, one can obtain $(d + R)/\sigma_i = /\sigma_m + R/\sigma_d$ and

$$\sigma_m \approx \sigma_i \,(d/R)k/(k - 1) \quad (2.3)$$

where $k = \sigma_d/\sigma_i$ is the electrical conductivity contrast.

The typical values of σ_i and k for different fruit and vegetable tissues (Bazhal et al. 2003a) are presented in Table 2.2. At $\sigma_i \approx 0.02$–0.08 S/m, $R = 50$ μm, and $k \gg 1$, the membrane conductivities of plant tissues may be estimated as $\sigma_m \approx \sigma_i \,(d/R) \approx 2$–$8.10^{-6}$ S/m.

TABLE 2.2

Tissue Characteristics for Different Fruits and Vegetables, Measured at a Temperature (T) of 293 K and a Frequency (f) of 1 kHz

Material	Cell Radius, R (μm)	Intact Conductivity, σ_i (S/m)	Contrast, $k = \sigma_d/\sigma_i$
Apple	35 ± 5	0.022 ± 0.007	10 ± 3
Banana	39 ± 13	0.082 ± 0.018	5.4 ± 0.9
Aubergine	–	0.051 ± 0.009	–
Carrot	30 ± 3	0.059 ± 0.019	4.5 ± 0.6
Courgette	30 ± 4	0.029 ± 0.009	11.9 ± 3.1
Cucumber	–	0.032 ± 0.005	–
Potato	47 ± 6	0.044 ± 0.014	13 ± 3
Pear	–	0.032 ± 0.005	–
Orange	59 ± 9	0.063 ± 0.009	1.26 ± 0.23

Source: Lebovka, N.I. et al., *Innov Food Sci Emerg*, 2, 113, 2001; Bazhal, M. et al., *Biosyst Eng*, 86, 339, 2003; and Ben Ammar, J. et al., *J Food Sci*, 76, E90, 2011.

Note: The presented data correspond mean ± SD.

Note that for plant tissues, the electroporation factor e can be noticeably smaller than 1 and be dependent on the contrast ratio k (Figure 2.3). Thus, it can be expected that the threshold electric field strength value E_t, required for noticeable electroporation, will be high for plant tissues with small electrical conductivity contrast k and small electroporation factor e. This conclusion was recently supported by a comparison of electroporation efficiency for fruit and vegetable tissues with different conductivity contrasts (Ben Ammar et al. 2011).

2.2.1.3 Resealing of Cells and Mass Transfer Process

Lebovka et al. (2000, 2001) have put forward a hypothesis explaining how PEF treatment affects the structure of cellular tissues. They considered the PEF effect as a correlated percolation that is governed by two processes: (i) resealing of cells and (ii) moisture mass transfer inside the cellular structure, which is sensitive to PEF treatment repetitions. At a low enough electric field intensity, electroporation is reversible as far as the resealing process is quick enough to repair the membranes immediately after the termination of PEF treatment. At moderate PEF treatment, some of the cells lose their permeability, but others may reseal (Lebovka et al. 2001). It was demonstrated that the insulating properties of the cell membrane (e.g., in potato, apple, and fish tissues) can be recovered within several seconds after pulse termination (Angersbach et al. 2000). The reversible permeabilization of potato cells was confirmed by transient changes in the viscoelastic properties after PEF application with a single 10^{-5}, 10^{-4}, or 10^{-3} s rectangular pulse at electric field strength E ranging from 30 to 500 V/cm (Pereira et al. 2009). According to calorimetric data, PEF application resulted in a strong metabolic response of potato tissue dependent on the pulsing conditions (Galindo et al. 2008a,b,c; Galindo et al. 2009a,b). The PEF-specific metabolic responses 24 h after the application of PEF (one 1 ms

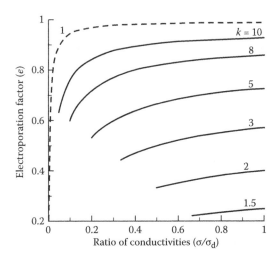

FIGURE 2.3 Electroporation factor e versus σ/σ_i (Equation 2.2). The curves, $k = \sigma_d/\sigma_i$, were obtained from Equations 2.2 through 2.3 at $R = 50$ μm (for plant tissues). Curve 1 was calculated for $\sigma_m = 3 \times 10^{-7}$ S/m, $\sigma_d = 0.3$ S/m, and $R = 5$ μm (for microbial cell).

rectangular pulse at $E = 30–500$ V/cm) may involve degradation of starch and ascorbic acid (Galindo et al. 2008b). High-intensity PEF treatment causes an irreversible damage to the cell membrane. Long-term changes in conductivity after the application of PEF treatment can also be related to osmotic flow and moisture redistribution inside the sample (Lebovka et al. 2001).

2.2.2   QUANTIFICATION OF PEF-INDUCED DISINTEGRATION

The damage degree Z can be defined as the volume fraction of the damaged cells. However, the experimental determination of Z and quantification of PEF-induced disintegration is not an easy task, although many experimental techniques have been tested thus far.

2.2.2.1   Microscopic Study

Optical microscopy was used for the study of PEF-treated aqueous suspensions of Chinese hamster ovary cells (Valic et al. 2003). Microscopic observations evidenced that the degree of electropermeabilization may be dependent on the anisotropy of cells. Visual observations evidenced that elongated cells became electropermeabilized more intensively when the longest axis of the cell was parallel to the electric field. The same conclusion was reached for apple tissues. It was shown that lower electric fields were required for permeabilization of anisotropic apple cells when the electric field was applied parallel to the longest axes of the cells (Chalermchat et al. 2010).

Optical microscopy was used for *in situ* visualization of PEF-induced color changes in onion epidermis stained with neutral red (Fincan and Dejmek 2002). The final electrical conductivity increase was directly proportional to the number of permeabilized cells. Microscopic studies showed that intact cell architecture was preserved, while membrane damage was confirmed by free colorant diffusion inside electroporated cells. Thus, these experiments evidenced that PEF did not noticeably affect the structure of cell walls. This important conclusion was supported by scanning electron microscope images, where a similarity in cell wall structure, and area and morphology of starch granules between untreated and PEF-treated potato tissues was observed (Ben Ammar et al. 2010).

Microscopic observation is the most direct way for the visual determination of the fraction of damaged cells, Z. However, the application of this method for estimation of Z in plants is not simple, accounting for the difficulties related with sample preparation, pH sensitivity of the method, and conductivity of the solution used in mounting the epidermis. That is why this method is not widely used for characterization of PEF-induced damage in plant tissues.

2.2.2.2   Electrical Conductivity

The simplest way for characterization of Z is based on electrical conductivity measurements, because the average electrical conductivity of a tissue increases with the degree of its damage. The electrical conductivity disintegration index seems Z_C can be defined as (Rogov and Gorbatov 1974; Lebovka et al. 2002)

$$Z_C = (\sigma - \sigma_i)/(\sigma_d - \sigma_i) \tag{2.4}$$

where σ is the electrical conductivity value measured at low frequency (≈ 1 kHz), and indexes i and d refer to the conductivities of intact and totally damaged tissue, respectively. This equation gives $Z_C = 0$ for the intact tissue and $Z_C = 1$ for the totally disintegrated material.

The procedure is simple and can be easily applied for continuous monitoring of the damage degree during PEF treatment (Figure 2.4). This method requires knowledge of the value of σ_d. This value can be estimated by measurement of the electrical conductivity of the freeze–thawed tissue. Another way to estimate σ_d is based on PEF treatment at high electric field strengths ($E \approx 1000$ V/cm) and long treatment durations ($t_{PEF} \approx 0.1$–1 s) (Bazhal et al. 2003a; Lebovka et al. 2004a). However, the value of σ_d determined in such a way is not well defined because freeze–thawing or strong PEF treatment can affect the structure of cell walls and influence σ_d.

Other methods are based on electrical conductivity measurements at low and high frequencies, and assume validity of some bioimpedance models for plant tissues (Angersbach et al. 2002; Pliquett 2010). For example, the conductivity disintegration index Z can be estimated as (Angersbach et al. 2002)

$$Z_C = (\alpha\sigma^0 - \sigma_i^0)/(\sigma_i^\infty - \sigma_i^0) \tag{2.5}$$

where $\alpha = \sigma_i^\infty/\sigma^\infty$ and the indexes 0 and ∞ refer to the low (≈ 1 kHz) and high (3–50 MHz) frequency conductivity limits, respectively (Figure 2.4b).

However, any method based on electrical conductivity should be applied with caution (Pliquett 2010). The electrical conductivity of tissues is sensitive to the spatial redistribution of air and moisture content inside the tissue, membrane resealing, and other factors (Lebovka et al. 2001). As a result, the transient behavior of

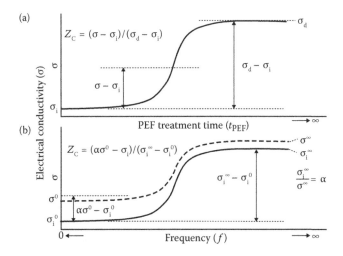

FIGURE 2.4 Estimation of electrical conductivity disintegration index Z_C from (a) PEF treatment time t_{PEF} and (b) frequency f dependencies of tissue electrical conductivity σ.

electrical conductivity σ after PEF treatment with time constants from seconds to hours is rather typical (Angersbach et al. 2002).

Electric impedance measurements and methods based on frequency dependency of the phase shift in the range of 500 Hz–10 MHz were also applied for estimation of electroporation effects in PEF-treated mash from wine grapes (Sack et al. 2009). Good correlations were observed between measurements of the complex impedance and color intensity of the must.

Finally, we noted that all conductivity-based methods for Z estimation are straightforward and may be useful for the rough estimation of the impact of PEF on plant tissues and colloidal biosuspensions (Lebovka et al. 2000; El Zakhem et al. 2006a,b; Vorobiev et al. 2006).

2.2.2.3   Diffusion Coefficient

Similarly, the diffusion coefficient disintegration index Z_D can be defined as (Jemai and Vorobiev 2001; Lebovka et al. 2007a)

$$Z_D = (D - D_i)/(D_d - D_i) \tag{2.6}$$

where D is the measured apparent diffusion coefficient and subscripts i and d refer to the values for intact and totally destroyed material, respectively.

The apparent diffusion coefficient D can be determined from solute extraction or convective drying experiments. Unfortunately, diffusion techniques are indirect and invasive for biological objects and can influence the structure of tissues. Moreover, there exists an equivalent problem with determination of D_d. Drying experiments with potato tissue have shown that the D_d value of freeze–thawed tissue is noticeably higher than the D_d value of PEF-disintegrated tissue with a high Z_C index (Lebovka et al. 2007a). Evidently it reflects cell wall damage after the freeze–thawing treatment.

2.2.2.4   Textural Characteristics

Some attempts in using textural methods for characterization of PEF-treated tissues were done (Fincan and Dejmek 2003; Lebovka et al. 2004a). The pressure–displacement and displacement–time (stress relaxation) curves were compared for untreated and PEF-treated tissues (Grimi 2009; Grimi et al. 2009a). Differences in the pressure–displacement curves (P–ε) for PEF-treated and untreated tissues were usually observed (Lebovka et al. 2004a; Chalermchat and Dejmek 2005; Bazhal et al. 2003b,c). Textural investigations (stress–deformation and relaxation tests) have shown that tissues (carrot, potato, and apple) lose a part of their textural strength after PEF treatment, and both the elasticity modulus and fracture stress decrease with increase in the damage degree (Lebovka et al. 2004a). For PEF-treated apples, linear dependency was observed between fracture pressure and the value of Z_C (Bazhal et al. 2003b,c, 2004). These data were confirmed by investigations on the textural and solid–liquid expression of PEF-treated potato tissues (Chalermchat and Dejmek 2005). The effects of PEF on the compression and solid–liquid expression of different vegetable tissues were also extensively studied (Lebovka et al. 2004a; Jemai and Vorobiev 2006; Grimi et al. 2007; Praporscic et al. 2007a,b).

The compression-to-failure and stress–relaxation measurements of apple, carrot, and potato tissues treated by PEF with different durations of treatment (t_{PEF}) were reported by Lebovka et al. (2004a). After a rather high-intensity, long-duration ($E = 1.1$ kV/cm, $t_{PEF} = 0.1$ s) PEF treatment, the tissues partially lose their initial strength. However, changes both in the elasticity modulus G_m and the fracture stress P_F were significantly smaller than changes observed for the freeze–thawed and thermally ($T = 45°C$, 2 h) pretreated tissues. Thus, tissue structure seems to be less affected by PEF treatment than by freeze–thawing or heating. This conclusion was later confirmed by textural studies on PEF-treated sugar beet tissue (Shynkaryk et al. 2008).

PEF treatment also accelerated the stress relaxation of tissues (Fincan and Dejmek 2003; Lebovka et al. 2004a, 2005a; De Vito et al. 2008). The relaxation behavior reflected the degree of membrane damage, but it was also sensitive to the state of the cell walls and the turgor pressure. Note that freeze–thawed tissues usually demonstrate faster relaxation than tissues treated by strong PEF (Lebovka et al. 2004, 2005a).

Note that the results of textural tests may depend on the mode of experiments—for example, they may be different for experiments with uniaxial (1d) and three-dimensional (3d) pressing. In 1d compression experiments, PEF treatment usually leads to depression of P–ε curves—that is, $\Delta P = P_{PEF} - P_i < 0$ (here P_{PEF} and P_i are the pressures for PEF-treated and intact tissues, respectively) for the same level of deformation ε (Lebovka et al. 2004a; Chalermchat and Dejmek 2005). The negative value of ΔP for 1d pressing was explained by the softening of tissue texture after PEF treatment followed by unconstrained liquid expression through the sidewalls (Lebovka et al. 2004a; Chalermchat and Dejmek 2005). However, another behavior was observed in experiments with 3d pressing (Grimi et al. 2009a), where the difference $\Delta P = P_{PEF} - P_i$ increases with the increase in deformation, ε. The positive value of ΔP in this case was explained by constrained filtration through the filter cake and higher stiffness of the network of cell walls saturated by intracellular liquid in PEF-treated tissues. Moreover, the fracture pressure P_c in 3d pressing experiments was approximately the same for untreated and PEF-treated potato samples, $P_c \approx 4.5 \pm 0.4$ MPa. This value is noticeably larger than the fracture pressure ($P_c \approx 1.5$–1.6 MPa) of potato samples used in 1d pressing experiments (Chalermchat and Dejmek 2005).

It can be concluded that the textural parameters of plant tissues may indefinitely reflect the PEF-induced changes in a complex form, although definitive relations between these parameters and cell damage degree remain unknown. However, textural experiments are rather useful for qualitative characterization of PEF-induced changes.

2.2.2.5 Acoustic Measurements

The acoustic technique is widely used for characterization of the quality of agricultural products (García-Ramos et al. 2005). For example, its applications to apple (Abbott et al. 1995), pineapple (Chen and De Baerdemaeker 1993), pear fruit (De Belie et al. 2000), avocado (Galili et al. 1998), watermelon (Yamamoto et al. 1980), and tomato (Schotte et al. 1999) tissues have been reported.

This technique allows measuring of the index of firmness F that shows good correlations with the quality and maturity of fruits and vegetables (Chen and Sun 1991). The index F (or stiffness coefficient) is a dynamic characteristic defined as $f^2 m^{2/3} \rho^{1/3}$, where f is the frequency corresponding to the maximum amplitude (A) in the acoustic spectrum, m is the mass of the sample, and ρ is the density of the tested tissue (Abbott et al. 1995; García-Ramos et al. 2005).

The successful application of the acoustic technique for characterization of PEF-treated tissues was recently reported (Grimi et al. 2010). The acoustic disintegration index Z_A was defined as

$$Z_A = (F - F_i)/(F_d - F_i) \tag{2.7}$$

where F is the measured index of firmness and subscripts i and d refer to the indices of firmness of the intact (untreated) and completely damaged tissues, respectively. Completely damaged tissue was obtained after freezing–thawing of the sample. Note that the definition of Z_A (Equation 2.7) is in clear analogy with the definitions of Z_C (Equation 2.4) and Z_D (Equation 2.6).

Note that the advantages of the acoustic technique for characterization of PEF-induced effects may be important when fruits and vegetables are processed as whole unpeeled samples. Examples of PEF application to whole samples have been demonstrated for sugar beet (Sack et al. 2005) and potato (Jaeger et al. 2008). PEF can also be attractive for treatment of other fruits and vegetables. The application of other methods (e.g., microscopy, electrical conductivity, or diffusion coefficient measurements) requires cutting and special preparation of samples; thus, they are destructive and may be dependent on local tissue characteristics.

2.2.2.6 Correlations between Z Values Estimated by Different Techniques

Although the question of correlations between the values of damage degree estimated by different techniques is rather important, it is not practically discussed in the literature. Figure 2.5 presents Z_C versus Z_D and Z_C versus Z_A dependencies obtained from the data of PEF treatment experiments with potato and apple, respectively (Lebovka et al. 2007a; Grimi et al. 2010). Note that the protocol of PEF treatment was the same for the same product. The observed dependencies Z_C (Z_D) and Z_C (Z_A) were nonlinear and were close to the power laws (i.e., $Z_C = Z_D^{m_D}$ and $Z_C = Z_A^{m_A}$), where $m_D = 1.68 \pm 0.04$ for potato and $m_A = 3.77 \pm 0.26$ for apple. The phenomenological theory (Archie 1942) predicts nonlinear dependence between the conductivity disintegration index Z_C and real damage degree Z (i.e., $Z_C = Z^m$), and the estimated values of m fall within the range of 1.8–2.5 for different plant tissues (apple, carrot, potato) (Lebovka et al. 2002). It was assumed that the acoustic disintegration index Z_A is better adapted for characterization of damage degree characterization than the conductivity disintegration index Z_C (Grimi et al. 2010). Use of Z_C results in a systematic underestimation of the damage degree. It is rather important because at high values of real damage degree, $Z \approx Z_A \approx 0.8–0.9$, when further PEF treatment is not efficient and gives no increase in Z value, the apparent conductivity disintegration index seems to be small, $Z_C \approx 0.4–0.6$ (Figure 2.4). However, PEF overtreatment is not desirable and may result in excessive power consumption.

2.2.3 KINETICS OF DAMAGE

The kinetics of biological material damage under PEF processing is governed by the mechanism of cell membrane electroporation.

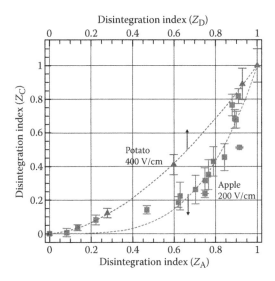

FIGURE 2.5 Dependencies of Z_C versus Z_D and Z_C versus Z_A for potato and apple, respectively. The pulse protocols were as follows: $E = 400$ V/cm, $t_i = 10^{-4}$ s (potato) and $E \approx 300$ V/cm, $t_i = 10^{-4}$ s (apple). The dashed lines correspond to the least square fitting of the experimental data to power equations $Z_C = Z_D^{m_D}$ and $Z_C = Z_A^{m_A}$ with $m_D = 1.68 \pm 0.04$ for potato and $m_A = 3.77 \pm 0.26$ for apple. (Compiled from Lebovka, N.I. et al., *J Food Eng*, 78, 606–613, 2007a, and Grimi, N. et al., *Biosyst Eng*, 105, 266, 2010.)

The time dependence of the membrane damage may be approximated by the first-order kinetic equation (Weaver and Chizmadzhev 1996)

$$Z = \exp(-t/\tau) \tag{2.8}$$

where τ is the damage time dependent on the transmembrane potential u_m and characteristics of membrane (τ_∞, Q, u_o)

$$\tau_m = \tau_\infty \exp(Q/(1 + (u_m/u_o)^2)) \tag{2.9}$$

The last equation follows from the fluctuation theory of electroporation, and we can refer as an example the typical values of $\tau_\infty \approx 3.7 \times 10^{-7}$ s, $u_o \approx 0.17$ V, and $Q \approx 109$, experimentally estimated at 293 K for lipid membranes (Lebedeva 1987).

Cell membranes in food tissues or in suspensions are exposed to highly inhomogeneous electric fields. Thus, the experimentally estimated time dependence of the damage degree Z may reflect the complexity of electric field distribution on the membrane surface, which is related to distribution of cell sizes, cell shape anisotropy, peculiarities of tissue structure, concentration of cells in suspension, and others (Lebovka et al. 2002). The kinetics of material disintegration during PEF treatment may be also influenced by mass transport and resealing processes (Lebovka et al. 2001; Knorr et al. 2001) and may be dependent on the PEF treatment protocol (Lebovka et al. 2001).

Different empirical equations were used for approximation of experimental dependencies in PEF damage kinetics (Barbosa-Canovas et al. 1998; Wouters and Smelt 1997)—for example, Hulsheger's equation (Hulsheger et al. 1983)

$$Z = (t/\tau)^{-(E-E_c)/k} \tag{2.10}$$

Weibull's equation (van Boekel and Martinus 2002)

$$Z = 1 - \exp(-t/\tau)^k \tag{2.11}$$

or the transition equation (Bazhal et al. 2003a)

$$Z = 1/(1 + (\tau/t)^k) \tag{2.12}$$

Here τ, E_c, and k are the empirical parameters.

Equations 2.10 through 2.12 fulfill the conditions $Z = 1$ at $t = 0$ and $Z = 1$ at $t = \infty$.

2.2.3.1 Characteristic Damage Time

Equation 2.12 was successfully used for the approximation of damage evolution in fruit and vegetable tissues (Bazhal et al. 2003a). It follows from Equation 2.12 that $Z = 0.5$ at $t = \tau$. Here τ is the characteristic damage time, which is defined as a time necessary for half-damage of material (i.e., $Z = 0.5$) (see inset in Figure 2.6). This measure is useful for crude characterization of damage kinetics, when the strict law is unknown, yet it obviously differs from the first-order kinetics law described by Equation 2.8. Figure 2.6 presents examples of characteristic time τ versus electric field strength E for different vegetable and fruit samples (Grimi 2009). These data were obtained for PEF-treated (by square wave pulses, duration $t_i = 100$ μs) whole products. Onions and oranges have stronger resistance to PEF treatment and require longer treatment time or higher electric field strength. In contrast, tomatoes and apples have demonstrated weaker resistance to PEF than all the other tested products, and their τ values reach a minimum at $E \geq 400$ V/cm. The effects observed for PEF-treated whole products reflect the specific structure of cellular materials, differences in the size of their cells, and differences in the relative electrical conductivities of the product and the aqueous medium. Similar $\tau(E)$ dependencies for cut cubic (1 cm^3) apple, potato, cucumber, aubergine, pear, banana, and carrot samples were reported by Bazhal et al. (2003a).

2.2.3.2 Synergy of Simultaneous Electrical and Thermal Treatments

An obvious synergy of simultaneous electrical and thermal treatments of food products is usually observed (Vorobiev and Lebovka 2008; Lebovka et al. 2005a,b, 2007a). This synergy is most evident for electroprocessing at a moderate electric field strength ($E < 100$ V/cm) under ambient conditions, or only thermal processing at a moderate temperature ($T < 50°C$). The thermal damage of a biomaterial under

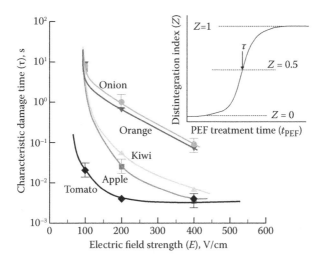

FIGURE 2.6 Characteristic time τ versus electric field strength E for different vegetable and fruit samples. Data were obtained from the measurements of acoustic disintegration index of PEF-treated samples in tap water. (Compiled from data presented in Grimi, N., PhD dissertation, University of the Technology of Compiègne, Compiègne, 2009.) The inset shows schematic Z versus t dependence; here τ is the characteristic damage time, defined as the time necessary for half-damage of material (i.e., $Z = 0.5$).

ambient conditions is noticeable only if the duration of treatment exceeds 10^5 s and could be accelerated only by increasing the temperature above 50°C.

Moreover, a rather complex kinetics with an intermediate saturation step (when disintegration index Z reaches a plateau, $Z = Z_s$) was often observed for long-duration PEF treatment at a moderate electric field ($E < 300$ V/cm) and a moderate temperature ($T < 50$°C) (Lebovka et al. 2001, 2007a). For example, the maximal disintegration index Z_s was of the order of 0.75 at $E = 100$ V/cm for sugar beet tissue (Lebovka et al. 2007a, 2008). The step-like behavior of $Z(t)$ was also observed for inhomogeneous tissues such as red beetroots (Shynkaryk 2007; Shynkaryk et al. 2008). Such saturation at the level of $Z = Z_s$ possibly reflects the presence of a wide distribution of cell survivability, related with different cell geometries and sizes. It was experimentally observed that the saturation level Z_s increases with increase of both electric field strength E (Lebovka et al. 2001) and temperature T (Lebovka et al. 2007b). For tissues with relatively homogeneous structures (potatoes, apples, chicory, etc.), this saturation behavior is less pronounced and not practically observed at higher electric fields ($E > 500$ V/cm). If PEF treatment stops at the saturation level, the scenario of the further evolution can be different depending on the type of material and the level of its disintegration. The cells can partially reseal at a very small level of disintegration (Knorr et al. 2001). However, a higher level of disintegration usually results in further increase of Z after a relatively long time (Lebovka et al. 2001; Angersbach et al. 2002).

The synergy of simultaneous PEF and thermal treatment with increase in temperature T or electric field strength E (or both) was evidently demonstrated by the presence of a drastic drop of the characteristic damage time by many orders of magnitude (Lebovka et al. 2005a,b, 2007b). Moreover, the electroporation activation energy W of tissues was a decreasing function of electric field strength E as a result of electrothermal synergy (Loginova et al. 2010). This synergism of tissue damage possibly reflects the existence of softening transitions in membranes at temperatures within 20–55°C (Exerova and Nikolova 1992; Mouritsen and Jørgensen 1997). A noticeable drop of the breakdown transmembrane voltage u_m of a single membrane was experimentally observed near the region of thermal softening (≈50°C) (Zimmermann 1986). The fluidity and domain structure of the cell membrane exert a noticeable influence on electropermeabilization of cells (Kandušer et al. 2008).

The general relations between characteristic damage time τ, electric field strength E, and temperature T may be rather complex. These relations were studied in detail for potato tissues. The following equation was used for the fitting of experimental data (Lebovka et al. 2005a)

$$\tau_m = \tau_\infty \exp(W/kT(1 + (E/E_o)^2)) \tag{2.13}$$

Here τ_∞, W, and E_o are adjustable empirical parameters. Note that that this equation is fully empirical and resembles the form of Equation 2.1.

Interesting synergetic effects of simultaneous electrical and thermal treatments were also observed in ohmic heating experiments (Lebovka et al. 2005a,b, 2007b). A direct method based on experimental observations of electrical conductivity changes during the ohmic heating was proposed for monitoring of electroporation changes, and it was shown that ohmic heating at electric field strength E of the order of 20–80 V/cm induced, inside the tissue, structural changes related to loss of membrane barrier functions.

2.2.4 INFLUENCE OF PULSE CONTROL

Sale and Hamilton (1967) concluded that two main relevant parameters determine the efficiency of PEF damage: the electric field strength (E) and the total time of PEF (t_{PEF}). Typically, higher electric field strengths lead to better damage efficiency (Bazhal 2001; Bouzrara 2001; Praporscic 2005; Toepfl 2006; Shynkaryk 2007); however, electrical power consumption and ohmic heating also become essential at high electric fields. More detailed experiments have also shown that electroporation efficiency may depend on the parameters of the pulse, such amplitude (or electric field strength E), shape, duration t_i, number of repetitions n, and intervals between pulses Δt (Canatella et al. 2001, 2004). A typical PEF protocol for bipolar pulses of a near-rectangular shape is presented in Figure 2.7. Such complex protocol with adjustable long pause between pulse trains allows fine regulation of the disintegration index Z without noticeable temperature elevation during the PEF treatment.

2.2.4.1 Waveforms of PEF Pulses

The waveforms of pulses commonly used in PEF generators are exponential decay, oscillatory, triangular, square, or more complex waveforms (Miklavcic and Towhidi

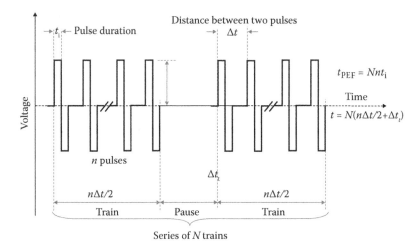

FIGURE 2.7 The typical PEF protocol. Bipolar square waveform pulses are presented. A series of N pulses (train) is shown. Each separate train consists of n pulses with pulse duration t_i, pause between pulses Δt, and pause Δt_t after each train. The total time of PEF treatment is regulated by variation of the number of series N and is calculated as $t_{PEF} = nNt_i$.

2010). The exponential decay, triangular, and square pulses may be either monopolar or bipolar. Square-wave generators are more expensive and require more complex equipment than exponential decay generators. However, experiments with inactivation of microbial cells have shown (Zhang et al. 1994) that application of square-wave pulses resulted in better energy performance and higher disintegrating efficiency than exponential decay pulses. The superiority of square-wave pulses over exponential decay pulses was explained by the better uniformity of electric field strength during each pulse application (Barbosa-Canovas and Altunakar 2006).

Bipolar pulses seem to be more advantageous as they cause additional stress in membrane structure, allow avoiding asymmetry of membrane damage in the cell, and result in more efficient electroporation responses (Saulis 1993, 2010; Fologea et al. 2004; Talele et al. 2010). Moreover, application of bipolar pulses offers minimum energy consumption, with reduced deposition of solids on electrodes and smaller food electrolysis (Chang 1989; Qin et al. 1994; Wouters and Smelt 1997).

2.2.4.2 Pause between Pulses

The pause between pulses Δt may be an essential parameter affecting PEF electroporation efficiency (Kinosita and Tsong 1979). It was shown that relaxation of the conductivity of membranes was complete for a relatively long pause ($\Delta t > 1$ s); however, it was incomplete for high repetition frequency (above 1 kHz), and the initial level of membrane conductivity for consecutive pulses increased. These results can be explained by the existence of short- and long-lived transient ("transport") membrane pores (Pavlin et al. 2008). The influence of distance between pulses Δt on disintegration of the apple tissue (Lebovka et al. 2001) and on inactivation of *Escherichia coli* cells (Evrendilek and Zhang 2005) was also discussed. For example, it was shown that a protocol with a longer pause between pulses at fixed values of E and t_{PEF} allowed acceleration of the disintegration kinetics of apple tissue. The results were explained, accounting for the moisture transport processes inside the cell structure. However, the impact of pause between pulses on PEF-induced effects is still ambiguous and requires a more detailed investigation in the future.

2.2.4.3 Pulse Duration

The impact of pulse duration t_i on PEF-induced effects in treatment of plant tissues and microbial species was also observed (Martin-Belloso et al. 1997; Wouters et al. 1999; Mañas et al. 2000; Raso et al. 2000; Aronsson et al. 2001; Abram et al. 2003; Sampedro et al. 2007; De Vito et al. 2008). Some authors demonstrated that inactivation of microbes was more efficient at higher pulse width, subject to invariable quantity of applied energy (Martin-Belloso et al. 1997; Abram et al. 2003), while others observed little effect of pulse width on inactivation at equal energy inputs (Raso et al. 2000; Mañas et al. 2000; Sampedro et al. 2007; Fox et al. 2008). The effect of pulse width on microbial inactivation seems to vary depending on electric field strength; still, the obtained results are controversial (Wouters et al. 1999; Aronsson et al. 2001). A critical review of the effect of pulse duration on electroporation efficiency in relation to therapeutic applications was recently published (Teissié et al. 2008).

A distinct correlation between pulse duration and damage efficiency was recently observed in PEF treatment experiments with apples (De Vito et al. 2008). The theory predicts deceleration of the membrane charging processes in materials with large cell sizes (Kotnik et al. 1998). An efficient PEF treatment requires application of relatively long pulses. To reach the maximum transmembrane voltage, the pulse duration t_i should be larger than membrane charging time t_c. The experimental data supported this conclusion and clearly demonstrated the influence of pulse duration t_i (10–1000 µs) on the efficiency of PEF treatment of grapes, apples, and potatoes (De Vito et al. 2008; Grimi 2010). Longer pulses were found to be more effective, and their effect was particularly pronounced at room temperature and moderate electric fields ($E = 100–300$ V/cm) (De Vito et al. 2008).

2.2.5 POWER CONSUMPTION

The power consumption Q (mass density of the energy input) during PEF treatment can be estimated from the following equation

$$Q = \int_0^t \sigma(t)E^2 \, dt/\rho \tag{2.14}$$

Here ρ is the density of material.

It is usually assumed that electrical conductivity $\sigma(t)$ is a complex function of time, owing to the development of two processes during PEF treatment: damage of material and temperature increase (related to ohmic heating). Both of these processes result in increase in the value of $\sigma(t)$.

The power consumption Q is the most important measure for estimation of industrial attractiveness of any electrotechnology, and Q values have been reported for PEF inactivation and extraction-oriented experiments.

The theoretical estimations predict that the product τE^2, as well as the power consumption Q, goes through a minimum with increase of the electric field strength E (Lebovka et al. 2002). Hence, there exists some optimum value of electric field strength $E \approx E_o$, which corresponds to minimum power consumption, and this prediction was supported by experimental data obtained for different fruit and vegetable tissues (Bazhal et al. 2003a). It was shown that an increase of E above E_o resulted in progressive increase of power consumption, but gave no additional increment to the conductivity disintegration index Z. For some vegetable and fruit tissues (apple, potato, cucumber, aubergine, pear, banana, and carrot), the typical values of E_o were within 200–700 V/cm and PEF treatment times required for effective damage, t_{PEF}, were within 1000 µs–0.1 s (Bazhal et al. 2003a). However, for grape skins, efficient PEF-induced damage was observed at higher electric fields (1–10 kV/cm) for PEF treatment times within 5–100 µs (López et al. 2008a). Note that the specific power consumption may be roughly estimated from Equation 2.14 as

$$Q \sim \sigma_d E^2 t_{PEF} / \rho \qquad (2.15)$$

where σ_d is the electrical conductivity of the totally damaged tissue (Lebovka et al. 2002).

Putting $\sigma_d = 0.1$ S/cm and $\rho = 0.8 \times 10^3$ kg/m³ (these are the typical values for apples (Lebovka et al. 2000), we obtain approximately the same value, $Q \approx 3$ kJ/kg, both for treatment by moderate electric field ($E = 500$ V/cm and $t_{PEF} = 10000$ µs) and by high electric field ($E = 5000$ V/cm and $t_{PEF} = 100$ µs).

However, in the general case, estimations of the values of E_o and Q require more thorough accounting of the tissue structure, tissue heterogeneity, cell geometry, and other factors (Ben Ammar et al. 2011). An approximate proportionality in behavior of characteristic damage time τ and power consumption Q was observed for different fruit and vegetable tissues (Figure 2.8). Comparisons of theory and experiment have shown that the optimal values of E_o and power consumption Q may be critically dependent on electrical conductivity contrast—that is, the difference between electrical conductivities of intact (σ_i) and completely damaged (σ_d) tissues (Figure 2.8).

Figure 2.9a displays power consumption Q ($Z_C = 0.8$) versus electric field strength E for two values of electrical conductivity contrast $k = \sigma_i/\sigma_d$ as predicted by Monte Carlo simulations (Ben Ammar et al. 2011). Theory predicts the increase of E_{opt} and $Q(E_{opt})$ values with the decrease in electrical conductivity contrast, $k = \sigma_i/\sigma_d$. The same tendency was experimentally observed in PEF-treated potato and orange (Figure 2.9b). With increase of E from 400 V/cm to 1000 V/cm, the power consumption Q, required for attaining of the given level of disintegration, $Z_D = 0.8$, increased for potato (high k, $k \approx 14$) and decreased for orange (small k, $k \approx 1.3$).

The experimentally estimated power consumptions Q for PEF-treated tissues were found to be rather low and typically lying within 1–15 kJ/kg. For example, they were 6.4–16.2 kJ/kg ($E = 0.35$–3.0 kV/cm) for potato (Angersbach et al. 1997), 0.4–6.7 kJ/kg ($E = 2$–10 kV/cm) for grape skin (López et al. 2008a), 2.5 kJ/kg (7 kV/cm)

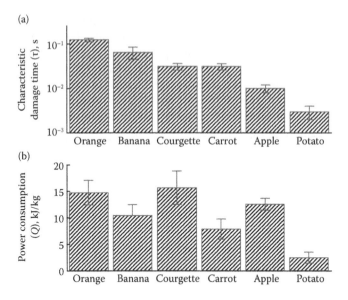

FIGURE 2.8 Power consumption Q ($Z_C = 0.8$) versus electric field strength E at different values of $k = \sigma_i/\sigma_d$: (a) results of Monte Carlo simulations and (b) experimentally estimated values for potato and orange. (Compiled from the data presented by Ben Ammar, J. et al., *J Food Sci*, 76, E90–E97, 2011.)

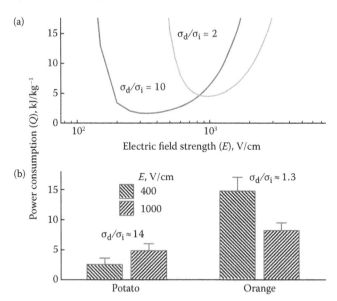

FIGURE 2.9 Correlations between characteristic damage time τ and power consumption Q for different fruit and vegetable tissues. The value of Q was estimated at a relatively high level of disintegration ($Z_C = 0.8$) for PEF treatment at $E = 400$ V/cm with 1000 µs bipolar pulses of near-rectangular shape. (Compiled from the data presented by Ben Ammar, J. et al., *J Food Sci*, 76, E90–E97, 2011.)

for red beetroot (López et al. 2009a), 3.9 kJ/kg (7 kV/cm) for sugar beet (López et al. 2009b), and 10 kJ/kg (400–600 V/cm) for chicory root (Loginova et al. 2010).

Thus, from the standpoint of power consumption, the PEF method is practically ideal for the production of damaged plant tissues as compared to other methods of treatment such as mechanical (20–40 kJ/kg), enzymatic (60–100 kJ/kg), and heating or freezing–thawing (>100 kJ/kg) (Toepfl et al. 2006). However, bacterial inactivation and food preservation requires high electric field strengths ($E_0 = 15$–40 kV/cm), and it naturally results in a noticeably higher specific power consumption, of the order of 40–1000 kJ/kg (Toepfl et al. 2006). Such power consumption is also typical for HVED (Boussetta et al. 2009). Hydroxide treatment at a moderate electric field (typically 20–80 V/cm) requires high power consumption, typically comparable with heating or freezing–thawing (20–40 kJ/kg).

2.3 PEF-ASSISTED EXTRACTION

2.3.1 VEGETABLE AND FRUIT TISSUES

In raw food plants, valuable compounds are initially enclosed in cells, which have to be damaged for facilitation of intracellular matter recovery. Conventional cell damage techniques, such as fine mechanical fragmentation and thermal, chemical, and enzymatic treatments, lead to more severe disintegration of the tissue components, including cell walls and cell membranes. PEF treatment, which is less destructive than conventional methods, can be used for a more selective extraction of cell components.

2.3.1.1 Potato

Potato was used as a model system in many electrically assisted experiments for testing electroporation effects in plant tissues (Angersbach et al. 1997, 2000; Lebovka et al. 2005a,b, 2006, 2007a; Galindo et al. 2008a,b; Pereira et al. 2009). The presence of reversible electroporation has been reported for potato tissue (Angersbach et al. 2000; Galindo 2008; Pereira et al. 2009). This process involved formation of pores after 0.7 μs of membrane charging; however, the vitality and metabolic activity of potato cells were recovered within seconds after electric field shutdown (Angersbach et al. 2000). The transient processes in the viscoelastic behavior of potato during PEF application at $E = 30$–500 V/cm followed by recovery of cell membrane properties and turgor were attributed to consequences of electroporation (Pereira et al. 2009). The reversibility of electropermeabilization was dependent on PEF parameters and different stress-induced effects, and metabolic responses were observed (Lebovka et al. 2008; Galindo et al. 2008a,b; Pereira et al. 2009; Galindo 2009). It was shown that the effects of low-temperature permeabilization of potato at 50°C were stimulated by preliminary electric field treatment (Lebovka et al. 2008). Mild PEF treatment allows the recovery of the functional properties of membranes, and metabolic responses may arise in a time scale of seconds (Galindo et al. 2008a,b). Isothermal calorimetry, electrical resistance, and impedance measurements have shown that 24 h after PEF treatment, the metabolic response of potato tissue involved oxygen-consuming pathways (Galindo et al. 2009a,b). The metabolic responses were strongly dependent on the PEF protocol and were independent of total permeabilization. It was shown that even mild electrical treatment of potato permeabilizes tissue, and this effect could be seen in electrical conductivity behavior 24 h after the treatment (Kulshrestha and Sastry 2010).

The effects of PEF treatment on the textural and compressive properties of potato were studied in detail (Fincan and Dejmek 2003; Lebovka et al. 2004a; Chalermchat and Dejmek 2005; Grimi et al. 2009a). It was shown that relatively strong PEF treatment resulted in decrease in the stiffness of potato tissues to levels similar to hyperosmotically treated samples (Fincan and Dejmek 2003). 1D force textural investigations (stress–deformation and relaxation tests with unconfined potato samples) have shown that the tissue loses a part of its textural strength after PEF treatment, and both the elasticity modulus and fracture stress decrease with increase in damage degree (Lebovka et al. 2004a). These data were confirmed by textural and solid–liquid expression investigations of PEF-treated potato tissues (Chalermchat and Dejmek 2005). It was also shown that the application of PEF treatment only was not sufficiently effective for complete elimination of textural strength; however, mild thermal pretreatment at 45–55°C allowed to increase PEF efficiency (Lebovka et al. 2004). 3d textural investigations (of confined potato samples) with the applied pressure varying within 0.5–4 MPa showed that fracture pressure was approximately the same for both PEF-treated and untreated specimens, but PEF-treated tissues displayed higher stiffness than untreated ones (Grimi et al. 2009a). The critical pressure P_m, at which the time the pressure-induced cell rupture should be of the same order of magnitude as the time of fluid expression from the damaged cells, was estimated as $P_m \approx 6$ MPa.

Potato tissue was used for detailed studies of the effects of temperature and the PEF protocol on the characteristic damage time (Lebovka et al. 2002, 2005a), dehydration (Arevalo et al. 2004), freezing (Jalté et al. 2009; Ben Ammar et al. 2009, 2010), and drying (Lebovka et al. 2007a). However, despite the numerous fundamental studies of PEF effects in potato, attempts in extraction-oriented application of PEF treatment are still rare. One can refer to the work by Propuls GmbH (Bottrop, Germany) on PEF application for facilitation of starch extraction from potato (Loeffler 2002; Topfl 2006). It was shown that PEF treatment also allowed the enhancement of the extractability of an anthocyanin-rich pigment from purple-fleshed potato (Topfl 2006). Note that the effects of PEF on the structure of potato starch were recently revealed (Han et al. 2009). Starch granules lost their shape after PEF treatment at 30–50 kV/cm: dissociation, denaturation, and damage of potato starch granules were observed. However, sequential PEF (at 400 V/cm) and osmotic pretreatments of potato tissue resulted in starch granules with a rougher surface. A noticeable disordering of the surface morphology of starch granules inside potato cells in the freeze-dried potatoes after sequential PEF and osmotic pretreatment was also observed (Ben Ammar et al. 2009, 2010).

2.3.1.2 Sugar Beet

The extraction technology conventionally used in the sugar industry is a power-consuming hot water technique. It involves diffusion of sugar from sliced sugar beet cossettes at 70–75°C. A relatively high temperature is required for tissue denaturation by heat. Unfortunately, treatment by heat also causes alterations in cell wall structure through hydrolytic degradation reactions (molecular chain breakage, detachment of polysaccharide fragments) (Van der Poel et al. 1998). The elastic properties of tissues can be strongly affected by thermal treatment. Moreover, cell components other than sugar, such as pectin, pass into the juice during extraction, thus affecting juice purity. In addition, formation of some colorants such as melanoidins is promoted by thermal diffusion. This results in the necessity for the application of a complex multistage process (preliming, liming, first and second carbonation, several filtrations, and sulfitation) and lime discharge (3–3.2 kg of limestone for 100 kg of beetroot) for juice purification (Van der Poel et al. 1998).

The diffusion process in the sugar extraction technology can be intensified by electric field treatment. Early investigations show that application of low-gradient alternating electric fields (<100 V/cm) can enhance the juice extraction process (Zagorulko 1958) and increase the soluble matter diffusion coefficient in the sugarbeet (Bazhal et al. 1983; Katrokha and Kupchik 1984; Jemai 1997). Therefore, PEF is a potential alternative method to conventional thermal technology.

Previous experimental studies have shown in principle the possibility of sugar extraction by cold or moderately heated water (Jemai and Vorobiev 2003). The strong dependence of damage efficiency on temperature and pulse protocol parameters was recently demonstrated in sugar beets (Lebovka et al. 2007b). The Arrhenius form of the characteristic damage time versus inverse temperature dependencies was observed both for electrical and thermal processes with activation energies within $W = 149–166$ kJ/mol. Such large activation energies correspond to the membrane damage of sugar beet cells. The thermal and cold diffusion kinetics and diffusion coefficients D_{eff} for untreated and PEF-treated ($E = 400$ V/cm, $t_{PEF} = 0.1$ s) sugar beet slices (1.5 mm × 10 mm × 10 mm) were compared (Lebovka et al. 2007b). PEF treatment noticeably accelerated the aqueous diffusion of sugar from the sliced cossettes. The Arrhenius dependencies of the effective diffusion coefficient D_{eff} versus inverse temperature were observed with activation energies of 21 ± 2 kJ/mol and 75 ± 5 kJ/mol for the PEF-treated and untreated sugar beet slices, respectively (Figure 2.10). The activation energy of PEF-treated slices was close to that characteristic for sugar diffusion in aqueous solutions, $W \approx 22$ kJ/mol (Lysjanskii 1973). The larger activation energy for the untreated slices possibly reflects interrelations of the effects of restricted diffusion and thermally induced damage (Lebovka et al. 2007b). At 70°C, the effective diffusion coefficient D_{eff} was nearly the same for untreated and PEF-pretreated slices ($1 \times 10^{-9}–1.5 \times 10^{-9}$ m^2/s). The difference in D_{eff} values of the untreated and PEF-pretreated tissues increased significantly for less heated tissue. For instance, the values of D_{eff} were nearly the same for sugar diffusion from untreated tissue at 60°C and from PEF-pretreated tissue at 30°C. The purest juice was obtained after cold diffusion. However, even after thermal diffusion at 70°C, juice purity was higher for slices pretreated by PEF than for untreated slices.

Diffusion experiments with PEF-treated cossettes prepared from sugar beet using industrial knives were recently done under electric field strength E varying between 100 and 600 V/cm and total time of PEF treatment $t_{PEF} = 50$ ms (Loginova et al. 2011). Figure 2.11 presents the temperature dependencies of diffusion juice purity P and sucrose concentration S in experiments with untreated and PEF-treated sugar

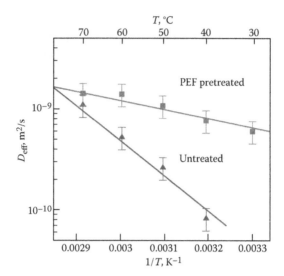

FIGURE 2.10 Arrhenius plots of the effective diffusion coefficient D_{eff} for the untreated and PEF-pretreated sugar beet slices. (From Lebovka, N. et al., *J Food Eng*, 80, 639–644, 2007b. With permission.)

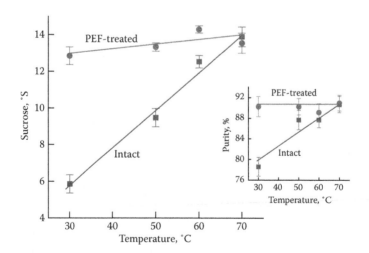

FIGURE 2.11 Temperature dependencies of diffusion juice purity P and sucrose concentration S in experiment with untreated and PEF-treated sugar beet cossettes. PEF treatment was done at $E = 600$ V/cm; the pulse duration t_i was 100 μs; and the total time of PEF treatment t_{PEF} was 50 ms, which corresponded to 5.4 kW·h/t of power consumption. (From Loginova, K.V. et al., *J Food Eng*, 102, 340–347, 2011. With permission.)

beet cossettes. For the untreated cossettes, the sucrose content in the cold diffusion juice was very low (<6% at 30°C) and the purity of cold juice was <80%. For the PEF-treated cossettes, the soluble solids and the sucrose content in the cold diffusion juice (30°C) were slightly lower than in the hot diffusion juice (70°C). However, the purity of the cold diffusion juice (30°C) was not lower than that of the hot (70°C) juice (Figure 2.11a). Some studies even showed that the purity of the cold diffusion juice is higher than that of the hot juice (Lebovka et al. 2007b). It was also shown (Loginova et al. 2011) that sugar beet pulp could be well exhausted by cold or mild thermal extraction of PEF-treated cossettes, and the pulp obtained by cold extraction of PEF-treated cossettes had a noticeably higher (>30%) dryness than that of the pulp obtained by conventional hot water extraction. The estimated energy surplus for cold extraction with a temperature reduction from 70°C to 30°C (i.e., by $\Delta T = 40°C$) was ≈ 46.7 kW·h/t, and was noticeably higher than the power consumption required for PEF treatment, ≈ 5.4 kW·h/t. Loginova et al. (2011) concluded that such power consumption can even be reduced by further optimization of PEF parameters and minimization of the liquid-to-solid ratio during PEF treatment.

The different aspects of PEF-assisted pressing and aqueous extraction from sugar beets were recently studied (Bouzrara and Vorobiev 2000, 2001, 2003; Eshtiaghi and Knorr 2002; El-Belghiti and Vorobiev 2004, 2005a,b; El-Belghiti 2005; Vorobiev et al. 2005; Praporscic et al. 2005; Jemai and Vorobiev 2006). The efficiency of the so-called cold pressing of PEF-treated sugar beet cossettes was demonstrated (Bouzrara and Vorobiev 2000, 2003; Eshtiaghi and Knorr 2002). Jemai and Vorobiev (2006) reported that up to 82% of the overall yield could be achieved by a two-stage pressing with an intermediate PEF application ($E = 400$ V/cm, $t_{PEF} = 0.1$ s). Initial pressurization served to assure good electrical contact between slices. In addition, juices obtained after PEF application (the so-called second juices) systematically had higher sugar content and better color. PEF treatment was successfully applied in scale-up experiments aimed at developing a novel process of cold juice extraction from sugar beet cossettes (Jemai and Vorobiev 2006). The processing scheme consisted of two initial pressing steps with an intermediate PEF treatment, followed by one or more washing steps and a final pulp pressing (pressure of 5–15 bar; particles filling of 4.5–15 kg). The cold juices expressed from sugar beet gratings after an intermediate PEF treatment had higher purity (95–98%) than those obtained before PEF application (90–93%). The losses of sugar in the pulp could be significantly reduced to about 3% of the initial sugar content through some washing and final pressing operations. The quantity of pectin was noticeably lower, and the color of juice was 3–4 times less intensive than the color of factory juices. In addition, significant amounts of potassium, sodium, and α-amino nitrogen were found to remain in PEF-treated particles, which explains why better-purity juices were obtained after PEF treatment (Jemai and Vorobiev 2006). These results demonstrating significant amelioration of juice quality open new interesting prospects of cold PEF-enhanced expression from the sugar beets.

El-Belghiti et al. (2005a,b) studied the influence of PEF intensity and duration on the static and centrifugal aqueous extraction from coarse sugar beet slices (1.5 mm in thickness) obtained on a 6 mm grater. The solute yield was significantly increased with PEF treatment at $E = 670$ V/cm: it was about 40% for untreated slices, and a yield of 93% was attained after PEF treatment and 2 h of extraction at ambient temperature. Further increasing the PEF intensity up to 800 V/cm was not effective. Centrifugal diffusion of slices was done with a liquid-to-solid mass ratio (L/S) of 3, and under different centrifugal accelerations (150–9660 × g) and temperatures (18–35°C). The extraction kinetics was much faster in the centrifugal field. For instance, the solute yield after PEF treatment reached 97% after 60 min of extraction even at a low centrifugal acceleration (14 × g) and at a temperature of 25°C. At high centrifugal acceleration (150 × g), a solute concentration of 97% was reached after 25 min of aqueous extraction at 25°C and just after 15 min of aqueous extraction at 35°C.

2.3.1.3　Sugar Cane

The effect of PEF pretreatment at 4–5 kV/2.5 cm on the disintegration of sugar cane was recently investigated (Kuldiloke et al. 2008). Sugar canes were cut into 10- to 15-cm-length cylinders, placed in a treatment chamber containing tap water, treated with PEF, cut into small pieces of approximately 3 × 3 mm length and 1 mm thickness, and then pressed. It was shown that PEF pretreatment at room temperature is a suitable method for permeabilization of sugar cane. The juice yield of PEF-pretreated samples was higher (74.5%) than that of the heat-treated (73.2%, 20 min at 70°C) and untreated (65.5%) sugar cane, and the power consumption of PEF (17 kJ/ kg) was 10 times less than that of heat treatment (171 kJ/ kg).

2.3.1.4　Red Beet

Red beets (*Beta vulgaris* L.) are widely used for the industrial production of natural water-soluble betalain pigments based on red-violet betacyanins (betanine) and yellow-orange betaxanthins (Stinzing and Carle 2004, 2007, 2008). Betalains exhibit good antiviral and antimicrobial activities and may be considered useful cancer-preventive agents, while betaxanthins may be used as a source of essential dietary amino acids (Delgado-Vargas and Paredes-López 2003). Betalains have potent antioxidant activity, conferring protection against degenerative diseases (Azeredo 2009). These pigments are usually produced by long-term solid–liquid extraction at room temperature (Delgado-Vargas et al. 2000; Delgado-Vargas and Paredes-López 2003) or by pressing grinded red beetroots followed by pasteurization of the resulting juice (Stinzing and Carle 2004, 2007, 2008). However, the stability of betalains is rather sensitive to temperature, presence of metals, pH, water activity, light exposure, enzymes, and oxygen (Saguy et al. 1978; Herbach et al. 2004; Stinzing and Carle 2004, 2007, 2008).

All of these factors complicate the extraction process, hinder extraction kinetics, and reduce the yield of colorants from the red beet. PEF treatment allows overcoming the restrictions typical for thermally sensitive bioproducts. A high degree of extractability of colorants from the red beetroot was observed after PEF treatment at the field strength 1 kV/cm, when samples released about 90% of their total red pigment following 1 h of aqueous extraction (Fincan et al. 2004; Chalermchat et al. 2004). The mechanism of electroporation was found to be responsible for acceleration of dye extraction from the red beet by moderate electric field treatment (Kulshrestha and Sastry 2003). A possibility of DC-assisted extraction and separation of the red beet pigment was also demonstrated (Zvitov et al. 2003; Zvitov and Nussinovitch 2005).

Lopez et al. (2009a) showed that application of the PEF treatment at 7 kV/cm enabled increasing the maximum yield of betanine by a factor of 4.2 compared to samples not subjected to PEF treatment, and achieved an almost complete betanine release. A combination of PEF at 7 kV/cm and pressing at 14 kg/cm² shortened the time of extraction by 18-fold. It can be noted that effective electroporation of the red beet tissue at ambient temperatures can be attained even at lower electric fields of 400–600 V/cm, and it allows acceleration of drying of the red beet tissue (Shynkaryk et al. 2008).

2.3.1.5 Carrot

Carrots are vegetables containing both water-soluble (mainly soluble sugars) and water-insoluble (carotenoids) components. They are rich in sugars, as indicated by their sweetness. In addition, the main components of the cellular juice (°Brix$_j$) include soluble sugars such as sucrose (56.9%), glucose (24.6%), and fructose (18.5%) (Rodríguez-Sevilla et al. 1999). There exist many examples of PEF application for the expression (Knorr et al. 1994; Bouzrara 2001; Praporscic et al. 2007b) and extraction (El-Belghiti and Vorobiev 2005a,b; El-Belghiti et al. 2007; Grimi et al. 2007) of juice from carrots.

The carrot juice yield increased from 30–50% to 70–80% as a result of PEF pre-treatment (Knorr et al. 1994; Rastogi et al. 1999). PEF treatment (at E = 220–1600 V/cm) accelerated the osmotic dehydration of carrots, and the effective diffusion coefficients of water and solute increased exponentially with electric field strength. It was demonstrated that a large juice yield could be attained even at a rather low voltage gradient of 360 V/cm, and juice expressed after PEF treatment are more transparent and less turbid than that from untreated carrots (Bouzrara 2001). Moreover, the °Brix values instantly increased after PEF treatment (Praporscic et al. 2007b). The effects of PEF treatment (at 25 and 30 kV/cm) on the release of carotenoids in an orange–carrot juice mixture (80:20, v/v) was studied by Torregrosa et al. (2005). Liquid chromatography was used for quantification of carotenoids, and PEF processing was shown to result in a significant increase in carotenoid and vitamin A concentrations.

The kinetics of extraction from carrot slices obtained by grating carrots in a 6 mm grater (1.5-mm-thick coarse slices) or in a 2 mm grater (0.5 mm thick fine slices) were studied by El-Belghiti and Vorobiev (2005a,b). They showed that fine and coarse slices demonstrated almost the same extraction kinetics after PEF treatment at E = 550 V/cm. This confirms the attractiveness of PEF treatment especially for coarse particles. In the absence of PEF pretreatment, only 45% of solute was obtained from the coarse slices after 8 h of extraction at 18°C. The increase in stirring speed up to 250 per minute enhanced extraction kinetics. An energy input of 9 kJ/kg was considered as optimal and was maintained to optimize the diffusion parameters (duration, temperature, and stirring velocity).

The centrifugal aqueous extraction of solute (accelerations from $14 \times g$ to $5434 \times g$) from the carrot gratings treated by PEF at 670 V/cm (300 pulses of 100 μs) was investigated at temperatures within 18–35°C (El-Belghiti et al. 2005, 2007). It was shown that an increase of the centrifugal acceleration up to $150 \times g$ enhanced extraction kinetics, and tissue preheating at 50°C allowed easier electrical permeabilization of the cell membranes.

Grimi et al. (2007) studied carrot juice extraction using laboratory filter-press chamber with different combinations of pressing and washing operations. For the smallest slices (0.078 × 0.078 × 2 mm), the washing–pressing procedure gave the highest Brix of the juice and PEF provided no additional effect on the juice yield and soluble matter content. However, the resulting juice was highly clouded; it contained cell wall residues and was rich in submicrometer suspended particles. For the largest slices (7 × 2 × 30 mm), PEF application noticeably improved the yield and soluble matter content of the juice, but left most of the carotenoids inside the press cake. An example of PEF application to extract juice from large slices of carrots evidenced the possibility of selective extraction of water-soluble components (soluble sugars) and the production of a "sugar-free" concentrate rich in vitamins and carotenoids, which can be used as an additive in dietary foods (Grimi et al. 2007).

PEF also influence carrot drying and rehydration (Gachovska et al. 2008; Gachovska et al. 2009a). Carrots pretreated at $E = 1000–1500$ V/cm showed a higher drying rate; however, the rehydration rate of PEF-pretreated carrots was lower than that of blanched carrots. It was also shown that PEF pretreatment reduced peroxidase activity by 30–50%, while blanching completely inactivated the enzyme (>95%).

2.3.1.6 Apple

The effects of electropermeabilization on apple tissues have been studied by different investigators (Angersbach et al. 2000; Bazhal and Vorobiev 2000; Lebovka et al. 2001, 2002; Bazhal et al. 2003c,d; De Vito et al. 2008; Chalermchat et al. 2010; Grimi et al. 2010). Models of dielectric breakage were developed and tested using experimental data obtained from PEF-treated apple tissues (Lebovka et al. 2001, 2002). These models accounted for the resealing of cells and the moisture transfer processes that occur inside the tissues. A mathematical model was also developed to describe the elastic properties of PEF-treated apples (Bazhal et al. 2004). The effects of apple tissue anisotropy and orientation with respect to the applied electric field on electropermeabilization were reported (Grimi et al. 2010; Chalermchat et al. 2010). It was shown that elongated cells (taken from the inner region of the apple parenchyma) responded to the electric field in a different manner, while no field orientation dependence was observed for round cells (taken from the outer region of the parenchyma) (Chalermchat et al. 2010). The effects of PEF treatment on the textural and biomechanical properties of apple tissues were also intensively discussed (Bazhal et al. 2003d; Wu and Guo 2009). A linear dependency was observed between failure stress and conductivity disintegration index (Bazhal et al. 2003d). It was shown that PEF treatment (at 1000 V/cm) decreased bulk density, decreased volume shrinkage, and increased porosity of air-dried apple tissues (Bazhal et al. 2003b,c,d). The size of the PEF-induced pores was comparable with the cell wall thickness. The textural relaxation data suggest a higher damage efficiency of longer pulse durations on apple tissues (De Vito et al. 2008). Textural tests have shown that shear strength, compression yield strength, firmness, and elastic moduli of the PEF-treated apple samples are lower than those of the untreated control group. The synergy of PEF and the effect of thermal treatment on the textural properties of apple tissues and apple juice expression were demonstrated (Lebovka et al. 2004a). It was shown that mild thermal treatment allows to increase the damage efficiency of PEF treatment, and apple tissue preheated at 50°C and treated by PEF at $E \approx 500$ V/cm exhibited a noticeable enhancement of juice extraction by pressing (Lebovka et al. 2004a,b).

Jemai and Vorobiev (2001, 2002) studied the apparent diffusion coefficient D in both thermally and electrically ($E = 100–500$ V/cm, $t_{PEF} = 0.1$s) treated apple discs (Golden Delicious) and demonstrated that detectable enhancement of the diffusion kinetics started at field intensities within 100–150 V/cm. Further increase both in the field intensity and pulse duration led to further enhancement of diffusion kinetics. It was shown that for the thermally treated samples, the temperature variation of the diffusion coefficient D was of Arrhenius type with two diffusion regimens: (i) without thermal pretreatment (E_a ~28 kJ/mol) and (ii) after thermal denaturation (E_a ~13 kJ/mol). Only one regimen with intermediate activation energy (E_a ~20 kJ/mol) was observed for electrically treated samples. Furthermore, it was found that electrical pretreatment with moderate temperature elevation (10–15°C) combined with a low temperature treatment significantly enhanced the diffusion coefficient D compared with the reference values (Jemai and Vorobiev 2002). It indicates that electrical treatment has a greater effect on the structure and permeability of apple tissue than thermal treatment.

The effects of PEF on drying and osmotic dehydration of apples have been widely discussed in the literature (Taiwo et al. 2003; Arevalo et al. 2004; Amami et al. 2005, 2006; Liu and Guo 2009). PEF treatment can decrease the drying rate, improve the quality of dried product, and reduce the power consumption (Liu and Guo 2009). PEF was used for acceleration of osmotic dehydration; it was reported that PEF treatment at 0.5–2.0 kV/cm improved mass transfer during osmotic dehydration. The vitamin C content of dried apples was reduced at higher field strengths and longer immersion times (Taiwo et al. 2003). PEF application at 900 V/cm increased both convection and diffusion rates and resulted in decreased sugar concentration in the osmotic solution and higher solid content in apples (Amami et al. 2005). The increase in solute (sucrose) concentration and PEF treatment resulted in the acceleration of osmotic dehydration. The PEF-treated apples exhibited higher water loss and higher solid gain than the untreated apples; the effect of PEF was more pronounced for water loss than for solid gain (Amami et al. 2006).

Different research groups have studied the influence of PEF treatment on juice expression from apples (Bazhal and Vorobiev 2000; Bazhal et al. 2001; Jemai and Vorobiev 2002; Lebovka et al. 2003, 2004b; Praporscic et al. 2007b; Schilling et al. 2007; Vorobiev et al. 2007; Turk et al. 2010a). It was observed that in fine-cut apples, a combination of pressing and PEF treatment gives optimum results, significantly enhances the yield of juice, and improves juice quality (Bazhal and Vorobiev 2000). Enhancement of the juice yield Y after PEF application was accompanied with a noticeable decrease in absorbance and increase in Brix value of the juice (Praporscic et al. 2007b). The reported improvement of the juice yield was more or less significant, probably due to the different process conditions employed in different studies (degree of particle fragmentation, PEF parameters, and compression pressure).

PEF application allowed increase of juice yield from apple slices (Praporscic et al. 2007a) and apple mash (Schilling et al. 2007). The time of PEF application and the size of slices noticeably affected juice characteristics (Praporscic et al. 2007b). The size of particles and the method of fragmentation (slicing, grinding, or milling) can be essential for the improvement of juice yield. Recently, it was demonstrated (Grimi 2009; Grimi et al. 2011) that the yield of juice, obtained from $2 \times 3.5 \times 55$ mm slices of Golden Delicious apples, increased after PEF treatment ($E = 400$ V/cm, $t_{PEF} = 0.1$ s) by 28%, but only by 5% when the apples were more finely sliced ($1 \times 1.9 \times 55$ mm). Finely slicing the apples caused most of the cells to be disrupted mechanically, and the additional effect of PEF on the total juice yield from such material is rather limited. On the contrary, when particles are coarse, the percentage of electrically damaged cell membranes increases, but the cell wall structure is less affected. It might be the cause of the more transparent and less cloudy apple juices obtained after PEF treatment of coarse particles (Bazhal and Vorobiev 2000; Praporscic et al. 2007b).

No apparent change in pH value and total acidity of juice was detected. Moreover, the content of many nutritionally valuable compounds was retained or even enhanced (Jaeger et al. 2008). Schilling et al. (2008a,b) reported a comparative study of apple juice production using PEF treatment and enzymatic maceration of mash on a pilot scale. It was shown that the chemical compositions, sensory properties, and total yield (~85%) of juice were similar for PEF and enzymatic maceration–assisted processes; however, PEF treatment resulted in an enhanced release of nutritionally valuable phenolics into the juice and retained genuine pectin quality. It allows sustainable pomace utilization and offers additional commercial benefits (Schilling et al. 2008a,b). It was shown that the overall composition of juice (pH; total soluble solids; total acidity; density; sugar, malic acid, and pectin contents; and nutritive value with respect to polyphenol contents and antioxidant capacities) obtained by pressing of PEF-treated ($E = 1–5$ kV/cm) mash did not significantly differ from that of the untreated control samples (Schilling et al. 2007). The effects of PEF treatment at $E = 450$ V/cm and the size of apple mash on juice yield, polyphenolic compounds, sugars, and malic acid were recently reported by Turk et al. (2010a). Juice yield Y increased significantly after PEF treatment of the large mash ($Y = 71.4\%$) as compared with the small mash (45.6%, control). The acid–sweetness balance was not altered by PEF for the large mash; however, a decrease in native polyphenol yield after PEF treatment was observed (control, 9.6%; treated, 5.9% for the small mash).

The effects of PEF treatment (at 400 V/cm) on the characteristics of apple juice (turbidity, polyphenolic content, and antioxidant capacities) were recently studied (Grimi et al. 2011). PEF pretreatment was accompanied by a noticeable improvement in juice clarity, an increase of the total soluble matter and polyphenol contents, and intensification of the antioxidant capacities of juice. Most of these effects (juice clarity and antioxidant contents) were more pronounced for the treated whole apples than for untreated whole apples and PEF-treated apple slices. Moreover, the evolution of apple browning before and after PEF treatment was more pronounced in whole samples (Grimi et al. 2011).

The possibility of PEF application (at 15 kV/cm) for improvement of pectin extraction from apple pomace was recently reported (Yin et al. 2009). The PEF technique was compared with use of chemical additives and ultrasonic and microwave-assisted extraction techniques; it was concluded that the PEF-assisted technique gave the highest yield (14.12% pectin) and was the most effective method for pectin extraction from apple pomace.

2.3.1.7 Grapes

Electroporation of wine grapes is an alternative nonthermal process leading to prudent extraction of colorants and valuable constituents. Praporscic et al. (2007a) investigated the quantitative (juice yield) and qualitative (absorbance and turbidity) characteristics of juices during the expression of white grapes (Muscadelle, Sauvignon, and Semillon). The experiments were carried out at an expression pressure of 5 bars, using a laboratory compression chamber equipped with a PEF treatment system. A PEF with strength $E = 750$ V/cm and total treatment duration $t_{PEF} = 0.3$ s was applied. The PEF treatment resulted in an increase in the final juice yield (Y_f) of up to 73–78% as compared with $Y_f \approx 49$–54% for the untreated grapes. A rather noticeable decrease of absorbance and turbidity was observed as a result of PEF treatment for all the studied white grape varieties. Later on, Grimi et al. (2009b) showed that PEF treatment enhanced the compression kinetics and extraction of polyphenols from Chardonnay grapes.

The effects of PEF pretreatment of grape skins on the evolution of color intensity, anthocyanin content, and total polyphenolic index during vinification of red grapes (Tempranillo, Garnacha, Mazuelo, Graciano, Cabernet Sauvignon, Syrah, and Merlot) was investigated. It was shown that these parameters increased in the final wine when the electric field strength was increased from 2 to 10 kV/cm (López et al. 2008a,b; Puértolas et al. 2010a). Lopez et al. (2009c) studied the application of PEF treatment (5 kV/cm, 50 pulses) to destemmed, crushed, and slightly compressed grape pomace (skins, pulp, and seeds) of Cabernet Sauvignon grapes. The application of PEF treatment to the pomace before the process of vinification produced freshly fermented wine that has richer color intensity, more anthocyanin and tannin contents, and showed better visual characteristics. PEF treatment permitted to reduce the maceration time during vinification of Cabernet Sauvignon grapes from 268 to 72 h. PEF treatment was shown to be more effective in terms of preserving color intensity and phenolic content than the enzymatic method (Puértolas et al. 2009). Phenolic extraction during fermentation of the red grapes with continuous PEF system was tested at the pilot-plant scale. The obtained data evidenced the attractiveness of this PEF-assisted technology at the commercial scale (Puértolas et al. 2010a,b,c,d,e). Moreover, it was demonstrated that PEF processing of Cabernet Sauvignon grapes allowed bottling of wines with better characteristics (Folin–Ciocalteu index, color intensity, polyphenol concentrations) using shorter maceration times (Puértolas et al. 2010b,c). It was also reported that PEF application increased the antioxidant activity of extracts from grape by-products (i.e., two-fold higher than in the control extraction) (Corrales et al. 2008). The other aspects and potential applications of PEF technology in the winemaking industry were recently reviewed by Puértolas et al. (2010e)

2.3.1.8 Oil- and Fat-Rich Plants

Guderjan et al. (2005, 2007) reported the application of PEF treatment for improvement of recovery and quality of oils extracted from oil-rich plants. PEF treatment was tested on maize, olives, soybeans, and rapeseeds. A modified PEF-assisted (at $E = 0.6$ kV/cm) process scheme was used, which allowed obtaining a high yield of maize germ oil (up to 88.4%) with increased amounts of phytosterols (up to 32.4%). The oil yield of fresh olives increased by 6.5–7.4%, and the amount of genistein and daidzein isoflavonoids in soybeans increased by 20–21% in comparison to the reference samples (Guderjan et al. 2005). The higher oil yield and higher concentrations of tocopherols, polyphenols, total antioxidants, and phytosterols were obtained for the oil extracted from PEF-treated (at 3 kV/cm) rapeseeds (Guderjan et al. 2007).

The impact of PEF application at field strength E = 100–2500 V/cm on cell permeabilization of coconut was reported by Ade-Omowaye et al. (2000). They showed that optimal PEF treatment resulted in 20% increase in milk yield, and a combination of PEF and centrifugation steps resulted in approximately 22% reduction of the drying time as compared with the untreated samples.

2.3.1.9   Other Vegetable and Fruit Tissues

PEF treatment was also applied to other tissues, such as red bell pepper and paprika (Ade-Omowaye et al. 2001, 2003), fennel (El-Belghiti et al. 2008), chicory (Loginova et al. 2010), alfalfa (Gachovska et al. 2006, 2009b), and red cabbage (Gachovska et al. 2010). Ade-Omowaye et al. (2001) studied the impact of PEF treatment at 1.7 kV/ cm on yield and quality parameters (pH, soluble solids [Brix], total dry matter, color, total carotenoids [as β-carotene], and vitamin C) of juice obtained from paprika. The results were compared with those obtained for juice from enzymatically treated or untreated paprika mash. It was shown that the quality of juice from the PEF-treated paprika was well comparable with that of the enzyme-treated or untreated juice, and both PEF and enzymatic treatments resulted in approximately the same (about 9–10%) increase in juice yield. However, the amount of β-carotene extracted into the juice was more than 60% for PEF treatment as compared to about 44% for enzymatic treatment (Ade-Omowaye et al. 2001).

The benefits of PEF application for the enhancement of soluble matter extraction from chicory were recently demonstrated (Loginova et al. 2010). Chicory (*Cichorium intybus*) roots contain many useful components, such as sucrose, proteins, and inulin. Note that inulin is used for the production of high-quality dietary foods or sugar substitute in tablets (Franck 2006). The thermally accelerated extraction of soluble solids from chicory roots is very similar to that used for sugar production from sugar beets (Berghofer et al. 1993). PEF treatment with a field strength 100–600 V/cm, duration of 10^{-3}–50 s, and temperature of 20–80°C was applied (Loginova et al. 2011). The activation energy of thermal damage was rather high ($W_\tau \approx 263$ kJ/mol); however, it could be noticeably reduced to $W_\tau = 30$–40 kJ/mol by application of PEF treatment. Moderate electric power consumption ($Q < 10$ kJ/kg) at room temperature demands application of a relatively high electric field strength ($E = 400$–600 V/cm); however, the value of E noticeably decreases as temperature increases. PEF pretreatment noticeably accelerated diffusion even at low temperatures within 20–40°C. The proposed technique appears to be promising for future industrial applications of "cold" soluble matter extraction from chicory roots (Loginova et al. 2010).

PEF-assisted juice extraction from alfalfa mash was studied in Gachovska et al. (2006). Each PEF treatment at 1500 V/cm was followed by pressure application. PEF treatment significantly increased extraction of juice (by 38%), dry matter, protein, and mineral contents compared with untreated samples. It was reported that

PEF technology can be efficiently applied for extraction of anthocyanins from red cabbage (Gachovska et al. 2010). PEF treatment of red cabbage at $E = 2.5$ kV/cm enhanced by 2.15 times the total anthocyanin extracted in water (Gachovska et al. 2010) and yielded higher proportions of nonacylated anthocyanins (Gachovska et al. 2010).

2.3.2 BIOSUSPENSIONS

Disruption of some microorganisms is a very important step in industrial extraction of valuable proteins, cytoplasmic enzymes, and polysaccharides, which are present inside the cells. Moreover, it is possible to extract the valuable proteins in cultures of recombinant host cells (*Saccharomyces cerevisiae, E. coli*) containing foreign genes. Important medical materials can be synthesized in such cells (Ohshima 1992; Rokkones et al. 1994). However, extraction of intracellular proteins is not an easy task and requires application of special techniques for cell disruption and for purification.

2.3.2.1 Cell Disruption Techniques

The existing cell disruption techniques are based on the application of different treatments: mechanical (high-pressure homogenization [HPH], wet milling), chemical (organic solvents, enzymes, detergents), and physical (sonification, freeze–thawing, electrically-assisted treatment). The successful recovery of intracellular products involves the preservation of their contents and removal of cell debris (Engler 1985; Harrison 1991; Chisti 2007; Peternel and Komel 2010).

Thermal treatment at $T > 50°C$ may result in damage of yeast membranes, and also causes denaturation and degradation of intracellular proteins and DNA (Chisti 2007). Mechanical methods are most appropriate for the large-scale disruption of cells and allow high recovery of intracellular material. However, they are restricted by temperature elevation, they require high power consumption and multiple passes with supplementary cooling, and their final products contain large quantities of cell debris (Brookman 1974; Engler 1985; Lovitt et al. 2000; Middelberg 2000; Wuytack et al. 2002). Cryogenic grinding at $-196°C$ is a promising technology for protein release, which has allowed a nearly 100% release of soluble protein from yeasts (*S. cerevisiae*), with a small degree of protein denaturation ($\approx 18\%$); however, this method was inefficient for DNA release (Singh et al. 2009).

Ultrasonification-assisted methods are restricted by heat generation, high costs, and extraction yield variability, as well as by generation of free radicals (Bar 1987; Riesz and Kondo 1992). Chemical methods are rather expensive, usually result in low recovery of intracellular material, cause protein degeneration, and require additional purification in the downstream processes (Harrison et al. 1991; Tamer et al. 1998). Detergent-based methods are very sensitive to cell type, pH, ionic strength, and temperature, and may denature proteins or destroy their activity and functions (Gough 1988; Cordwell 2008; Patel et al. 2008). Biological, autolysis, and enzymatic methods are rather expensive and may affect protein stability. Moreover, they present a potential problem in that the susceptibility of cells to the enzyme can be dependent on the state of the cells (Salazar and Asenjo 2007; Chisti 2010).

2.3.2.2 PEF Application for Killing and Disruption of Microorganisms

Nowadays there exist many examples of PEF application for killing and disruption of microorganisms (Grahl and Märkl 1993; Barbosa-Cánovas et al. 1998). Numerous studies have investigated the effect of PEF application on electrofusion of cells and transport of nanoparticles or biopolymers across the cell wall into the recipient cells (Van Wert and Saunders 1992; Jen et al. 2004; Pakhomov et al. 2010). Electroporation-assisted extraction from biocells is expected to be highly selective with respect to low and high molecular weight intracellular components (Ohshima et al. 2000), and promising for the recovery of homogeneous and heterogeneous intracellular proteins having wide biotechnological applications (Ganeva et al. 1999, 2001, 2003; Suga et al. 2006, 2007). The supplementary attractivity of the PEF-assisted method is related to the fact that this method is nonthermal and is expected to have a small influence on the cell wall.

However, the efficiency of PEF-assisted extraction in its application to biosuspensions may be dependent on multiple factors. The efficiency of electroporation in suspensions may be governed by cell shape and orientation, cell type and strain, physiological state of cells (age of culture and temperature of cultivation), and state of cell aggregation (Wouters and Smelt 1997), as well as suspension properties such as electrical conductivity, salinity, and pH; the presence of surfactants; cell density; and others (Barbosa-Cánovas et al. 1998; Susil et al. 1998; Pavlin et al. 2002).

It was shown that leakage of cytoplasmic ions during PEF application influences the ionic concentration of the medium and its electrical conductivity (Eynard et al. 1992; Kinosita and Tsong 1997; El Zakhem et al. 2006a,b). The leakage of the intracellular components after PEF application was accompanied by decrease in sizes of *S. cerevisiae* and *E. coli* cells (El Zakhem et al. 2006a,b). The conductometric approach was used for continuous monitoring of the degree of cell damage (*S. cerevisiae* and *E. coli* cells); it was applied for studying the effects of temperature and surfactant on inactivation efficiency (El Zakhem et al. 2006a,b, 2007).

At high concentration of cells in suspension, PEF disruption efficiency was found to be affected by formation of large aggregates (Zhang et al. 1994; El Zakhem et al. 2006b; Calleja 1984). The possibility of formation of a "pearl chain," in which the cells are in very close contact with each other, was described by Zimmermann et al. (1986, 1992). It was experimentally demonstrated (El Zakhem et al. 2006b) in *S. cerevisiae* suspensions that intact cells have a negative charge, as compared with the positive charge of the damaged cells. Thus, PEF treatment can induce an electrostatic attraction between intact and damaged cells, and the formation of large aggregates. In principle, this effect may facilitate the PEF-induced damage due to the formation of "equivalent cells" of larger volume, or may protect cells against PEF-induced damage (Zhang et al. 1994). However, incomplete damage of cells inside the clusters is also possible; it can occur because of the formation of low-conductive cores (consisting of damaged cells) enveloping a surface of intact cells inside a floc. Moreover, theoretical calculations predict the dependence of induced transmembrane potential on cell density and arrangement (Susil et al. 1998; Pavlin et al. 2002), and that higher voltage amplitude or longer pulse duration is required to cause the same poration effects if cells are in a cluster (Joshi et at. 2008).

The efficiency of electroporation of PEF-treated cells may be increased by the addition of supplementary chemical reagents and nanoparticles. Improvement of the damage efficiency in suspensions by the addition of surfactants, peptides, dimethyl sulfoxide, or polylysine has been previously reported (Melkonyan et al. 1996; Diederich et al. 1998; Tung et al. 1999; El Zakhem et al. 2007). The use of nanotubes for enhancement of cell electroporation was recently discussed by several investigators (Rojas-Chapana et al. 2004; Yantzi and Yeow 2005; Raffa et al. 2009). Owing to the so-called lightning rod effect, the nanotubes have the ability to strongly enhance the electric field at the tube ends, which makes them ideal for localized electroporation. It was demonstrated that nanotubes can be used as nanotools, enabling electropermeabilization of cells at rather low electric fields (40–60 V/cm) (Raffa et al. 2009). The pulsing protocol (electric field strength, pulse shape, pulse length, total time of treatment, temperature) is very important. Note that, typically, inactivation and disruption of microorganisms requires high critical electric fields ($E > 2$–5 kV/cm). It presumably reflects relatively small cell sizes—for example, between 2 and 15 μm for $S.$ $cerevisiae$ (near spherical shape) and 0.4–0.6 μm diameter and 2–4 μm length for $E.$ $coli$ (rod-like shape) (Bergey 1986).

2.3.2.3 Yeasts

The commonly reported values of field strength E needed for disintegration of membranes in $S.$ $cerevisiae$ yeast cells by short pulses of microsecond duration are rather high, typically $E > 7.5$ kV/cm (Zhang et al. 1994). However, smaller electric fields can also affect the structure of yeast cells at a long duration of PEF treatment. For instance, a noticeable damage in yeast cells at $E < 7.5$ kV/cm was observed at a long-duration PEF treatment (>1 s) (El Zakhem et al. 2006a,b).

The release of proteins in a PEF-treated aqueous suspension of $S.$ $cerevisiae$ cells was observed for relatively low (below 10 kV/cm) PEF (Ohshima et al. 1995). No cell wall damage related to PEF application was observed by scanning electron microscopy, and the concentration of protein was found to increase with the increase in electric field strength E (0–18 kV/cm) during treatment. However, the maximal yield of PEF-assisted extraction was only 5% of that obtained using glass bead homogenization. It was concluded that some intracellular proteins could be released through the pores (induced by PEF treatment) selectively, depending on the PEF protocol.

PEF application (at $E = 3$–4.5 kV/cm) to yeast suspensions resulted in a high extraction yield of intracellular proteins and enzymes, with their functional activities preserved (Ganeva and Galutzov 1999; Ganeva et al. 2001, 2003). The specific activities of the electroextracted enzymes were higher than those of enzymes obtained by mechanical disintegration or enzymatic lysis (Ganeva et al. 2003). The highest extraction yield of proteins, glutathione reductase, 3-phosphoglycerate kinase, and alcohol dehydrogenase was observed for supplementary pretreatment by dithiothreitol (a reducing agent), and maximal yield was observed 3–8 hours after PEF application (Ganeva and Galutzov 1999). Electropulsing (4–4.5 kV/cm and 2 ms pulse duration) allowed effective extraction of the enzyme β-galactosidase from the yeast $Kluyveromyces$ $lactis$ with 75–80% yield within 8 h after PEF application (Ganeva et al. 2001). The extraction efficiency was strongly dependent on the growth phase of yeast cells and salinity of the solution in the postpulse incubation period (Ganeva and Galutzov 1999; Ganeva et al. 2001). It was shown that high yields of intracellular enzymes from yeast can be obtained through PEF treatment of the flowing suspensions. The maximal yield of enzymes (hexokinase, 3-phosphoglycerate kinase, and glyceraldehyde-3-phosphate dehydrogenase) from *S. cerevisiae* and of β-galactosidase from *K. lactis* was reached within 4 h. The proposed flow method permitted treatment of large volumes and treatment of at least 20% wet weight suspensions (Ganeva et al. 2003).

PEF treatment of *S. cerevisiae* at 5 kV/cm allowed attaining a high conductivity disintegration index, $Z \approx 1$, with higher amounts of released peptides and proteins than nucleic acid bases (El Zakhem et al. 2006b). However, in PEF-treated suspensions of wine yeast cells (*S. cerevisiae bayanus* strain DV10), a relatively small release of proteins was observed even at a high index of $Z \approx 0.8$ (Shynkaryk et al. 2009). Moreover, it was demonstrated that high levels of membrane disintegration ($Z > 0.8$) in these yeasts require a rather strong PEF treatment (at $E = 10$ kV/cm using 2×10^5 pulses of 100 μs). Thus, the efficiency of PEF-assisted extraction was dependent on the yeast strain. It can reflect the presence of hard cell walls in addition to cell membranes that can restrict extraction of intracellular compounds. More effective extraction of high molecular weight contents (e.g., proteins) from electrically resistant strain requires more powerful mechanical disintegration of cell walls, which is provided by high-voltage electrical discharges (HVED) and HPH techniques. In principle, HPH permitted better extraction than HVED (Loginov et al. 2009; Liu et al. 2010). However, a synergistic enhancement of protein release from yeasts can be attained using combined disruption techniques. It was shown that a combination of HVED and HPH techniques allowed reaching a high level of protein extraction from wine yeast cells (*S. cerevisiae bayanus*, strain DV10) at lower pressures or smaller number of passes through the homogenizer (Shynkaryk et al. 2009).

2.3.2.4 *Escherichia coli*

The commonly reported values of field strength E needed for disintegration of membranes in *E. coli* cells by short pulses of microsecond durations are even higher than for *S. cerevisiae*, which typically require $E = 10–35$ kV/cm (Grahl and Märkl 1996; Aronsson et al. 2001; Aronsson and Rönner 2005; Amiali 2006; Bazhal et al. 2006). However, a noticeable permeabilization of the membranes in *E. coli* cells was observed at significantly smaller fields ($E = 1.25–3.75$ kV/cm) (Eynard et al. 1998). Note that PEF-induced orientation of rod-like cells in external electric fields can facilitate their electropermeabilization (Eynard et al. 1998).

PEF treatment (10 kV/cm, with a needle-plate electrode geometry) of genetically engineered *E. coli* suspension allowed the effective release of β-glucosidase and α-amylase (Ohshima 2000). It was noted that PEF treatment could easily disrupt the outer membrane, but it was difficult to disrupt the cytoplasmic membrane simultaneously, and it was concluded that PEF treatment is useful for easy selective release of periplasmic proteins (Ohshima 2000).

Experimental studies on PEF treatment of the flowing concentrated aqueous suspensions of *E. coli* (1 wt.%) at $E = 5–7.5$ kV/cm and medium temperatures within 30–50°C were done by El Zakhem et al. (2007, 2008). A noticeable disruption of cells was observed at PEF treatment time (t_{PEF}) within 0–0.2 s and thermal treatment time (t_T) within 0–7000 s. It was shown that disruption of *E. coli* was accompanied by decrease in cell size and release of intracellular components. Absorbance analysis of supernatant solutions evidenced the leakage of nucleic acids. The electrical conductivity disintegration index Z was monitored in a continuous mode in the course of PEF–thermal treatment through electrical conductivity measurements (Figure 2.12).

Thermal treatment alone at $T = 30$–$50°C$ was ineffective for disruption of *E. coli* cells and required a long treatment time ($t_T \gg 1$ h). For example, 1 h of thermal treatment resulted in increases in electrical conductivity disintegration index Z of up to ≈ 0.02, ≈ 0.28, and ≈ 0.66 at temperatures of $30°C$, $40°C$, and $50°C$, respectively (Figure 2.12). Moreover, there was an evident synergism between the simultaneously applied electrical and thermal treatments. The electrical conductivity disintegration index Z after 1 h of PEF and thermal treatments reached ≈ 0.22, ≈ 0.83, and ≈ 0.99 at temperatures of $30°C$, $40°C$, and $50°C$, respectively (Figure 2.12). The observed behavior can be explained by the increased fluidity of cell membranes and possible phase transitions inside them (Stanley 1991). A synergy between PEF and thermal treatments was also observed in *E. coli* in inactivation experiments with higher electric fields (Zhang et al. 1995; Pothakamury et al. 1996; Aronsson et al. 2001; Bazhal et al. 2006).

It was shown that surfactant additives (Triton X-100) additionally improved disruption of cells in *E. coli* suspensions (El Zakhem et al. 2007, 2008). The influence of the surfactant on *E. coli* disruption efficiency was explained by changes in the membrane fluidity properties and changes in the state of cell aggregation in

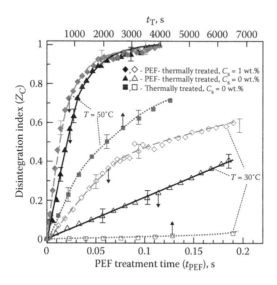

FIGURE 2.12 The electrical conductivity disintegration index Z_C versus effective PEF treatment time (t_{PEF}) and thermal treatment time (t_T) at different temperatures T. C_s is the surfactant concentration (wt.%). PEF treatment was done at electric field strength $E = 5$ kV/cm and pulse duration $t_i = 10^{-3}$ s. (From El Zakhem, H. et al., *Int J Food Microbiol* 120, 259–265, 2007. With permission.)

suspension (El Zakhem et al. 2007). Addition of a surfactant resulted in enhanced aggregation and formation of an "equivalent cell" of larger size that can enhance the PEF damage efficiency (Zimmermann et al. 1986). The disruption efficiency of cells in *E. coli* suspensions was also noticeably improved by the addition of organic (citric, malic, and lactic) acids in small concentrations (≤0.5 g/L), and an 8 log cycle reduction was reached by using 0.375 g/L of lactic acid (El Zakhem et al. 2008). Lactic acid can be efficiently used in food-related applications, and it is well known as an effective permeabilizer and disintegrating agent of outer membranes in gram-negative bacteria, including *E. coli* (Alakomi et al. 2000; Theron and Lues 2011). It was assumed that lactic acid may act as a very effective potentiator of PEF effects in membranes of *E. coli* cells (El Zakhem et al. 2008).

2.4   PEF PILOT-SCALE EXPERIMENTS AND APPLICATIONS

The recent applications of PEF treatment in the food industry are mainly restricted by attempts on gentle microbial inactivation and pasteurization of pumpable foods (e.g., milk, fruit juices) (Lelieveld et al. 2007) and extraction of cellular constituents from the tissues (Vorobiev and Lebovka 2008).

2.4.1   Some Examples of Related Recent Patents

Doevenspeck (1991) has patented an electric-impulse method for treating substances located in an electrolyte, and Bushnell et al. (2000) have patented a pumpable serial-electrode treatment system for deactivating organisms in a food product. Eshtiaghi and Knorr (1999) and Arnold et al. (2010) have patented methods for treating sugar beets. Vorobiev et al. (2000) have patented a PEF-assisted process for acceleration of extraction from tissues, where PEF treatment is combined with mechanical pressing. Ngadi et al. (2009) have presented the invention of a PEF-assisted method for enhancement of extraction of phytochemicals from plant materials, wherein PEF treatment and pressing are applied and the PEF treatment could be accomplished in a unique treatment chamber.

2.4.2   Pasteurization and Regulation of Microbial Stability

For fluid food products, several fundamental works were done for elucidation of the association between laboratory-, pilot plant–, and commercial-scale applications. The pilot scale continuous scheme was used for the study of the relationship between PEF protocol parameters (power consumption 0–300 kJ/kg, electric field strength 25–70 kV/cm, square wave pulse width 0.05–3 µs, and initial product temperature 4–20°C) and efficiency of *Salmonella enteritidis* inactivation in aqueous solutions (Korolczuk et al. 2006). A 3d computational model of fluid dynamics in a pilot-scale PEF system with colinear electrodes was developed by Buckow et al. (2010). Note that PEF is a more energy-efficient process than thermal pasteurization, and it would add only US$0.03–US$0.07/L to the final food costs (Ramaswamy et al. 2008).

Different types of pilot plant–scale PEF systems were developed for pasteurization and regulation of microbial stability in pumpable foods such as yogurt, milk, and juices (Barbosa-Canovas et al. 2000). A pilot plant–scale PEF continuous processing system integrated with an aseptic packaging machine was used as a nonthermal tool for effective microbial inactivation of fresh orange juice at a flow rate of 75–150 liters/h (Qiu 1997, 1998). The PEF-treated and aseptically packaged fresh orange juice demonstrated the feasibility of use of the PEF technology to extend product shelf lives with very little loss of flavor, vitamin C, and color (Qiu et al. 1997). A synergistic effect of temperature and PEF inactivation was also observed in a pilot plant PEF unit with the flow rate of 200 liters/h (Wouters et al. 1999). The applications of pilot plant PEF facilities (at 16.4–37.3 kV/cm) in the batch and continuous flow modes for inactivation of microorganisms capable of secreting lipases in milk and dairy products were discussed by Bendicho et al. (2002). PEF processing of yogurt-based products by using the OSU-2C pilot plant scale system was studied by Evrendilek et al. (2004). Mild heat (at 60°C for 30 s) combined with PEF treatment (at 30 kV/cm electric field strength and 32 µs total treatment time) did not affect the main characteristics (color, pH, and °Brix) of the product, and prevented the growth of microorganisms and decreased the total mold and yeast count in yogurt-based products during their storage at 4°C and 22°C.

A pilot plant–sized PEF treatment (at electric field strength 25–37 kV/cm, pulse width 1.84 µs, power consumption 11.9 J/ml per pulse, and total treatment time varying within 54–478 µs) was applied for nonthermal preservation of liquid whole eggs (Góngora-Nieto et al. 2003). Citric acid (CA) additives were used as color stabilizers and also for increasing the effectiveness of PEF treatment. It was shown that the maximum shelf life of PEF-treated liquid whole eggs (at 4°C) was 20 days, and almost 30 days with 0.15% CA and 0.5% CA, respectively.

The efficiency of PEF treatment for pasteurization of apple sauces was demonstrated in a pilot plant scale by Jin et al. (2009). A system for continuous flow PEF treatment followed by high-temperature, short-time processing integrated with an aseptic packaging machine was tested. The PEF treatment system included a co-field continuous flow tubular chamber (inner diameter, 0.635 cm), with boron carbide electrodes (gap distance between electrodes, 1.27 cm) and high voltage pulse generator (OSU-6; Diversified Technology Inc., Bedford, MA, USA). The generator provided bipolar square pulses with 60-kV voltage, maximum peak current of 750 A, maximum frequency of 2000 Hz, and pulse width of 2–10 µs.

The pilot-scale PEF treatment (45.7 µs at 34 kV/cm) combined with mild heating (24 s at 67.2–73.6°C) was applied to salad dressing inoculated with *Lactobacillus plantarum* 8014, and more than 7 log inactivation was achieved. It was reported that no *L. plantarum* 8014 was recovered in the model salad dressing at room temperature for at least 1 year (Li et al. 2007).

The pilot plant–scale PEF treatment (94 µs mean total treatment time at 35 kV/cm) was applied for the study of inactivation of *E. coli* O157:H7 and evaluation of shelf life of aseptically packaged apple juice and cider (Evrendilek et al. 2000). It was shown that PEF treatment improved the microbial shelf life of the apple cider and did not alter its natural food color and vitamin C content. A portable pilot-scale PEF processing machine was constructed and evaluated in the pasteurization of apple cider (Jin and Zhang 2005). Different PEF pilot systems for microbial inactivation and pasteurization were developed in the Eastern Regional Research Center,

Wyndmoor, PA, USA, and the first commercial application of PEF for pasteurization of apple cider was reported by Ravishankar (2008).

2.4.3 EXTRACTION

An industrial prototype for starch extraction from potatoes was developed by Propuls GmbH, Bottrop, Germany (Loeffler 2002; Topfl 2006). The automated flow of potatoes came from a feeding funnel with two cross electrodes. After passing the water-filled electrode section, the electrically treated potatoes were separated from water with a screw conveyer for further treatment.

A commercial pilot plant–scale PEF mobile device (Karlsruher Elektroporations Anlage, KEA-Tec, Germany) was contracted for effective treatment of large specimens (e.g., entire sugar beets) in a continuous mode (Schultheiss et al. 2003; Sack et al. 2005). It consisted of a 300-kV Marx generator operating at 10 Hz and delivering pulses to a cylindrical reaction chamber with maximal electric field strength up to 60 kV/cm. This device was used for demonstration of the advantages of PEF treatment for sugar production. Encouraging results were obtained by several research groups. They revealed an industrial interest in PEF pretreatment, and a semi-industrial–scale equipment was built for PEF-assisted extraction (Sack et al. 2010). The equipment allowed handling a throughput of up to 1 t/h, with a power consumption about 15 kW·h/t. Both red and white wine grapes were processed using this equipment (Sack et al. 2010).

The efficiency of PEF treatment (at $E = 400$ V/cm and total treatment time of 50 ms) for sugar extraction from sugar beets was justified using a pilot countercurrent section extractor (Loginova et al. 2011). Cossettes were prepared from sugar beets by using industrial knives, and the temperature was varied between 30°C and 70°C. The possibility of PEF-assisted cold (at 30°C) and moderate thermal (50–60°C) sugar extraction was shown.

The good industrial potential of PEF-assisted apple juice expression was confirmed on laboratory and pilot scales using belt-press equipment (Jaeger et al. 2008; Grimi et al. 2008). Figure 2.13 shows a scheme (a) and a photo (b) of a pilot belt press recently used for PEF-assisted expression of sugar beets (Grimi et al. 2008; Grimi 2009). The pilot experiments were done for untreated and PEF-treated sugar beet slices of different sizes: S1 (0.045 mm³), S2 (47.5 mm³), S3 (280 mm³), and S4 (1050 mm³). The obtained results confirmed the amelioration of juice yield and purity on application of PEF pretreatment. It was concluded that the size of particles treated by PEF should be optimized for attaining the maximal yield and better purity of the juice.

A pilot plant–scale PEF treatment (at $E = 2$, 5, and 7 kV/cm) for improving extraction of anthocyanins and phenols from red grapes (Cabernet Sauvignon, Syrah, and Merlot) during the maceration-fermentation step was investigated by Puértolas et al. (2010a). PEF treatment was done in a colinear continuous treatment chamber, and the maximum PEF treatment capacity was about 1000 kg/h (Figure 2.14). The PEF generator (Modulator PG; ScandiNova, Uppsala, Sweden) provided square waveform pulses at 30 kV voltage, maximum peak current of 200 A, maximum frequency of 300 Hz, and pulse width of 3 μs. The reported energy requirements of the process were rather low (6.76–0.56 kJ/kg), and the PEF-assisted technology allowed to decrease the duration of maceration during vinification or to increase the quantity

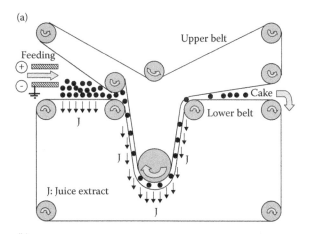

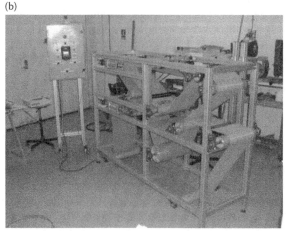

FIGURE 2.13 (a) A scheme and (b) a photo of a pilot belt press recently used for PEF-assisted expression from the sugar beets. (From Vorobiev, E. and Lebovka, N., *Food Eng Rev* 2, 95–108, 2010. With permission.)

of anthocyanins and phenolic compounds in the wine. For example, in experiments with PEF treatment of Cabernet Sauvignon at 5 kV/cm, the maximum concentrations of anthocyanins and total phenols were 34% and 40% higher than in untreated control samples, respectively (Puértolas et al. 2010a). Similar pilot plant–scale studies of the influence of PEF treatment of grape berries on the evolution of chromatic and phenolic characteristics of Cabernet Sauvignon red wines were done by Puértolas et al. (2010c). Better chromatic characteristics and higher phenolic content were observed in PEF-treated wine samples during aging in American oak barrels and subsequent storage in bottles. It evidenced that PEF-assisted processing is a promising enological technology for the production of aged red wines with high phenolic content (Puértolas et al. 2010b).

Comparative laboratory- and pilot plant–scale studies of PEF and thermal processing were done for apple juice (Schilling et al. 2008a) and apple mash (Schilling

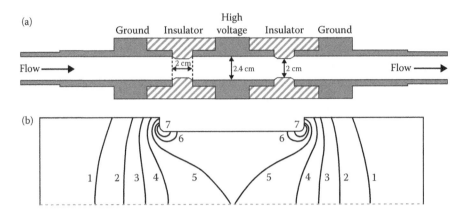

FIGURE 2.14 The colinear treatment chamber used at a pilot plant for PEF processing of red grapes. (a) The treatment chamber consisted of three cylindrical electrodes (stainless steel) separated by two methacrylate insulators. The central electrode was connected to high voltage and two others were grounded. (b) The distribution of the electric field strength E was not uniform. An example of E distribution simulated by method of finite elements for 14.2 kV input voltage is shown. The value of E changes from the weakest (1 kV/cm) to the strongest (7 kV/cm). (From Puértolas, E. et al., *J Food Eng*, 98, 120–125, 2010a. With permission.)

et al. 2008b). It was shown that juice composition was not affected by PEF treatment; however, PEF treatment of apple mash enhanced the release of nutritionally valuable phenolics into the juice (Schilling et al. 2008b). The observed browning of PEF-treated juices provided evidence of residual enzyme activities. The different combinations of preheating and PEF treatment had a synergistic effect on peroxidase and polyphenoloxidase deactivation. For example, a 48% deactivation of polyphenoloxidase activity was achieved on a plant scale on preheated (to 40°C) and PEF-treated (at 30 kV/cm, 100 kJ/kg) juices (Shilling et al. 2008a).

PEF treatment (1000 V/cm, 200 Hz, and 100 µs pulse duration) was applied to French cider apple mash pumped into a collinear treatment chamber at the flow rate of 280 kg/h (Turk at al. 2009). Juices were recovered continuously under a single belt press. PEF treatment of mash increased the juice yield by approximately 4%. Juice from the treated mash had a better color than that from the control. The overall chemical composition of the treated juices showed no differences from their respective controls.

Recently, Turk and colleagues (Turk 2010; Turk et al. 2010b) confirmed their results obtained with French cider apple mash on an industrial scale (flow rate of 4500 kg/h). PEF (E = 650 V/cm and t_{PEF} = 23.2 ms) application permitted a 5.2% increase in juice yield. The energy provided (3.5 W·h/kg of mash) contributed to the increase in dry matter of the marc from 19.8% to 22.5%. The reduction in the quantity of water to be evaporated during the drying process was estimated as 12.1 W·h/kg. Consequently, the total energy savings from pressing/drying would be approximately 8.6 W·h/kg of mash. The juices treated by PEF were significantly less turbid and had more intense odor, savor, and flavor.

2.4.4 FOOD SAFETY ASPECTS

PEF treatment of foods at a relatively moderate electric field strength (<50 kV/cm) can affect the integrity of cell membranes; however, such electric fields are still rather low and do not influence covalent chemical bonds, and do not cause protein alternation and gelatinization of starch. Many experimental works confirmed the retention of freshness, color, and nutrients in food products after PEF treatment. The use of PEF treatment is subject to US and EU food regulations; however, its associated electrochemical processes are still insufficiently understood (Smith 2007; Knorr et al. 2008). Electrochemical reactions at the electrode surfaces may introduce undesirable effects, which can be avoided by selection of suitable electrodes and PEF protocols (Morren et al. 2003; Roodenburg et al. 2005a,b; Roodenburg 2007). The food safety aspects associated with the application of PEF-assisted technologies for processing of foods were recently discussed by the working group "Food technology and safety" of the Deutsche Forschungsgemeinschaft Senate Commission on Food Safety (Knorr et al. 2008) and reviewed by Smith (2007).

2.5 CONCLUSIONS

PEF is a novel method in food processing that is based on disruption of very thin (≈0.5 nm) membranes in biological cells. The selectivity of disruption was shown to be very high, and this method practically retains the integrity of cell walls, and the color, flavor, vitamin C content, and important nutrients of food materials. The PEF method is energetically cost-efficient and nonthermal—that is, the elevation of temperature related to ohmic heating may be unessential.

However, disruption of cell membranes results in the acceleration of mass exchange processes in food materials. Recent laboratory experiments demonstrated many promising examples of the PEF-assisted extraction of juices, sugars, colors, polyphenolic substances, and oils from solid foods (sugar beets, apples, grapes, etc.). PEF treatment also poses unique possibilities for extraction of valuable proteins, cytoplasmic enzymes, and polysaccharides from cells in suspensions, and opens new prospects for the production of valuable proteins in cultures of recombinant host cells (*S. cerevisiae*, *E. coli*). The first important steps for practical implementations of PEF-assisted technologies at the pilot plant scale have already been done. Future technologies will allow overcoming the difficulties associated with mechanical, thermal, or chemical pretreatments presently used in food processing, and provide a potential method for production of foods with excellent sensory and nutritional qualities.

ACKNOWLEDGMENTS

The authors thank Dr. N. S. Pivovarova for her help with manuscript preparation.

REFERENCES

Abbott, J.A., Massie D.R., Upchurch, B.L., Hruschka, W.R. 1995. Nondestructive sonic firmness measurement of apples. *Trans ASAE* 38: 1461–1466.

Abram, F., Smelt, J.P.P.M., Bos, R., Wouters, P.C. 2003. Modelling and optimization of inactivation of *Lactobacillus plantarum* by pulsed electric field treatment. *J Appl Microbiol* 94: 571–579.

Ade-Omowaye, B.I.O., Angersbach, A., Eshtiaghi, N.M., Knorr, D. 2000. Impact of high intensity electric field pulses on cell permeabilisation and as pre-processing step in coconut processing. *Innov Food Sci Emerg* 1: 203–209.

Ade-Omowaye, B.I.O., Angersbach, A., Taiwo, K.A., Knorr, D. 2001. The use of pulsed electric fields in producing juice from paprika (*Capsicum annuum* L.). *J Food Process Preserv* 25: 353–365.

Ade-Omowaye, B.I., Rastogi, N.K., Angersbach, A., Knorr, D. 2003. Combined effects of pulsed electric field pre-treatment and partial osmotic dehydration on air drying behaviour of red bell pepper. *J Food Eng* 60: 89–98.

Alakomi, H.-L., Skytta, E., Saarela, M., Mattila-Sandholm, T., Latva-Kala, K., Helander, I.M. 2000. Lactic acid permeabilizes gram-negative bacteria by disrupting the outer membrane. *Appl Environ Microbiol* 66: 2001–2005.

Altunakar, B., Gurram, S.R., Barbosa-Cánovas, G.V. 2007. Applications of pulsed electric fields processing of food, in *Food Preservation by Pulsed Electric Fields: From Research to Application*, H.L.M. Lelieveld, S. Notermans, and S.W.H. de Haan (Eds.), pp. 266–293. Cambridge: Woodhead Publishing Limited.

Amami, E., Vorobiev, E., Kechaou, N. 2005. Effect of pulsed electric field on the osmotic dehydration and mass transfer kinetics of apple tissue. *Dry Technol* 23: 581–595.

Amami, E., Vorobiev, E., Kechaou, N. 2006. Modelling of mass transfer during osmotic dehydration of apple tissue pre-treated by pulsed electric field. *LWT-Food Sci Technol* 39: 1014–1021.

Amiali, M. 2006. Inactivation of *Escherichia coli* O157:H7 and *Salmonella enteritidis* in liquid egg products using pulsed electric field. PhD dissertation. Montreal: McGill University.

Anderson, A.K., Finkelstein, R. 1919. A study of electro-pure process of treating milk. *J Diary Sci* 2: 374–406.

Angersbach, A., Heinz, V., Knorr, D. 1997. Elektrische Leitfähigkeit als Maß des Zellaufschlußgrades von Zellulären Materialien durch Verarbeitungsprozesse. *LVT* 42: 195–200.

Angersbach, A., Heinz, V., Knorr, D. 2000. Effects of pulsed electric fields on cell membranes in real food systems. *Innov Food Sci Emerg* 1: 135–149.

Angersbach, A., Heinz, V., Knorr, D. 2002. Evaluation of process-induced dimensional changes in the membrane structure of biological cells using impedance measurement. *Biotechnol Prog* 18: 597–603.

Anglim, T.H. 1923. Method and apparatus for pasteurization milk. *U.S. Patent* 1468 871.

Archie, G.E. 1942. The electrical resistivity log as an aid in determining some reservoir characteristics. *Trans AIME* 146: 54–62.

Arevalo, P., Ngadi, M.O., Bazhal, M.I., Raghavan, G.S.V. 2004. Impact of pulsed electric fields on the dehydration and physical properties of apple and potato slices. *Dry Technol* 22: 1233–1246.

Arnold, J., Frenzel, S., Michelberger, T., Scheuer, T. 2010. Extraction of constituents from sugar beet chips. *U.S. Patent* 7695566.

Aronsson, K., Lindgren, M., Johansson, B.R., Rönner, U. 2001. Inactivation of microorganisms using pulsed electric fields: The influence of process parameters on *Escherichia coli*, *Listeria innocua*, *Leuconostoc mesenteroides* and *Saccharomyces cerevisiae*. *Innov Food Sci Emerg* 2: 41–54.

Aronsson, K., Rönner, U., Borch, E. 2005. Inactivation of *Escherichia coli*, *Listeria innocua* and *Saccharomyces cerevisiae* in relation to membrane permeabilization and subsequent leakage of intracellular compounds due to pulsed electric field processing. *Int J Food Microbiol* 99: 19–32.

Azeredo, H.M.C. 2009. Betalains: Properties, sources, applications, and stability—A review. *Int J Food Sci Technol* 44: 2365–2376.

Ball, C.O. 1937. Apparatus for pasteurization milk. *U.S. Patent* 2091263.

Bar, R. 1987. Ultrasound enhanced bioprocesses. *Biotechnol Eng* 32: 655–663.

Barbosa-Cánovas, G.V., Pothakamury, U.R., Palou, E., Swanson, B. 1998. *Nonthermal Preservation of Foods*. New York: Marcel Dekker.

Barbosa-Canovas, G.V., Pierson, M.D., Zhang, Q.H., Schaffner, D.W. 2000. Pulsed electric fields. *J Food Sci* 65: 65–79.

Barbosa-Cánovas, G.V., Cano, M.P. 2004. *Novel Food Processing Technologies*. New York: Marcel Dekker.

Barbosa-Cánovas, G.V., Altunakar, B. 2006. Pulsed electric fields processing of foods: An overview, in *Pulsed Electric Field Technology for Food Industry: Fundamentals and Applications*, J. Raso and V. Heinz (Eds.), pp. 153–194, New York: Springer.

Bazhal, M. 2001. Etude du mécanisme d'électropermeabilisation des tissus végétaux. Application à l'extraction du jus des pommes. PhD dissertation. Compiègne: University of the Technology of Compiègne.

Bazhal, I.G., Gulyi, I.S., Bobrovnik, L.D. 1983. Extraction of sugar from sugar beet in a constant electric field. *Izv Vuz Pishch Tekh* 5: 49–51 (in Russian).

Bazhal, M.I., Vorobiev, E.I. 2000. Electrical treatment of apple cossettes for intensifying juice pressing. *J Sci Food Agric* 80: 1668–1674.

Bazhal, M.I., Lebovka, N.I., Vorobiev, E.I. 2001. Pulsed electric field treatment of apple tissue during compression for juice extraction. *J Food Eng* 50: 129–139.

Bazhal, M., Lebovka, N., Vorobiev, E. 2003a. Optimisation of pulsed electric field strength for electroplasmolysis of vegetable tissues. *Biosyst Eng* 86: 339–345.

Bazhal, M.I., Ngadi, M.O., Ragavan, G.S.V. 2003b. Synergy between pressure and pulsed electric field in compression of vegetable tissue. *Elektronnaya Obrabotka Materialov* 3: 59–66.

Bazhal, M.I., Ngadi, M.O., Raghavan, G.S.V., Nguyen, D.H. 2003c. Textural changes in apple tissue during pulsed electric field treatment. *J Food Sci* 68: 249–253.

Bazhal, M.I., Ngadi, M.O., Raghavan, V.G.S. 2003d. Influence of pulsed electroplasmolysis on the porous structure of apple tissue. *Biosyst Eng* 86: 51–57.

Bazhal, M.I., Ngadi, M.O., Raghavan, G.S.V. 2004. Modeling compression of cellular systems exposed to combined pressure and pulsed electric fields. *Trans Am Soc Agric Eng* 47: 165–171.

Bazhal, M.I., Ngadi, M.O., Raghavan, G.S.V., Smith, J.P. 2006. Inactivation of *Escherichia coli* O157:H7 in liquid whole egg using combined pulsed electric field and thermal treatments. *LWT-Food Sci Technol* 39: 420–426.

Beattie, J.M. 1914. Electrical treatment of milk for infant feeding. *Br J State Med* 24: 97–113.

Ben Ammar, J., Van Hecke, E., Lebovka, N., Vorobiev, E., Lanoisellé, J.-L. 2009. Pulsed electric fields improve freezing process, in *Proceedings of BFE*, E. Vorobiev, N. Lebovka, E. Van Hecke, and J.-L. Lanoisellé (Eds.), C16 (1–6). Compiègne: University of the Technology of Compiègne.

Ben Ammar, J., Lanoisellé, J.-L., Lebovka, N.I., Van Hecke, E., Vorobiev, E. 2010. Effect of a pulsed electric field and osmotic treatment on freezing of potato tissue. *Food Biophys* 5: 247–254.

Ben Ammar, J., Lanoisellé, J.-L., Lebovka, N.I., Van Hecke, E., Vorobiev, E. 2011. Impact of a pulsed electric field on damage of plant tissues: Effects of cell size and tissue electrical conductivity. *J Food Sci* 76: E90–E97.

Bendicho, S., Estela, C., Giner, J., Barbosa-Cánovas, G.V., Martín, O. 2002. Effects of high intensity pulsed electric field and thermal treatments on a lipase from *Pseudomonas fluorescence*. *J Dairy Sci* 85: 19–27.

Berghofer, E., Cramer, A., Schmidt, U., Veigl, M. 1993. Pilot-scale production of inulin from chicory roots and its use in foodstuffs, in *Inulin and Inulin-Containing Crops. Studies in Plant Science 3*, A. Fuchs (Ed.), pp. 77–84. Amsterdam: Elsevier.

Bergey L. 1986–1989. *Manual of Systematic Bacteriology*. v 1–4. Baltimore: Williams and Wilkins.

Boussetta, N., Lebovka, N.I., Vorobiev, E.I., Adenier, H., Bedel-Cloutour, C., Lanoisellé, J.-L. 2009. Electrically assisted extraction of soluble matter from chardonnay grape skins for polyphenol recovery. *J Agric Food Chem* 57: 1491–1497.

Bouzrara, H. 2001. Amélioration du pressage de produits végétaux par Champ Electrique Pulsé. Cas de la betterave à sucre. PhD dissertation. Compiègne: University of the Technology of Compiègne.

Bouzrara, H., Vorobiev, E.I. 2000. Beet juice extraction by pressing and pulsed electric fields. *Int Sugar J* CII: 194–200.

Bouzrara, H., Vorobiev, E.I. 2001. Non-thermal pressing and washing of fresh sugarbeet cossettes combined with a pulsed electrical field. *Zuckerindustrie* 126: 463–466.

Bouzrara, H., Vorobiev, E.I. 2003. Solid/liquid expression of cellular materials enhanced by pulsed electric field. *Chem Eng Process* 42: 249–257.

Brookman, J.S.G. 1974. Mechanism of cell disintegration in a high pressure homogenizer. *Biotechnol Bioeng* 16: 371–383.

Buckow, R., Schroeder, S., Berres, P., Baumann, P., Knoerzer, K. 2010. Simulation and evaluation of pilot-scale pulsed electric field (PEF) processing. *J Food Eng* 101: 67–77.

Bushnell, A.H., Dunn, J.E., Clark, R.W., Lloyd, S.W. 2000. High-strength-electric-field pumpable-food-product treatment in a serial-electrode treatment cell. *U.S. Patent* 6110423.

Calleja, G.B. 1984. *Microbial Aggregation*. Boca Raton: CRC Press.

Canatella, P.J., Karr, J.F., Petros, J.A., Prausnitz, M.R. 2001. Quantitative study of electroporation mediated uptake and cell viability. *Biophys J* 80: 755–764.

Canatella, P.J., Black, M.M., Bonnichsen, D.M., McKenna, C., Prausnitz, M.R. 2004. Tissue electroporation: Quantification and analysis of heterogeneous transport in multicellular environments. *Biophys J* 86: 3260–3268.

Chalermchat, Y., Fincan, M., Dejmek, P. 2004. Pulsed electric field treatment for solid–liquid extraction of red beetroot pigment: Mathematical modelling of mass transfer. *J Food Eng* 64: 229–236.

Chalermchat, Y., Dejmek, P. 2005. Effect of pulsed electric field pretreatment on solid–liquid expression from potato tissue. *J Food Eng* 71: 164–169.

Chalermchat, Y., Malangone, L., Dejmek, P. 2010. Electropermeabilization of apple tissue: Effect of cell size, cell size distribution and cell orientation. *Biosyst Eng* 105: 357–366.

Chang, D.C. 1989. Cell poration and cell fusion using an oscillating electric field. *Biophys J* 56: 641–652.

Chen, P., Sun, Z. 1991. A review of non-destructive methods for quality evaluation and sorting of agricultural products. *J Agr Eng Res* 49: 85–98.

Chen, H., De Baerdemaeker, J. 1993. Modal analysis of the dynamic behavior of pineapples and its relation to fruit firmness. *Trans ASAE* 36: 1439–1444.

Chisti, Y. 2007. Strategies in downstream processing, in *Bioseparation and Bioprocessing: A Handbook*, vol. 1, G. Subramanian (Ed.), pp. 29–62. New York: Wiley-VCH.

Chisti, Y. 2010. Fermentation technology, in *Industrial Biotechnology: Sustainable Growth and Economic Success*, W. Soetaert and E.J. Vandamme (Eds.), pp. 149–171. New York: Wiley-VCH.

Corrales, M., Toepfl, S., Butz, P., Knorr, D., Tauscher, B. 2008. Extraction of anthocyanins from grape by-products assisted by ultrasonic, high hydrostatic pressure or pulsed electric fields: A comparison. *Innov Food Sci Emerg* 9: 85–91.

Cordwell, S.J. 2008. Sequential extraction of proteins by chemical reagents. *Methods Mol Biol* 424: 139–146.

de Alwis, A.A.P., Fryer, P.J. 1990. The use of direct resistance heating techniques in the food industry. *J Food Eng* 11: 3–27.

De Belie, N., Schotte, S., Lammertyn, J., Nicolai, B., De Baerdemaeker, J. 2000. Firmness changes of pear fruit before and after harvest with the acoustic impulse response technique. *J Agric Eng Res* 77: 183–191.

Delgado-Vargas, F., Jiménez, A.R., Paredes-López, O. 2000. Natural pigments: Carotenoids, anthocyanins, and betalains—Characteristics, biosynthesis, processing, and stability. *CRC Cr Rev Food Sci* 40: 173–289.

Delgado-Vargas, F., Paredes-López, O. 2003. *Natural Colorants for Food and Nutraceutical Uses*. Boca Raton: CRC Press.

De Vito, F., Ferrari, G., Lebovka, N.I., Shynkaryk, N.V., Vorobiev, E. 2008. Pulse duration and efficiency of soft cellular tissue disintegration by pulsed electric fields. *Food Bioprocess Technol* 1: 307–313.

Diederich, A., Bhr, G., Winterhalter, M. 1998. Influence of polylysine on the rupture of negatively charged membranes. *Langmuir* 14: 4597–4605.

Doevenspeck, H. 1961. Influencing cells and cell walls by electrostatic impulses. *Fleischwirtschaft* 13: 968–987.

Doevenspeck, H. 1991. Electric-impulse method for treating substances and device for carrying out the method. *U.S. Patent* 4994160.

Donsì, F., Ferrari, G., Pataro, G. 2010. Applications of pulsed electric field treatments for the enhancement of mass transfer from vegetable tissue. *Food Eng Rev* 2: 109–130.

El-Belghiti, K. 2005. Effets d'un champ électrique pulsé sur le transfert de matière et sur les caractéris-tiques végétales. PhD dissertation. Compiègne: University of the Technology of Compiègne.

El-Belghiti, K., Vorobiev, E.I. 2004. Mass transfer of sugar from beets enhanced by pulsed electric field. *Trans IChemE* 82: 226–230.

El-Belghiti, K., Vorobiev, E.I. 2005a. Kinetic model of sugar diffusion from sugar beet tissue treated by pulsed electric field. *J Sci Food Agric* 85: 213–218.

El-Belghiti, K., Vorobiev, E.I. 2005b. Modelling of solute aqueous extraction from carrots subjected to a pulsed electric field pre-treatment. *Biosyst Eng* 90: 289–294.

El-Belghiti, K., Rabhi, Z., Vorobiev, E.I. 2005. Effect of the centrifugal force on the aqueous extraction of solute from sugar beet tissue pretreated by a pulsed electric field. *J Food Process Eng* 28: 346–358.

El-Belghiti, K., Rabhi, Z., Vorobiev, E.I. 2007. Effect of process parameters on solute centrifugal extraction from electropermeabilized carrot gratings. *Food Bioprod Process* 85: 24–28.

El-Belghiti, K., Moubarik, A., Vorobiev, E.I. 2008. Aqueous extraction of solutes from fennel (*Foeniculum vulgare*) assisted by pulsed electric field. *J Food Process Eng* 31: 548–563.

El Zakhem, H., Lanoisellé, J.-L., Lebovka, N.I., Nonus, M., Vorobiev, E.I. 2006a. Behavior of yeast cells in aqueous suspension affected by pulsed electric field. *J Colloid Interf Sci* 300: 553–563.

El Zakhem, H., Lanoisellé, J.-L., Lebovka, N.I., Nonus, M., Vorobiev, E.I. 2006b. The early stages of *Saccharomyces cerevisiae* yeast suspensions damage in moderate pulsed electric fields. *Colloids Surf B* 47: 189–197.

El Zakhem, H., Lanoisellé, J.-L., Lebovka, N.I., Nonus, M., Vorobiev, E.I. 2007. Influence of temperature and surfactant on *Escherichia coli* inactivation in aqueous suspensions treated by moderate pulsed electric fields. *Int J Food Microbiol* 120: 259–265.

El Zakhem, H., Lanoiselle, J.-L., Lebovka, N., Nonus, M., Allali, H., Vorobiev, E. 2008. Combining moderate pulsed electric fields with temperature and with organic acids to inactivate *Escherichia coli* suspensions. *IUM Eng J* 9: 1–8

Engler, C.R. 1985. Disruption of microbial cells in comprehensive biotechnology, in *Comprehensive Biotechnology*, M. Moo-Young (Ed.), pp. 305–324. Oxford: Pergamon Press.

Eshtiaghi, M.N., Knorr, D. 1999. Method for treating sugar beet. *International Patent* WO 99/6434.

Evrendilek, G.A., Jin, Z.T., Ruhlman, K.T., Qiu, X., Zhang, Q.H., Richter, E.R. 2000. Microbial safety and shelf-life of apple juice and cider processed by bench and pilot scale PEF systems. *Innov Food Sci Emerg* 1: 77–86.

Evrendilek, G.A., Yeom, H.W., Jin, Z.T., Zhang, Q.H. 2004. Safety and quality evaluation of yogurt-based drink processed by a pilot plant PEF system. *J Food Proc Eng* 27: 197–212.

Evrendilek, G.A., Zhang, Q.H. 2005. Effects of pulse polarity and pulse delaying time on pulsed electric fields-induced pasteurization of *E. coli* O157:H7. *J Food Eng* 68: 271–276.

Exerova, D., Nikolova, A. 1992. Phase transitions in phospholipid foam bilayers. *Langmuir* 8: 3102–3108.

Eynard, N., Sixou, S., Duran, N., Teissié, J. 1992. Fast kinetics studies of *Escherichia coli* electrotransformation. *Eur J Biochem* 209: 431–436.

Eynard, N., Rodriguez, F., Trotard, J., Teissié, J. 1998. Electrooptics studies of *Escherichia coli* electropulsation: Orientation, permeabilization, and gene transfer. *Biophys J* 75: 2587–2596.

Fettermann, J.C. 1928. Electrical conductivity method of processing milk. *Agric Eng* 9: 107–108.

Fincan, M., Dejmek, P. 2002. In situ visualization of the effect of a pulsed electric field on plant tissue. *J Food Eng* 55: 223–230.

Fincan, M., Dejmek, P. 2003. Effect of osmotic pretreatment and pulsed electric field on the viscoelastic properties of potato tissue. *J Food Eng* 59: 169–175.

Fincan, M., De Vito, F., Dejmek, P. 2004. Pulsed electric field treatment for solid–liquid extraction of red beetroot pigment. *J Food Eng* 64: 381–388.

Flaumenbaum, B.L. 1949. Electrical treatment of fruits and vegetables before extraction of juice. *Trudy OTIKP* 3: 15–20 (in Russian).

Flaumenbaum, B.L. 1968. Anwendung der Elektroplasmolyse bei der Herstellung von Fruchtsäften. *Flüssiges Obst* 35: 19–22.

Fologea, D., Vassu, T., Stoica, I., Csutak, O., Sasarman, E., Smarandache, D., Ionescu, R. 2004. Efficient electrotransformation of yeast using bipolar electric pulses. *Rom Biotechnol Lett* 9: 1505–1510.

Franck, A. 2006. Inulin, in *Food Polysaccharides and Their Application*, A.M. Stephen, G.O. Phillips, and P.A. Williams (Eds.), pp. 335–351. Boca Raton: CRC Press.

Foster, K.R., Schwan, H.P. 1989. Dielectric properties of tissues—A review. *CRC Crit Rev Bioeng* 17: 25–104.

Fox, M.B., Esveld, D.C., Mastwijk H., Boom, R.M. 2008. Inactivation of *L. plantarum* in a PEF microreactor. The effect of pulse width and temperature on the inactivation. *Innov Food Sci Emerg* 9: 101–108.

Gachovska, T.K., Ngadi, M.O., Raghavan, G.S.V. 2006. Pulsed electric field assisted juice extraction from alfalfa. *Can Biosyst Eng/Le Genie des biosystems au Canada* 48: 3.33–3.37.

Gachovska, T.K., Adedeji, A.A., Ngadi, M., Raghavan, G.V.S. 2008. Drying characteristics of pulsed electric field-treated carrot. *Dry Technol* 26: 1244–1250.

Gachovska, T.K., Marian, V., Ngadi, M.O., Raghavan, G.S.V. 2009a. Pulsed electric field treatment of carrots before drying and rehydration. *J Sci Food Agric* 89: 2372–2376.

Gachovska, T.K., Adedeji, A.A., Ngadi, M. 2009b. Influence of pulsed electric field energy on the damage degree in alfalfa tissue. *J Food Eng* 95: 558–563.

Gachovska, T., Cassada, D., Subbiah, J., Hanna, M., Thippareddi, H., Snow, D. 2010. Enhanced anthocyanin extraction from red cabbage using pulsed electric field processing. *J Food Sci* 75: E323–E329.

Galindo, F.G. 2008. Reversible electroporation of vegetable tissues—Metabolic consequences and applications. *Rev Bol Quím* 25: 30–35.

Galindo, F.G., Vernier, P., Dejmek, P., Vicente, A., Gundersen, M. 2008a. Pulsed electric field reduces the permeability of potato cell wall. *Bioelectromagnetism* 29: 296–301.

Galindo, F.G., Wadsö, L.P., Vicente, A., Dejmek, P. 2008b. Exploring metabolic responses of potato tissue induced by electric pulses. *Food Biophys* 3: 352–360.

Galindo, F.G., Dejmek, P., Lundgren, K., Rasmusson, A.G., Vicente, A., Moritz, T. 2009a. Metabolomic evaluation of pulsed electric field-induced stress on potato tissue. *Planta* 230: 469–479.

Galindo, F.G., Wadso, L., Vicente, A., Dejmek, P. 2009b. Gross metabolic responses of potato tissue subjected to pulsed electric fields, in: *Proceedings of the 5th International Symposium on Food Processing, Monitoring Technology in Bioprocesses and Food Quality Management*, pp. 601–613. Potsdam, Germany.

Galili, N., Shmulevich, I., Benichou, N. 1998. Acoustic testing of avocado for fruit ripeness evaluation. *Trans ASAE* 41: 399–407.

Ganeva, V., Galutzov, B. 1999. Electropulsation as an alternative method for protein extraction from yeast. *FEMS Microbiol Lett* 174(2): 279–284.

Ganeva, V., Galutzov, B., Eynard, N., Teissié, J. 2001. Electroinduced extraction of β-galactosidase from *Kluyveromyces lactis*. *Appl Microbiol Biotechnol* 56: 411–413.

Ganeva, V., Galutzov, B., Teissié, J. 2003. High yield electroextraction of proteins from yeast by a flow process. *Anal Biochem* 315: 77–84.

García-Ramos, F.J., Valero, C., Homer, I., Ortiz-Cañavate, J., Ruiz-Altisent, M. 2005. Non-destructive fruit firmness sensors: A review. *Span J Agric Res* 3: 61–73.

Getchell, B.E. 1935. Electric pasteurization of milk. *Agric Eng* 16: 408–410.

Góngora-Nieto, M.M., Pedrow, P.D., Swanson, B.G., Barbosa-Cánovas, G.V. 2003. Energy analysis of liquid whole egg pasteurized by pulsed electric fields. *J Food Eng* 57: 209–216.

Gough, N.M. 1988. Rapid and quantitative preparation of cytoplasmic RNA from small numbers of cells. *Anal Biochem* 173: 93–95.

Grahl, T., Märkl, H. 1996. Killing of microorganisms by pulsed electric fields. *Appl Microbiol Biotechnol* 45: 148–157.

Grimi, N. 2009. Vers l'intensification du pressage industriel des agroressources par champs électriques pulsés: Étude multi-échelles (Towards the intensification of industrial pressing of agricultural resources by pulsed electric fields: Multiscale study). PhD dissertation. Compiègne: University of the Technology of Compiègne.

Grimi, N., Praporscic, I., Lebovka, N., Vorobiev, E. 2007. Selective extraction from carrot slices by pressing and washing enhanced by pulsed electric fields. *Separ Purific Technol* 58: 267–273.

Grimi, N., Vorobiev, E., Vaxelaire, J. 2008. Juice extraction from sugar beet slices by belt filter press: Effect of pulsed electric field and operating parameters, in *Proceedings of* 14th *World Congress of Food Science and Technology,* TS24-29, 554. Shanghai, China.

Grimi, N., Lebovka, N.I., Vorobiev, E., Vaxelaire, J. 2009a. Compressing behavior and texture evaluation for potatoes pretreated by pulsed electric field. *J Text Stud* 40: 208–224.

Grimi, N., Lebovka, N.I., Vorobiev, E., Vaxelaire, J. 2009b. Effect of a pulsed electric field treatment on expression behavior and juice quality of chardonnay grape. *Food Biophys* 4: 191–198.

Grimi, N., Mamouni, F., Lebovka, N., Vorobiev, E., Vaxelaire, J. 2010. Acoustic impulse response in apple tissues treated by pulsed electric field. *Biosyst Eng* 105: 266–272.

Grimi, N., Boussetta N., Mamouni, F., Lebovka, N., Vorobiev, E., Vaxelaire, J. 2011. Impact of apple processing modes on extracted juice quality: Pressing assisted by pulsed electric fields. *J Food Eng* 103: 52–61.

Guderjan, M., Elez-Martínez, P., Knorr, D. 2007. Application of pulsed electric fields at oil yield and content of functional food ingredients at the production of rapeseed oil. *Innov Food Sci Emerg* 8: 55–62.

Guderjan, M., Töpfl, S., Angersbach, A., Knorr, D. 2005. Impact of pulsed electric field treatment on the recovery and quality of plant oils. *J Food Eng* 67: 281–287.

Gulyi, I.S., Lebovka, N.I., Mank, V.V., Kupchik, M.P., Bazhal, M.I., Matvienko, A.B., Papchenko, A.Y. 1994. *Scientific and Practical Principles of Electrical Treatment of Food Products and Materials.* Kiev: UkrINTEI (in Russian).

Han, Z., Zeng, X.A., Yu, S.J., Zhang, B.S., Chen, X.D. 2009a. Effects of pulsed electric fields (PEF) treatment on physicochemical properties of potato starch. *Innov Food Sci Emerg* 10: 481–485.

Harrison, S.T. 1991. Bacterial cell disruption: A key unit operation in the recovery of intracellular products. *Biotechnol Adv* 9: 217–240.

Harrison, S.T.L., Dennis, J.S., Chase, H.A. 1991. Combined chemical and mechanical processes for the disruption of bacteria. *Bioseparation* 2: 95–105.

Herbach, K.M., Stinzing, F.C., Carle, R. 2004. Impact of thermal treatment on color and pigment pattern of red beet (*Beta vulgaris* L.) preparations. *J Food Sci C* 69: C491–C498.

Hulsheger, H., Potel, J., Niemann, E.G. 1983. Electric field effects on bacteria and yeast cells. *Radiat Environ Biophys* 22: 149–162.

Jaeger, H., Balasa, A., Knorr, D. 2008. Food industry applications for pulsed electric fields, in *Electrotechnologies for Extraction from Food Plants and Biomaterials*, E. Vorobiev and N. Lebovka (Eds.), pp. 181–217. New York: Springer.

Jaeger, H., Knorr, D. 2010. Pulsed electric fields (PEF): Mass transfer enhancement. In *Encyclopedia of Agricultural, Food, and Biological Engineering*, D.R. Heldman and C.I. Moraru (Eds.), pp. 1391–1394. London: Taylor & Francis.

Jalté, M., Lanoisellé, J.-L., Lebovka, N.I., Vorobiev, E. 2009. Freezing of potato tissue pretreated by pulsed electric fields. *LWT-Food Sci Technol* 42: 576–580.

Jemai, A.B. 1997. Contribution a l'etude de l'effet d'un traitement electrique sur les cossettes de betterave a sucre. Incidence sur le procede d'extraction. PhD dissertation. Compiègne: Université de Technologie de Compiègne.

Jemai, A.B., Vorobiev, E. 2001. Enhancement of the diffusion characteristics of apple slices due to moderate electric field pulses (MEFP), in *Proceedings of the 8th International Congress on Engineering and Food*, J. Welti-Chanes, G.V. Barbosa-Cánovas, and J.M. Aguilera (Eds.), pp. 1504–1508. Pennsylvania: Technomic Publishing Co.

Jemai, A.B., Vorobiev, E. 2002. Effect of moderate electric field pulse (MEFP) on the diffusion coefficient of soluble substances from apple slices. *Int J Food Sci Technol* 37: 73–86.

Jemai, A.B., Vorobiev, E. 2003. Enhancing leaching from sugar beet cossettes by pulsed electric field. *J Food Eng* 59: 405–412.

Jemai, A.B., Vorobiev, E. 2006. Pulsed electric field assisted pressing of sugar beet slices: Towards a novel process of cold juice extraction. *Biosyst Eng* 93: 57–68.

Jen, C.P., Chen, Y.H., Fan, C.S., Yeh, C.S., Lin, Y.C., Shieh, D.B., Wu, C.L., Chen, D.H., Chou, C.H. 2004. A nonviral transfection approach in vitro: The design of a gold nanoparticle vector joint with microelectromechanical systems. *Langmuir* 20: 1369–1374.

Jin, T., Zhang, H. 2005. Development of pilot scale fluid handling system with energy recovery for pulsed electric field processing, in *AIChE Annual Meeting, Conference Proceedings*, pp. 12599–12605. Cincinnati, OH.

Jin, Z.T., Zhang, H.Q., Li, S.Q., Kim, M., Dunne, C.P., Yang, T., Wright, A.O., Venter-Gains, J. 2009. Quality of apple sauces processed by pulsed electric fields and HTST pasteurization. *Int J Food Sci Technol* 44: 829–839.

Jones, F. 1897. Apparatus for electrically treating liquids. *U.S. Patent* 592 735.

Joshi, R.P., Mishra, A., Schoenbach, K.H. 2008. Model assessment of cell membrane breakdown in clusters and tissues under high-intensity electric pulsing. *IEEE Trans Plasma Sci* 36: 1680–1688.

Kandušer, M., Šentjurc, M., Miklavčič, D. 2008. The temperature effect during pulse application on cell membrane fluidity and permeabilization. *Bioelectrochemistry* 74: 52–57.

Katrokha, I.M., Kupshik, M.P. 1984. Intensification of sugar extraction from sugar-beet cossettes in an electric field. *Sakharnaya Promyshlennost* 7: 28–31 (in Russian).

Kinosita Jr., K., Tsong, T.Y. 1977. Hemolysis of human erythrocytes by transient electric field. *Proc Nat Acad Sci U S A* 74: 1923–1927.

Knorr, D., Geulen, W., Grahl, T., Sitzmann, W. 1994. Food application of high electric field pulses. *Trends Food Sci Technol* 5: 71–75.

Knorr, D., Angersbach, A., Eshtiaghi, M.N., Heinz, V., Lee, D.-U. 2001. Processing concepts based on high intensity electric field pulses. *Trends Food Sci Technol* 12: 129–135.

Knorr, D., Engel, K.-H., Vogel, R., Kochte-Clemens, B., Eisenbrand, G. 2008. Statement on the treatment of food using a pulsed electric field. *Mol Nutr Food Res* 52: 1539–1542.

Kogan, F.I. 1968. *Electrophysical Methods in Canning Technologies of Foodstuff*. Kiev: Tehnika (in Russian).

Korolczuk, J., Mc Keag, J.R., Fernandez, J.C., Baron, F., Grosset, N., Jeantet, R. 2006. Effect of pulsed electric field processing parameters on *Salmonella enteritidis* inactivation. *J Food Eng* 75: 11–20.

Kotnik, T., Miklavčič, D., Slivnik, T. 1998. Time course of transmembrane voltage induced by time-varying electric fields: A method for theoretical analysis and its application. *Bioelectrochem Bioenerg* 45: 3–16.

Kuldiloke, J., Eshtiaghi, M.N., Neatpisarnvanit, C. 2008. Application of high electric field pulses for sugar cane processing, in *Proceedings of 34th Congress on Science and Technology of Thailand (STT34), Science and Technology for Global Challenges*, H_H0015:1–6. Bangkok, Thailand.

Kuldiloke J., Eshtiaghi M.N., Neatpisarnvanit C., Uan-On, T. 2008. Application of high electric field pulses for sugar cane processing. *KMITL Sci Technol J* 8: 75–83.

Kulshrestha, S., Sastry, S. 2003. Frequency and voltage effects on enhanced diffusion during moderate electric field (MEF) treatment. *Innov Food Sci Emerg* 4: 189–194.

Kulshrestha, S.A., Sastry, S.K. 2010. Changes in permeability of moderate electric field (MEF) treated vegetable tissue over time. *Innov Food Sci Emerg* 11: 78–83.

Lebedeva, N.E. 1987. Electric breakdown of bilayer lipid membranes at short times of voltage effect. *Biol Membr* 4: 994–998 (in Russian).

Lebovka, N.I., Bazhal, M.I., Vorobiev, E.I. 2000. Simulation and experimental investigation of food material breakage using pulsed electric field treatment. *J Food Eng* 44: 213–223.

Lebovka, N.I., Bazhal, M.I., Vorobiev, E.I. 2001. Pulsed electric field breakage of cellular tissues: Visualization of percolative properties. *Innov Food Sci Emerg* 2: 113–125.

Lebovka, N.I., Bazhal, M.I., Vorobiev, E.I. 2002. Estimation of characteristic damage time of food materials in pulsed-electric fields. *J Food Eng* 54: 337–346.

Lebovka, N.I., Praporscic, I., Vorobiev, E.I. 2003. Enhanced expression of juice from soft vegetable tissues by pulsed electric fields: Consolidation stages analysis. *J Food Eng* 59: 309–317.

Lebovka, N.I., Praporscic, I., Vorobiev, E.I. 2004a. Effect of moderate thermal and pulsed electric field treatments on textural properties of carrots, potatoes and apples. *Innov Food Sci Emerg* 5: 9–16.

Lebovka, N.I., Praporscic, I., Vorobiev, E.I. 2004b. Combined treatment of apples by pulsed electric fields and by heating at moderate temperature. *J Food Eng* 65: 211–217.

Lebovka, N.I., Praporscic, I., Ghnimi, S., Vorobiev, E. 2005a. Temperature enhanced electroporation under the pulsed electric field treatment of food tissue. *J Food Eng* 69: 177–184.

Lebovka, N.I., Praporscic, I., Ghnimi, S., Vorobiev, E. 2005b. Does electroporation occur during the ohmic heating of food. *J Food Sci* 70: 308–311.

Lebovka, N.I, Shynkaryk, M.V., Vorobiev, E. 2006. Drying of potato tissue pretreated by ohmic heating. *Dry Technol* 24: 1–11.

Lebovka, N.I., Shynkaryk, N.V., Vorobiev, E.I. 2007a. Pulsed electric field enhanced drying of potato tissue. *J Food Eng* 78: 606–613.

Lebovka, N.I., Shynkaryk, M.V., El-Belghiti, K., Benjelloun, H., Vorobiev, E.I. 2007b. Plasmolysis of sugarbeet: Pulsed electric fields and thermal treatment. *J Food Eng* 80: 639–644.

Lebovka, N.I., Kupchik, M.P., Sereda, K., Vorobiev, E. 2008. Electrostimulated thermal permeabilization of potato tissues. *Biosyst Eng* 99: 76–80.

Lebovka, N., Vorobiev, E. 2010. Food and biomaterials processing assisted by electroporation, in *Advanced Electroporation Techniques in Biology and Medicine*, A.G. Pakhomov, D. Miklavcic, and M.S. Markov (Eds.), pp. 463–490. Boca Raton: CRC Press.

Lelieveld, H.L.M., Notermans, S., de Haan S.W.H. (Eds.). 2007. *Food Preservation by Pulsed Electric Fields: From Research to Application*. Boca Raton: CRC Press.

Li, S.-Q., Zhang, H.Q., Jin, T.Z., Turek, E.J., Lau, M.H. 2007. Elimination of *Lactobacillus plantarum* and achievement of shelf stable model salad dressing by pilot scale pulsed electric fields combined with mild heat. *Innov Food Sci Emerg* 6: 125–133.

Liu, D., Savoire, R., Vorobiev, E., Lanoisellé, J.-L. 2010. Effect of disruption methods on the dead-end microfiltration behavior of yeast suspension. *Separ Sci Technol* 45: 1042–1050.

Liu, Z., Guo, Y. 2009. BP neural network prediction of the effects of drying rate of fruits and vegetables pretreated by high-pulsed electric field. *Nongye Gongcheng Xuebao/Trans Chin Soc Agric Eng* 25: 235–239.

Loeffler, M.J. 2002. Commercial pulsed power applications in Germany, in *Papers of Technical Meeting on Plasma Science and Technology, IEE Japan*, PST-02: pp. 95–100.

Loginov, M., Lebovka, N., Larue, O., Shynkaryk, M., Nonus, M., Lanoisellé, J.-L., Vorobiev, E. 2009. Effect of high voltage electrical discharges on filtration properties of *Saccharomyces cerevisiae* yeast suspensions. *J Membr Sci* 346: 288–295.

Loginova, K.V., Shynkaryk, M.V., Lebovka, N.I., Vorobiev, E. 2010. Acceleration of soluble matter extraction from chicory with pulsed electric fields. *J Food Eng* 96: 374–379.

Loginova, K.V., Vorobiev, E., Bals, O., Lebovka N.I. 2011. Pilot study of countercurrent cold and mild heat extraction of sugar from sugar beets, assisted by pulsed electric fields. *J Food Eng* 102: 340–347.

López, N., Puértolas, E., Condón, S., Álvarez, I., Raso, J. 2008a. Effects of pulsed electric fields on the extraction of phenolic compounds during the fermentation of must of Tempranillo grapes. *Innov Food Sci Emerg* 9: 477–482.

López, N., Puértolas, E., Condón, S., Álvarez, I., Raso, J. 2008b. Application of pulsed electric fields for improving the maceration process during vinification of red wine: Influence of grape variety. *Eur Food Res Technol* 227: 1099–1107.

López, N., Puértolas, E., Condón, S., Álvarez, I. 2009a. Enhancement of the extraction of betanine from red beetroot by pulsed electric fields. *J Food Eng* 90: 60–66.

López, N., Puértolas, E., Condón, S., Raso, J., Álvarez, I. 2009b. Enhancement of the solid-liquid extraction of sucrose from sugar beet (*Beta vulgaris*) by pulsed electric fields. *LWT-Food Sci Technol* 42: 1674–1680.

López, N., Puértolas, E., Hernández-Orte, P., Álvarez, I., Raso, J. 2009c. Effect of a pulsed electric field treatment on the anthocyanins composition and other quality parameters of Cabernet Sauvignon freshly fermented model wines obtained after different maceration times. *LWT-Food Sci Technol* 42: 1225–1231.

Lovitt, R.W., Jones, M., Collins, S.E., Coss, G.M., Yau C.P., Attouch, C. 2000. Disruption of baker's yeast using a disrupter of simple and novel geometry. *Process Biochem* 36: 415–421.

Lysjanskii V.M. 1973. *The Extraction Process of Sugar from Sugarbeet: Theory and Calculations (Process ekstractzii sahara iz svekly: Teorija i raschet)*. Moscow: Pischevaja Promyshlennost (in Russian).

Mañas, P., Barsotti, L., Cheftel, J.C. 2000. Microbial inactivation by pulsed electric fields in a batch treatment chamber: Effects of some electrical parameters and food constituents. *Innov Food Sci Emerg* 2: 239–249.

Martín-Belloso, O., Vega-Mercado, H., Qin, B.L., Chang, F.J., Barbosa-Cánovas, G.V., Swanson, B.G. 1997. Inactivation of *Escherichia coli* suspended in liquid egg using pulsed electric fields. *J Food Proc Preserv* 21: 193–208.

Matov, B.I., Reshetko, E.V. 1968. *Electrophysical Methods in Food Industry*. Kishinev: Kartja Moldavenjaske (in Russian).

Melkonyan, H., Sorg, C., Klempt, M. 1996. Electroporation efficiency in mammalian cells is increased by dimethyl sulfoxide (DMSO). *Nucleic Acids Res* 24: 4356–4357.

Middelberg, A.P. 2000. Microbial cell disruption by high-pressure homogenization, in *Downstream Processing of Proteins*, vol. 9. M.A. Desai (Ed.), pp. 11–21. Totowa: Humana Press.

Miklavcic, D., Towhidi, L. 2010. Numerical study of the electroporation pulse shape effect on molecular uptake of biological cells. *Radiol Oncol* 44: 34–41.

Morren, J., Roodenburg, B., de Haan, S.W.H. 2003. Electrochemical reactions and electrode corrosion in pulsed electric field (PEF) treatment chambers. *Innov Food Sci Emerg* 4: 285–295.

Mouritsen, O.G., Jørgensen, K. 1997. Small-scale lipid-membrane structure: Simulation versus experiment. *Curr Opin Struct Biol* 7: 518–527.

Neumann, E., Rosenheck, K. 1972. Permeability changes induced by electric impulses in vesicular membranes. *J Membr Biol* 10: 279–290.

Ngadi, M., Raghavan, V., Gachovska, T. 2009. Pulsed electric field enhanced method of extraction. *U.S. Patent* 20090314630.

Ohshima, T., Zhang, X.L., Iijima, S., Kobayashi, T. 1992. Control of gene expression from SUC2 promoter of *Saccharomyces cerevisiae* using two carbon sources. *Kagaku Kogaku Ronbunshu* 18: 693–700.

Ohshima, T., Sato, M., Saito, M. 1995. Selective release of intracellular protein using pulsed electric field. *J Electrostat* 35: 103–112.

Ohshima, T., Hama, Y., Sato, M. 2000 Releasing profiles of gene products from recombinant *Escherichia coli* in a high-voltage pulsed electric field. *Biochem Eng J* 5: 149–155.

Pakhomov, A.G., Miklavcic, D., Markov M.S. (Eds.). 2010. *Advanced Electroporation Techniques in Biology and Medicine*. Boca Raton: CRC Press.

Patel, N., Solanki, E., Picciani, R., Cavett, V., Caldwell-Busby, J.A., Bhattacharya, S.K. 2008. Strategies to recover proteins from ocular tissues for proteomics. *Proteomics* 8: 1055–1070.

Pavlin, M., Pavselj, N., Miklavčič, D. 2002. Dependence of induced transmembrane potential on cell density, arrangement, and cell position inside a cell system. *IEEE Trans Biomed Eng* 49: 605–612.

Pavlin, M., Kotnik, T., Miklavčič, D., Kramar, P., Lebar, A.M. 2008. Electroporation of planar lipid bilayers and membranes, in *Advances in Planar Lipid Bilayers and Liposomes*, L.A. Leitmanova (Ed.), pp. 165–226. Amsterdam: Elsevier.

Pereira, R.N., Galindo, F.G., Vicente, A.A., Dejmek, P. 2009. Effects of pulsed electric field on the viscoelastic properties of potato tissue. *Food Biophys* 4: 229–239.

Peternel, Š., Komel, R. 2010. Isolation of biologically active nanomaterial (inclusion bodies) from bacterial cells. *Microb Cell Fact* 9: 66.

Pliquett, U. 2010. Bioimpedance: A review for food processing. *Food Eng Rev* 2: 74–94.

Pothakamury, U.R., Vega, H., Zhang, Q., Barbosa-Cánovas, G.V., Swanson, B.G. 1996. Effect of growth stage and processing temperature on the inactivation of *Escherichia coli* by pulsed electric fields. *J Food Process* 59: 1167–1171.

Praporscic, I. 2005. Influence du traitement combiné par champ électrique pulsé et chauffage modéré sur les propriétés physiques et sur le comportement au pressage de produits végétaux. PhD dissertation. Compiègne: University of the Technology of Compiègne.

Praporscic, I., Ghnimi, S., Vorobiev, E.I. 2005. Enhancement of pressing sugar beet cuts by combined ohmic heating and pulsed electric field treatment. *J Food Proc Preserv* 29: 378–389.

Praporscic, I., Lebovka, N.I., Vorobiev, E.I., Mietton-Peuchot, M. 2007a. Pulsed electric field enhanced expression and juice quality of white grapes. *Sep Purif Technol* 52: 520–526.

Praporscic, I., Shynkaryk, M.V., Lebovka, N.I., Vorobiev, E.I. 2007b. Analysis of juice colour and dry matter content during pulsed electric field enhanced expression of soft plant tissues. *J Food Eng* 79: 662–670.

Prescott, S.C. 1927. The treatment of milk by an electrical method. *Am J Public Health* 17: 221–223.

Prochownick, L., Spaeth, F. 1890. Über die keimtötende Wirkung des galvanischen Stroms. *Dtsch Med Wochenschr* 26: 564–565.

Puértolas, E., Saldaña, G., Condón, S., Álvarez, I., Raso, J. 2009. A comparison of the effect of macerating enzymes and pulsed electric fields technology on phenolic content and color of red wine. *J Food Sci C* 74: C647–C652.

Puértolas, E., López, N., Saldaña, G., Álvarez, I., Raso, J. 2010a. Evaluation of phenolic extraction during fermentation of red grapes treated by a continuous pulsed electric fields process at pilot-plant scale. *J Food Eng* 98: 120–125.

Puértolas, E., Saldaña, G., Álvarez, I., Raso, J. 2010b. Effect of pulsed electric field processing of red grapes on wine chromatic and phenolic characteristics during aging in oak barrels. *J Agr Food Chem* 58: 2351–2357.

Puértolas, E., Saldana, G., Condon, S., Alvarez, I., Raso, J. 2010c. Evolution of polyphenolic compounds in red wine from Cabernet Sauvignon grapes processed by pulsed electric fields during aging in bottle. *Food Chem* 119: 1063–1070.

Puértolas, E., Hernández-Orte, P., Sladaña, G., Álvarez, I., Raso, J. 2010d. Improvement of winemaking process using pulsed electric fields at pilot-plant scale. Evolution of chromatic parameters and phenolic content of Cabernet Sauvignon red wines. *Food Res Int* 43: 761–766.

Puértolas, E., López, N., Condón, S., Álvarez, I., Raso, J. 2010e. Potential applications of PEF to improve red wine quality. *Trends Food Sci Technol* 21: 247–255.

Qin, B.L., Zhang, Q., Swanson, B.G., Pedrow, P.D. 1994. Inactivation of microorganisms by different pulsed electric fields of different voltage waveforms. *Inst Electrical Electronics Eng Trans Ind Appl* 1: 1047–1057.

Qiu, X., Tuhela, L., Zhang, Q.H. 1997. Application of pulsed power technology in non-thermal food processing and system optimization, in *Digest of Technical Papers—11th IEEE International Pulsed Power Conference* 1: 85–90.

Qiu, X., Sharma, S., Tuhela, L., Jia, M., Zhang, Q.H. 1998. An integrated PEF pilot plant for continuous nonthermal pasteurization of fresh orange juice. *Trans Am Soc Agric Eng* 41: 1069–1074.

Raffa, V., Ciofani, G., Cuschieri, A. 2009. Enhanced low voltage cell electropermeabilization by boron nitride nanotubes. *Nanotechnology* 20: 075104 (5pp).

Ramaswamy, R., Jin, T., Balasubramaniam, V.M., Zhang, H. 2008. Pulsed electric field processing, in *Fact Sheet for Food Processors*. The Ohio State University: Department of Food Science and Technology. Available at http://ohioline.osu.edu/fse-fact/pdf/0002 .pdf.

Raso, J., Álvarez, I., Condón, S., Sala-Trepat, F.J. 2000. Predicting inactivation of *Salmonella senftenberg* by pulsed electric fields. *Innov Food Sci Emerg* 1: 21–29.

Raso, J., Heinz, V. (Eds.). 2006. *Pulsed Electric Fields Technology for the Food Industry. Fundamentals and Applications*. New York: Springer.

Rastogi, N.K., Eshtiaghi, M.N., Knorr, D. 1999. Accelerated mass transfer during osmotic dehydration of high intensity electrical field pulse pretreated carrots. *J Food Sci* 64: 1020–1023.

Ravishankar, S., Zhang, H., Kempkes, M.L. 2008. Pulsed electric fields. *Food Sci Technol Int* 14: 429–432.

Riesz, P., Kondo, T. 1992. Free radical formation induced by ultrasound and its biological implications. *Free Radic Biol Med* 13: 247–270.

Rodríguez-Sevilla, M.D., Villanueva-Suárez, M.J., Redondo-Cuenca, A. 1999. Effects of processing conditions on soluble sugars content of carrot, beetroot and turnip. *Food Chem* 66: 81–85.

Rojas-Chapana, J.A., Correa-Duarte, M.A., Ren, Z., Kempa, K., Giersig, M. 2004. Enhanced introduction of gold nanoparticles into vital *Acidothiobacillus ferrooxidans* by carbon nanotube-based microwave electroporation. *Nano Lett* 4: 985–988.

Rokkones, E., Kareem, B.N., Olstad, O.K., Hogset, A., Schenstrom, K., Hansson, L., Gautvik, K.M. 1994. Expression of human parathyroid hormone in mammalian cells, *Escherichia coli* and *Saccharomyces cerevisiae*. *J Biotechnol* 338: 293–306.

Roodenburg, B., Morren, J., Berg, H.E. (Iekje), de Haan, S.W.H. 2005a. Metal release in a stainless steel Pulsed Electric Field (PEF) system: Part I. Effect of different pulse shapes; theory and experimental method. *Innov Food Sci Emerg* 6: 327–336.

Roodenburg, B., Morren, J., Berg, H.E. (Iekje), de Haan, S.W.H. 2005b. Metal release in a stainless steel pulsed electric field (PEF) system: Part II. The treatment of orange juice; related to legislation and treatment chamber lifetime. *Innov Food Sci Emerg* 6: 337–345.

Roodenburg, B. 2007. Electrochemistry in pulsed electric field treatment chambers, in *Food Preservation by Pulsed Electric Fields: From Research to Application*, H.L.M. Lelieveld, S. Notermans, and S.W.H. de Haan (Eds.), pp. 94–107, Boca Raton: CRC Press.

Rogov, I.A. 1988. *Electrophysical Methods of Foods Product Processing*. Moscow: Agropromizdat (in Russian).

Rogov, I.A., Gorbatov, A.V. 1974. *Physical Methods of Foods Processing*. Moscow: Pischevaja Promyshlennost (in Russian).

Sack, M., Schultheiss, C., Bluhm, H. 2005. Triggered Marx generators for the industrial-scale electroporation of sugar beets. Industry applications. *IEEE Trans* 41: 707–714.

Sack, M., Eing, C., Stangle, R., Wolf, A., Muler, G., Sigler, J., Stukenbrock, L. 2009. Electric measurement of the electroporation efficiency of mash from wine grapes. *IEEE Trans Diel Electr Insul* 16: 1329–1337.

Sack, M., Sigler, J., Frenzel, S., Eing, C., Arnold, J., Michelberger, T., Frey, W., Attmann, F., Stukenbrock, L., Müller, G. 2010. Research on industrial-scale electroporation devices fostering the extraction of substances from biological tissue. *Food Eng Rev* 2: 147–156.

Saguy, I., Kopelman, I.J., Mizrahi, S. 1978. Thermal kinetic degradation of betanin and betalamic acid. *J Agric Food Chem* 26: 360–362.

Salazar, O., Asenjo, J.A. 2007. Enzymatic lysis of microbial cells. *Biotechnol Lett* 29: 985–994.

Sale, A., Hamilton, W. 1967. Effect of high electric fields on microorganisms. I. Killing of bacteria and yeast. *Biochim Biophys Acta* 148: 781–788.

Sampedro, F., Rivas, A., Rodrigo, D., Martínez, A., Rodrigo, M. 2007. Pulsed electric fields inactivation of *Lactobacillus plantarum* in an orange juice–milk based beverage: Effect of process parameters. *J Food Eng* 80: 931–938.

Saulis, G. 1993. Cell electroporation. Part 3. Theoretical investigation of the appearance of asymmetric distribution of pores on the cell and their further evolution. *Bioel Bioen* 32: 249–265.

Saulis, G. 2010. Electroporation of cell membranes: The fundamental effects of pulsed electric fields in food processing. *Food Eng Rev*: 52–73.

Schilling, S., Alber, T., Toepfl, S., Neidhart, S., Knorr, D., Schieber, A., Carle, R. 2007. Effects of pulsed electric field treatment of apple mash on juice yield and quality attributes of apple juices. *Innov Food Sci Emerg* 8: 127–134.

Schilling, S., Schmid, S., Jäger, H., Ludwig, M., Dietrich, H., Toepfl, S., Knorr, D., Neidhart, S., Schieber, A., Carle, R. 2008a. Comparative study of pulsed electric field and thermal processing of apple juice with particular consideration of juice quality and enzyme deactivation. *J Agric Food Chem* 56: 4545–4554.

Schilling, S., Toepfl, S., Ludwig, M., Dietrich, H., Knorr, D., Neidhart, S., Schieber, A., Carle, R. 2008b. Comparative study of juice production by pulsed electric field treatment and enzymatic maceration of apple mash. *Eur Food Res Technol* 226: 1389–1398.

Schultheiss, C., Bluhm, H., Mayer, H.-G., Kern, M., Michelberger, T., Witte, G. 2003. Processing of sugar beets with pulsed-electric fields. *IEEE Trans Plasma Sci* 30: 1547–1551.

Schwan, H.P. 1957. Electrical properties of tissue and cell suspensions, in *Advances in Biological and Medical Physics*, J.H. Lawrence and A. Tobias (Eds.), pp. 147–209. New York: Academic Press.

Singh, M.R., Roy, S., Bellare, J.R. 2009. Influence of cryogenic grinding on release of protein and DNA from *Saccharomyces cerevisiae*. *Int J Food Eng* 5: 1–25.

Schotte, S., Belie, N.D., Baerdemaeker, J.D. 1999. Acoustic impulse-response technique for evaluation and modeling of firmness of tomato fruit. *Postharvest Biol Technol* 17: 105–115.

Shynkaryk, M.V. 2007. Influence de la perméabilisation membranaire par champ électrique sur la performance de séchage des végétaux. PhD dissertation. Compiègne: University of the Technology of Compiègne.

Shynkaryk, M.V., Lebovka, N.I., Vorobiev, E. 2008. Pulsed electric fields and temperature effects on drying and rehydration of red beetroots. *Dry Technol* 26: 695–704.

Shynkaryk, M.V., Lebovka, N.I., Lanoisellé, J.-L., Nonus, M., Bedel-Clotour, C., Vorobiev, E.I. 2009. Electrically-assisted extraction of bio-products using high pressure disruption of yeast cells (*Saccharomyces cerevisiae*). *J Food Eng* 92: 189–195.

Smith, M. 2007. Regulatory acceptance of pulsed electric fields processing of foods, in *Food Preservation by Pulsed Electric Fields: From Research to Application*, H.L.M. Lelieveld, S. Notermans, and S.W.H. de Haan (Eds.), pp. 352–357. Bosa Roca: CRC Press.

Stampfli, R. 1958. Reversible electrical breakdown of the excitable membrane of a Ranvier node. *Ann Acad Brasil Ciens* 30: 57–63.

Stanley, D.W. 1991. Biological membrane deterioration and associated quality losses in food tissue. *Crit Rev Food Sci Nutr* 30: 487–553.

Stinzing, F.C., Carle, R. 2004. Functional properties of anthocyanins and betalains in plants, food, and in human nutrition. *Trends Food Sci Technol* 15: 19–38.

Stinzing, F.C., Carle, R. 2007. Betalains—Emerging prospects for food scientists. *Trends Food Sci Technol* 18: 514–525.

Stinzing, F.C., Carle, R. 2008. Betalains in food: Occurrence, stability, and postharvest modifications, in *Food Colorants. Chemical and Functional Properties*, C. Socaciu (ed.), pp. 277–299. Boca Raton: CRC Press.

Stone, G.E. 1909. Influence of electricity on micro-organisms. *Bot Gaz* 48: 359–379.

Suga, M., Goto, A., Hatakeyama, T. 2006. Control by osmolarity and electric field strength of electro-induced gene transfer and protein release in fission yeast cells. *J Electrostat* 64: 796–801.

Suga, M., Goto, A., Hatakeyama, T. 2007. Electrically induced protein release from *Schizosaccharomyces pombe* cells in a hyperosmotic condition during and following a high electropulsation. *J Biosci Bioeng* 103: 298–302.

Susil, R., Semrov, D., Miklavčič, D. 1998. Electric field induced transmembrane potential depends on cell density and organization. *Electro Magnetobiol* 17: 391–399.

Taiwo, K.A., Angersbach, A., Knorr, D. 2003. Effects of pulsed electric field on quality factors and mass transfer during osmotic dehydration of apples. *J Food Process Eng* 26: 31–48.

Talele, S., Gaynor, P., van Ekeran, J., Cree, M.J. 2010. Modelling single cell electroporation with bipolar pulse: Simulating dependence of electroporated fractional pore area on the bipolar field frequency, in *Technological Developments in Education and Automation*, M. Iskander, V. Kapila, and M.A. Karim (Eds.), pp. 355–359. New York: Springer.

Tamer, I.M., Moo-Young, M., Chisti, Y. 1998. Disruption of *Alcaligenes latus* for recovery of poly(beta-hydroxybutyric acid): Comparison of high-pressure homogenization, bead milling, and chemically induced lysis. *Ind Eng Chem Res* 37: 1807–1814.

Teissié, J., Golzio, M., Rols, M.P. 2005. Mechanisms of cell membrane electropermeabilisation: A minireview of our present (lack of?) knowledge. *Biochim Biophys Acta* 1724: 270–280.

Teissié, J., Escoffre, J.M., Rols, M.P., Golzio, M. 2008. Time dependence of electric field effects on cell membranes. A review for a critical selection of pulse duration for therapeutical applications. *Radiol Oncol* 42: 196–206.

Tewari, G., Juneja V.K. (Eds.). 2007. *Advances in Thermal and Non-Thermal Food Preservation*. Oxford: Wiley-Blackwell Publications.

Theron, M.M., Lues, J.F.R. 2010. *Organic Acids and Food Preservation*. Boca Raton: CRC Press.

Toepfl, S., Heinz, V., Knorr, D. 2005. Overview of pulsed electric field processing of foods, in *Emerging Technologies for Food Processing*, D.-W. Sun (Ed.), pp. 67–97. Oxford: Elsevier.

Toepfl, S. 2006. Pulsed electric fields (PEF) for permeabilization of cell membranes in food- and bioprocessing—Applications, process and equipment design and cost analysis. PhD dissertation. Berlin: Institut für Lebensmitteltechnologie und Lebensmittelchemie.

Toepfl, S., Knorr, D. 2006. Pulsed electric fields as a pretreatment technique in drying processes. *Stewart Postharvest Rev* 4: 1–6.

Toepfl, S., Mathys, A., Heinz, V., Knorr, D. 2006. Potential of high hydrostatic pressure and pulsed electric fields for energy efficient and environmentally friendly food processing. *Food Rev Int* 22: 405–423.

Toepfl, S., Heinz, V., Knorr, D. 2007a. History of pulsed electric field application, in *Preservation of Food by Pulsed Electric Fields*, H. Lelieveld, S. Notermans, and S.W. de Haan (Eds.), pp. 9–39. Cambridge: Woodhead.

Toepfl, S., Heinz, V., Knorr, D. 2007b. High intensity pulsed electric fields applied for food preservation. *Chem Eng Process* 46: 537–546.

Toepfl, S. Heinz, V. 2010. Pulsed electric field assisted extraction—A case study, in *Textbook Food Preservation*, G. Barbosa-Canovas and H. Zhang (Eds.). Weinheim: Wiley-VCH Verlag.

Torregrosa, F., Cortés, C., Esteve, M.J., Frígola, A. 2005. Effect of high-intensity pulsed electric fields processing and conventional heat treatment on orange–carrot juice carotenoids. *J Agric Food Chem* 53: 9519–9525.

Tung, L., Troiano, G.C., Sharma, V., Raphael, R.M., Stebe, K.J. 1999. Changes in electroporation thresholds of lipid membranes by surfactants and peptides. *Ann N Y Acad Sci* 888: 249–265.

Turk, M. 2010. Vers une amélioration du procédé industriel d'extraction des fractions solubles de pommes à l'aide de technologies électriques. PhD dissertation. Compiègne: Université de Technologie de Compiègne, France.

Turk, M., Baron, A., Vorobiev, E. 2009. Pulsed electric field assisted pressing of apple mash on a continuous pilot scale plant: Extraction yield and qualitative characteristics of cider juice, in *Proceedings of BFE*, E. Vorobiev, N. Lebovka, E. Van Hecke, and J.-L. Lanoisellé (Eds.), pp. 114–119. Compiègne: University of the Technology of Compiègne.

Turk, M.F., Baron, A., Vorobiev, E. 2010a. Effect of pulsed electric fields treatment and mash size on extraction and composition of apple juices. *J Agric Food Chem* 58: 9611–9616.

Turk, M., Baron, A., Vorobiev, E., Lanoiselle, J.L. 2010b. How to improve apple juice extraction and quality by pulsed electric field on an industrial scale, in *Proceedings IUF0st, 15th World Congress of Food Science and Technology*. Cape Town, South Africa.

Valic, B., Golzio, M., Pavlin, M., Schatz, A., Faurie, C., Gabriel, B., Teissié, J., Rols, M.-P., Miklavčič, D. 2003. Effect of electric field induced transmembrane potential on spheroidal cells: Theory and experiment. *Eur Biophys J* 32: 519–528.

van Boekel, Martinus A.J.S. 2002. On the use of the Weibull model to describe thermal inactivation of microbial vegetative cells. *Int J Food Microbiol* 74: 139–159.

Van der Poel, P.W., Schiweck, H., Schwartz, T. 1998. *Sugar Technology Beet and Cane Sugar Manufacture*. Denver: Beet Sugar Development Foundation.

Van Wert, S.L., Saunders, J.A. 1992. Electrofusion and Electroporation of Plants. *Plant Physiol* 99: 365–367.

Vega-Mercado, H., Gongora-Nieto, M.M., Barbosa-Cánovas, G.V., Swanson, B.G. 2007. Pulsed electric fields in food preservation, in *Handbook of Food Preservation*, M. Shafiur Rahman (Ed.), pp. 783–813. Boca Raton: CRC Press.

Vorobiev, E., Andre, A., Bouzrara, H., Bazhal, M. 2000. Procédé d'extraction de liquide d'un matériau cellulaire, et dispositifs de mise en œuvre du dit procédé (Method for extracting liquid from a cellular material and devices therefore). *De brevet en France* 002159, WO/2001/062482.

Vorobiev, E.I., Jemai, A.B., Bouzrara, H., Lebovka, N.I., Bazhal, M.I. 2005. Pulsed electric field assisted extraction of juice from food plants, in *Novel Food Processing Technologies*, G. Barbosa-Cánovas, M.S. Tapia, and M.P. Cano (Eds.), pp. 105–130. New York: CRC Press.

Vorobiev, E.I., Lebovka, N.I. 2006. Extraction of intercellular components by pulsed electric fields, in *Pulsed Electric Field Technology for the Food Industry. Fundamentals and Applications*, J. Raso and V. Heinz (Eds.), pp. 153–194. New York: Springer.

Vorobiev, E., Praporscic, L., Lebovka, N. 2007. Pulsed electric field assisted solid/liquid expression of agro-food materials: Towards a novel environmentally friendly technology. *Filtration* 7: 45–49.

Vorobiev, E.I., Lebovka, N.I. (Eds.). 2008. *Electrotechnologies for Extraction from Food Plants and Biomaterials*. New York: Springer.

Vorobiev, E., Lebovka, N. 2010. Enhanced extraction from solid foods and biosuspensions by pulsed electrical energy. *Food Eng Rev* 2: 95–108.

Weaver, J.C., Chizmadzhev, Y.A. 1996. Theory of electroporation: A review. *Bioelectrochem Bioenergetics* 41: 135–160.

Wouters, P.C., Smelt, J.P.P.M. 1997. Inactivation of microorganisms with pulsed electric fields: Potential for food preservation. *Food Biotechnol* 11: 193–229.

Wouters, P.C., Dutreux, N., Smelt, J.P.P.M., Lelieveld, H.L.M. 1999. Effects of pulsed electric fields on inactivation kinetics of *Listeria innocua*. *Appl Environ Microbiol* 65: 5364–5371.

Wu, Y., Guo, Y. 2009. Effect of high pulsed electric field on biomechanical properties of fruits and vegetables. *Nongye Gongcheng Xuebao/Trans Chin Soc Agric Eng* 25: 336–340.

Wuytack, E.Y., Diels, A.M., Michiels, C.W. 2002. Bacterial inactivation by high-pressure homogenisation and high hydrostatic pressure. *Int J Food Microbiol* 77: 205–212.

Yamamoto, H., Iwamoto, M., Haginuma, H. 1980. Acoustic impulse response method for measuring natural frequency of intact fruits and preliminary applications to internal quality evaluation of apples and watermelons. *J Text Stud* 11: 117–136.

Yantzi, J.D., Yeow, J.T.W. 2005. Carbon nanotube enhanced pulsed electric field electroporation for biomedical applications, in *Proceedings of Mechatronics Automation, IEEE International Conference* vol. 4, pp. 1872–1877.

Yin, Y.-G., Fan, X.-D., Liu, F.-X., Yu, Q.-Y., He, G.-D. 2009. Fast extraction of pectin from apple pomace by high intensity pulsed electric field. *Jilin Daxue Xuebao (Gongxueban)/J Jilin Univ (Eng Technol Ed)* 39: 1224–1228.

Zagorulko, A.Ja. 1958. Technological parameters of beet desugaring process by the selective electroplosmolysis, in *New Physical Methods of Foods Processing*, pp. 21–27. Moscow: Izdatelstvo GosINTI (in Russian).

Zhang, Q., Monsalve-Gonzalez, A., Qin, B.L., Barbosa-Cánovas, G.V., Swanson, B.G. 1994. Inactivation of *Saccharomyces cerevisiae* in apple juice by square-wave and exponential—Decay pulsed electric fields. *J Food Process Eng* 17: 469–478.

Zhang, B., Qin, L., Barbosa-Cánovas, G.V., Swanson, B.G. 1995. Inactivation of *Escherichia coli* for food pasteurization by high-strength pulsed electric fields. *J Food Process Preserv* 19: 103–118.

Zimmermann, U. 1986. Electrical breakdown, electropermeabilization and electrofusion. *Rev Physiol Biochem Pharmacol* 105: 175–256.

Zimmermann, U. 1992. Electric field-mediated fusion and related electrical phenomena. *Biochim Biophys Acta* 694: 227–277.

Zvitov, R., Schwartz, A., Nussinovitch, A. 2003. Comparison of betalain extraction from beet (*Beta vulgaris*) by low DC electrical field versus cryogenic freezing. *J Text Stud* 34: 83–94.

Zvitov, R., Nussinovitch, A. 2005. Low DC electrification of gel-plant tissue "sandwiches" facilitates extraction and separation of substances from *Beta vulgaris* beetroots. *Food Hydrocolloids* 19: 997–1004.

3 Microwave-Assisted Extraction

María Dolores Luque de Castro
and Feliciano Priego-Capote

CONTENTS

3.1 INTRODUCTION

Proper use of equipment and techniques requires a prior sound knowledge of their principles, their operational characteristics, and the variables influencing their performance. Because microwave energy is no exception to this rule, a description of these factors is in order before addressing microwave-assisted extraction.

3.1.1 FUNDAMENTALS OF MICROWAVE ENERGY AND ITS INTERACTION WITH MATTER

Microwave radiation consists of electromagnetic waves, and hence of an electric field and a magnetic field that are normal to each other. Microwave energy is a nonionizing type of radiation that causes molecular motion through migration of ions and rotation of dipoles without altering molecular structure.

The microwave region of the electromagnetic spectrum lies between far infrared light and radio frequencies, and spans the frequency range from 300 MHz to 300 GHz, which corresponds to wavelengths from 1 m to 1 cm. Wavelengths from 1 to 25 cm are extensively used for RADAR (RAdio Detection And Ranging) transmissions, and those in the remaining range are used for telecommunications. Four specific frequencies (viz. 915 ± 25, 2450 ± 13, 5800 ± 75, and $22{,}125 \pm 125$ MHz) are used for industrial and scientific microwave heating. These frequencies were reserved for industrial, scientific, and medical use by the U.S. Federal Communications Commission and conform to the International Radio Regulations adopted in Geneva in 1959 (Radio Regulations to the International Telecommunication Convention of Geneva, December 21, 1959).

A material can be heated by applying energy to it in the form of electromagnetic waves. The origin of the heating effect is the ability of an electric field to exert a force on charged particles. If the particles present in a substance can move freely through it as a result, then a current has been induced. On the other hand, if the charge carriers in the substance are bound to certain regions, they will move until a counterforce balances them, the net result being dielectric polarization. Both conduction and dielectric polarization are sources of microwave heating. The heating effect of microwaves depends on their frequency and on the applied power. Unlike microwave spectroscopy, however, the effect does not result from well-spaced discrete quantized energy states.

The dielectric properties of materials, which play a key role on the heating effect of microwaves, are defined by two parameters, namely the *dielectric constant* and the *dielectric loss*. The former, symbolized by ε, describes the ability of a molecule to be polarized by the electric field. At low frequencies, ε peaks at the maximum amount of energy that can be stored by the material. The dielectric loss, ε', measures the efficiency with which the energy of the electromagnetic radiation can be converted into heat. The dielectric loss moves toward a maximum as the dielectric constant decreases (Kingston and Haswell 1997). The *dissipation factor*, $\tan \delta$, is the ratio of the dielectric loss of a material, also called the "loss factor," to its dielectric constant, $\tan \delta = \varepsilon'/\varepsilon$.

When microwave energy penetrates a sample, the energy is absorbed at a rate dependent on its dissipation factor. Penetration is considered to be infinite in materials that are transparent to microwave energy and zero in reflective materials such as metals. The dissipation factor is a finite amount for absorptive samples. Because the energy is rapidly absorbed and dissipated as microwaves pass into the sample, the greater the dissipation factor of the sample is, the less microwave energy at a given frequency will penetrate into it. One effective way to characterize penetration is via the half-power depth or the distance from the sample surface at which the power density is reduced to one-half that at the surface (Smith 1984).

The half-power depth varies with the dielectric properties of the sample and also, roughly, with the inverse of the square root of the frequency. Thus, Lambert's expression for power absorption gives

$$P = P_0\, e^{-2\alpha' d} \tag{3.1}$$

where P_0 is the incident power, P is the power at the penetration depth, d is the penetration depth, and α' is the attenuation factor. Therefore, if the half-power depth is the distance from the surface of a material where the power is one-half of the incident power,

$$P/P_0 = 1/2 \tag{3.2}$$

$$e^{-2\alpha' d} = 1/2 \tag{3.3}$$

solving for d,

$$d = 0.347/\alpha' \tag{3.4}$$

Upon application to the sample, microwave energy is transformed into heat via two mechanisms, ionic conduction and dipole rotation, which occur simultaneously in many practical applications of microwave heating.

Ionic conduction is the conductive migration of dissolved ions in the applied electromagnetic field. This ion migration is a flow of current that results in I^2R losses (heat production, where I is current intensity and R is resistance) through resistance to ion flow. All ions in solution contribute to the conduction processes; however, the fraction of current carried by any given species is determined by its relative concentration and its inherent mobility in the medium. Therefore, the losses due to ionic migration depend on the size, charge, and conductivity of the dissolved molecules (Decareau 1985). Ion conduction is influenced by ion concentration, ion mobility, and solution temperature. Every ionic solution contains at least two ionic species, each of which conducts current according to its concentration and mobility. The dissipation factor of an ionic solution changes with temperature, which affects ion mobility and concentration.

Dipole rotation refers to the alignment, by effect of the electric field, of molecules in the sample having permanent or induced dipole moments. As the electric field of microwave energy increases, molecules become polarized; as the field decreases, thermally induced disorder is restored.

Temperature largely dictates the relative contribution of each energy conversion mechanism (dipole rotation and ionic conduction). With small molecules such as water and other solvents, the dielectric loss to a sample due to the contribution of dipole rotation decreases as the sample temperature increases. By contrast, the dielectric loss due to ionic conduction increases as the temperature is increased. Therefore, as an ionic sample is heated by microwave energy, the dielectric loss to the sample is initially dominated by the contribution of dipole rotation, and, as the temperature increases, the dielectric loss becomes dominated by ionic conduction.

The relative contribution of these two heating mechanisms depends on the mobility and concentration of the sample ions and on the relaxation time of the sample. If the ion mobility and concentration of the sample ions are both low, heating of the sample will be entirely dominated by dipole rotation. On the other hand, as the mobility and concentration of the sample increase, microwave heating will be dominated by ionic conduction, and the heating time will be independent of the relaxation time of the solution. As the ionic concentration increases, the dissipation factor will increase and the heating time will decrease. The heating time depends not only on the dielectric absorptivity of the sample but also on the particular microwave system design and sample size.

It is worth emphasizing the difference between conductive and microwave heating: because the vessels used for conductive heating are usually poor heat conductors, they take time to generate heat and transfer such heat to the solution. Also, because the liquid vaporizes at the surface, a thermal gradient is established by convection currents, so only a small portion of the fluid is at the temperature of the heat applied to the outside of the vessel. Therefore, with conductive heating, only a small portion of the fluid is above the boiling point of the solution. On the other hand, microwaves heat all of the sample fluid simultaneously without heating the vessel. As a result, microwave heating causes the solution to reach its boiling point very rapidly. Because heating is so much faster, substantial localized superheating can occur (Kingston and Jassie 1998).

3.1.2 Variables Affecting Microwave-Assisted Processes

Performance in microwave-assisted processes depends on a number of variables, including the microwave power output, exposure time, pressure, sample viscosity, sample size, and solvent. In specific treatments, performance is additionally dependent on other factors.

Microwave power and *time of exposure* to radiation have opposite effects: for a given process, the use of a high microwave power affords a decreased exposure time; on the other hand, the use of a low power requires irradiating the sample for a longer time to apply the same amount of energy.

Temperature is a key variable in most analytical processes. In microwave-assisted processes, it plays a prominent role and affects the rate of some reactions, the degradation of thermolabile species, and the solubilization of some substances, among others. A number of microwave devices have been developed for monitoring or even controlling the temperature, some of which are commented on in the following section.

Pressure is a highly influential factor in closed-vessel systems. The development of vessels capable of withstanding pressures over 50 bar has enabled digestion with pure or mixed liquid acids at very high temperatures, thereby dramatically increasing the digestion efficiency and decreasing the exposure time. Some such devices allow the pressure to be both monitored and controlled.

The *viscosity of a sample* reflects its ability to absorb microwave energy inasmuch as it affects molecular rotation. The effect of viscosity is best illustrated by considering ice water. When water is frozen, the water molecules become locked in a crystal lattice. This greatly restricts molecular mobility and makes it difficult for the molecules to align with the microwave field. Thus, the dielectric dissipation factor of ice is low (2.7×10^{-4} at 2450 MHz). When the temperature of water is increased to 27°C, the viscosity decreases and the dissipation factor increases to a much higher value (12.2).

Concerning *sample size*, the input microwave frequency affects the penetration depth of microwave energy; at a given input frequency, however, the greater the dissipation factor of a sample is, the less it will be penetrated by microwave energy. In large samples with high dissipation factors, heating beyond the penetration depth of the microwave energy is due to thermal conductance through molecular collisions. Therefore, temperatures at or near the surface will be higher as a result.

Although the small sample size used in most analytical processes has some advantages, it also has at least one disadvantage: the amount of energy absorbed decreases with decreasing sample size. With small sample sizes, a substantial amount of energy is not absorbed but reflected, and reflected energy can damage the magnetron. Thus, in using small samples for analytical work, it is advisable to employ microwave systems designed to protect the magnetron from reflected power.

The role of *solvents* in microwave-assisted processes is crucial for optimal development. When selecting a solvent, consideration should be given to its microwave-absorbing properties, the solvent–sample matrix interaction, and the solubility of the target compound(s) in the solvent. In solid–liquid extraction processes (also known as "leaching" or "lixiviation"), the solvent should have a high selectivity toward the target compound(s) and exclude unwanted matrix components. One key requirement is compatibility of the extractant with the analytical method used in the final analysis step. The optimal extractants here need not coincide with those used in conventional procedures. If the extractant is unable to absorb microwave energy, then it will not be heated and no effective extraction will occur. When one of these extractants must be used, special bars of chemically inert fluoropolymers, which absorb microwave energy and transfer heat to the surrounding medium, are recommended (Hummert et al. 1996).

In general, the heating process during extraction may occur via a number of mechanisms depending on the particular extractant: thus, the sample can be immersed in a single solvent or in a mixture capable of strongly absorbing microwave energy (mechanism I); the sample can be extracted into a combination of solvents with both high and low dielectric losses mixed in variable ratios (mechanism II); a sample with a high dielectric loss can be extracted with a microwave transparent solvent (mechanism III); or both the sample and the extractant are transparent to microwaves, and special bars of chemically inert fluoropolymers are used (mechanism IV). Usually, extractions and partitioning of solutes occur via one of these mechanisms or a combination thereof. Table 3.1 shows the effect of microwave heating on various solvents. The temperature reached by each solvent on application of a given amount of energy can differ widely, as does its effect on the efficiency of the target microwave-assisted process.

The microwave-assisted extraction (MAE) of nonpolar compounds usually requires a compromise since, although these compounds are more readily dissolved in nonpolar solvents such as *n*-hexane, the interaction of microwaves with the solvent

TABLE 3.1

Dielectric Constant, Dipole Moment, and Temperature Reached by Various Solvents Upon Heating from Room Temperature Using Microwave Energy of 2450 MHz and 560 W for 1 min

Solvent	ε	Dipole Moment, D	T (°C)	Boiling Point (°C)
Water	80.1	1.87	81	100
Methanol	32.7	2.87	65	65
Ethanol	24.5	1.66	78	78
1-Propanol	20.3	3.09	97	97
1-Butanol	17.5	1.75	109	117
1-Pentanol	13.9	1.70	106	137
1-Hexanol	13.3	1.80	92	158
Acetic acid	6.2	1.74	110	119
Ethyl acetate	6.0	1.88	73	77
Chloroform	4.8	1.15	49	61
Acetone	20.7	2.69	56	56
Dimethylformamide	36.7	3.86	131	153
Dimethyl ether	5.0	1.30	32	35
Hexane	1.88	0.08	25	68
Heptane	1.92	0.00	26	98
Carbon tetrachloride	2.20	0.00	26	77

depends on its dielectric constant (the greater ε, the stronger will be the interaction); this requires the use of a mixture of solvents of different polarity in many cases. For example, the most suitable solvent for the MAE of organic compounds such as polycylic aromatic hydrocarbons and polychlorinated biphenyls is a mixture of hexane and acetone. In specific cases such as the extraction of organometal compounds, the use of methanol acidified with acetic acid allows organometals to be extracted in a rapid, efficient manner with no degradation.

The solvents of choice for digestion (microwave-assisted digestion included) are solutions of both oxidizing (e.g., HNO_3, H_2SO_4) and nonoxidizing acids (e.g., HCl, HF). The acid or mixture of acids should be chosen for its efficiency in decomposing the matrix. Combinations of acids are also frequently used to digest some sample, the particular choice being dictated by the characteristics of the sample matrix (Johnson and Maxwell 1981).

3.1.3 Aim of This Chapter

Microwave-assisted treatments are frequently applied to liquids (e.g., for elution [Chee et al. 1997], thermospraying [Ding et al. 2000], ozonization [Jiang et al. 1997], desolvation [Gras et al. 1999], distillation [Conte et al. 1996]), and solids (e.g., for drying [Isengard and Walter 1998]; protein hydrolysis [Reichelt et al. 1999]; and, mainly, digestion and leaching, solid–liquid extraction, or MAE [Luque de Castro and Luque-García 2002]). One patented variant of MAE is the microwave-assisted process (MAP™), developed by Environmental Canada, which covers both analytical-scale methods and industrial processes (Paré et al. 1994). This chapter is mainly devoted to MAE (or MAP at the analytical scale) as this is the authors' field of expertise, and aimed at giving an overview of the potential and limitations of this type of extraction in the food field.

3.2 TYPES OF MICROWAVE DEVICES

Microwave systems for sample preparation can be of two different types depending on the way microwave energy is applied to the sample, namely

1. Multimode systems, in which microwave radiation is allowed to disperse randomly in a cavity and the sample it contains is evenly irradiated, as shown in Figure 3.1.
2. Single-mode or focused systems, in which microwave radiation is focused on a restricted zone where the sample is subjected to a much stronger electrical field than in multimode systems.

Usually, multimode systems are of the closed-vessel type and focused systems of the open-vessel type, the former being defined as systems where the microwave treatment is conducted at a high pressure, and the latter as systems where microwaves are applied at atmospheric pressure and no overpressure occurs as a result. However, this mutual association is incorrect since some commercial devices that operate at a high pressure use focused microwaves (Matusiewicz et al. 1991), and domestic ovens have been used to couple microwave treatment to detection using multimode sample irradiation, but at atmospheric pressure (Cuesta et al. 1998). The latter application has aroused much confusion as many authors consider these systems, which are usually flow injection manifolds, to be of the closed-vessel type (Pérez-Jordán et al. 1998). In a different context, this definition is quite correct and causes no confusion since the sample, once inserted into the flow injection system, actually circulates

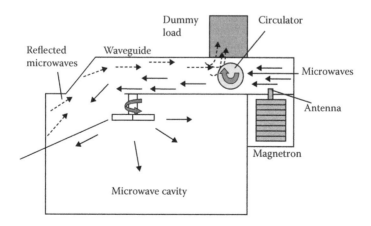

FIGURE 3.1 Major components of a typical multimode microwave system.

through a series of tubes and devices isolated from the outside. To avoid confusion, however, we shall use the designation "closed vessel" in this chapter only to refer to systems operating above atmospheric pressure.

3.2.1 COMPONENTS OF A MICROWAVE DEVICE

Both multimode and focused microwave devices comprise four major basic components: (i) the microwave generator, usually called the "magnetron," which produces the microwave energy; (ii) the waveguide, which is used to propagate the microwaves from the magnetron to the microwave cavity; (iii) the applicator, where the sample is placed; and (iv) the circulator, which allows microwaves to pass in the forward direction only.

The *magnetron* consists of a number of identical small cavities or resonators arranged in a cylindrical pattern around an also cylindrical cathode, normal to the cross-section of which a permanent magnet produces a strong magnetic field. The anode, which consists of a series of circuits tuned to oscillate at a specific frequency, is kept at a high voltage relative to the cathode. Electrons emitted from the cathode are accelerated toward the anode block, but the presence of the magnetic field produces a force in the azimuthal direction that causes the electron trajectory to be deflected in the same direction. The deflected electrons pass through the resonator gaps and induce a small charge into the circuit that causes the resonator to oscillate.

The *waveguide* is a rectangular or circular channel constructed of reflective material such as sheet metal and is designed to direct the microwaves generated by the magnetron to the cavity without mismatch. Waveguides with a rectangular cross-section are the more widely used types. Depending on size, they can be used for frequencies from 320 MHz to 333 GHz.

The *applicator* can be a multimode cavity where microwaves are randomly dispersed or the waveguide itself, in which case the sample vessel is placed directly inside it to focus the microwave radiation onto the sample. Uniform distribution of microwave radiation in a microwave cavity can be achieved using a mode stirrer, which is a fan-shaped blade used to reflect and mix the energy entering the microwave cavity from the waveguide. A mode stirrer assists in distributing the incoming energy, thus homogeneizing sample heating.

Commercial equipment is endowed with other additional units such as those for temperature and pressure monitoring and control. Typical of MAE closed-vessel devices are both pressure units and a turntable for the simultaneous treatment of several samples (in addition to other electronic components, the number and nature of which depend on the particular model).

3.2.2 CLOSED-VESSEL MICROWAVE SYSTEMS

The earliest microwave devices for analytical purposes were closed-vessel systems with a multimode cavity. The primary improvement over domestic units was the addition of safety mechanisms such as isolation and ventilation of the cavity to prevent acid fumes from attacking the electronics. Ever since the first commercial laboratory microwave unit (model MDS-81 D from CEM Corporation) was introduced in 1985, a number of manufacturers have delivered gradually improved versions with more uniform microwave fields, the ability to control the microwave power (and hence the temperature inside the vessels) and the pressure, and, most importantly, increased safety. Closed vessels themselves have evolved greatly and now span three generations. The first generation is represented by an all-Teflon (polytetrafluoroethylene [PTFE]) design with low-pressure limits that decreased through stress from previous treatments as the vessel aged. The second generation was that of jacketed vessels, which typically consisted of a Teflon liner and cap on a polymer case (usually of polyetherimide) that afforded inner pressures in the vessels up to 20 atm. The third generation of vessels, also lined, are thoroughly redesigned models that can withstand pressures as high as 30–150 atm and are offered as workstations for total control over the automated sample preparation process. Figure 3.2a depicts one of the most recent (third-generation) designs of high-pressure vessels marketed by CEM under the name MARS system, which is offered in two configurations aimed at (i) high-throughput processing of up to 40 samples simultaneously with accurate temperature internal control and (ii) preparation of large sample volumes (up to 100 mL) with capability for 14 simultaneous treatments with accurate temperature and pressure control. Figure 3.2b illustrates one of the recent extractors commercially available from Milestone—the Ethos EX labstation, which is highly flexible. This system can be configured with three different rotors depending on the intended application: (i) routine analysis with capacity for 12 samples (100 mL),

(a) MARS system

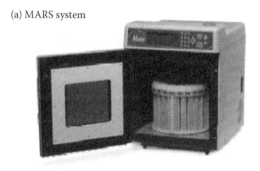

(b) Ethos EX labstation

FIGURE 3.2 Commercially available closed-vessel systems from (a) CEM Corporation and (b) Milestone.

(ii) high-throughput analysis with capacity for 24 samples (100 mL), and (iii) analysis of larger samples with capacity for 6 samples (270 mL). Interestingly, the material currently used to construct closed vessels is a function of the type of treatment to be applied and the subsequent type of analysis to be performed on the sample. The web pages of the different manufacturers provide useful information about the most advisable type depending on the target application. For guidance, Table 3.2 provides a summary of recent closed-vessel microwave-assisted devices from major manufacturers and their most salient features.

Although most commercially available closed-vessel microwave systems are based on multimode microwaves, the advantages of high-pressure vessels and focused microwave heating have led to the development of laboratory systems that combine both assets. These so-called focused high-pressure, high-temperature microwave systems (Matusiewicz 1994) consist of an integrated closed vessel and a focused microwave-heated system operating at a very high pressure and temperature. An extension of the system allows the simultaneous treatment of up to six samples in a gas-pressurized metal chamber, while measuring and controlling the pressure in one vessel. Pressures up to 130 bar and temperatures up to 320°C can thus be reached (Matusiewicz 1999). These laboratory systems have had recent commercial counterparts designed by CEM and were named Discover systems. This group of devices implements a patented focused single-mode cavity for highly efficient application of microwaves in a closed system. This technology enables the wattage magnetron to be reduced up to 300 W with performance similar to that of conventional closed systems having a magnetron power of up to 850 W. These devices, initially proposed to assist synthesis reactions, can also be used for sample preparation based on digestion or extraction.

Some dynamic systems for high-pressure microwave treatment appeared much later than open-vessel systems, possibly because operating under a high pressure reduces the flexibility afforded by working at atmospheric pressure. The highly complex first design required solid samples to be slurried and the assistance of a nitrogen bomb to supply N_2 at a constant pressure up to 5 bar (Pichler et al. 1999). A subsequent design avoided the need for the nitrogen bomb and enabled the direct introduction of solid samples, through which fresh extractant was continuously pumped, and connection to a subsequent step of the analytical process (Ericsson and Colmsjö 2000). Recently, a commercial system based on the Discover platform has been developed by implementing a module named Voyager. This module is an automated flow system intended to scale up microwave-assisted treatments for both continuous-flow and stop-flow processing. Thus, the same parameters used with the Discover system can be used to scale up from milligram amounts to approximately 1 kg with identical results. A dynamic stirring device ensures homogeneous, uniform mixing.

3.2.3 OPEN-VESSEL MICROWAVE SYSTEMS

The first completely reengineered laboratory focused microwave system was introduced by Prolabo in 1986. Most commercial open-vessel microwave systems manufactured since then are of the focused microwave type. According to Prolabo (and later to CEM after acquisition of this Prolabo section), the coupling efficiency of

TABLE 3.2

Summary of Recent Closed-Vessel Microwave-Assisted Devices and Their Most Salient Features

Manufacturer	Device	Configuration	Notes
CEM Corporation	MARS	Open and closed configurations	Programmable up to 1600 W
			Possibility of adapting reflux condensers, reagent addition, overhead stirring, etc.
			Up to 40 vessels (75 mL) in closed format and one 5 mL vessel in open format
			50–300°C depending on the vessel, at pressures up to 34 bar
	Discover series	Open and closed configurations	Focused microwaves up to 300 W with high efficiency
			Single-mode cavity with vessel capacity up to 300 mL
			Coupling of autosampler from 12 to 96 positions
			−80–300°C at pressures up to 21 bar
			Dynamic operation under continuous or stop-flow regimens
Milestone	Ethos EX Lab	Closed configuration	From 1 to 100 g
			Programmable up to 1600 W
			Modular rotors for routine, high-throughput, and large sample analysis
			Magnetic stirring to ensure homogeneous mixing, solvent evaporation, and recovery after treatment
			Pressure control up to 35 bar
	Ethos Digestion Lab	Open and closed configurations	Suitable for digestion protocols
			Maximum operation pressure 100 bar
			Different rotor configurations depending on application
Anton Paar	Multiwave 3000	Closed configuration	Programmable up to 1400 W
			Agitation unit and fast cooling system
			Processing of up to 48 samples simultaneously
			Controlled evaporation of the solvent to dry the extract
Aurora Biomed	Transform 800	Closed configuration	Processing of up to 10 samples
			Maximum conditions at 250°C and 55 bar
			Direct control of pressure and temperature
Sineo	MDS-8	Closed configuration	Programmable up to 1200 W
			Processing of up to 10 samples
			Maximum conditions at 300°C and 80 bar
	MDS-10	Closed configuration	Programmable up to 1800 W
			Processing of up to 15 samples
			Maximum conditions at 300°C and 150 bar

these approaches is increased by a factor of 10 relative to a multimode cavity. A constraint of the former devices commercialized by Prolabo—to allow for the use of only one flask at a time—was overcome with more recent developments involving the use of four flasks at a time by symmetrically splitting the microwave energy among the flasks at the end of each waveguide. The above-described CEM systems (MARS and Discover instruments) additionally include the choice to operate as an open system by continuous venting during sample preparation. Milestone markets another digestion system suitable for operation in the open configuration mode by continuous removal of gas to maintain a low pressure.

Unlike multimode cavity devices, which typically use pulsed power, these open devices use a continuously adjustable percentage of the maximum power over a preset period of time. The maximum power is in the 200–800 W range, depending on the type of microwave system used (Kingston and Haswell 1997).

In most open-vessel treatments, it is desirable to achieve as much refluxing as possible to reduce the need for continual additions of solvent to maintain the volume. Refluxing is generally better in open-vessel microwave treatments because the vessel is secondarily heated as heat from the solution is dissipated, and remains cooler as a result. Therefore, provided air can flow freely through the cavity, as is indeed the case with analytical microwave systems, refluxing will be more efficient.

As noted earlier, not all open-vessel systems (viz. those operating under atmospheric conditions) are of the focused type. In fact, a number of reported applications use a household multimode oven to process samples for analytical purposes, usually with a view to coupling the microwave treatment to another step of the analytical process.

Open-vessel microwave-assisted online sample treatment. Most online procedures involving microwaves that are conducted to couple a microwave treatment with another step of the analytical process (usually detection, with or without prior cleanup) use either a household oven (Crespín et al. 2000; Serrano and Gallego 2006) or a commercial focused system (Luque-García and Luque de Castro 2004; Morales-Muñoz et al. 2004a,b). Figure 3.3 shows two such systems using a household oven for the simultaneous treatment of several samples (Figure 3.3a) (Burguera et al. 1988), or for processing one sample at a time (Figure 3.3b) (Crespín et al. 2000). A very recent design for continuous MAE uses up to three in-series household ovens as shown in Figure 3.3c (Terigar et al. 2010). The main advantages of online designs using household ovens are their low cost, automatability, reduced delay between sample delivery and analysis, and isolation of samples from the environment. On the other hand, their main problem is the inherent inhomogeneity of microwave power distribution within the cavity: with only a small area occupied by the reaction coil, a high proportion of the microwave power within the cavity is not absorbed.

Unlike household ovens, where solid samples are usually introduced as slurries, in commercial focused microwave devices the sample is placed directly in the vessel (Fernández-Pérez et al. 2000) or in an extraction cartridge that is in turn placed in the vessel (Luque-García et al. 2000). Figure 3.4 depicts two such systems in which the microwave device is not coupled to a subsequent step of the analytical process (Figure 3.4a); rather, the dynamic approach used allows several consecutive extraction cycles to be performed to ensure quantitative removal of the target compounds (Fernández-Pérez et al. 2000). The extract provided by each cycle is aspirated through a cellulose filter to retain solid particles. The use of the pump itself as an interface between the extractor and a flow injection manifold (Figure 3.4b) allows MAE to be coupled to filtration, cleanup, preconcentration, individual chromatographic separation, and detection, and hence allows the whole analytical process to be automated (Luque-García et al. 2002).

Special open-vessel microwave systems. The most outstanding of these systems are the microwave–ultrasound combined extractor and the focused microwave-assisted Soxhlet extractor.

The first design of a *microwave–ultrasound combined extractor* was constructed by Lagha et al. in 1999 from a Prolabo Maxidigester and a cup-horn Branson Sonifier at the base of the microwave oven for indirect ultrasonic agitation of the sample under focused microwaves. A dramatic shortening of the MAE time in the presence of ultrasonic agitation (from 3 to 1 h) was widely observed. A subsequent, complex design of Chemat et al. (2004) was simplified and commercialized as the result of a COST project (Domini et al. 2009).

The *focused microwave-assisted Soxhlet extractor* was a prototype designed by the authors' group that was constructed by Prolabo in 1997 (Societé Prolabo et al. 1997). The device combines the advantages of the Soxhlet extractor with those of MAE. As can be seen from Figure 3.5, the prototype is based on the same principles as a conventional Soxhlet extractor modified to facilitate accommodation of the sample cartridge compartment in the irradiation zone of a microwave oven. This enables focused microwave-assisted Soxhlet extraction (FMASE) while retaining the advantages of conventional Soxhlet extraction (viz. the sample is repeatedly brought into contact with fresh portions of extractant to facilitate displacement of the transfer equilibrium and no filtration is thus required) and overcoming problems such as long extraction times and nonquantitative extraction of strongly retained analytes, thanks to the easier cleavage of analyte–matrix bonds by effect of interactions with focused microwave energy or the difficulty of automating the processing of large volumes of organic solvent waste—unlike the conventional Soxhlet extractor, an FMASE allows up to 75–80% of the total extractant volume to be recycled. In addition, FMASE affords solvent distillation by electrical heating irrespective of the extractant polarity, thus avoiding the greatest problem of commercial focused microwave devices such as those of the Soxwave series from Prolabo. Some drawbacks of the original prototype have been circumvented in subsequent versions of the FMASE. Thus, the glassware has been reduced to facilitate the use of high-boiling-point extractants such as water, an optical sensor for the extractant level at the siphon, and a solenoid valve at the bottom of the siphon to empty the sample vessel at preset intervals (prototype MIC II from SEV, Puebla, México). Also, the incorporation of two extraction units has enabled the simultaneous processing of two samples for replicate extraction and automation with an optical sensor, a solenoid valve, and microprocessor software control (prototype MIC V, also from SEV). Finally, a new, more compact prototype called Accesox (Medicarin, Barcelona, Spain) is currently being developed for a wide market. This device has the additional choice of the maximum temperature to be reached in the sample–extractant medium during microwave irradiation; in this way, the temperature of the leaching process can be effectively controlled, which can be essential with thermolabile compounds.

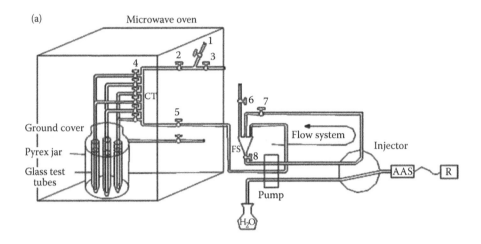

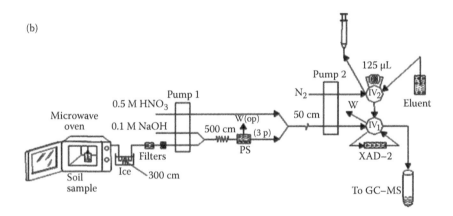

FIGURE 3.3 (a) Assembly for the simultaneous treatment of up to 6 samples. 1–8, open–close valves; AAS, atomic absorption spectrometer; CT, sample collector tube; FS, flowing sample collector; R, recorder. (From Burguera, M. et al., *Anal Chim Acta*, 214, 421–427, 1988. With permission.) (b) Online development of leaching, liquid–liquid extraction, and sorption/cleanup with manual transportation to the GC–MS equipment. IV, injection valve; PS, membrane phase separator; W, waste; XAD–2, sorbent material. (From Crespín, M.A. et al., *J Chromatogr A* 897, 279–293, 2000. With permission.)

Worthy of a special note here are the attempts by the Chemat team to develop a microwave-assisted extractor, "microwave-integrated Soxhlet extractor," which they deem similar to a Soxhlet extractor but in fact differs markedly from it in operational terms. Thus, there is no contact of the sample with fresh extractant and no siphoning of the extract; also, the extractant is heated by microwaves (similarly to the Soxwave-100) and a filtration step is required. Low-polar and nonpolar extractants are heated to their boiling points using microwaves while stirring with a Weflon magnetic stirrer to absorb microwave radiation. In this way, solvent vapors penetrate through the sample and are condensed on arrival at the condenser, where

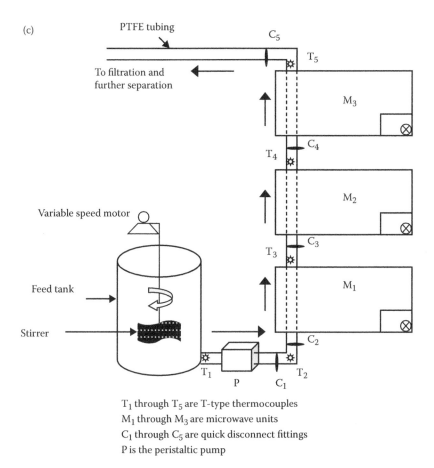

(c)

PTFE tubing

C_5

T_5

To filtration and further separation

M_3

C_4

T_4

M_2

Variable speed motor

C_3

T_3

M_1

Feed tank

Stirrer

C_2

T_1

P

C_1

T_2

T_1 through T_5 are T-type thermocouples
M_1 through M_3 are microwave units
C_1 through C_5 are quick disconnect fittings
P is the peristaltic pump

FIGURE 3.3 (Continued) (c) Schematic depiction of the continuous microwave system. (From Terigar, B.G. et al., *Bioresour Technol* 101, 2466–2471, 2010. With permission.)

the condensate is dropped down onto the sample. Obviously, this operation is not based on the Soxhlet principle, which exploits contact between the sample and fresh extractant in each leaching cycle, so displacing the partitioning equilibrium to complete extraction is impossible as a result. Extraction must inevitably be followed by filtration to separate the remaining solid matrix from the extract. Despite the name used by the authors, the device does not integrate Soxhlet and microwaves (Virot et al. 2007, 2008).

3.2.4 Working Modes in Microwave-Assisted Extraction (Discontinuous/Continuous/Semicontinuous)

In analytical jargon, the words "continuous" and "discrete" are used in connection with extraction and are used to refer to the continuous or batchwise, respectively, obtainment of an extract from a sample rather than to the continuous or

(a)

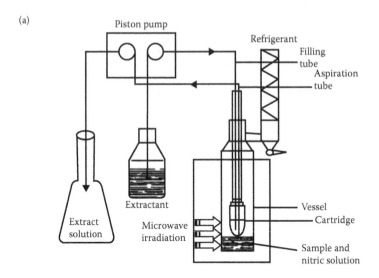

(b)

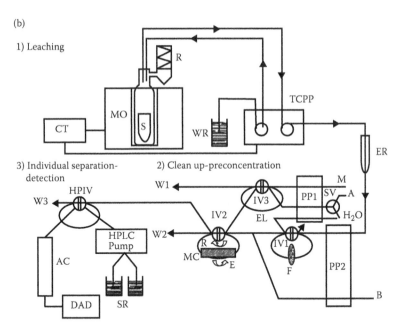

FIGURE 3.4 (a) Dynamic focused microwave-assisted extractor. (b) Experimental setup used to integrate microwave-assisted extraction with the subsequent steps of the analytical process. Leaching: CT, controller; ER, extract reservoir; MD, microwave digestor; R, refrigerant; S, sample; TCPP, two-channel piston pump; WR, water reservoir. Clean up-preconcentration: A, air; B, buffer; E, elution direction; EL, elution loop; F, filter; M, methanol; MC, microcolumn; PP, peristaltic pump; R, retention direction; SV, switching valve; IV, injection valve; W, waste. Individual separation-detection: AC, analytical column; DAD, diode array detector; HPIV, high-pressure injection valve; SR, solvent reservoirs. (From Fernández-Pérez, V. et al., *Analyst* 125, 317–322, 2000; Luque-García, J.L. et al., *Chromatographia* 55, 117–122, 2002. With permission.)

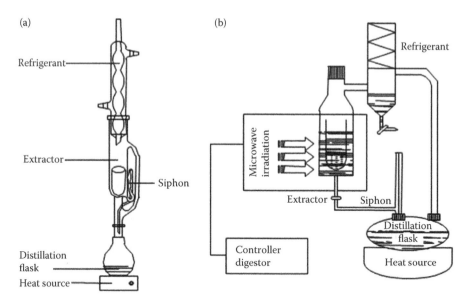

FIGURE 3.5 (a) Comparison of the performance of a conventional Soxhlet extractor and (b) the early prototype of focused microwave-assisted Soxhlet extractor from Prolabo. (From Luque de Castro, M.D. and Luque-García, J.L., *Acceleration and Automation of Solid Sample Treatment.* Amsterdam: Elsevier, 2002. With permission.)

batchwise feeding of the extractor with sample. The principal MAE modes of the continuous type are described below. Performance in the dynamic modes using microwaves to facilitate extraction does not depend on the multimode or focused nature of the irradiation source, but rather on the design of the sample container and peripherals.

Closed-vessel systems only allow discrete, batch operation with no connection to other steps of the analytical process: transfer to the following step must be made manually, so some user intervention is always necessary. The time required for cooling before vessels can be opened depends on the temperature reached during the extraction or digestion step and on the nature of evolved gases. The potential for simultaneously processing a number of samples and the wide commercial development for adaptation to any type of sample and use are the principal reasons for preferring closed-vessel systems to open-vessel systems.

By contrast, open-vessel systems are more flexible. With them, multimode approaches allow the simultaneous treatment of a number of samples, dependent on the dimensions of the irradiated cavity (see Figure 3.3a), and truly continuous extraction as in the in-series household oven arrangement proposed by Terigar et al. (2010).

Focused microwave approaches usually allow only one sample to be processed at a time and simultaneous treatment of several requires splitting the waveguide, which clearly complicates the system.

The most salient advantage of open-vessel systems is their ease of online connection to other steps of the analytical process via a simple or complex dynamic manifold (see Figure 3.4a and b, respectively) for fully automatic development of the process (Figure 3.4b).

The FMASE allows a special type of dynamic extraction that could be called a "semicontinuous regimen." In this mode, the FMASE is operated as a conventional Soxhlet device (except for the accelerated leaching provided by focused microwaves): the sample, held in a cartridge, is extracted with a preset volume of extractant that is dictated by the dimensions of the siphon or is adjusted by switching of the unloading valve at preset times, depending on the particular prototype (García-Ayuso and Luque de Castro 1999; Fernández-Pérez et al. 2000; García-Ayuso et al. 2000a,b; Luque-García and Luque de Castro 2001; Priego-Capote et al. 2003); this is followed by dropping of a new portion of clean extractant in each cycle. This extractor can also be online connected to a dynamic manifold for semicontinuous monitoring of a leaching process (García-Ayuso et al. 2000b) or for online development of other steps of the analytical process (Luque-García et al. 2002; Morales-Muñoz et al. 2004a,b).

3.2.5 ADVANTAGES AND DISADVANTAGES OF MICROWAVE DEVICES

The earliest microwave systems adapted for laboratory work were of the closed-vessel type; however, many were superseded by open-vessel systems developed to circumvent their shortcomings. In any case, each type of system has specific advantages and disadvantages, so neither can be said to be the better choice for all microwave-assisted treatments. The advantages of closed-vessels can be summarized as follows:

(a) They can reach higher temperatures than open-vessel systems because the boiling points of the solvents used are raised by the increased pressure inside the vessel. The higher temperatures in turn decrease the time needed for microwave treatment.
(b) Losses of volatile substances during microwave irradiation are almost completely avoided by virtue of the absence of vapor losses.
(c) Less solvent is required. Because no evaporation occurs, there is no need to continuously add solvent to maintain the volume. Also, the risk of contamination is avoided as a result.
(d) The fumes produced during an acid microwave treatment are contained within the vessel, so no provision for handling potentially hazardous fumes need be made.

On the other hand, the use of closed-vessel systems is subject to several shortcomings, namely

(a) The high pressures used pose safety (explosion) risks derived from the production of hydrogen in acid treatments of metals and alloys.
(b) The amount of sample that can be processed is limited (usually less than 100 g); by exception, the dynamic version commercially available by CEM can treat approximately 1 kg of sample.

(c) The usual constituent material of the vessels, PTFE, cannot withstand high solution temperatures. Other materials can be designed for applications at extremely high temperatures.

(d) These systems are rarely suitable for organic compounds.

(e) The single-step procedure excludes the addition of reagents or solvents during operation.

(f) The vessel must be cooled down before it can be opened after treatment.

(g) The use of porous PTFE can result in memory effects.

(h) Unlike digestion, leaching involves mass transfer equilibrium of the analytes between the sample matrix and the leachant, which hinders quantitative removal of the target species.

Atmospheric pressure (open-vessel) microwave sample preparation can be as effective method as a closed-vessel microwave sample preparation methods or even more so. The use of atmospheric pressure provides substantial advantages over pressurized vessels, namely

(a) Increased safety resulting from operating at atmospheric pressure with open vessels containing, for example, gas-forming species.

(b) The ability to add reagents at any time during treatment, which enables sequential acid attacks, if required.

(c) The ability to use vessels made of various materials, including PTFE, glass, and quartz.

(d) The ability to operate at high temperatures with quartz material when using sulfuric acid near its boiling point to destroy organic compounds.

(e) The ease with which excess solvent can be removed to ensure complete dryness of the digest or extract.

(f) The ability to process large samples.

(g) The lack of need for cooling down or depressurization.

(h) The low cost of the equipment required.

(i) The ability to develop several leaching cycles until quantitative removal of the target species is achieved.

In addition to these advantages, the following specifications would make open-vessel systems even more useful:

(a) Highly efficient transfer and precise control of the energy deposited into the sample.

(b) Fully automatic operation.

(c) Digestion of samples equal to, or higher than, 10 g—especially those with a high carbon content—produces large amounts of gas and vapors. In open vessels, these are released by the reaction mixture and continuously swept from the headspace above the sample. Thus, in contrast to closed vessels, completion of gas-forming reactions is favored as per Le Chatelier's principle.

(d) Open-vessel operation is better suited to thermolabile species (e.g., organo-metals) since it uses low temperatures relative to closed-vessel systems.

(e) Easy connection of the sample extraction step to other steps of the analytical process.

Despite their many advantages, open-vessel systems are also subject to several shortcomings, namely

(a) The ensuing methods are usually less precise than those developed using closed-vessel systems.

(b) The sample throughput is lower since most open systems cannot process many samples simultaneously—closed-vessel systems can handle up to 40 samples at once.

(c) The operation times required to obtain results similar to those of closed-vessel systems are usually longer.

(d) Digestion is especially cumbersome with some samples owing to the difficulty of reaching the drastic conditions they require.

3.3 TYPE OF MICROWAVE-ASSISTED SAMPLE TREATMENTS FOR FOOD

The two major types of sample treatment of food assisted by microwaves are digestion and extraction (leaching). Both are widely applied in food analysis, and the use of one or the other is dictated by the characteristics of the matrix–analyte couple and the subsequent steps of the analytical process to be carried out. In fact, leaching is more desirable whenever possible. Leaching a sample with complete removal of the target analytes provides less complex liquids and the ability to avoid interferences. However, leaching is not specific, but only more selective than digestion because it maintains most matrix interferences in the solid. It is therefore common to use a cleanup step after leaching to remove any species behaving like the target analytes in the leaching step. In any case, complete dissolution of the analytes must be ensured if they are to be accurately quantified. Special care must be taken in applying a method developed for spiked matrices to samples with natural content of the target analytes.

Most available MAE methods are based on a conventional existing method for the same or a similar matrix—and analyte—and using the same extractant that is simply reoptimized for microwave-related variables via a univariate approach, even in the presence of strongly correlated quantities (Kovács et al. 1998; Martín-Calero et al. 2009; Yu et al. 2009). Recent contributions in this field tend to use multivariate optimization designs (Fuentes et al. 2008, 2009; Yang and Zhai 2010), however.

The usual inclusion in the overall method of one—but sometimes two or more—cleanup steps shows that the strong conditions used to remove analytes favor undesirable leaching of other sample components. Filtration of the extract through glass wool, a glass microbore filter, or a membrane syringe filter and centrifugation of the solid–liquid system resulting from leaching provide simple and effective cleanup (Eskilsson and Björklund 2000). However, the most common choice for cleanup after MAE is solid-phase extraction using one (Kerem et al. 2005; Yang and Zhai 2010) or several serially arranged cartridges for exhaustive removal of interferents (Dévier et al. 2010); equally effective for this purpose is solid-phase microextraction (Wang et al. 1997). Liquid–liquid extraction has also provided good cleanup results (Sing et al. 2004). It would be of interest to develop methods where the absence of interferences could be used as a response variable of the multivariate optimization approach. In this way, a more comprehensive optimization step would result in a shorter overall method, thanks to the avoidance of cleanup steps, which would be of great importance for routine analysis methods.

One less frequent but interesting microwave-assisted sample treatment is steam distillation, also known as solvent-free microwave extraction. This treatment, especially indicated for the removal of essential oils from aromatic plants, has been applied to hard, dry plant materials such as bark, roots, and seeds (Périno-Issartier et al. 2010); aromatic plants such as basil (*Ocimum basilicum* L.), garden mint (*Mentha crispa* L.), and thyme (*Thymus vulgaris* L.) (Lucchesi et al. 2004) or oregano (Bayramoglu et al. 2008); and mainly to flowers (Sahraoui et al. 2008; Chunzhu Jiang et al. 2010). The dramatic shortening of the extraction time as compared with conventional steam distillation has been evidenced in all instances.

One field of such a growing interest and expansion as metabolomics—also expanding in food analysis—has also profited from the use of MAE (Kim and Verpoorte 2010). One primary objective of metabolomics is to measure as many metabolites as possible. This entails exercising some caution to avoid degradation and the spurious results it may produce for metabolic processes. Multivariate optimization of the MAE step, including potential degradation products as a response variable, would be of great interest here as well.

One very uncommon sample preparation step assisted by microwaves is liquid–liquid extraction. To the authors' knowledge, only Fuentes et al. (2008, 2009) have reported on it; they used multivariate optimization designs but failed to compare the results with conventional methods or the tested method in the absence of microwaves.

3.4 TYPE OF SAMPLES

Solid samples are typical targets for MAE. In closed-vessel devices, solid samples are simply brought into contact with a suitable extractant and the resulting solid–liquid system is irradiated with microwaves under appropriate power and time conditions. The most flexible open-vessel devices allow management of solid samples brought into contact with the extractant and subjected to either multimode or focused microwaves similarly to the process in closed vessels, except for pressure. However, open vessels additionally afford online coupling of extraction with other steps of the analytical process; thus, the extract is frequently aspirated through a filter to prevent solid particles from reaching the dynamic system.

Continuous MAE circulates *slurry samples* through a dynamic manifold, where they are subjected to microwave irradiation (Terigar et al. 2010). This procedure could also be used with *semisolid samples* such as yogurt, which would open up prospects for the continuous treatment of these common food samples.

While solid, slurry, and semisolid samples can be processed with common MAE devices, *liquid samples* require special extraction approaches such as that proposed by Fuentes et al. for the extraction of pesticides from oil. The first design reported by these authors in 2008 was improved in 2009 as shown in Figure 3.6 and used to extract organophosphorous pesticides from oils. Although the system operates at atmospheric pressure, the authors used a Milestone high-pressure microwave oven extraction system equipped with an exhaust module, using a 50-mL Erlenmeyer flask as sample container (with ground socket) where the oil was diluted with hexane and brought into contact with the extractant (acetonitrile). Following connection of an air-cooler condenser, the glass system was placed in the microwave oven and heated under preset time and power conditions. After cooling, the inner wall of the condenser was rinsed with acetonitrile and removed from the flask, the upper layer being carefully transferred with a Pasteur pipette to a test tube for cleanup by solid-phase extraction or low-temperature precipitation before gas chromatography–flame ionization detection and gas chromatography–mass spectrometry (GC–MS) analysis. The authors failed to compare the microwave-assisted process with a conventional method using no microwaves; thus, the improvement, if any, was not quantified.

The nature of the matrix to which the target analytes are bound can have a great effect on the efficiency of the microwave-assisted process in general and in MAE in particular. This influence has been widely demonstrated in spiking experiments where solid samples were spiked with the analytes and the results were compared with those for the extraction of samples natively containing the analytes (Eskilsson and Björklund 2000). The recoveries of freshly added compounds are usually higher

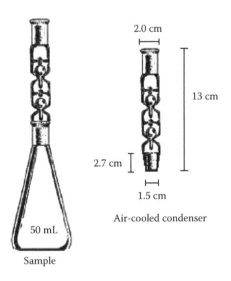

FIGURE 3.6 Scheme of a glass system used for atmospheric pressure microwave-assisted liquid–liquid extraction. (From Fuentes, E. et al., *J Chromatogr A* 1207, 38–45, 2008. With permission.)

than those from aged samples, the latter decreasing as the aging time increases (Luque-García and Luque de Castro 2003). Interestingly, this aging effect is also present in other extraction techniques and the origin of the restricted application of supercritical fluid extraction (Erickson 1998).

In general, the performance of a method for a given analyte or family of analytes is a function of the matrix characteristics. The dependence was widely studied by López-Ávila et al. (1995) in organic compounds in soils, and has also been demonstrated for pesticides in crops (lettuces and tomatoes), where some compounds were extracted three times more efficiently from tomatoes than from lettuce, thus confirming the dependence of the performance on the particular type of sample (Pylyw et al. 1997).

One key influence on the extraction efficiency is that of matrix moisture, the effect of which clearly depends on the extractant used. Except for increasing the polarity of the extractant, water, whether added or naturally present in the sample, will always affect its microwave-absorbing ability and facilitate heating. Water may also cause the matrix to swell and/or influence analyte–matrix interactions, making the analytes more readily available to the extractant (García-Ayuso and Luque de Castro 1999). These effects, which have been widely demonstrated in the extraction of contaminants in soils (Eskilsson and Björklund 2000), have also been observed in matrices such as biomass and tissues for the extraction of taxane and drugs, respectively. Freeze-drying to a moisture level below 10% and additional soaking in water before extraction in the former type of matrix provided full extraction of taxane (Mattina et al. 1997); on the other hand, extraction of a drug from swine tissue was more efficient with freeze-dried samples than with wet samples (Akhtar et al. 1998).

Concerning the need for prior manipulation of food to be subjected to microwave-assisted sample preparation, fresh (Yu et al. 2009), cooked (Sing et al. 2004), and manufactured food (Zhang et al. 2010) have been irradiated with this type of energy to facilitate removal of the target analytes. The efficacy of microwaves has always been a function of the matrix–analyte combination and type of extractant used.

3.5   TYPE OF LEACHABLE COMPOUNDS IN FOOD

Some of the compounds to be removed for food analyses (e.g., richness) include major components of natural foods such as fat, proteins, and carbohydrates, the richness of which needs to be known; food additives (e.g., dyes, stabilizers, sweeteners, gelling agents), the nature or concentration of which should be known for effective control; food supplements or nutraceuticals (e.g., antioxidants, vitamins; Japón-Luján and Luque de Castro 2006) with preventive or curative effects; and contaminants (e.g., herbicides, dioxins, toxic metals, degradation products, can coatings), the contents of which should not exceed their maximum allowed level. Effective MAE methods for all these types of compounds have been reported, and all are much more expeditious than conventional methods for the same or a similar matrix–analyte combination.

Easily oxidized analytes such as antioxidants must be protected from degradation under certain working extraction conditions (e.g., air and elevated temperatures, released oxidative enzymes). These problems have been solved by developing low-temperature vacuum MAE for the highly efficient extraction of oxidizable compounds (Wang et al. 2008). The principle behind this technique is that the boiling point of the extraction solvent in vacuum is lower than that at normal air pressure. Yu et al. (2009) used a nitrogen atmosphere to prevent the oxidation of ascorbic acid in its open-vessel MAE from fruit and vegetables, thereby reducing analyte losses.

The enthusiasm aroused by MAE in some users has led them to compare its performance with methods using other types of auxiliary energy (e.g., ultrasound). In most cases, the improvement achieved using a commercial MAE system specially designed for this purpose and equipped with a variety of devices for enhanced performance was compared with the use of a low-price ultrasound bath designed for cleaning and degassing, which usually exhibits power decay with time and heterogeneous distribution of ultrasonic energy (Zhang et al. 2010). Their results are therefore unreliable toward comparing the potential of the two types of energy for the given application. Tests with both types of energy in appropriate systems have shown them to provide similar results (Smythe and Wakeman 2000; Bermejo-Barrera et al. 2001) or even revealed that ultrasound energy may be more convenient, especially with thermolabile, hard to oxidize compounds (Luchini et al. 2000; Ruiz-Jiménez et al. 2003; Wang et al. 2003).

3.5.1 FOOD COMPONENTS

This section only deals with those MAE uses for some typical major components of food, the extraction of which from various matrices is substantially expedited by microwaves.

One of the most common components of a food is fat (as the combination of fatty acids and expressed as triglycerides or as the content of each lipid component). Fat is frequently determined in food analysis laboratories. Traditionally, fat is removed by Soxhlet extraction, which is the basis for a number of official methods despite its slowness.

FMASE has exhibited excellent performance in the removal of fat from various matrices such as olive drupes (García-Ayuso and Luque de Castro 1999) and oily seeds (sunflower, rape, and soybean) (García-Ayuso et al. 2000a), cheese (García-Ayuso et al. 1999a), milk (García-Ayuso et al. 1999b), fried and prefried foods (Luque-García et al. 2002), and sausage products (Priego-López et al. 2003), among others. In all instances, FMASE was much more expeditious than reference methods for fat extraction. For example, the extraction of fat from fried and prefried foods takes 55 min with FMASE and 8 h with the reference method. Also, FMASE reduces the time needed to extract fat from seeds from 8 h to 20–25 min, and the procedure is less labor intensive than the official method, which, for example, requires halting the process twice to grind the sample (García-Ayuso et al. 2000a). With cheese, FMASE reduces the extraction time from 6 h to 40 min; with milk (García-Ayuso and Luque de Castro 1999), from 10 h to 50 min. In addition, FMASE dramatically shortens the prehydrolysis step required for digestion of dairy products (from 1 h to 10 min) and avoids the need for subsequent neutralization. Moreover, FMASE provides cleaner extracts, possibly as a result of its shorter operational times. For example, milk fat extracted by FMASE exhibits less marked chemical transformation of triglycerides during extraction (García-Ayuso and Luque de Castro 1999).

The increased extraction efficiency of FMASE has been confirmed by some kinetics studies. Figure 3.7a illustrates the kinetics of extraction of fat from two bakery products by FMASE with and without microwave assistance (solid and dashed lines, respectively) (Priego-Capote and Luque de Castro 2005). As can be seen from the figure, seven cycles were required for complete fat isolation from a snack sample by FMASE. In the absence of microwaves, fat was extracted by 8.21% from snacks versus 26.22% with FMASE (data obtained by gravimetry). Therefore, microwave irradiation under the optimum conditions had a substantial influence on the outcome. Figure 3.7b further illustrates the kinetics of conventional Soxhlet extraction with the isolation of fat from the same type of sample. The kinetics study was conducted over periods of 4 and 20 h to determine the time needed for complete extraction. Clearly, the extraction time was much longer in the absence of microwaves than in their presence, even if the duration of each FMASE cycle was identical with that of a conventional Soxhlet cycle. The most salient result was that the time required for total fat isolation by FMASE was only 35 min for snack samples versus 8 h with Soxhlet extraction (Priego-Capote and Luque de Castro 2005). This confirms the boosting effect of microwaves on the extraction efficiency.

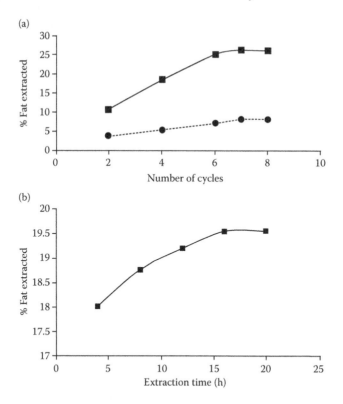

FIGURE 3.7 Extraction kinetics of fat from bakery products as performed by (a) FMASE (solid line) and without microwave assistance (dashed line) and (b) with the classic Soxhlet technique for the same target sample. (Adapted from Priego-Capote, F. and Luque de Castro, M.D., *Talanta* 65, 81–86, 2005.)

The same method used for the bakery products was employed to determine *trans* fatty acids as a contribution to the initiative of adding *trans* fat content to the label of processed foods, either by legislation or as a quality index to be implemented in the near future. FMAS extracts can be used to quantify the content in *trans* fatty acids by middle infrared spectrometry (MIR) (Priego-Capote et al. 2004) and the profiles of individual compounds by GC–MS (Priego-Capote et al. 2007). The FMASE results were compared with those provided by the Folch reference extraction method, which uses mild temperatures. Both extraction methods provided similar results in terms of total content in *trans* isomers and also of individual isomer concentrations. Therefore, a method based on FMASE sample preparation and subsequent MIR or GC–MS analysis has the potential to become an effective, expeditious alternative to the Folch method for routine analyses. Two other very common determination methods for food components are also worth discussing here on the grounds that they dramatically reduce the time needed to release analytes under microwave irradiation, namely Kjeldahl extraction and protein hydrolysis.

The determination of total nitrogen in vegetables, which takes 2–4 h with the conventional Kjeldahl method, is of special interest in agrifood analysis laboratories. The early attempts by Alvarado et al. (1988) using closed test tubes for Kjeldahl digestion of various foods in a household microwave oven, which expedited digestion considerably, were followed by the use of an open-vessel focused microwave system combined with a flow injection manifold for microwave-enhanced color development. In this way, Mason et al. (1999) achieved a throughput of 60 samples/h.

Acid hydrolysis of proteins has been the rate-determining step in amino acid analysis since commercial high-performance liquid chromatography (HPLC) amino acid analyzers became available in the late 1960s. Microwave-assisted hydrolysis is an alternative to conventional hydrolysis that has changed this situation dramatically. Thus, protein hydrolysates cannot be prepared in less time than a single chromatographic run without compromising accuracy or precision. Microwave-assisted protein hydrolysis is particularly advantageous in those cases where two or more hydrolysis reactions are needed for optimum recovery of labile and refractory amino acids. For example, microwave-assisted hydrolysis runs of 20 and 40 min duration can be performed without altering the chemistry of amino acids instead of 1 and 2 days of conventional hydrolysis.

Although, initially, microwave-assisted hydrolysis of proteins was done in household ovens, a number of laboratory microwave ovens operating in various methods were soon made commercially available. At present, microwave-assisted hydrolysis systems equipped with temperature regulators enabling quality control of protein products are available. With these devices, hydrolysis is usually performed in sealed containers where the sample is either brought into contact with the concentrated acid used for hydrolysis or subjected to vapor–gas microwave hydrolysis.

One major concern in the hydrolysis process is the potential racemization of amino acids. Peter et al. (1993) used both microwave and conventional hydrolysis to determine the amino acid composition of three synthetic peptides and found microwave-assisted hydrolysis to result in reduced racemization and in higher recovery of sensitive amino acids than conventional heating with hydrochloric acid.

3.5.2　Food Additives

Food additives are substances added to food to preserve flavor or enhance its taste and appearance. The microwave-assisted methods described in this section are used for either (i) extraction of the target compounds from the raw material for subsequent use in foods or (ii) extraction of the target compounds from the food itself to determine its content for labeling or content checking.

The most typical additives used to improve food appearance are colorants, which have traditionally been of synthetic nature. Recent research has exposed long-term problems with synthetic colorants (e.g., that they increase hyperactivity in children with attention deficit hyperactivity disorder [Nair et al. 2006]), which has led to their banning and raised the need for fast, accurate methods for their detection and quantitation, and also for the development of more effective methods for extraction of natural colorants from plants than those previously available. One recent method for fast extraction of colorants is that developed by Yang and Zhai (2010) for MAE of anthocyanins from purple corn; these compounds are usually extracted by conventional solvent extraction, which takes a long time, uses high extractant volumes, is inefficient, and causes loss of color followed by the formation of brownish degradation products and insoluble compounds (Castillo-Sánchez et al. 2006). By contrast, the MAE method requires only 19 min for complete extraction and exhibits much higher efficiency than conventional extraction.

Pectin, a structural heteropolysaccharide contained in the primary cell walls of terrestrial plants, is mainly extracted from citrus fruits and used as a gelling agent for foods (particularly jams and jellies). In addition, pectin is used in fillings, sweets, as a stabilizer in fruit juices and milk drinks, and as a source of dietary fiber. Fishman et al. (2006) reported the MAE of pectin from lime. They studied changes in molar mass, viscosity, radius of gyration, and hydrated radius, which determine its usefulness as a food additive, under variable working conditions (particularly irradiation time). By adjusting the variables of the microwave device, they managed to optimize the properties of pectins for addition to a specific type of food.

One example of MAE use for sample preparation in the determination of food additives is the method of Barroso et al. (2008) for maltol, an enhancer of food flavors in cakes. Using an appropriate extractant (1:1 hexane–acetone) ensured virtually complete extraction of the target analyte in 19 min versus more than 2 h with the conventional method.

3.5.3　Food Supplements and Nutraceuticals

Although "food supplements" (or "functional foods") and "nutraceuticals" are vague terms, most people do not distinguish between the two. A supplemented food, functional food, or medicinal food is "any fresh or processed food that is considered to have properties to favor health and/or prevent diseases, in addition to the nutritional function of providing nutrients." On the other hand, a nutraceutical is "a food extract with medicinal effects on health with a prescribed capsule, tablet, or powder format." In addition to their basic nutritional functions, nutraceuticals provide a proven physiological benefit and/or reduce the risk of a chronic disease (Proestos and Komaitis

2008). The increasing interest in these two types of compounds (plant secondary metabolites, mainly) has raised the need to expand and modify available conventional protocols for their extraction.

Compound classes or families of well-known capabilities as antioxidants such as simple phenols and polyphenols are highly demanded both to enrich food and as nutraceuticals (Luque de Castro et al. 2006). This warrants the development of fast methods for their joint or individual extraction from a variety of raw materials.

Thus, the authors' group has developed a number of batch and continuous methods using auxiliary energy (e.g., ultrasound, microwaves, superheated liquids) (Luque de Castro and Luque García 2002; Luque de Castro and Priego-Capote 2006) for extraction of food enriched in phenols from by-products of the Mediterranean agrifood industry (viz. olives, olive oil, vine, wine). The products thus obtained include oleuropein, hydroxytyrosol, and luteolin, from olive, and malvidin-3-glucoside, myricetin, quercetin, quercetin-3-b-glucoside, caffeic acid, and p-coumaric acid from vine. MAE methods based on focused microwaves have been applied to raw materials such as olive tree leaves (Japón-Luján et al. 2006) and branches (Japón-Luján and Luque de Castro 2007), and compared with other methods based on ultrasound or superheated liquids (Japón-Luján and Luque de Castro 2006). The data in Table 3.3 testifies to the extraction-boosting effect of auxiliary energy (ultrasound, microwaves, or a high temperature and pressure).

One recent contribution by the authors' group to the exploitation of residues from the Mediterranean agrifood industry is an MAE method for obtaining phenol compounds from lees (Pérez-Serradilla and Luque de Castro 2010). The method was optimized using a multivariate approach for phenols from wine lees based on the total phenol index, oxygen radical absorbance capacity (ORAC), and extraction yield as response variables. Under optimum working conditions, the proposed MAE method provides better extraction efficiency, in a much shorter time (17 min), than the conventional extraction method for phenols (24 h). The liquid extract obtained by MAE was spray dried. The types of excipients used and their contents, as well as the spray-drying temperature, were optimized to minimize oxidation of phenols and maximize the yield of the spray-drying process. The total phenol index in the dried extract thus obtained was 36.8% (expressed as gallic acid), and the ORAC was 3930 μmol Trolox equivalents per gram of extract. Additionally, malvidin-3-glucoside,

---

TABLE 3.3

Comparison of Extraction Time, Ethanol–Water Ratio, and Extractant Volume for the Isolation of Olive Phenols from Leaves with Different Methods Using Auxiliary Energies

Variable	Superheated Liquid Extraction	Ultrasound-Assisted Extraction	MAE
Extraction time (min)	13	25	8
Ethanol–water ratio	70:30	59:41	80:20
Extractant volume (mL)	11	15	24

myricetin, quercetin, quercetin-3-b-glucoside, and caffeic and *p*-coumaric acids were quantified in the dry extract by HPLC–diode array detection. The results show that antioxidant extracts from wine lees can provide an effective, economical alternative to those from grape seeds or skin.

Isoflavones, which are another type of antioxidant, have been extracted from soybean flour with the continuous MAE system described in Section 3.2.4 (Terigar et al. 2010). Also, Chukwumah et al. (2007) compared MAE, sonication, stirring, Soxtec, and MAE + sonication for the extraction of selected isoflavones and *trans*-resveratrol from peanuts. They found the MAE and Soxtec extracts to contain significantly greater amounts of the target compounds, and the methods based on stirring or sonication at near ambient temperature to cause no hydrolysis, which afforded discrimination between aglycons and glycoside conjugates present in the raw material.

Saponins and their aglycons, termed sapogenins, are compounds of controversial health-promoting properties, which have enjoyed massive commercial promotion as dietary supplements and nutraceuticals. Sapogenins are lipophilic triterpene derivatives that, similarly to phenols, protect plants against microbes, fungi, and other hostile organisms. The MAE of saponins from chickpea (*Cicer arietinum* L.) has been demonstrated by Kerem et al. (2005), who justified its choice for isolation of these compounds in terms of the volume of extractant required (an ethanol–water mixture or a butanol–water immiscible system, which is more selective), and time and energy expended. Subsequent separation of the extract components by thin layer chromatography revealed the presence of two major saponins (as confirmed by ^{1}H and ^{13}C nuclear magnetic resonance spectroscopy), one of which, containing a reducing sugar moiety, was endowed with unique characteristics, including antioxidant and antifungal capacities in addition to sweet taste. Interestingly, the heat-sensitive sugar moiety was preserved under MAE conditions. In addition to their use as colorants, anthocyanins possess pharmacological properties (e.g., they protect against a myriad of human diseases such as liver dysfunction, hypertension, vision disorders, microbial infections, or diarrhea); therefore, the MAE method of Yang and Zhai (2010) warrants inclusion here.

3.5.4 FOOD CONTAMINANTS

Food contaminants are a matter of concern, the importance of which depends on the toxicity of the particular compound. They often require efficient extraction and cleanup in addition to resolutive individual separation and sensitive detection owing to their low maximum allowed levels in food. Conventional extraction methods for contaminants are time consuming and labor intensive; also, they use large amounts of organic solvents. The need to determine widely different contaminants in large numbers of samples with highly diverse matrices has promoted the development of methods in which the central step (dissolving the target analytes) can be dramatically shortened using auxiliary energy such as that of microwaves.

While pesticides, dioxins, toxic metals, and organometal compounds are the most common contaminants in any type of food (fresh, cooked, or manufactured), packed food contains special contaminants as a result of the manufacturing process or further contamination from the packaging material. Some selected MAE methods for sample preparation before the determination of both sources of contamination are described below.

Polycylic aromatic hydrocarbons are ubiquitous contaminants in all types of food matrices. The cumulative characteristics of these compounds in seafood ashes has been emphasized by Peréz-Gregorio et al. (2010), and the use of microwaves to facilitate their removal from food matrices has also been reported by the same group (García-Falcón et al. 2000).

Singh et al. (2004) used a household oven for MAE of thiamethozam, imidacloprid, and carbendazim residues in fresh and cooked vegetables such as cabbage, tomatoes, chillies, potatoes, and peppers that were spiked with the target analytes but were not subjected to aging. The disparate physical characteristics of the insecticides made selection of an appropriate extractant difficult. Thus, the analyte recoveries obtained using acetone for 30 s ranged from 68% to 106% (in the absence of microwaves, recoveries ranged from 37.2% to 61.4% for an identical extraction time). Using spiked but unaged samples is of little interest since the analytes are not bound to the sample matrix; therefore, no microwave assistance is needed and the method, as proposed, is inapplicable to samples containing the analytes since before plant collection.

Reyes et al. (2008) applied eight different analytical extraction procedures commonly used to extract mercury species from biological samples to Tuna Fish Tissue Certified Reference Material (ERM-CE464) for the content of total mercury and methylmercury. They used speciated isotope dilution mass spectrometry (SIDMS; U.S. Environmental Protection Agency's method 6800) to evaluate and effectively compensate for potential errors during measurement and accurately quantify mercury species using all the extraction methods. SIDMS was used to accurately evaluate species transformations during sample pretreatment, preparation, and analysis protocols. The extraction methods tested in this research were alkaline extraction with KOH solution or tetramethylammonium hydroxide; acid leaching with HCl, HNO_3, or CH_3COOH; extraction with L-cysteine hydrochloride; and enzymatic digestion with protease XIV in the presence of auxiliary energy (microwaves or ultrasound). Detection of total mercury and mercury species from all extraction methods was carried out by inductively coupled plasma mass spectrometry (ICP-MS) and HPLC–ICP-MS, respectively. MAE and USAL (ultrasound-assisted leaching) were found to be the most efficient alkaline digestion protocols and also the methods causing the lowest transformation of mercury species (6% or less). Extraction with 5 M HCl or enzymatic digestion with protease resulted in the second-highest extraction efficiency, with relatively lower transformation of methylmercury to inorganic mercury (3% and 1.4%, respectively). Despite the frequent use of acid leaching for the extraction of mercury species from tuna fish samples, the lowest extraction efficiencies and highest mercury species transformations were provided by MAE with 4 M HNO_3 or CH_3COOH. Transformations as high as 30% have been reported; however, all extraction methods tested by Reyes et al. allowed accurate quantitation when corrected in accordance with the SIDMS method as standardized in the U.S. Environmental Protection Agency's method 6800.

One uncommon example of MAE described in a previous section is the liquid–liquid extraction of organophosphorous pesticides from oils proposed by Fuentes et al. (2008, 2009).

Metal cans used to store many preserved foods (e.g., fruits, vegetables, soft drinks, milk powder, and coffee) are lined with an inner coating to avoid direct food contact with metals, prevent electrochemical corrosion, and increase the shelf life of canned food. Epoxy resins and vinylic organosols (polyvinylchloride) have been widely used as inner varnishes for food cans. The former are synthetized mainly from bisphenol A and bisphenol A diglycidyl ether (BADGE). For polyvinylchloride organosol resins, BADGE, bisphenol F diglycidyl ether (BFDGE), or some other Novolac glycidyl ether (NOGE) is usually added to scavenge the hydrochloric acid released during heating of the cans. As a result, residues of BFDGE and BADGE monomers can migrate into the preserved food and form diverse chlorohydroxy compounds. In addition, epoxy residues can form monohydrated and dihydrated products. Several studies have shown that these compounds may have genotoxic effects, which has led the European Union (EU) to legislate on NOGE, BADGE, and their derivatives in foodstuffs or food simulants (Commission of the European Communities 2002, 2005). The EU legislation and that which is bound to be passed in other countries promote the development of methods such as that of Zhang et al. (2010) based on MAE with simultaneous individual separation by LC with fluorescence detection. These authors optimized the MAE variables of their CEM system using a univariate approach and found the best temperature for fast extraction without analyte degradation to be 105°C, at which 20 min was enough for complete extraction of samples previously spiked with the target analytes but were not subjected to aging. A chaotic comparison of the MAE results with those provided by conventional liquid–liquid extraction and USAL does not warrant their conclusions, however, since the three methods used a different extractant, volume, and time. In addition, USAL was implemented with an unsuitable device (a conventional ultrasonic cleaning bath). Therefore, any users interested in this method should check by themselves the allegedly increased extraction efficiency of MAE under these conditions (20–50% improvement).

3.6 SAFETY CONSIDERATIONS ON THE USE OF MICROWAVE ENERGY

Anyone using microwave equipment should be aware of the effects of microwave radiation exposure and the specific safety features of the microwave device used. Standards, limits, and tolerance ranges for microwave radiation exposure have been established in most of the industrial world. Military and governmental bodies in developed countries, and international organizations, have all established safety standards. An underlying reason for the large number of exposure standards in existence today is the varying manners in which they are defined (e.g., by the electromagnetic energy frequency, duration of exposure, body mass, and time or periodicity of exposure).

Research into the biological effects of microwave radiation exposure has been extensively discussed in several reviews dealing with the scientific, industrial, and medical applications of microwaves (Thuery 1992). Overall, the effects on human tissue are thermal in nature and relate to overheating of exposed tissue. The underlying protective principle of several standards is derived from data on the amount of energy required to raise human skin and tissue temperatures to biologically significant levels. Exposure to energy such as sunlight is basically a surface phenomenon; however, microwave energy penetrates the skin up to the subcutaneous tissue, thereby also raising the temperature level of tissue and blood.

Most commercial systems include specific safety features such as rupture membranes for extraction vessels. The membranes are designed to burst at pressures exceeding 200 psi (ca. 14 bar). Other safety features include a solvent vapor detector (which interrupts the supply of microwaves when detecting traces of solvent), an exhaust fan to evacuate air from the instrument cavity, and an isolator to divert reflected microwave energy into a dummy load to reduce the microwave energy within the cavity. Some manufacturers offer equipment with resealable vessels that are secured with a calibrated torque wrench; thus, if the pressure exceeds the vessel limits, a spring device allows the vessel to open and close quickly to release excess pressure. Another safety feature is the movable wall, which prevents the door from being blown away: the door moves in and out to release pressure from the microwave cavity. Most manufacturers offer several types of vessels differing in material, volume, or pressure-withstanding capabilities. Although the pressure in a closed-vessel system is typically below 20 bar, today's technology can raise it up to 150 bar.

Proper usage of laboratory microwave equipment is the responsibility of laboratory personnel. The safety devices of many commercial systems and vessels can be rendered ineffective by carelessness or misuse. It is the analyst's responsibility to follow good laboratory practices and the manufacturer's instructions when assembling, using, and maintaining microwave equipment. For example, by placing a microwave system inside a fume hood, where exhausted acid fumes may become circulated around the unit, the designed physical isolation of the electronics from the cavity is defeated and accelerated corrosion of the electronics, including safety interlock mechanisms and control circuits, can result. Chemical vapors should always be transported away from the unit, or the cavity air be swept away to an exhaust hose, a fume extraction or neutralization system, or a hood. Deterioration of the microwave guide, door seals, or cavity walls can provide leak paths for microwave radiation to escape and result in degradation of the equipment.

The literature concerning stability of food components subjected to MAE is scant. An interesting study on this subject was developed by Liazid et al. (2007), who used grape skin and grape seeds as raw materials subjected to MAE for extraction of their phenol contents. Twenty-two phenol compounds were monitored in the extracts obtained at different MAE conditions with temperatures from 50°C to 175°C. The authors concluded that (i) instability increases with the number of substituents in the aromatic ring; (ii) for a similar number of substituents in the ring, hydroxyl groups confer to the molecule more instability than methoxyl groups; and (iii) working temperatures of 100°C and extraction times of 20 min should not be surpassed to ensure the absence of degradation.

3.7 PROSPECTS OF MAE

Microwave assistance has proved efficient toward improving a number of processes, including analytical and industrial extraction of foods (Kingston and Haswell 1997).

This is bound to result in an increasing use of microwaves to accelerate sample preparation in existing methods. Also, new types of manufactured and cooked foods, new pesticides for crop protection, and new environmental contaminants to which plants are exposed are bound to require the development of appropriate extraction methods based on existing technologies.

In fact, MAE will with time be one of the most widely demanded sample preparation, its expansion being facilitated by the large number of commercially available devices spanning virtually any requirement in this field and the proved efficiency of microwaves for extraction. Nevertheless, users should be aware that MAE is not the panacea, so occasionally—and only occasionally—other types of energy may be more effective. A good professional would know the alternatives and adopt them when necessary, without prejudices.

ACKNOWLEDGMENTS

Spain's Ministerio de Ciencia e Innovación (MICINN) and the European FEDER Program are thanked for financial support through project CTQ2009-07430. F.P.C. is also grateful to MICINN for award of a Ramón y Cajal contract (RYC-2009-03921).

REFERENCES

Akhtar, M.H., Wong, M., Crooks, S.R.H., Sauve, A. 1998. Extraction of incurred sulfamethazine in swine tissue by microwave assisted extraction and quantification without clean up by high performance liquid chromatography following derivatization with dimethylaminobenzaldehyde. *Food Addit Contamin* 15: 542–549.

Alvarado, J., Márquez, M., León, L.E. 1988. Determination of organic nitrogen by the Kjeldahl method using microwave acid digestion. *Anal Lett* 21: 357–365.

Barroso, M.F., Sales, M.G.F., Almeida, S.A.A., Vaz, M.C.V.F., Delerue-Matos, C. 2008. Maltol determination in food by microwave assisted extraction and electrochemical detection. *J Food Drug Anal* 16: 30–36.

Bermejo-Barrera, P., Moreda-Piñeiro, A., Bermejo-Barrera, A. 2001. Sample pre-treatment methods for the trace elements determination in seafood products by atomic absorption spectrometry. *Talanta* 57: 969–974.

Burguera, M., Burguera, J.L., Alarcón, O.M. 1988. Determination of zinc and cadmium in small amounts of biological tissues by microwave-assisted digestion and flow injection atomic absorption spectrometry. *Anal Chim Acta* 214: 421–427.

Castillo-Sánchez, J.J., Mejuto J., Garrodo, J., García-Falcón, S. 2006. Influence of winemaking protocol and fining agent on the evolution of the anthocyanin content, color and general organoleptic quality of Vinhao wines. *Food Chem* 97: 130–136.

Chee, K.K., Wong, H.K., Lee, H.K. 1997. Membrane solid-phase extraction with closed vessel microwave elution for the determination of phenolic compounds in aqueous matrices. *Mikrochim Acta* 126: 97–104.

Chemat, S., Lagha, A., Amar, H.A., Chemat, F. 2004. Ultrasound assisted microwave digestion. *Ultrasonics Sonochem* 11: 5–8.

Chukwumah, I.C., Walter, L.T., Verghese, M., Bokanga, M., Ogutu, S., Alphonse, K. 2007. Comparison of extraction methods for the quantification of selected phytochemicals in peanuts (*Arachis hypogaea*). *J Agric Food Chem* 55: 285–290.

Commission of the European Communities EC/16/2002. 2002. *Off J Eur Commun* L51/27.

Commission of the European Communities EC/1895/2005. 2005. *Off J Eur Commun* L302/28.

Conte, E.D., Shen, C.Y., Perschbacher, P.W., Miller, D.W. 1996. Determination of geosmin and methylisoborneol in catfish tissue (*Ictalarus punctatus*) by microwave-assisted distillation–solid phase adsorbent trapping. *J Agric Food Chem* 44: 829–835.

Crespín, M.A., Gallego, M., Valcárcel, M. 2000. Continuous microwave-assisted extraction, solvent changeover and preconcentration of monophenols in agricultural soils. *J Chromatogr A* 897: 279–293.

Cuesta, A., Todolí, J.L., Mora, J., Canals, A. 1998. Rapid determination of chemical oxygen demand by a semi-automated method based on microwave sample digestion, chromium(VI) organic solvent extraction and flame atomic absorption spectrometry. *Anal Chim Acta* 372: 399–409.

Decareau, R.V. 1985. *Microwaves in the Food Processing Industry*. New York: Academic Press.

Dévier, M.H., Labadie, P., Togola, A., Budzinski, H. 2010. Simple methodology coupling microwave-assisted extraction to SPE/GC/MS for the analysis of natural steroids in biological tissues: Application to the monitoring of endogenous steroids in marine mussels *Mytilus* sp. *Anal Chim Acta* 657: 28–35.

Ding, L., Liang, F., Huan, Y.F., Cao, Y.B., Zhang, H.Q., Jin, Q.H. 2000. A low-powered microwave thermospray nebulizer for inductively coupled plasma atomic emission spectrometry. *J Anal At Spectrom* 15: 293–296.

Domini, C., Vidal, L., Cravotto, G., Canals, A. 2009. A simultaneous, direct microwave/ultrasound-assisted digestion procedure for the determination of total Kjeldahl nitrogen. *Ultrasonics Sonochem* 16: 564–569.

Erickson, B. 1998. Fattening up SFE sales. *Anal Chem* 70: 333A–336A.

Ericsson, M., Colmsjö, A. 2000. Dynamic microwave-assisted extraction. *J Chromatogr A* 877: 141–151.

Eskilsson, C.S., Björklund, E. 2000. Analytical-scale microwave-assisted extraction. *J Chromatogr A* 902: 227–250.

Fernández-Pérez, V., García-Ayuso, L.E., Luque de Castro, M.D. 2000. Focused microwave Soxhlet device for rapid extraction of mercury, arsenic and selenium from coal prior to atomic fluorescence detection. *Analyst* 125: 317–322.

Fishman, M.L., Chau, H.K., Hoagland, P.D., Hotchkiss, A.T. 2006. Microwave-assisted extraction of lime pectin. *Food Hydrocolloids* 20: 1170–1177.

Fuentes E., Báez, M.E., Quiñones, A. 2008. Suitability of microwave-assisted extraction coupled with solid-phase extraction for organophosphorus pesticide determination in olive oil. *J Chromatogr A* 1207: 38–45.

Fuentes, E., Báez, M.E., Díaz, J. 2009. Microwave-assisted extraction at atmospheric pressure coupled to different clean-up methods for the determination of organophosphorous pesticides in olive and avocado oils. *J Chromatogr A* 1216: 8859–8866.

García-Ayuso, L.E., and Luque de Castro, M.D. 1999. A multivariate study of the performance of a microwave-assisted Soxhlet extractor for olive seeds. *Anal Chim Acta* 382: 309–316.

García-Ayuso, L.E., Velasco, J., Dobarganes, M.C., Luque de Castro, M.D. 1999a. Double use of focused microwave irradiation for accelerated matrix hydrolysis and lipid extraction in milk samples. *Int Dairy J* 9: 667–674.

García-Ayuso, L.E., Velasco, J., Dobarganes, M.C., Luque de Castro, M.D. 1999b. Accelerated extraction of the fat content in cheese using a focused microwave-assisted Soxhlet device. *J Agric Food Chem* 47: 2308–2314.

García-Ayuso, L.E., Velasco, J., Dobarganes, M.C., Luque de Castro, M.D. 2000a. Determination of the oil content of seeds by focused microwave-assisted Soxhlet extraction. *Chromatographia* 52: 103–109.

García-Ayuso, L.E., Luque-García, J.L., Luque de Castro, M.D. 2000b. Approach for independent-matrix removal of polycyclic aromatic hydrocarbons from solid samples based on microwave-assisted Soxhlet extraction with on-line fluorescence monitoring. *Anal Chem* 72: 3627–3634.

García-Falcón, M.S., Simal-Gándara, J., Carril-González-Barros, S.T. 2000. Analysis of benzo[*a*]pyrene in spiked fatty foods by second derivative synchronous spectrofluorimetry alter microwave-assisted treatment of samples. *Food Addit Contam* 17: 957–964.

Gras, L., Mora, J., Todoli, J.L., Canals, A., Hernandis, V. 1999. Microwave desolation for acid simple introduction in inductively coupled plasma atomic emission spectrometry. *Spectrochim Acta B* 54: 469–480.

Hummert, K., Vetter, W., Lucas, B. 1996. Fast and effective sample preparation for determination of organochlorine compounds in fatty tissue of marine mammals using microwave extraction. *Chromatographia* 42: 300–304.

Isengard, H.D., Walter, M. 1998. Can the water content of dairy products be determined accurately by microwave drying? *Lebensm Unters Forsch* 207: 377–380.

Japón-Luján, R., Luque de Castro, M.D. 2006. Superheated liquid extraction of oleuropein and related biophenols from olive leaves. *J Chromatogr A* 1136: 185–191.

Japón-Luján, R., Luque-Rodríguez, J.M., Luque de Castro, M.D. 2006. Multivariate optimization of the microwave-assisted extraction of oleuropein and related biophenols from olive leaves. *Anal Bioanal Chem* 385: 753–759.

Japón-Luján, R., Luque de Castro, M.D. 2007. Small branches of olive tree: A source of biophenols complementary to olive leaves. *J Agric Food Chem* 55: 4584–4588.

Jiang, W., Chalk, S.J., Kingston, H.M. 1997. Ozone degradation of residual carbon in biological samples using microwave irradiation. *Analyst* 122: 211–216.

Jiang, C., Sun, Y., Zhu, X., Gao, Y., Wang, L. Jian, W., Wu, L., Song, D. 2010. Solvent-free microwave extraction coupled with headspace single-drop microextraction of essential oils from flower of *Eugenia caryophyllata* Thunb. *J Sep Sci* 33: 2784–2790.

Johnson, W.M., Maxwell, J.A. 1981. *Rock and Mineral Analysis.* New York: Wiley & Sons.

Kerem, Z., German-Shashoua, H., Yarden, O. 2005. Microwave-assisted extraction of bioactive saponins from chickpea (*Circer arietinum* L). *J Sci Food Agric* 85: 406–412.

Kim, H.K., Verpoorte, R. 2010. Sample preparation for plant metabolomics. *Phytochem Anal* 21: 4–13.

Kingston, H.M., Haswell, S.J. 1997. *Microwave-Enhanced Chemistry. Fundamentals, Sample Preparation and Applications.* Washington, DC: American Chemical Society.

Kingston, H.M., Jassie, L.B. 1998. *Introduction to Microwave Sample Preparation.* Washington, DC: American Chemical Society.

Kovács, A., Ganzler, K., Simon-Sarkadi, L. 1998. Microwave-assisted extraction of free amino acids from foods. *Z Lebensm Unters Forsch A* 207: 26–30.

Liazid, A., Palma, M., Brigui, J., Barroso, C.G. 2007. Investigation on phenolic compounds stability during microwave-assisted extraction. *J Chromatogr A* 1140: 29–34.

López-Ávila, V., Young, R., Benedicto, J., Ho, P., Kim, R. 1995. Extraction of organic pollutants from solid samples using microwave energy. *Anal Chem* 67: 2996–2102.

Lucchesi, M.E., Chemat, F., Smadja, J. 2004. Solvent-free microwave extraction of essential oil from aromatic herbs: Comparison with conventional hydrodistillation. *J Chromatogr A* 1043: 323–327.

Luchini, L.C., Peres, T.B., Andrea, M.M. 2000. Monitoring of pesticide residues in a cotton crop soil. *J Environ Sci Health* 35: 51–57.

Luque de Castro, M.D., Luque-García, J.L. 2002. *Acceleration and Automation of Solid Sample Treatment.* Amsterdam: Elsevier.

Luque de Castro, M.D., Luque-Rodríguez, J.M., Japón-Luján, R. 2006. Exploitation of residues from vineyards, olive groves and wine and oil production to obtain phenolic compounds of high-added value. In *Methods of Analysis for Functional Foods and Nutraceuticals*, Ed. J. Hurst, New York, Taylor & Francis.

Luque de Castro, M.D., Priego-Capote, F. 2006. *Analytical Applications of Ultrasound.* Amsterdam: Elsevier.

Luque-García, J.L., Morales-Muñoz, S., Luque de Castro, M.D. 2000. Microwave-assisted extraction of acid herbicides from soils coupled to continuous filtration, pre-concentration, chromatographic separation and UV detection. *Chromatographia* 55: 117–122.

Luque-García, J.L., Luque de Castro, M.D. 2001. Water Soxhlet extraction assisted by focused microwaves: A clean approach. *Anal Chem* 73: 5903–5908.

Luque-García, J.L., Morales-Muñoz, S., Luque de Castro, M.D. 2002. Microwave-assisted water extraction of acid herbicides from soils coupled to continuous filtration, pre-concentration, chromatographic separation and UV detection. *Chromatographia* 55: 117–122.

Luque-García, J.L., Velasco, J., Dobarganes, M.C., Luque de Castro, M.D. 2002. Fast quality monitoring of oil from prefried and fried foods by focused microwave-assisted Soxhlet extraction. *Food Chem* 76: 241–248.

Luque-García, J.L., Luque de Castro, M.D. 2003. Extraction of polychlorinated biphenyls from soils by automated focused microwave-assisted Soxhlet extraction. *J Chromatogr A* 998: 21–29.

Luque-García, J.L., Luque de Castro, M.D. 2004. Focused microwave-assisted Soxhlet extraction: Devices and applications. *Talanta* 64: 571–577.

Martín-Calero, A., Pino, V., Ayala, J.H., González, V., Afonso, A.M. 2009. Ionic liquids as mobile phase additives in high-performance liquid chromatography with electrochemical detection: Application to the determination of heterocyclic aromatic amines in meta-based infant foods. *Talanta* 79: 590–597.

Mason, C.J., Coe, G., Edwards, M., Riby, P.G. 1999. The use of microwaves in the acceleration of digestion and colour development in the determination of total Kjeldahl nitrogen in soil. *Analyst* 124: 1719–1726.

Mattina, M.J.I., Berger, W.A.I., Denson, C.L. 1997. Microwave-assisted extraction of taxanes from Taxus biomass. *J Agric Food Chem* 45: 4691–4696.

Matusiewicz, H. 1994. Development of a high pressure/temperature focused microwave heated Teflon bomb for sample preparation. *Anal Chem* 66: 751–755.

Matusiewicz, H. 1999. Development of a high-pressure asher focused microwave system for sample preparation. *Anal Chem* 71: 3145–3149.

Matusiewicz, H., Sturgeon, R.E., Bermann, S.S. 1991. Vapour-phase acid digestion of inorganic and organic matrices for trace element analysis using a microwave heated bomb. *J Anal At Spectrom* 6: 283–287.

Morales-Muñoz, S., Luque-García, J.L., Luque de Castro, M.D. 2004a. Screening method for linear alkylbenzene sulfonates in sediments based on water Soxhlet extraction assisted by focused microwaves with on-line preconcentration/derivatization/detection. *J Chromatogr A* 1026: 41–46.

Morales-Muñoz, S., Luque-García, J.L., Luque de Castro, M.D. 2004b. Continuous microwave-assisted extraction coupled with derivatization and fluorimetric monitoring for the determination of fluoroquinolone antibacterial agents from soil samples. *J Chromatogr A* 1059: 25–31.

Nair, J., Ehimare, U., Beitman, B.D., Nair, S.S., Lavin, A. 2006. Clinical review: Evidence-based diagnosis and treatment of ADHD in children. *Mol Med* 103: 617–621.

Paré, J.R.J., Bélanger, J.M.R., Stafford, S.S. 1994. Microwave-assisted process (MAP™): A new tool for the analytical laboratory. *Trends Anal Chem* 13: 176–184.

Peréz-Gregorio, M.R., García-Falcón, M.S., Martínez-Carballo, E., Simal-Gándara, J. 2010. Removal of polycyclic aromatic hydrocarbons from organic solvents and ashes wastes. *J Hazard Mater* 178: 273–281.

Pérez-Jordán, M.Y., Salvador, A., de la Guardia, M. 1998. Determination of Sr, K, Mg, and Na in human teeth by atomic spectrometry using a microwave-assisted digestion in a closed flow system. *Anal Lett* 31: 867–877.

Pérez-Serradilla, J.A., Japón-Luján, R., Luque de Castro, M.D. 2007. Simultaneous microwave-assisted solid–liquid extraction of polar and nonpolar compounds from alperujo. *Anal Chim Acta* 602: 82–88.

Pérez-Serradilla, J.A., Luque de Castro, M.D. 2010. Microwave-assisted extraction of phenolic compounds from wine lees and spray-drying of the extract. *Food Chem*. DOI information: 10.1016/j.foodchem.2010.07.046.

Peter, A., Laus, G., Tourwe, D., Gerlo, E., Van Binst, G. 1993. An evaluation of microwave heating for the rapid hydrolysis of peptide samples for chiral amino acid analysis. *Peptide Res* 6: 48–52.

Périno-Issartier, S., Abert-Vian, M., Petitcolas, E., Chemat, F. 2010. Microwave turbo hydrodistillation for rapid extraction of the essential oil from *Schinus terebinthifolius* Raddi berries. *Chromatographia* 72: 347–350.

Pichler, U., Haase, A., Knapp, G. 1999. Microwave-enhanced flow system for high-temperature digestion of resistant organic materials. *Anal Chem* 71: 4050–4055.

Priego-Capote, F., Luque-García, J.L., Luque de Castro, M.D. 2003. Automated fast extraction of nitrated polycyclic aromatic hydrocarbons from soil by focused microwave-assisted Soxhlet extraction prior to gas chromatography–electron-capture detection. *J Chromatogr A* 994: 159–167.

Priego-Capote, F., Ruiz Jiménez, J., García Olmo, J., Luque de Castro, M.D. 2004. Fast method for the determination of total fat and trans fatty-acids content in bakery products based on microwave-assisted Soxhlet extraction and medium infrared spectroscopy detection. *Anal Chim Acta* 517: 13–20.

Priego-Capote, F., Luque de Castro, M.D. 2005. Focused microwave-assisted Soxhlet extraction: A convincing alternative for total fat isolation from bakery products. *Talanta* 65: 81–86.

Priego-Capote, F., Ruiz Jiménez, J., Luque de Castro, M.D. 2007. Identification and quantification of trans fatty acids in bakery products by gas chromatography-mass spectrometry after focused microwave Soxhlet extraction. *Food Chem* 100: 859–867.

Priego-López, E., Velasco, J., Dobarganes, M.C., Ramis-Ramos, G., Luque de Castro, M.D. 2003. Focused microwave-assisted Soxhlet extraction: An expeditive approach for the isolation of lipids from sausage products. *Food Chem* 83: 143–149.

Proestos, C., Komaitis, M. 2008. Application of microwave assisted extraction to the fast extraction of plant phenolic compounds. *J Food Qual* 31: 402–414.

Pylyw, H.M.J., Arsenault, T.L., Thetford, C.M., Incorvia, M.M.J. 1997. Suitability of microwave-assisted extraction for multiresidue pesticide analysis of produce. *J Agric Food Chem* 45: 3522–3528.

Reichelt, M., Hummert, C., Luckas, B. 1999. Hydrolysis of microcystins and nodularin by microwave radiation. *Chromatographia* 49: 671–677.

Reyes, L.H., Mizanur-Rahman, G.M., Fahrenholz, T., Skip Kingston, H.M. 2008. Comparison of methods with respect to efficiencies, recoveries, and quantitation of mercury species interconversions in food demonstrated using tuna fish. *Anal Bioanal Chem* 390: 2123–2132.

Ruiz-Jiménez, J., Luque García, J.L., Luque de Castro, M.D. 2003. Dynamic ultrasound-assisted extraction of cadmium and lead from plants prior to electrothermal atomic absorption spectrometry. *Anal Chim Acta* 480: 231–238.

Serrano, A., Gallego, M. 2006. Continuous microwave-assisted extraction coupled on-line with liquid–liquid extraction: Determination of aliphatic hydrocarbons in soil and sediments. *J Chromatogr A* 1104: 323–330.

Sing, S.B., Foster, G.D., Khan, S.U. 2004. Microwave-assisted extraction for the simultaneous determination of thiamethoxam, imidacloprid, and carbendazim residues in fresh and cooked vegetable samples. *J Agric Food Chem* 52: 105–109.

Smith, R.D. 1984. Microwave power in industry. Final report. *Elec Power Res Inst* EM-3465: A-8.

Smythe, M.C., Wakeman, R.J. 2000. The use of acoustic fields as a filtration and dewatering aid. *Ultrasonics* 38: 657–662.

Societé Prolabo, Luque de Castro, M.D., García-Ayuso, L.E. 1997. *PCT application n°PCT/FR97/00883* (published under WO 97/44109).

Terigar, B.G., Balasubranian, S., Boldor, D., Xu, Z., Lima, M., Sabriov, C.M. 2010. Continuous microwave-assisted isoflavone extraction system: Design and performance evaluation. *Bioresour Technol* 101: 2466–2471.

Thuery, J. 1992. *Microwaves: Industrial, Scientific and Medical Applications.* Norwood, MA: Artech House.

Virot, M., Tomao, V., Colnagui, G., Visinoni, F., Chemat, F. 2007. New microwave-integrated Soxhlet extraction: An advantageous tool for the extraction of lipids from food products. *J Chromatogr A* 1174: 138–144.

Virot, M., Tomao, V., Ginies, C., Visinoni, F., Chemat, F. 2008. Green procedure with a green solvent for fats and oils' determination: Microwave-integrated Soxhlet using limonene followed by microwave Clevenger distillation. *J Chromatogr A* 1196–97: 147–152.

Wang, Y., Bonilla, M., McNair, H.M. 1997. Solid phase microextraction associated with microwave assisted extraction of food products. *J High Resolut Chromatogr* 20: 213–216.

Wang, T., Jia, X., Wu, J. 2003. Direct determination of metals in organics by inductively coupled plasma atomic emission spectrometry in aqueous matrices. *J Pharm Biomed Anal* 33: 639–642.

Wang, J.X., Xiao, X.H., Li, G.K. 2008. Study of vacuum microwave-assisted extraction of polyphenolic compounds and pigments from Chinese herbs. *J Chromatogr A* 1198–1199: 45–53.

Yang, Z., Zhai, W. 2010. Optimization of microwave-assisted extraction of anthocyanins from purple corn (*Zea mays L.*) cob and identification with HPLC–MS. *Innovative Food Sci Emerging Technol* 11: 470–476.

Yu, Y., Chen, B., Xie, M., Duan, H., Li, Y., Duan, G. 2009. Nitrogen-protected microwave-assisted extraction of ascorbic acid from fruit and vegetables. *J Sep Sci* 32: 4227–4233.

Zhang, H., Xue, M., Lu, Y., Dai, Z., Wang, H. 2010. Microwave-assisted extraction for the simultaneous determination of Novolac glycidyl ethers, bisphenol A diglycidyl ether and its derivatives in canned food using HPLC with fluorescence detection. *J Sep Sci* 33: 235–243.

4 Ultrasonically Assisted Diffusion Processes

Zbigniew J. Dolatowski and
Dariusz M. Stasiak

CONTENTS

4.1 INTRODUCTION

Diffusion processes are among the most important technical operations in various industries, especially chemical, pharmaceutical, and food production. In the diffusion model, it is assumed that viscous forces and permeability barriers within the microstructure of the biological material prevent coherent velocity flow of water, whose moisture movement occurs by random molecular diffusive motion. Cellular material approximates to the diffusive model because the cell membranes and cell walls act as permeability barriers, preventing coherent fluid flow. The complicated system of tissue, cell and its components, and substructural division into cell parts results from the evolution of cellular organisms. Cells are able to exchange nutrients and molecules with their environment through the cell membrane, which encloses cell materials. The process rate is determined by the physical properties and chemical composition of the system (Brennan 2006; Sun 2006).

4.2 DIFFUSION

The term mass transfer is used to denote the transfer of a component in a mixture from a region of higher concentration to a region of lower concentration. Diffusion is part of the transport phenomenon. It can result from the random movements of molecules (molecular diffusion) or from the circulating currents present in turbulent gas, vapor, or liquid (eddy diffusion) (Brennan 2006).

Industrial methods of separation of mixture into its components are based on differences in physical properties. The differences in solubility of the gases in a selective absorbent are the underlying principles in gas absorption, and the selectivity of an immiscible liquid solvent for the target ingredient in liquid–liquid extraction. Fractional distillation is based on the differences in the volatility of components. The rate of the process is dependent on both driving force (i.e., difference in concentrations) and mass transfer resistance. Mass transfer takes place across the phase boundary until the concentration on both sides is different. In a state of equilibrium, transfer still occurs but it does not lead to change in concentration (e.g., aroma in the air in a closed chamber). When a transfer process is accompanied by a chemical reaction, the overall transfer rate depends on both the mass transfer resistance and on the chemical kinetics of the reaction (Coulson et al. 1999; Brennan 2006).

The simplest example of diffusion is a mixing process in a chamber divided by a partition into two compartments containing different fluids. After removing the partition, the fluids start to mix and the mixing continues at a constantly decreasing rate until the whole system acquires a uniform composition. The process is attributable solely to the random motion of the molecules—the so-called Brownian motion. The rate of diffusion in a mixture of two gases A and B, assumed ideal, is governed by Fick's law and expresses the mass transfer rate as a linear function of the molar concentration gradient. In steady state, diffusion may be written as (Coulson et al. 1999)

$$N_A = -D_{AB} \frac{dC_A}{dy} \tag{4.1}$$

where N_A is a molar flux of A, C_A is a concentration of A, D_{AB} is a diffusion coefficient for A in B, and y is a distance in the direction of transfer.

Equation 4.1 is applicable to a stationary medium or a fluid in streamline flow. The description of the molecular mechanism in a turbulent fluid, where circulating or eddy currents are present, will be reinforced, and the total mass transfer rate may be written as (Coulson et al. 1999)

$$N_A = -(D + E_P) \frac{dC_A}{dy} \tag{4.2}$$

where D is a physical property of the system, which includes composition, pressure, and temperature, and E_p is an eddy diffusivity that depends on the flow pattern and varies with position.

Molecular diffusivity increases with increase in molecular velocity and in the mean free path of the molecules in fluid. Therefore, the use of high temperature and low pressure, especially in gas, is a means to enhance diffusivity. Extremely high temperatures and variable pressures are accompanied by ultrasonic cavitation and may affect the diffusion process (Coulson et al. 1999; Brenan 2006).

Mass transfer across a phase boundary is a specific phenomenon in many important applications such as vapor and liquid distillation, gas absorption, liquid–liquid extraction, and crystal dissolution. A number of mechanisms have been suggested to represent conditions in the region of the phase boundary. Whitman's two-film theory suggests that the resistance to transfer in each phase possibly lie in a thin film near the interface. The transfer is regarded as a steady-state process of molecular diffusion. The turbulence in the bulk fluid dissipates at the interface of the films. In Higbie's penetration theory, it is assumed that the eddies in the fluid bring an element of fluid to the interface where it is exposed to the second phase for a fixed period; however, Danckwerts suggested that the exposure lasts for a random amount of time. Subsequently, the film-penetration theory of Toor and Marchello incorporates some of the principles of both the two-film theory and the penetration theory, and is more general (Coulson et al. 1999).

In food production, pharmaceutical, and chemical industries, there are many technologies in which diffusion is a basic phenomenon and determines the possibility of the production of particular goods. Most often, the process is run in laboratory conditions under the control of operators; however, in certain conditions, the process can be run beyond the operator's control or run spontaneously, and can cause undesired changes in raw materials and products. The following technologies are based on the process of diffusion, especially with the application of ultrasound (Povey and Mason 1998; Mason and Lorimer 2002; Knorr et al. 2004; Patist and Bates 2008):

- Extraction involves using a liquid solvent to remove target liquid component from a liquid mixture or a target soluble component from a solid (e.g., production of sugar, oil, instant foods, dyes, and pigments), the driving power being the chemical potential difference.
- Crystallization and dissolution, in which the substance undergoes transition between a liquid and a solid (crystal) phase (e.g., sugar crystallization, water freezing), the driving power being the chemical potential difference defined as the degree of solution saturation.
- Salt curing, in which salt and the ingredients of the curing mixture are added into solid foods (e.g., salted cheese, cured meat).
- Drying, in which water molecules pass through the border of phases from solid or liquid into the surrounding gas (e.g., fruit and vegetable drying, drying meat), the driving force being the difference in vapor pressure near the surface of the material and in the surrounding gas.

- Smoking, in which the particles of smoke (gaseous phase) penetrate the product (e.g., smoking sausages or cheeses).
- Osmosis, which is based on the movement of molecules and ions through semipermeable membranes toward a solution of lower concentration.

4.3   PROPERTIES OF ULTRASOUNDS

Ultrasounds are pressure waves, or areas of medium particles oscillating around a position of equilibrium, that spread into a material. Through the oscillation movement toward the medium, energy is transmitted. The energy level can be very large or insignificant, depending on the intensity of the ultrasounds. In this sense, ultrasounds are exceptional means of energy transport resulting from elastic deformations due to pressure changes in the medium rather than due to movement of substances. Mechanical vibrations spreading in gaseous, liquid, and solid elastic materials assume frequencies in the range from 0 to 10^{13} Hz. This spectrum is divided into three basic ranges: the frequency range below 16 Hz, defined as infrasounds; the oscillations from 16 Hz to 16×10^4 Hz, audible for the human ear, called sounds; and the frequencies above 16×10^4 Hz, named ultrasounds. Furthermore, frequencies more than 10^9 Hz are referred to as hypersounds, which are natural heat waves related to the oscillation of molecules and atoms in a crystal network (Povey and Mason 1998; Mason and Lorimer 2002).

The ultrasounds moving in an elastic medium create a wave surface that assumes a spherical, flat, cylindrical, or an unspecified shape. Because of the direction of wave propagation, the means of medium stimulation to vibration, boundary conditions defined by the measures of the vibrating particles in the substance, and physical properties of the environment such as viscosity and density, two main types of waves are defined: longitudinal and transverse waves. They can also be of surface, dilatation, bend, and torsional type. In elastic, solid bodies and extremely viscous liquids, ultrasounds propagate perpendicularly (longitudinal waves) and tangentially (transverse waves) to the wave surface. In liquids of lower viscosity and in gases, ultrasounds propagate as longitudinal waves in forms of substance thickening and thinning. Concerning the liquids and longitudinal waves only, the phenomenon of ultrasound travel can be examined using the basic characteristics of wave movement or by observation of the medium in which wave travel takes place (Mason and Lorimer 2002).

The propagation of ultrasounds is subject to typical mechanical wave phenomena of reflection, diffusion, and absorption resulting from the viscosity, conduction, and heat radiation and molecular relaxation of the surroundings. Biological medium anisotropy particularly influences the irreversibility of real acoustic processes accompanying vibration spreading, which is connected to wave absorption and amplitude reduction as a result of physical (e.g., energy dissipation, cavitation), chemical (e.g., sonochemical reactions), and biological (e.g., microflows, cell disintegration, biostimulation) phenomena. For the purpose of analyzing ultrasound wave movement in a medium, simplified linear descriptions are most frequently employed. Detailed descriptions of the phenomena of nonlinear characteristics are more complicated.

In most cases, the phenomena caused by ultrasounds, particularly in biological sub-stances, are assessed differently in theory and practice (Mason and Lorimer 2002).

The use of ultrasounds in research and the modification of biological material are most often considered in the aspect of low–high frequency and low–high intensity. Ultrasounds of high frequency (5–10 MHz) and low intensity (under 1 W·cm^{-2}) are commonly used for diagnostic purposes; texture examination; and analysis of chemical composition, viscosity, and other characteristics. Vibrations of low frequency (20–100 kHz) and high intensity (10–1000 W·cm^{-2}) are used for evoking physical and chemical changes in the acoustic field. The characteristics of the interactions between the ultrasounds and the environment depend primarily on their frequency and energy level. It is assumed that ultrasounds of high frequency are most useful for examination of materials. They are used for the examination of nondegradable materials in medical diagnostics and acoustic microscopy (Table 4.1).

The influence of ultrasounds of low frequencies and higher intensities on sub-stances and biological materials is related to the presence of

- Primary effects present in the time of operations (e.g., diffusion, disintegration, sonochemical reactions)
- Secondary effects as a delayed result of primary phenomena (e.g., specific biochemical reactions as an effect of untypical enzyme diffusion)

TABLE 4.1
Major Applications of Ultrasounds in the Food Industry

Main Effect of Ultrasound	Examples of Application
Sonochemical	Initiation of chemical reactions
	Free radical production
	Oxidation
	Denaturation
	Macromolecule degradation
Sonophysical	Destruction of physical structure
	Fragmentation
	Emulsification
	Acceleration of mass and heat transfer
	Control of crystallization
	Control of filtration and drying
	Control of diffusion and extraction
Sonobiological	Stimulation and inhibition processes in cells of organisms
	Stimulation and inhibition of enzymatic reactions
	Destruction of tissue

Source: Povey, M.J.W. and Mason, T. (Eds.), *Ultrasound in Food Processing*, Blackie Academic & Professional, London, 1998; Mason, T.J. and Lorimer, J.P. (Eds.), *Applied Sonochemistry. The Uses of Power Ultrasound in Chemistry and Processing*, Wiley-VCH, Weinheim, 2002; Knorr, D. et al., *Trends Food Sci Technol* 15, 261–266, 2004.

Ultrasounds cause three main effects in the treated material: mechanical (e.g., cavitation, thixotropy, coagulation, emulgation), thermal (increase of temperature), and physicochemical (e.g., polymerization, depolymerization, increase of ionization, increase of reaction rate). For these reasons, this chapter deals primarily with ultrasounds of low frequency, which can influence effective mass transfer processes, especially in substances of biological origin (Povey and Mason 1998).

4.4   ULTRASOUND AND CAVITATION

Ultrasonic cavitation, widely described in the scientific literature (Suslick 1990; Povey and Mason 1998; Mason and Lorimer 2002), is a physical phenomenon based on the creation of small vacuum and gas–steam bubbles, and their pulsation and implosion. It happens when liquid is under the influence of changeable pressure caused by ultrasound waves. When the intensity of ultrasounds exceeds the threshold of cavitation, the forces of liquid cohesion can be overcome. In the spaces of lower pressure, small vacuum bubbles are created. The bubbles increase their size in a short time, fill themselves with vapors of liquids and gases dissolved in the liquid, and, after some time, disappear.

Cavitation at the liquid–solid border causes fast liquid flows that create significant shear forces. They undergo physical changes, resulting in surface stripping, erosion, and particle disintegration (Mason and Lorimer 2002). Because of this phenomenon, new layers of the surface are exposed, which intensifies diffusion. Microscopic examinations of plant and animal tissues that underwent extraction with the aid of ultrasounds confirmed that there are two phenomena responsible for the process:

(a) Increased diffusion of elements through cellular membranes
(b) Extraction of the cellular structures of the cells damaged by cavitation (Povey and Mason 1998; Dolatowski 1999; Toma et al. 2001; Vinatoru 2001; Chen et al. 2007; Vilkhu et al. 2008; Ebringerová and Hromádkova 2010)

The microflow of liquid directed toward the surface of the solid that accompany the implosion of the cavitation bubble is also capable of causing damage to bacterial cells (Dolatowski 1999). The behavior of cavitation bubbles is different and its classification is well known (Povey and Mason 1998; Mason and Lorimer 2002). When cavitation bubble collapse, one bubble wall passes through another, thus forming a high speed microjet. This is shown in Figure 4.1.

The use of ultrasounds to assist the extraction of biological materials from cells presents certain difficulties because of the variations in the physical makeup, chemical composition, and rheological properties of substances. Moreover, materials demonstrate different vulnerabilities to ultrasounds, acoustic pressure, and cavitation. It has also been reported that the use of ultrasound waves of intensities below the threshold of cavitation also speeds up the diffusion processes in nonhomogeneous liquid systems (Dolatowski 1989; Sivakumar and Rao 2001; Stasiak 2005). Variable ultrasound pressure causes alternating rarefaction and compression of liquid and circulating or eddy currents at the film close to the interfacial area. The course of extraction also depends on the probability that the bubble is created at the phase

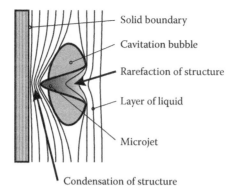

Solid boundary

Cavitation bubble

Rarefaction of structure

Layer of liquid

Microjet

Condensation of structure

FIGURE 4.1   Cavitation phenomenon at the solid phase boundary.

boundary and the microflow reaches the interface. The efficiency of extraction is also influenced by such factors as the plant tissue turgor and particle mobility (e.g., starch granules in the cellular cytoplasm). The impact of cavitation increases the number of particle collisions and intensity of liquid flow inside the medium (Povey and Mason 1998).

4.5   APPLICATIONS OF ULTRASOUNDS

4.5.1   EXTRACTION

The process of extraction is based on the transfer of one or more ingredients from a solution (or a solid) to another liquid phase (i.e., the solvent). Depending on the physical state of the raw material, there are two types of extraction: liquid–liquid system (when two components are liquids) and solid–liquid extraction (when one component is solid). In food production technology, we usually encounter solid–liquid types of systems. Extraction in a solid–liquid system means the selective removal of one or more components from a multicomponent solid. Usually, liquid solvent is used in which one component dissolves and is removed and others do not. The target substance in a porous inert may be the soluble solid or the solution that fills the pores. In the first step, diffusion must be preceded by the dissolution of the solid substance. In the second phase, the solution of target substance is transported (diffused) from the porous space of the inert matrix to the outside of the solid. Therefore, dissolution can be one of the stages in the extraction process (Coulson et al. 1999; Brennan 2006).

The course of the extraction process depends on the difference in concentrations between the solution in the pores and in the bulk liquid. From the kinetics point of view, the most favorable situation happens when the concentration in the bulk liquid equals zero; however, for technological reasons, such situations are unjustified because the aim of technological processes is to obtain a solution of the highest possible concentration. Having contacted the solid, the solvent usually increases its concentration in a continuous way, and consequently, the concentration at the solid

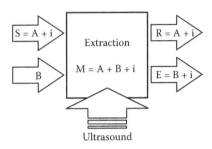

FIGURE 4.2 Principle of ultrasound-aided leaching (S, solid; A, solid matrix; i, solute; B, solvent; M, mixture; E, extract; R, residue).

surface consistently changes. The course of the process depends on the mutual phase impact (Coulson et al. 1999).

The extraction process has a fundamental importance in many branches of industry, and supporting the process with the use of ultrasounds (Figure 4.2) provides measurable economic and technological benefits—for example, in sugar production from sugar beets and oil production through extraction from oil plants. It is also used in the production of fruit and vegetable juices, aromatic oils, colorings, vitamins, enzymes, food concentrates, and other products (Sun 2006).

The desired component (i) contained in a solid component (A) is separated by using a liquid solvent (B) in which the solute (i) is soluble. The mixture (M) is treated with ultrasound. During the contact of the solvent (B) and the solid component (A), the solute (i) leaches from the solid (S) into the solvent (B). Thus, the phase compositions change. After the period of contact, the mixture (M) is divided into two streams: overflow (E) consisting of the solute (i) in the solvent (B) and a underflow (R) consisting of insoluble solid components (A) with some solution (i) adhering to it (Coulson et al. 1999).

The extraction of soluble substances from plant sources requires the right choice of solvent, and heating and stirring combination. Ultrasounds significantly speed up the process of mass exchange in the extraction system. This intensifying impact of the ultrasound field occurs after given threshold values of acoustic pressure are exceeded at which the flows resulting from nonlinear energy absorption are present. The phenomenon of mass transport stimulated by ultrasounds is then observed. Cavitation and microflows disrupt natural liquid layers system close the phase boundaries and consequently stimulate the process of mass transfer. Cavitation provides another advantage in the form of destruction of cellular structure, and consequently the release of cell contents into the surrounding solution.

4.5.1.1   Oil Extraction

Oil from flax seed contains many specific substances that are required for the normal functioning of the human body, especially unsaturated fatty acids that reduce the risk of cancer. Zhang et al. (2008) facilitated the extraction of linum oil by using an ultrasound device with a generator having a frequency of 20 kHz and power of 250 W. They achieved a reduction in the production time and the amount of solvent used. Process efficiency increased with the intensity of ultrasounds, with the highest efficiency reached within the first 30 min of extraction. Gas chromatographic analysis did not reveal any changes in chemical composition caused by ultrasounds (Zhang et al. 2008).

Extraction of oil from plant materials can be done in supercritical conditions. A research by Riera et al. (2010) showed that the extraction of almond oil, assisted with ultrasound (19 Hz), allowed for an efficiency of 19%. The method was tested with raw materials such as cocoa bean. Ultrasound extraction in supercritical conditions enabled an increase in efficiency by 43%. Consequently, with the use of ultrasounds (20 kHz, 100 W), the technological process of oil production is simplified (Li et al. 2004; Riera et al. 2004).

Cravotto et al. (2008) carried out extraction processes with a method that combined the use of ultrasounds and microwaves. Using diversified frequencies of ultrasounds within the range of 19–300 kHz, they achieved the highest extraction efficiency of soya and algae oils at the lowest frequency. Compared with conventional methods, extraction time was reduced by up to 10-fold and the yield increased by 50–500%. Chromatographic tests of extracts resulting from ultrasound-assisted processes showed insignificant changes in the chemical composition of the desired product. Slight or negligible differences in methyl ester profiles of oils extracted under high-intensity ultrasound and in a Soxhlet extractor were identified by gas chromatography (Cravotto et al. 2008).

The extraction of oil from ground olives takes less time if the olives are treated by low-frequency ultrasound (Jiménez et al. 2007). It is not significant whether the ultrasounds were emitted from the bottom or inside the container through a horn. In both cases, Jiménez et al. (2007) achieved an increase in extraction efficiency and effectiveness. The oil did not show any qualitative changes after the application of ultrasounds. However, the oil that is produced from the raw material with the aid of ultrasounds had more tocopherol, chlorophylls, and carotenoids; it did not present any sensory changes, especially in taste.

4.5.1.2 Sugar Extraction

Beet juice is obtained from the water extraction of cut beet. The juice is then cleaned, thickened, and crystallized to obtain sugar. The use of ultrasound allows shortening the time of extraction of sugar from beet slices. Under the influence of cavitation, the physical structure of cells is destroyed and microflows accompanying the bubble implosion speed up sugar diffusion (Povey and Mason 1998; Knorr et al. 2004). A study in this field was carried out by Stasiak (2005) and showed that ultrasound extraction at 25 kHz shortens the extraction time by fourfold, compared with the conventional static method. Higher sugar extraction from cut beet was also achieved. It allows the reduction of sugar loss in the beet pulp. Using a short application of ultrasound to the extraction of polysaccharides from different plant tissues resulted in a higher yield of product, with increased purity, compared with classical extraction (Li et al. 2007).

4.5.1.3 Extraction of Different Organic Compounds

The production of extracts from various plant parts necessitates the use of different solvent. The necessary step of cutting raw materials into smaller pieces before extraction and the effect of temperature during the extraction processes result in the loss of valuable compounds. The results of research from different laboratories worldwide show that with the use of ultrasounds, the efficiency of extraction can be significantly increased without negatively affecting the properties of obtained components, which is especially important in pharmaceutical production (Brennan 2006; Sun 2006).

Plant parts, especially herbs, contain many valuable ingredients that can be obtained by extraction. The effectiveness of ultrasound-aided extraction of biologically active substances was researched by Vinatoru et al. (1997) and Toma et al. (2001) on herbal plants such as menthe, camomile, marigold, and salvia. The extraction with ultrasounds was faster each time and allowed achieving higher process effectiveness, compared with the conventional method, especially for the components of lower molecular mass. Paniwnyk et al. (2001) achieved a 20% increase in rutin amount by using ultrasounds of 20 kHz with *Sophora japonica*. Boonkird et al. (2008) achieved a significant increase in speed of capasacinoid extraction from pepper fruit owing to the use of ultrasounds. The process was particularly intensive in its first stage, and after about 5 min the intensity rapidly decreased. Ultrasound-aided extraction at frequencies of 26 and 70 kHz proved that a higher yield of capasacinoids can be obtained at lower frequencies. The process efficiency in the pilot industrial tests was slightly lower than in the case of conventional extraction based on thermal maceration of raw material. Ultrasound-assisted extraction can significantly improve the yield of bioactive substances, achieving higher efficiencies and shorter reaction times at lower temperatures and lower process costs (Boonkird et al. 2008; Ebringerová and Hromádková 2010).

Anthocyanins belong to a family of coloring components naturally contained in fruits, vegetables, and plants. Chen et al. (2007) performed chromatographic tests in anthocyanins extracted from raspberry fruits, using ultrasounds with a frequency of 22 kHz and a generator power of 650 W. In comparison with conventional solvent extraction, ultrasound-assisted extraction was more efficient and rapid in extracting anthocyanins from red raspberries, owing to the strong disruption of the structure of fruit tissues under ultrasonic acoustic cavitation, as observed under a scanning electron microscope. However, the compositions of anthocyanins in extracts by both methods—conventional and ultrasound assisted—were similar, as confirmed by their chromatographic profiles. The optimization of the generator's power and the duration of the process eliminated the possibility of chemical changes in anthocyanins during ultrasound extraction. One of the richest sources of natural carotenoids are marigold petals. Lutein is the most important coloring agent from marigold (Gao et al. 2009). With the use of ultrasounds, the supercritical extraction of this compound carried out in the state phase increased from 3.1×10^{-9} to 4.3×10^{-9} m·s^{-1}. The highest efficiency of lutein extraction was achieved with a frequency of 25 kHz and a generator power of 400 W at an extraction pressure of 32.5 MPa.

Polyphenols present in food products are considered valuable compounds that have the potential to prevent heart diseases and cancer. They demonstrate antioxidizing properties and are present in the rind of citrus fruits. Khan et al. (2010) extracted polyphenols from citrus rinds, supporting the process with ultrasounds of 25 kHz frequency and 150 W generator power. As a result, they achieved approximately 30%

higher polyphenol contents in the extract, a shorter time of extraction, and a higher efficiency than in conventional methods. Chromatographic tests of polyphenols extracted with ultrasounds showed no significant changes in the antioxidizing values of the extracts. In this way, the usefulness of the ultrasound-assisted extraction method for polyphenol production was proven (Khan et al. 2010).

The application of ultrasounds is also used in the production of instant tea, as one of the operations in the extraction of ingredients from tea leaves. After the extract is dried, a water-soluble powder is obtained. The use of ultrasounds result in about 20% increase in soluble compound content compared with the extraction process without the use of ultrasounds. This helps shorten the time of effective extraction, especially because the process is at its most intensive in the first few minutes of ultrasound operations. Balachadran et al. (2006) studied the influence of ultrasounds of the frequency of 20 kHz on supercritical extraction of ginger. The increase of extraction speed and process efficiency depended on the generator's power; however, the relations were not proportional. The authors found that a significant increase in the temperature of the treated material restricts the use of high-intensity ultrasound in some applications.

Ultrasounds can increase the efficiency of date syrup extraction at lower temperatures and allow the production of a juice of lighter color (20 kHz, 500 W) (Nazary et al. 2004). With the use of ultrasounds, a higher efficiency of polysaccharide extraction from salvia can be achieved (20 kHz, 1 W·cm^{-2}) (Hromádková et al. 1999), as well as that of tannins from willow tree (20 kHz, 50 W·cm^{-2}) (Mantysalo and Mantysalo 2000), antioxidants from rosemary (Albu et al. 2004) and bearberry leaves (Gribova et al. 2008), isoflavonoids from soya seed (24 kHz, 200 W) (Rostagno et al. 2003), volatile compounds from wines (Vila et al. 1999), select elements from vegetables (47 kHz) (Nascentes et al. 2001), and fukoidins from algae (Li et al. 2008). Interesting information concerning ultrasound-aided extraction was published by Ebringervá and Hromádková (2010). The most significant benefits are the reduction of extraction time, decrease of reagent use, and decrease of extraction temperature.

Soy protein is used in a variety of foods for emulsification and texturizing. The protein concentrate is achieved by lixiviating most of the soluble nonprotein compounds from dehulled, defatted soybean meal. The research by Moulton and Wang (1982) resulted in the concept of continuous soya protein extraction supported by ultrasounds, which can decrease the energy consumption of the process by about 70%. The water extraction yield of proteins was higher by 54% and by 23% in alkaline extraction (Moulton and Wang 1982). Ultrasounds also facilitate the extraction of meat proteins in meat processing: they cause the relaxation of myofibril structures and therefore the release of proteins, especially in the ultrasound-assisted process of meat tumbling (Dolatowski 1999).

The research on ultrasound-aided extraction indicates that the benefits of ultrasounds include the shortening of process time, increase of process efficiency, and decrease of solvent use. A shorter process time implies energy savings, which increase the installation efficiency. The conditions of ultrasound extraction allow decreasing inert substance loss, and consequently better utilization of raw materials. A higher concentration of the extract is also achieved, because of which smaller amounts of solvent are used. These advantages show that ultrasound-aided extraction has the features of an industrial pro-ecological technology that can preserve the natural environment.

4.5.2 CRYSTALLIZATION AND DISSOLUTION

Crystallization is the process of obtaining a solid component (in the form of crystals) from a solution, or creating a solid phase in crystal form as a result of precipitation of a liquid substance. The process takes place in conditions of simultaneous mass and heat transfer. In the saturated solution, in which the amount of the dissolved substance is higher than the saturation concentration, crystal nuclei are created and they grow further as a result of the diffusion of solute from the solution to the surface of the solid. Application of ultrasound in liquid and solid–liquid systems can produce chemical and physical effects that enhance mass transfer and reactant diffusion. The speed of crystallization and dissolution processes is determined by the diffusion process at the phase boundary (Coulson et al. 1999; Brennan 2006). The proper use of ultrasound can initiate seeding and control subsequent crystal growth in a supercooled or supersaturated medium (McCausland et al. 2001; Chow et al. 2003; Stasiak and Dolatowski 2007). On the one hand, ultrasonic cavitation causes nuclei formation; on the other hand, it causes the disruption of seeds (nuclei) already present within the medium, thus increasing the number of nuclei. The products formed by sonocrystallization are superior to the products obtained by conventional crystallization. Kakinouchi et al. (2006) described the effects of applying ultrasonic power to the crystallization of hen egg white lysozyme. Their results suggest that long-term ultrasonic (100 kHz, 100 W) treatment of a solution containing clusters of protein molecules that include crystal nuclei damage the clusters, while brief treatment (10 s) of a protein solution without clusters promotes nucleation and generates crystal nuclei (Kakinouchi et al. 2006). The study of Louhi-Kultanen et al. (2006) shows that continuous ultrasound (20 kHz, 24–122 $W \cdot kg^{-1}$ solution) treatment can be used as a size reduction method to produce glycine with a uniform crystal morphology. Moreover, ultrasound is also used to control the crystallization process of sugar, lactose, and fats, as well as freezing of food products and other solids (Povey and Mason 1998; Chow et al. 2003; Knorr et al. 2004; Bund and Pandit 2007).

4.5.3 SALTING AND CURING

The technology of meat curing is based on the addition of salt, which positively influences meat stability and is responsible for physicochemical and biological phenomena that determine sensory properties. Sodium chloride facilitates the extraction of myofibril proteins that give the proper physicochemical and biological features of the product. Meat salting is based on the phenomenon of diffusion and can be intensified with low-frequency ultrasounds (Dolatowski 1988; Mulet et al. 2003; Cárcel et al. 2007). Cárcel et al. (2007) show that to obtain a significant increase in salt diffusion in meat, high-intensity ultrasounds (51 $W \cdot cm^{-2}$) must be used. However, Siró et al. (2009) achieved more than a twofold increase in diffusion rate by using a frequency of 20 kHz and an intensity of 4 $W \cdot cm^{-2}$. The higher the ultrasound power used, the faster the diffusion process was, and the relation was described by an exponential function. The course of salt diffusion also depends on the frequency, as indicated in previous studies (Sajas and Gorbatow 1978; Dolatowski 1999). The phenomenon of ultrasound propagation causes an increase in meat temperature during curing, which increases with the increase in field density and wave frequency. For this reason, optimal conditions for meat salting are ensured by using ultrasounds of low frequencies.

4.5.4 DRYING

Drying is one of the oldest methods of food preservation that uses heat and mass transfer to remove water or moisture from another solvent, by evaporation from a solid, semisolid, or liquid. The final product of drying is in solid form. The main industrial methods of drying use hot air, heating through a hot wall, superheated steam, or lyophilization. When water removal is considerable, the product is usually deformed and shrunken, except in a well-designed freeze-drying process. To achieve drying, there must be a source of heat and a moisture content difference. It requires continuous removal of vapor from the environment. The process consists of several stages: a period of constant drying, a phase of rate decrease, and a long period when the rate tends to the null value (Coulson et al. 1999; Sun 2006). The thin film close to the interface is affected by ultrasound waves. During the negative phase of the pressure cycle, the moisture is removed and it does not reenter during the positive pressure phase. The influence of ultrasound is particularly effective in the initial, constant-rate phase of drying, when the surface of the material is completely covered with moisture. Ultrasounds of relatively low frequencies (16–30 kHz) and high intensities (140–160 dB) decrease the thickness of the film and increases gas exchange at the phase boundary, which intensifies the rate of drying (Povey and Mason 1998). As a result, the time of ultrasound-aided drying is shortened, the process can be carried out at lower temperatures, and lower speeds of gas flow do not cause the blowout of small-grained materials. Consequently, the materials are less vulnerable to unfavorable changes in physicochemical properties, which are common when using conventional methods. Studies on acoustic drying showed that the best conditions for diffusion are created in tumble dryers, in devices equipped with a vibrating conveyor, and in fluid bed dryers (Povey and Mason 1998). Ultrasound converters are installed inside the dryer chamber in such a way that the directions of wave propagation and the movement of the dried material had opposite directions. Ultrasound energy is emitted to the gaseous environment (Gallego-Juarez et al. 1997). Figure 4.3 shows a diagram illustrating the operating principle of the dryers equipped with an ultrasound device.

In this type of device, material is conveyed through the dryer on a perforated conveyor. The heated air flows through the belt and the layer of material, upward at the beginning of the process and downward in the later stages (Sun 2006). High-intensity airborne ultrasounds introduce pressure variations at gas–liquid interfaces, and therefore increases the moisture evaporation rate. The use of a multistage conveyor during drying will provide higher efficiency, uniformity of drying, and better control of the process. A fluidized bed can be used instead of a conveyor belt.

Ultrasound vibrations can be directly emitted into the material. With the use of vibrations of 20 kHz and a power of about 100 W, de la Fuente-Blanco et al. (2006)

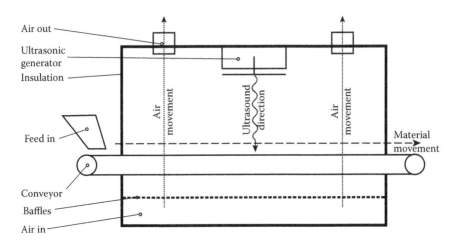

FIGURE 4.3 Principle of ultrasonically assisted drying process.

achieved the reduction of vegetable drying time. They used a direct-contact ultra-sonic dehydration system with rectangular vibrating plate. A forced-air generator was located above the samples. The results show the strong influence of acoustic intensity in drying of carrots.

4.5.5 MEMBRANE PROCESSES

Membrane operations in food production allow for selective removal of particular groups of ingredients or large-particle compounds without chemical or thermal pro-cesses. They are based on the use of selective membranes permeable for certain sub-stances only. Reverse osmosis removes large molecules and ions from solutions by applying pressure to the solution when it is on one side of a selective membrane. The result is that the solute is retained on the pressurized side of the membrane and the pure solvent is allowed to pass to the other side. Ultrafiltration is a kind of membrane filtration in which hydrostatic pressure forces a liquid against a semipermeable mem-brane. Suspended solids and solutes of high molecular weight are retained, while water and low molecular weight solutes pass through the membrane. Nanofiltration is a cross-flow filtration technology that ranges somewhere between ultrafiltration and reverse osmosis and takes place at lower transmembrane pressure, which signifi-cantly reduces the process cost (Sun 2006). Of primary interest in cross-flow filtra-tion are cavitation, rapid movement of fluids caused by variation of sonic pressure, and microstreaming. The enhancement of membrane filtration by ultrasonic waves is primarily a consequence of cavitation and is directly related to the parameters of the acoustic field. The relationship is described as follows:

$$I_x = I_0 \exp(-kf^2x) \tag{4.3}$$

where I_x is the intensity at distance x, I_0 is the initial intensity, f is the frequency, and k is the constant of proportionality (Povey and Mason 1998).

The degree of enhancement also depends on concentration, size and type of particle, and suspension viscosity. In a particular case, a film of concentrated solution is formed close to membrane. This may locally change the electrochemical potential and adversely affect the membrane properties (Sun 2006). Ultrasound intensification of the process is based on intensifying the mass transfer close to the membrane (Povey and Mason 1998).

4.5.6 OTHER MASS TRANSFER PROCESSES

Distillation is one of the methods used in industries for the separation of liquid solutions into individual components. The separation occurs because the gaseous stage obtained by partial vaporization of the liquid solution has a different composition from the initial solution. Distillation is based on the constant removal of vapor from a boiling mixture. The flow of liquid during distillation usually ensures quick mass exchange and diffusion in phase boundaries and in vapor (Brennan 2006; Sun 2006). In such circumstances, application of the ultrasound seems to be pointless. Similar conclusions may arise from the analysis of smoking processes. Penetration of the material by smoke components occurs in particular technological conditions. Utilizing ultrasounds during smoking can cause condensation of volatile smoke components. Consequently, the final quality of the product may be reduced because of smoke contamination. Therefore, there are no scientific reports indicating practical aspects of the use of ultrasounds for mass transfer support in the processes of distillation and smoking.

Ultrasounds speed up the process of dye diffusion in leather coloration, shortening the process by 25% and improving the leather quality at the same time by making the coloring more even (Sivakumar and Rao 2003).

4.6 ULTRASONIC EQUIPMENT

In control conditions of ultrasound-supported diffusion, two main kinds of devices are utilized—bath and horn emitters (Mason and Lorimer 2002). Differences in the makeup of the devices are due to the composition of transducers and methods of ultrasound application (Figure 4.4). Effective, ultrasound-supported diffusion requires the control of parameters vital for the process, in particular (Povey and Mason 1998; Knorr et al. 2004)

- Density of energy, $W \cdot dm^{-3}$, which is the energy per environment volume unit.
- Intensity of ultrasound source, $W \cdot cm^{-2}$, which is the power emitted by a transducer per a surface unit.
- Static pressure, Pa, which determines cavitation and its intensity; the influence of cavitation on various media is much more intensive at high pressures.
- Temperature, K, which influences the cavitation intensity; at higher temperatures, the cavitation bubble is filled with the vapor of liquid and the

(a)

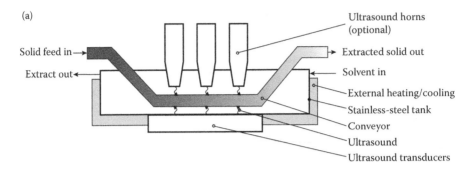

(b)

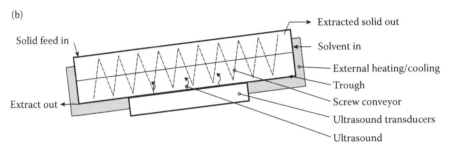

FIGURE 4.4 Diagram showing the principles behind extractors equipped with ultrasound: (a) with belt (bucket) conveyor and (b) with screw conveyor.

> intensity of the implosion is weakened. At the same time, higher liquid temperature helps mass exchange by convection.
>
> • Viscosity, Pa·s, which influences cavitation; high viscosity of the medium hinders the bubble creation and mass transfer processes.

Bath-type devices, where transducers are permanently fixed under a thin-walled, flat bottom of the bath, are the most common ultrasonic devices. Ultrasound energy is emitted directly from the bottom into the material filling the container. Extraction in smaller containers immersed into the bath can also be carried out. This solution is useful, in particular, in the case of small samples. The greatest efficiency of ultrasounds is acquired at the level of liquids, which allows for the creation of a standing wave at the liquid–air boundary. The standard bath devices create fields of low and medium intensity so as to avoid excessive erosion of the bath walls. Therefore, the density of the ultrasound field is low, owing to the large capacity of the device. Despite these limitations, bath-type devices are used in many laboratories and industries, mainly because of their universality (Povey and Mason 1998).

Much greater ultrasound intensity can be obtained when the transducer equipped with a concentrator is immersed in a liquid. Devices with horn transducers are used for processes involving samples of small volume, especially in laboratory conditions. A limitation is the cavitational erosion of the surface of the emitter and the risk of contamination of the material being processed (Povey and Mason 1998). Because of the high power of source and dissipation of energy, intense material heating is observed close to the emitter face. The control of the temperature of the processed material temperature is the main problem in applications. Excessive, uncontrolled temperature increase of the extracted biological material is undesirable because of the risk of changes in the physicochemical properties of the substance (Dolatowski 1999; Duck 2008). Most often, devices equipped with additional systems allow control of

- Ultrasound power
- Temperature of the processed material
- Duration of the process
- Sweeping the frequencies to increase the efficiency of cavitation

A simple extraction cell consists of a tank fitted with a false bottom that supports a bed of the solids to be extracted (Brennan 2006). Ultrasonic transducers can be attached to walls or immersed in the mixture. Single-extraction cells are used for laboratory and small-scale industrial applications. Continuous moving bed extractors equipped with ultrasonic transducers are shown in Figure 4.4.

The extractor (Figure 4.4a) consists of a conveyor (e.g., chain with perforated basket) immersed in the tank containing liquid (Brennan 2006). The ultrasound transducers are fitted to the wall of the container or immersed in the liquid. Another design of a continuous moving bed extractor consists of a trough set at a small angle to the horizontal containing screw conveyor with intermeshing flights (Figure 4.4b). The solid is fed in at the lower end of the trough and is moved up by conveyor countercurrent to the stream of solution. The ultrasound transducers are attached at the bottom of the trough. Ultrasonic cavitation and pressure oscillation increase the velocity and turbulence of the liquid as it flows over the solid particles, and the rate of the mass transfer coefficient increases. One of the components of the extractor is ultrasonic installation, which consists of transducers with a corresponding electronic generator. The generator transforms electrical power into mechanical vibrations of the maximum possible energy efficiency.

It is clear that the result of the ultrasound-assisted mass exchange process depends on the solid–liquid interface area, concentration gradient, and mass transfer coefficient on the one hand, and on the parameters of ultrasonication on the other. Therefore, optimization of the ultrasound-supported diffusion requires several experiments, which increases the implementation cost. As a consequence of considerable physicochemical diversity of materials processed by the food industry, such experiments are inevitable.

4.7 SAFETY OF ULTRASOUND APPLICATIONS

The first observations of the biological effects of ultrasound accompanied the genesis of the submarine detection method when fish were killed because of the rupture of their swim bladder. The hazards for humans associated with ultrasound are now quite well understood. Ultrasound causes thermal and nonthermal effects on medium (Povey and Mason 1990; Duck 2008). Dissipation of energy increases the temperature of tissues. Tissue–bone interfaces or tissue–air interfaces concentrate the ultrasound energy and may significantly increase local temperature. Acoustic cavitation in liquid gives rise to specific mechanical effects. The cells in the region of cavitation may be disrupted. In addition, radiation pressure exerts a force on the propagating medium. During the passage of ultrasound through a material, that material experiences local stress arising from energy density gradients (Duck 2008). The real health hazard for humans in proximity to an ultrasound device is insignificant. The ultrasound energy level emitted from the extraction device into the air is negligible because of the strong reflection of acoustic waves on the boundaries of the liquid–air and solid–air phases. The waves are almost completely reflected from the inner surface of the tank and the liquid surface. Only cavitation noise can be heard near the device (Mason and Lorimer 2002).

Ultrasound also induces sonochemical reactions in treated materials (Mason and Lorimer 2002). The chemical effects of ultrasound do not come from a direct interaction with molecular species. Three classes of sonochemical reactions exist: homogeneous sonochemistry of liquids, heterogeneous sonochemistry of liquid–liquid or solid–liquid systems, and, overlapping with the aforementioned, sonocatalysis. The natural components in foods (i.e., proteins, fats, and carbohydrates) may be also changed during sonochemical reactions. As a result of sonodegradation, undesirable and harmful chemical products are formed, especially free radicals (Suslick 1990; Mason and Lorimer 2002). Chemat et al. (2004), Patrick et al. (2004), and Schneider et al. (2006) studied properties of sonicated oils. During oil processing, a metallic and rancid odor is detected and some off-flavor compounds of oil are identified after treatment. Different edible oils (e.g., olive, sunflower, soybean) also show significant changes in their composition (chemical and flavor) owing to ultrasound treatment. Ultrasound causes changes in crystal structure and kinetics of oil crystallization. Patrick et al. (2004) conclude that the presence of off-flavors has prevented ultrasound from being considered a commercial technique for the crystallization of edible fats. However, the composition of the flaxseed oils was not affected significantly by the ultrasonically assisted extraction (Zhang et al. 2008).

High-power ultrasound is not a standardized extraction method and therefore needs to be developed and scaled up for industrial applications. Understanding of the complex physicochemical influence of high-power ultrasound and the relationship between the frequency of waves, intensity and duration of treatment, and their effects on the technological and functional properties of food and pharmaceutical materials would provide higher safety of food and pharmaceutical products generated using ultrasonically assisted mass transfer processes.

4.8   CONCLUSION

The ultrasonically assisted mass transfer processes typically utilize low-frequency waves in the range from 20 kHz to 100 kHz. Ultrasounds cause such physical phenomena as cavitation and acoustic pressure, which disturb the laminar flow in the direct neighborhood of phase boundaries and speed up the process of mass transfer. Ultrasound-assisted diffusion may be conducted in gaseous, liquid, and solid environments. In the case of biological materials, ultrasonic cavitation presents an additional benefit due to its ability to destroy cell structures. This allows for an easier release of the cell contents into the surrounding environment and enables the process of mass exchange. Consequently, diffusion-based technological processes, such as extraction, crystallization, drying, and osmosis, are sped up. Higher ultrasonic power is accompanied by higher diffusion efficiency.

The utilization of ultrasounds for diffusion in technological processes in the food industry improve the process efficiency, allow for the better use of materials, shorten the duration of the process, limit the use of solvents, and decrease energy consumption. Ultrasound-supported diffusion is used in the sugar industry for the extraction of sugar from sugar beets and in the oil industry for the extraction of oil from oil plants. The process is also utilized in the production of food concentrates, fruit and vegetable juices, essential oils, dyes, vitamins, enzymes, and other biological substances.

The use of ultrasound to aid diffusion, especially in biological materials (e.g., parts of plants) encounters difficulties due to the considerable diversity of the structure, chemical composition, and rheological properties of cells and tissues. Moreover, different materials have different degrees of susceptibility to the ultrasonic stimuli, which creates the need for an empirical selection of optimal extraction parameters (i.e., physical and chemical stimuli) in each case. This applies to the combination of frequency and intensity of ultrasounds, methods of their application, hydrostatic pressure, heating and mixing of the material, and the type of solvent used in extraction, in particular. Difficulties arising from the above mentioned factors have a role in the utilization of ultrasound in diffusion processes only in a laboratory or semi-technical scale.

The usefulness of ultrasound in supporting diffusion is undeniable, as indicated by its laboratory applications. At the same time, however, the relatively low number of industrial implementations indicates the need for adaptive works on the existing laboratory installations so that these can be efficient on an industrial scale. As shown in practice, this can be attained by the development of new, ultrasonic devices and fitting the existing ones with suitable ultrasonic installations.

REFERENCES

Albu, S., Joyce, E., Paniwnyk, L., Lorimer, J.P., Mason, T.J. 2004. Potential for the use of ultrasound in the extraction of antioxidants from *Rosmarinus officinalis* for the food and pharmaceutical industry. *Ultrason Sonochem* 11: 261–265.

Balachandran, S., Kentish, S.E., Mawson, R., Ashokkumar, M. 2006. Ultrasonic enhancement of the supercritical extraction from ginger. *Ultrason Sonochem* 13: 471–479.

Boonkird, S., Phisalaphong, C., Phisalaphong, M. 2008. Ultrasound-assisted extraction of capsaicinoids from *Capsicum frutescens* on a lab- and pilot-plant scale. *Ultrason Sonochem* 15: 1075–1079.

Brennan, J.G. (Ed.). 2006. *Food Processing Handbook.* Weinheim: Wiley-VCH.

Bund, R.K., Pandit, A.B. 2007. Sonocrystallization: Effect on lactose recovery and crystal habit. *Ultrason Sonochem* 14: 143–152.

Cárcel, J.A., Benedito, J., Bon, J., Mulet, A. 2007. High intensity ultrasound effects on meat brining. *Meat Sci* 76: 611–619.

Chemat, F., Grondin, I., Sing, A.S.C., Smadja, J. 2004. Deterioration of edible oils during food processing by ultrasound. *Ultrason Sonochem* 11: 13–15.

Chen, F., Sun, Y., Zhao, G., Liao, X., Hu, X., Wu, J., Wang, Z. 2007. Optimization of ultrasound-assisted extraction of anthocyanins in red raspberries and identification of anthocyanins in extract using high-performance liquid chromatography-mass spectrometry. *Ultrason Sonochem* 14: 767–778.

Chow, R., Blindt, R., Chivers, R., Povey, M. 2003. The sonocrystallization of ice in sucrose solutions: primary and secondary nucleation. *Ultrasonics* 41: 595–604.

Coulson, J.M., Richardson, J.F., Backhurst, J.R., Harker, J.H. 1999. *Chemical Engineering. Fluid Flow, Heat Transfer and Mass Transfer.* London: Elsevier.

Cravotto, G., Boffa, L., Mantegna, S., Perego, P., Avogadro, M., Cintas, P. 2008. Improved extraction of vegetable oils under high-intensity ultrasound and/or microwaves. *Ultrason Sonochem* 15: 898–902.

Dolatowski, Z.J. 1988. Ultraschall. 2. Einfluss von Ultraschall auf die Mikrostruktur von Nuskelgewebe bei der Poekehuig. *Fleischwirtschaft* 68: 1301–1303.

Dolatowski, Z.J. 1989. Ultraschall. 3. Einfluss von Ultraschall auf die Produktionstechnologie und Qualitaet von Kochschinken. *Fleischwirtschaft* 69: 106–111.

Dolatowski, Z.J. 1999. Influence of low frequency ultrasound on properties of meat. *Agricultural University in Lublin* No. 221. (In Polish).

Duck, F.A. 2008. Hazards, risk and safety of diagnostic ultrasound. *Med Eng Phys* 30: 1338–1348.

Ebringerová, A., Hromádková, Z. 2010. An overview on the application of ultrasound in extraction, separation and purification of plant polysaccharides. *Cent Eur J. Chem* 8: 243–257.

de la Fuente-Blanco, S., Riera-Franco de Sarabia, E., Acosta-Aparicio, V.M., Blanco-Blanco, A., Gallego-Juárez, J.A. 2006. Food drying process by power ultrasound. *Ultrasonics* 44: e523–e527.

Gallego-Juarez, J.A., Yang, T., Vazquez-Martinez, F., Galvez-Moraleda, J.C., Rodriquez-Corral, G. 1997. Ultrasound drying. *Trends Food Sci Technol* 8: 426.

Gao, Y., Nagy, B., Liu, X., Simandi, B., Wang, O. 2009. Supercritical CO_2 extraction of lutein esters from marigold (*Tagetes erecta* L.) enhanced by ultrasound. *J Supercrit Fluid* 49: 345–350.

Gribova, N.Yu., Filippenko, T.A., Nikolaevskii, A.N., Khizhan, E.I., Bobyleva, O.V. 2008. Effects of ultrasound on the extraction of antioxidants from bearberry (*Arctostaphylos adans*) leaves. *Pharma Chem J-USSR* 42: 593–595.

Hromádková, Z., Ebringerová, A., Valachowic, P. 1999. Comparison of classical and ultra-sound-assisted extraction of polysaccharides from *Salvia officinalis* L. *Ultrason Sonochem* 5: 163–168.

Jiménez, A., Beltrán, G., Uceda, M. 2007. High-power ultrasound in olive paste pretreatment. Effect on process yield and virgin olive oil characteristics. *Ultrason Sonochem* 14: 725–731.

Kakinouchi, K., Adachi, H., Matsumura, H., Inoue, T., Marakami, S., Mori, Y., Koga, Y., Takano, K., Kanaya, S. 2006. Effect of ultrasonics irradiation on protein crystallization. *J Cryst Growth* 292: 437–440.

Khan, M.K., Abert-Vian, M., Fabiano-Tixier, A.-S., Dangles, O., Chemat, F. 2010. Ultrasound-assisted extraction of polyphenols (flavanone glycosides) from orange (*Citrus sinensis* L.) peel. *Food Chem* 2: 851–858.

Knorr, D., Zenker, M., Heinz, V., Lee, D.-U. 2004. Applications and potential of ultrasonics in food processing. *Trends Food Sci Technol* 15: 261–266.

Li, B., Lu, F., Wei, X., Zhao, R. 2008. Fucoidan: Structure and bioactivity. *Molecules* 13: 1671–1695.

Li, H., Pordesimo, L., Weiss, J. 2004. High intensity ultrasound-assisted extraction of oil from soybeans. *Food Res Int* 37: 731–738.

Li, J., Ding, S., Ding, X. 2007. Optimization of the ultrasonically assisted extraction of polysaccharides from *Zizyphus jujuba* cv. *jinsixiaozao*. *J Food Eng* 80: 176–183.

Louhi-Kultanen, M., Karjalainen, M., Rantanen, J., Huhtanen, M., Kallas, J. 2006. Crystallization of glycine with ultrasound. *Int J Pharm* 320: 23–29.

Mason, T.J., Lorimer, J.P. (Ed.). 2002. *Applied Sonochemistry. The Uses of Power Ultrasound in Chemistry and Processing.* Weinheim: Wiley-VCH.

McCausland, L.J., Cains, P.W., Martin, P.D. 2001. Use the power of sonocrystallization for improved properties. *Chem Eng Prog* 97: 56–61.

Mantysalo, M., Mantysalo, E. 2000. Extraction and filtering in ultrasonic field: Finite element modeling and simulation of the processes. *Ultrasonics* 38: 723–726.

Moulton, J., Wang, C. 1982. A pilot plant study of continuous ultrasonic extraction of soybean protein. *J Food Sci* 47: 1127–1129.

Mulet, A., Cárcel, J.A., Sanjuán, N., Bon, J. 2003. New food drying technologies. Use of ultrasound. *Food Sci Technol Int* 9: 215–221.

Nascentes, C.C., Korn, M., Arruda, M.A.Z. 2001. A fast ultrasound-assisted extraction of Ca, Mg, Mn and Zn from vegetables. *Microchem J* 69: 37–43.

Nazary, S.H., Entezari, M., Haddad, M., Nazari, A.H. 2004. The effect of direct ultrasound on the extraction of date syrup. In *The Joint Agriculture and Natural Resources Symposium,* Tabriz–Ganja, May 14–16, 2004.

Paniwnyk, L., Beaufoy, E., Lorimer, P., Mason, J. 2001. The extraction of rutin from flower buds of *Sophora japonica. Ultrason Sonochem* 8: 299–301.

Patist, A., Bates, D. 2008. Ultrasonic innovations in the food industry: From the laboratory to commercial production. *Innov Food Sci Emerg* 9: 147–154.

Patrick, M., Blindt, R., Janssen, J. 2004. The effect of ultrasonic intensity on the crystal structure of palm oil. *Ultrason Sonochem* 11: 251–255.

Povey, M.J.W., Mason, T. (Ed.). 1998. *Ultrasound in Food Processing.* London: Blackie Academic & Professional.

Riera, E., Golas, Y., Blanco, A., Gallego, J.A., Blasco, M., Mulet, A. 2004. Mass transfer enhancement in supercritical fluids extraction by means of power ultrasound. *Ultrason Sonochem* 11: 241–244.

Riera, E., Blanco, A., Garcia, J., Benedito, J., Mulet, A., Gallego-Juarez, J.A., Blasco, M. 2010. High-power ultrasonic system for the enhancement of mass transfer in supercritical CO_2 extraction processes. *Ultrasonics* 50: 306–309.

Rostagno, M.A., Palma, M., Barroso, C.G. 2003. Ultrasound-assisted extraction of soy isoflavones. *J Chromatogr* 1012: 119–128.

Sajas, J.F., Gorbatow, W.M. 1978. The use of ultrasonics in meat technology. *Fleischwirtschaft* 6: 1009–1021.

Schneider, Y., Zahn, S., Hofmann, J., Wecks, M., Rohm, H. 2006. Acoustic cavitation induced by ultrasonic cutting devices: A preliminary study. *Ultrason Sonochem* 13: 117–120.

Siró, I., Vén, C.S., Balla, C.S., Jónas, G., Zeke, I., Friedrich, L. 2009. Application of an ultrasonic assisted curing technique for improving the diffusion of sodium chloride in porcine meat. *J Food Eng* 91: 353–362.

Sivakumar, V., Rao, P.G. 2003. Studies on the use of power ultrasound in leather dyeing. *Ultrason Sonochem* 10: 85–94.

Stasiak, D.M. 2005. Ultrasonically assisted extraction of sugar beet (in Polish). *Acta Sci Pol-Tech Agr* 4: 31–39.

Stasiak, D.M., Dolatowski, Z.J. 2007. Effect of sonication on the crystallization of honey. *Pol J Food Nutr Sci* 57: 133–136.

Sun, D.-W. (Ed.). 2006. *Emerging Technologies for Food Processing.* London: Elsevier Academic Press.

Suslick, K.S. 1990. Sonochemistry. *Science* 247: 1439–1445.

Toma, M., Vinatoru, M., Paniwnyk, L., Mason, T. 2001. Investigation of the effects of ultrasound on vegetal tissues during solvent extraction. *Ultrason Sonochem* 8: 137–142.

Vila, D.H., Mira, F.J.H., Lucena, R.B., Recamales, M.F. 1999. Optimization of an extraction method of aroma compounds in white wine using ultrasound. *Talanta* 50: 413–421.

Vilkhu, K., Mawson, R., Simons, L., Bates, D. 2008. Applications and opportunities for ultrasound assisted extraction in the food industry. A review. *Innov Food Sci Emerg* 9: 161–169.

Vinatoru, M. 2001. An overview of the ultrasonically assisted extraction of bioactive principles from herbs. *Ultrason Sonochem* 8: 303–313.

Vinatoru, M., Toma, M., Radu, O., Filip, P., Lazurca, D., Mason, T.J. 1997. The use of ultrasound for the extraction of bioactive principles from plant materials. *Ultrason Sonochem* 4: 135–139.

Zhang, Z.-S., Wang, L.-J., Li, D., Jiao, S.-S., Chen, X.D., Mao, Z.-H. 2008. Ultrasound-assisted extraction of oil from flaxseed. *Sep Purif Technol* 62: 192–198.

5 Pulsed Electrical Discharges: Principles and Application to Extraction of Biocompounds

*Nadia Boussetta, Thierry Reess,
Eugene Vorobiev, and Jean-Louis Lanoisellé*

CONTENTS

5.1 INTRODUCTION

Treatment of aqueous systems by a strong electric field has been studied for many years because of its importance for many practical applications in biology, chemistry, and electrochemistry. In particular, the method of pulsed electrical discharges

(PED) was developed for enhancing extraction of biocompounds. The general issues and questions regarding the role of electrical discharge processes in extraction of biocompounds include the following:

1. What are the fundamental reaction mechanisms and breakdown pathways of electrical discharge in water? How are electrical discharges initiated and how do they propagate from one electrode to another? What are the main chemical effects of the extreme electromagnetic and mechanical conditions induced by PED in the bulk of solution that influence cell disruption?
2. How can electrical discharges be applied in extraction of biocompounds? Are there any specific types of products more suitable for treatment by electrical discharge processes? Are there any alternative techniques utilizing electrical discharge processes in combination with conventional methods?

Answers to the aforementioned questions are subjects of the current study. The present review seeks for placing these questions and issues within the framework of what is known about electrical discharge processes in extraction of biocompounds.

5.2 PRINCIPLES AND MECHANISMS OF ELECTRICAL DISCHARGES IN WATER

There exists a considerable interest in using water gap discharge in engineering practices. It can be used for several purposes—for instance, water treatment for removal of organic chemical impurities, acoustic sources in medical or sonar applications, selective separation of solids, or plasma blasting. Depending on the discharge conditions, the electrical energy becomes transformed into light, heat, or mechanical energy (shock wave).

Two schools of thought try to explain the processes that lead to establishment of a conductive channel in water. The first hypothesis assumes development of a gaseous phase first, in which electronic avalanches take place. The second hypothesis posits that a gaseous phase is not required. It assumes that breakdown is governed by multiplication of the charge carriers caused by ionization of the liquid. The confrontation between the so-called bubble theory and direct impact ionization model is ongoing.

Whatever model is taken into account, it is well known that there exist two different ways for inducing the dielectric breakdown in a water gap: the pulsed corona electrohydraulic discharges (PCED) and the pulsed arc electrohydraulic discharges (PAED).

The aim of this chapter is to present experimental results on propagation of discharges generated by application of high electric fields (PCED treatment) and high-power switches (PAED treatment) in the water medium. Characteristic shock waves associated with PAED are also discussed.

The mechanisms of electrical discharge development in water and associated dynamic pressure waves were investigated in the Electrical Engineering Laboratory (LGE) of the University of Pau (Touya 2003). They used the experimental set-up for electrical and optical study of discharge propagation in water. The test device and the general measuring equipment of this setup are shown schematically in Figure 5.1. The point–plane electrode system was immersed in a cylindrical vessel filled with tap water at room temperature. A stainless-steel rod with a conical end was used as a point electrode. The tip radius R_t of the conical end could be varied from 100 μm to 15 mm. The electrodes were located in the center of the vessel, and the gap distance D between the point electrode and the plane could be adjusted within 3–20 mm. Water conductivity was 200 μS/cm (tap water).

Biexponential voltage impulses produced by the tank capacitor discharge (up to 40 kV–100 kJ) or generated by a Marx generator (120 kV–5 J) were applied to the point electrode in the water gap.

The voltage and the gap current were monitored using a North Star probe (40 kV–100 MHz) (Campbell, CA, USA) and a Pearson current monitor (50 kA–4 MHz) (Palo Alto, CA, USA), respectively.

An image converter (I.C.) (TSN 506; Thomson, Limoges, France) working in the frame mode recorded development of the discharge in water. The delay time t_{Fi} was defined as the time between the beginning of the voltage pulse and each ith frame. The I.C. was positioned behind an ultraviolet (UV)-transmitting quartz window of the vessel, and a 200 J xenon flash source was located in front of an orthogonal window to illuminate the discharge without saturation of the I.C. We will call it the "indirect light" recording.

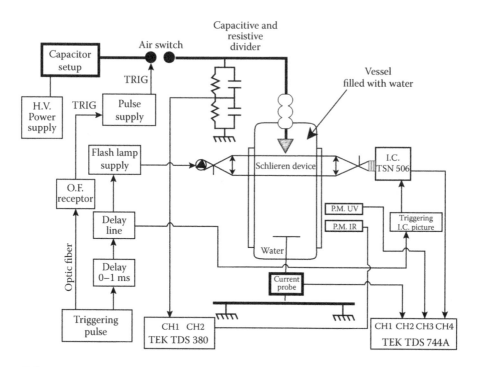

FIGURE 5.1 Experimental setup devoted to physical studies of discharges in water gaps (Electrical Engineering Laboratory of Pau University). H.V., high voltage; O.F., optic fiber.

Two photomultipliers (P.M.) located behind a quartz window were used for monitoring of the whole gap. They recorded UV or infrared (IR) irradiation emitted by the discharge.

A Schlieren device was used for analysis of the water refractive index variations. It gave qualitative results on water heating. The Schlieren device was connected with the I.C. or with a P.M. to improve its sensitivity.

The shock wave generated in water by electrical discharges was detected using a piezoelectric sensor developed by Dr. F. Bauer from the Franco-German Research Institute of Saint-Louis (France). To remove parasitic electromagnetic emissions, a differential measurement of the piezoelectric signal was especially developed for this device. The signal was recorded directly on an oscilloscope without preamplification. The maximum measurable pressure was about 100 bar, and the frequency of the sensor was 100 MHz.

5.2.1 PROPAGATION OF PCED

The PCED are associated on a submicrosecond time scale. This mode of water discharges is also called supersonic in accordance with the propagation velocity of streamers. A streamer corona develops in water when a very high electric field (from 200 kV/cm to more than MV/cm) is suddenly applied to the water gap. An avalanche of electrons seems to grow directly in water under the influence of a high amplitude field, leading to propagation of the water streamers. Even if both electronic and thermal mechanisms have been considered, the mechanism of the streamer formation in liquids is not yet understood (Katsuki et al. 2002). Initiation of streamers by high electric field application and their propagation in water has been recently studied experimentally (Kob et al. 2008). These authors discussed parameters that can be expected to influence PCED development in relation to the analogous models and theories describing the relevant mechanisms in gases.

The switching energies necessary for development of a breakdown in the water gap are weak (a few joules). Consequently, only weak shock waves are observed. However, the PCED generates UV light (Figure 5.2) and radicals, such as O_3 and OH, in narrow regions along the streamer corona discharge channel. Similarly, formation of chemical species and their effects on microorganisms were studied using PCED (Sunka et al. 1999). Various active species produced by PCED play an important role in the killing of bacteria or controlling water quality.

PCED have been studied at LGE (Touya 2003). A typical record of streamer propagation in water is presented in Figure 5.2. PCED consists of streamers developing from the needle electrode toward the grounded plane electrode. The record of UV light presents a single impulse of a few hundreds of nanoseconds (Figure 5.2). These photographs were taken with the I.C.: the point–plan gap was illuminated by the lamp flash; thus, not only light emitted by the discharge was recorded. The electric field near the point was high enough ($\approx$500 kV/cm) to develop such a discharge. In this case, no bubbles were optically visible on the I.C. In our experimental conditions, the streamer development speed was about 33 km/s.

The current data support the assumption that an initial low-density nucleation site or a gas-filled bubble assists initiation of the water streamers (Kolb et al. 2008).

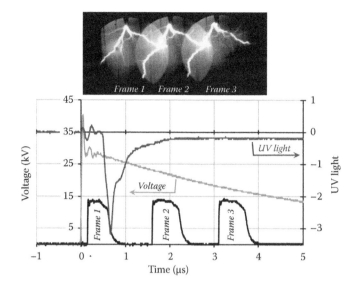

FIGURE 5.2 I.C. records in frame mode, associated voltage, and UV light records (electrode radius = 200 μm, water gap = 28 mm, U = 30 kV).

Our optical resolution does not allow showing of any gas bubble. Moreover, since PCED involves weak switching energy, such water discharges do not generate high amplitude shock wave.

5.2.2 PHYSICAL AND MECHANICAL EFFECTS OF PAED

The second mode of water discharge generation involves PAED. Long-lasting voltage application and, consequently, high energy (several tens of kilojoules) are used for inducing the electrical breakdown of water. This second mode of breakdown in water is also called subsonic discharges. In the first phase, long-lasting high-voltage pulse application between a pair of electrodes induces thermal effects and leads to development of bubbles, where the gas discharges take place. Aka-Ngnui and Beroual (2001) showed that conditions of gas bubble generation and its dimensions depend on the applied voltage. Mathematical models (Saniei et al. 2004) can be used for calculation of the local electric field in the water gap when bubbles develop. Moreover, the PAED are associated with emission of a powerful shock wave that propagates in radial directions in the water. The generated pressure waves have peak values in the range of 1 bar to 10 kbar (Madhavan et al. 2000). The bandwidths of the acoustic waves are wide, with frequencies of up to 10 MHz. Therefore, the PAED system is mainly applied to treatment of sludges, removal of foreign deposits from the pipe walls, or increase of intrinsic permeability of mortar (Maurel et al. 2010). Electrical shock waves can also be generated in water by fragmentation of rocks. Different electrical waveforms were examined to optimize PAED application. The PAED can also be used to generate high-power ultrasound for acoustic source applications (Mackersie et al. 2004).

Development of subsonic discharges in water subjected to the effect of voltage—with a long duration of voltage application—was investigated electrically, optically, and by transient pressure measurement (Touya et al. 2006) at LGE. Such discharges that are called subsonic because their speed of expansion does not exceed a few tens of meters per second develop from gas bubbles, which appear in the neighborhood of the high-voltage electrode (Figure 5.3).

Measurements using a Schlieren device (Touya et al. 2006) clearly point out the first phase, without any light emission, during which modification of the water refractive index around the high-voltage point can be detected. This modification is caused by local water heating. The vapor bubbles appear because of the conversion of electrical energy injected in the water into heat. Under our experimental conditions, when the energy injected into the water reaches 80 J, the heating of water led to development of bubbles, which can be detected by the optical device. The bubble ignition time coincides with the appearance of light emissions due to electrical discharges into the bubbles. The energy of these discharges is sufficiently high for emitting both UV and IR light.

The volume of the bubbles is directly related to the energy injected into the water. Consequently, a competition occurs between the development of bubbles and voltage decrease. If the time of voltage application is long enough to inject at least 200 J into the water, the bubbles can fill the entire interelectrode space and gap break-

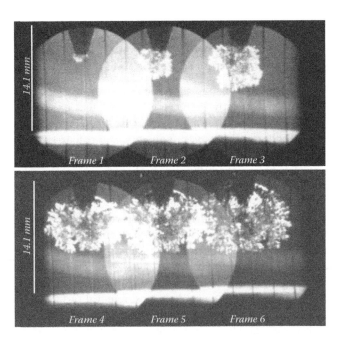

FIGURE 5.3 Typical I.C. records showing the development of bubbles into water (electrode radius = 1 mm, water gap = 10 mm, U_{max} = 19.5 kV; t_1 = 137 μs, t_2 = 246 μs, t_3 = 350 μs, t_4 = 523 μs, t_5 = 630 μs, and t_6 = 746 μs).

down occurs. Under such conditions, a fast and high-amplitude wave of pressure is associated with water gap discharge.

5.2.3 CHARACTERIZATION OF SHOCK WAVES ASSOCIATED WITH PAED

The generation of shock waves requires the occurrence of a dielectric breakdown in water, after which heat and expansion of plasma channel between the two electrodes produce a dynamic wave. Figure 5.4 presents typical records of the pressure waveform and associated frequency spectrum, corresponding to a breakdown in water performed in the previously described experimental conditions.

Under the conditions of our experiment, the dynamic pressure is generated by subsonic water discharges of biexponential form, which are characterized by a time increase of about 500 ns and a bandwidth increase of a few microseconds. The associated frequency spectrum reaches 300 kHz at –20 dB.

The general form of the applied pressure pulse can be described by the following relation (Equation 5.1):

$$P(t) = P_0(e^{-at} - e^{-bt}) \tag{5.1}$$

where P (Pa) is the pressure at time t (s), P_0 (Pa) denotes the peak pressure value, and a (s^{-1}) and b (s^{-1}) are constants under our experimental conditions.

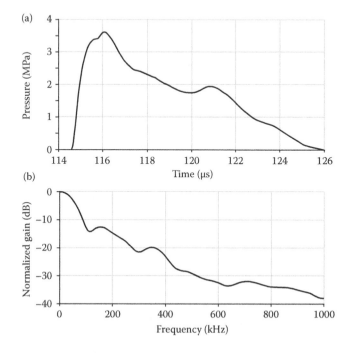

FIGURE 5.4 Example of (a) pressure versus time curve and (b) its associated Fourier transform.

It was pointed out in previous works (Touya et al. 2006) that the peak pressure P_0 (bar) depends on the energy E_B (kJ) remaining at the time of breakdown and can be approximated by Equation 5.2

$$P_0 = k \cdot E_B^\alpha \qquad (5.2)$$

where $E_B = \dfrac{1}{2} \cdot C \cdot U_B^2$ (U_B [V] is the breakdown voltage value, C [F] is the condenser capacity, and k [mm^{-1}] and α are parameters depending on the interelectrode geometry and the distance d between the pressure sensor and the plasma channel).

It is important to distinguish the maximum energy stored in the tank capacitor $E_M = \dfrac{1}{2} \cdot C \cdot U_M^2$ (where U_M [V] is the charging voltage) from the remaining energy E_B [J] at the time of breakdown t_B [s]). During the period of t_B, ionic conduction develops through the water to the ground electrode. Then, microdischarges occur inside the bubbles, thus reducing the energy available for dissipation in the plasma channel at the time of breakdown. Consequently, even for a high E_M value, a high t_B leads to low E_B and faint peak pressure values. For a given U_M value, enhancement of the peak pressure is associated with reduction in the time t_B. One can decrease the time of breakdown delay t_B without any modification of the electrode geometry: the higher is the applied voltage, the shorter is the time of breakdown delay.

Touya et al. (2006) showed that under the experimental conditions they have used ($D = 10$ mm, $R = 15$ mm), $\alpha = 0.35$ and $k = 9000/d$. Thus, it is possible to control the peak pressure value P_0 using the following relation (Equation 5.3):

$$P_0 = \frac{9000}{d} E_B^{0.35} \qquad (5.3)$$

with P_0 expressed in bar, E_B in kJ, and d in mm.

Consequently, the applied peak pressure can be adjusted and controlled by modification of the distance d or by adjusting the value of the energy remaining at time of breakdown.

5.2.4 APPLICATION OF PAED TO EXTRACTION OF VARIOUS BIOCOMPOUNDS

The arc discharge leads to generation of a hot localized plasma that strongly emits high-intensity UV light (Sun et al. 1998), produces shock waves (Touya et al. 2006; Vitkovitsky 1987), and generates hydroxyl radicals during water photodissociation (Joshi et al. 1995). UV light in the range of 200–400 nm is mutagenic to cells (Matsunaga et al. 1991), shock waves are known to cause mechanical rupture of cell membranes (Howard and Sturtevant 1997), and hydroxyl radicals lead to oxidative cell damage. As a result, PAED can find various applications, such as water treatment (Sato et al. 2001), wood treatment (Mikula et al. 1997), and extraction of biocompounds (Barskaya et al. 2000; Vorobiev and Lebovka 2010). The following sections are devoted to extraction of biocompounds from various products.

5.2.5 Extraction of Proteins from Wine Yeast

Wine yeast cells (*Saccharomyces cerevisiae*) contain valuable compounds such as proteins, polysaccharides, and cytoplasmic enzymes (Engler 1985). Disruption of these cells facilitates and accelerates extraction of intracellular biocompounds. Many techniques for cell disruption have been proposed: wet milling, high-pressure homogenization (HPH), sonification, and electrically assisted treatments. Mechanical treatment by HPH is the most suitable technique for large-scale disruption of biomaterial. However, this method requires multiple passes, which are usually accompanied by temperature increase and high cell debris content (Lovitt et al. 2000). To reduce these undesirable effects, combined disruption techniques can be applied, such as ultrasound and mechanical homogenization (Iida et al. 2008), or chemical and mechanical processing (Harrison et al. 1991). Electrical pretreatments can also be combined with the HPH technique (Shynkaryk et al. 2009). Indeed, electrotechnologies can damage cell membranes with a minimum temperature increase (Vorobiev and Lebovka 2008) and can efficiently enhance extraction of intracellular proteins (Loginov et al. 2009).

Shynkaryk et al. (2009) compared the efficiency of two pretreatments (pulsed electric fields [PEF] and PAED) for the enhancement of the release of bioproducts from yeast cells (*S. cerevisiae*, bayanus) by HPH. PAED (40 kV/cm, 10 kA, 1–150 pulses, 0.5 Hz, 25°C) were applied in a batch 1-L treatment chamber between a needle and a plane electrode. PAED were generated by electrical breakdown in the aqueous yeast suspension (1%, w/w). The temperature increase after PAED treatment was less than 2–4°C. PEF was applied between two plane electrodes in a plastic electroporation cuvette connected to a high voltage generator (3–10 kV/cm, 100 A, 200–200,000 pulses, 200 Hz, 25°C). The yeast suspension was pumped through the treatment chamber at a flow rate of 250 ml/min. The temperature increase after PEF treatment was less than 1–2°C. The HPH technique was applied to a yeast suspension at the flow rate of 10 l/h using successive passes (1–20) at fixed homogenization pressure in the range of 30–200 MPa. When all these treatments were applied separately, a high disintegration index (or cell permeability degree) ($Z = 0.8$) was obtained faster with PAED (100 pulses, 3 min) than with PEF (10 kV/cm, 200000 pulses, 100 min). With HPH, the same disintegration was obtained after 5 passes at 100 MPa. The spectroscopic analysis of the treated extracts showed an intermediate concentration of high molecular weight intracellular compounds (proteins and nucleic acids) in PAED-treated samples compared with PEF- and HPH-treated samples. Then combined PAED-HPH technique was applied to the yeast suspension. The results clearly evidence the benefits of PAED for the enhancement of yeast disruption (Figure 5.5).

For example, even PAED pretreatment with low disintegration degree ($Z = 0.15$) resulted in a noticeable acceleration of HPH disruption kinetics and decrease of the required number of passes. This proposed combined PAED–HPH method allows efficient disruption of yeast cells at low pressure and lower number of passes through the homogenizer.

Liu et al. (2010) have also studied the extraction of soluble matter from wine yeast cells (*S. cerevisiae*) with application of PAED and HPH. In this study, the

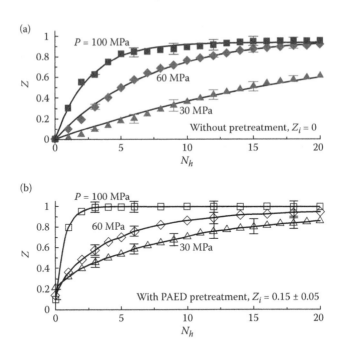

FIGURE 5.5 Conductivity disintegration index Z versus number of HPH passes N_h at different pressures P for yeast suspensions (a) without pretreatment, $Z_i = 0$, and (b) with discharges pretreatment (PAED), $Z_i = 0.15 \pm 0.05$. (From Shynkaryk, M.V. et al., *J Food Eng*, 92, 189–195, 2009. With permission.)

yeast suspension was more diluted (0.5%, w/w). PAED treatment (40 kV/cm, 10 kA, 0.5 Hz, 25°C) involved application of up to 500 pulses, and HPH (10 l/h, 1–20 passes) was performed at a high pressure of 80 MPa. Samples obtained after application of strong PAED (500 pulses, $Z \approx 0.8$) and HPH (20 passes, $Z \approx 1$) treatments were compared. The absorbance analysis of the supernatant showed that HPH allowed approximately 10-fold higher extraction of proteins than PAED. This result was in correlation with the turbidity measurements of the treated samples. Turbidity of both PAED- and HPH-treated suspensions increased with the disintegration index. At disintegration levels higher than 0.4, turbidity of HPH samples (≈ 1000 NTU) was about 10 times higher than that of PAED. These data also showed the release of fine cell debris during treatment. The authors have also studied the disruption effects of these two techniques by measurement of the particle size distribution in yeast suspensions. The maximal particle diameter in PAED-treated suspensions (4.3 ± 0.1 μm) and HPH-treated suspensions (3.5 ± 0.1 μm) were lower than that of untreated samples (4.7 ± 0.1 μm). PAED also resulted in the formation of cell aggregates. These data evidence disaggregation and damage of cells caused by both PAED and HPH techniques. However, differences in the particle diameter values in treated yeast suspensions reflect different mechanisms of disruption of the yeast cells. PAED generates pressure shock waves and air bubble cavitation that contribute to mechanical damage of cells, disintegration of the cell walls, and membrane rupture (Gros et al.

2003; Boussetta et al. 2009a). HPH treatment acts through mechanical breakage of cell walls, disintegration of cell aggregates, and suspension homogenization (Kleinig et al. 1998).

5.2.6 EXTRACTION OF TOTAL SOLUBLE MATTER FROM FENNEL

Fennel (*Foeniculum vulgare*) is known as an aromatic and medicinal herb that contains antioxidants (Gamiz-Gracia and Luque de Castro 2000). The methods commonly used for extraction of these compounds are based on thermal treatment (hydrodistillation and solid–liquid diffusion with hot water) (Miraldi 1999). However, heat can deteriorate thermosensitive compounds (such as antioxidants) and the product characteristics (freshness, vitamin content, color, etc.) (Damjanović et al. 2003). Nonthermal electrical treatments would thus be appropriate for enhancement of extraction at lower diffusion temperatures.

Moubarik et al. (2010) studied the effects of PAED on improvement of the aqueous extraction of solutes from fennel. PAED (40 kV/cm, 10 kA, 20–90 pulses, 0.5 Hz, 20°C) were applied to a suspension containing 400 g of distilled water and 200 g of fennel slices ($1.5 \times 0.75 \times 40$ mm^3; width × height × length). The temperature elevation during PAED treatment of fennel slices was rather small (≈5°C). After pretreatment, the diffusion of solutes (20°C, 80 min) was studied in a cylindrical cell under agitation (250 rpm). The solute extraction kinetics was studied by measurements of the Brix value in the samples taken during extraction. The authors showed that PAED-assisted diffusion resulted in increase in the final yield of solutes and acceleration of the extraction kinetics compared with diffusion without pretreatment (Figure 5.6).

For example, immediately after fennel treatment by 60 pulses PAED, the yield of solutes reached about 90%. The subsequent 80 min of diffusion allowed obtaining of

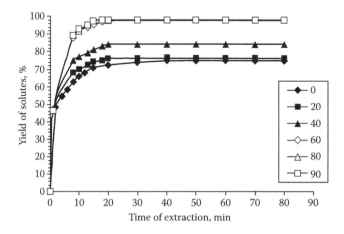

FIGURE 5.6 Yield of solute during extraction from slices treated at 40 kV and different numbers of discharges. (From Moubarik, A. et al., *Food Bioproducts Process*, in press, 2010. With permission.)

the final 99% yield of solutes. When the extraction was performed without pretreatment, the final yield reached only 74%. The results also showed that a number of pulses higher than 60 did not enhance extraction of solutes.

To describe such extraction kinetics, a two-exponential empirical model (So and Macdonald 1986) was applied (Equation 5.1). This model was previously applied for description of the kinetics of electrically assisted extraction from sugar beets and other products (El-Belghiti 2005; El-Belghiti et al. 2005, 2008). The model is based on the assumption that the process of extraction can be divided into two steps: the washing step with rapid solute transfer from the surface of the solid into the bulk of liquid, and the diffusion step with slower, prolonged transfer of solutes from the solid interior to the solid surface.

$$c^* = c_w^*(1 - e^{-k_w t}) + c_d^*(1 - e^{-k_d t}) \tag{5.4}$$

where $c^* = c/c_\infty$, c is the solute concentration in solution at any time during the process of extraction (g/L), c_∞ is the equilibrium solute concentration ($c_\infty = \text{Brix}_{\text{cellular juice}}/(n + 1)$) (g/L), n is the ratio of the extract volume (m^3) to the solid volume (m^3); $c_w^* = c_w/c_\infty$, c_w is the final solute concentration in solution after the washing stage; $c_d^* = c_d/c_\infty$, c_d is the final solute concentration in solution after the diffusion stage; k_w is the rate constant for the washing stage of extraction (min^{-1}); and k_d is the rate constant for the diffusion stage of extraction (min^{-1}).

The coefficients c_d^*, c_w^*, k_d, and k_w were determined for control of the extraction and PAED-assisted diffusion (Moubarik et al. 2010). These coefficients reached maximum values when the extraction was assisted by 60 pulses of PAED ($c_d^* = 0.42 \pm 0.02$, $c_w^* = 0.56 \pm 0.01$, $k_d = 0.280 \pm 0.020$, $k_w = 0.019 \pm 0.001$). Further increase in the number of pulses caused no variations in the measured values: all of them remained constant. The results evidence that the aqueous extraction of solutes was more efficient during the washing step than during the subsequent diffusion stage. Indeed, a higher amount of solutes was extracted during the washing step than during the diffusion stage ($c_w^* > c_d^*$), and the rate of solute extraction during the washing step was also faster ($k_w > k_d$).

5.2.7 Extraction of Glycosides from Stevia Leaves

Stevia rebaudiana leaves are characterized by a high glycoside content (glycosides are compounds used as sweeteners—e.g., steviosides, rebaudiosides, and dulcosides) (Zhang et al. 1999). The extraction of these compounds has gained a growing interest, as far as they are low-calorie products and are up to 300 times sweeter than sucrose at a concentration of 0.4% (Zhang et al. 1999). These sweetening substances could thus replace sucrose in many kinds of food. These compounds are usually extracted with water or alcohols at elevated temperatures.

Negm et al. (2009) studied the extraction of the total soluble matter from dried stevia leaves with the application of PAED. PAED (40 kV/cm, 10 kA, 50–125–200 pulses, 0.5 Hz) were applied to an aqueous suspension of stevia leaves (liquid-to-solid ratio of 16) at 20°C or preheated to 55°C. The PAED-treated suspension was then transferred to a cylindrical cell for the diffusion step (60 min at 20°C or 55°C) under agitation (170 rpm). The glycoside content was estimated by measurements of the Brix value of the suspension. Increasing the number of pulses resulted in improved extraction efficiency of the PAED pretreatment. For example, at room temperature, the yield of solutes increased to 48%, 62%, and 68% immediately after 50, 125, and 200 pulses, respectively. After 60 min of diffusion, equivalent 92% yields of solutes were reached in all the experiments. An increase in extraction temperature to 55°C improved the kinetics, and the final yields of solutes increased to 99%. Compared with a simple diffusion extraction, PAED (200 pulses)-assisted extraction accelerated the initial rate of solute extraction by 7.9 and 1.8 times at 20°C and 55°C, respectively. These data clearly show that PAED pretreatment can lead to higher yields of solutes at a reduced time of diffusion.

5.2.8 EXTRACTION OF OIL FROM LINSEEDS

Linseed contains 35–45% of oil, 10.5–31% of proteins, and 3.5–9.4% of mucilage (heterogeneous polysaccharide) (Oomah and Mazza 1993). Nowadays linseed oil is mainly used in industrial applications (paints, stains, or linoleum), notwithstanding the recent renewal of interest in the nutritional properties of oils and seeds (Oomah 2001). Industrial extraction of linseed oil is based on a technique involving a step of expression followed by a step of solvent extraction. However, because of ecological issues, use of solvents has to be minimized or excluded.

Gros et al. (2003) proposed a solvent-free process for oil extraction from linseeds, consisting of the following stages: crushing, expression, demucilagination (extraction of mucilage) by electrical discharges, centrifugal separation of mucilage and solid residue, enzymatic treatment of this residue, and final separation of oil, water, and solid fractions. As far as mucilage is a natural emulsifier that stabilizes the oil-in-water emulsion, its removal is required for recovery of oil. Extraction of mucilage was assisted by the application of PAED. Electrical discharges (40 kV/cm, 10 kA, 300 pulses, 0.5 Hz) were applied to 50 g of whole linseeds (Barbara variety) immersed in 500 ml of demineralized water (20°C). The residue was separated from solution by centrifugation (17,700 $\times$ g, 20°C, 10 min). Then the residue was treated again by PAED in fresh water under the same conditions. Aqueous extraction of mucilage (34°C, 360 min) under agitation (120 rpm) was also performed at different liquid-to-solid ratios (10:1, 15:1, 20:1). The results showed that three 10-min PAED treatments were effective and sufficient for almost complete extraction of mucilage. Three short (3 $\times$ 10 min) PAED treatments allowed extraction of about twice as much mucilage as in the control experiment with long extraction (liquid-to-solid ratio of 1:10, 360 min).

Li et al. (2009) also studied the PAED-assisted extraction of oil from different products (linseed meal, fruit of palm, and palm kernel). PAED (40 kV/cm, 10 kA, 50–600 pulses, 0.5 Hz, 20°C) were applied to three mixtures (linseed meal [30 g]/water; fruit of palm [20 g]/water; palm kernel [20 g]/water) with liquid-to-solid ratio of 10 and pH of 7. Then mixtures were subjected to centrifugation for separation of the liquid (oil-in-water emulsion) and solid residues. The mass of the extracted oil was studied as a function of the number of pulses. For all studied products,

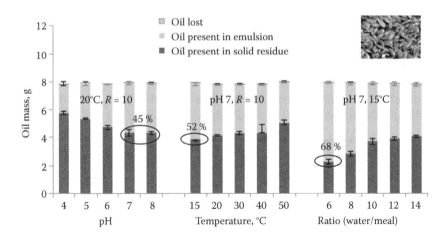

FIGURE 5.7 Optimization of oil extraction from linseed meal by PAED.

the increase in the number of PAED pulses resulted in the increase in the level of extracted oil in the emulsion and a parallel decrease of oil content in the solid residue. However, the oil mass in these two components remained constant when the number of pulses exceeded 500 for linseed meal, 100 for fruit of palm, and 300 for palm kernel. At these numbers of pulses, the oil extraction yields reached 45%, 11%, and 36% from linseed meal, fruit palm, and palm kernel, respectively. The oil initially present in the product was extracted after PAED treatment and broken up into small droplets owing to the cavitational and pressure effects of electrical discharges. Simultaneously, the mucilage present in the product was extracted, and it promoted formation of oil-in-water emulsion with a mean diameter of droplets of about 1 µm. The effects of temperature (15–50°C), pH (4–8), and liquid-to-solid ratio (6–14) (Figure 5.7) were studied for optimization of the operating parameters of PAED in linseed oil extraction.

Treatment in suspension at 15°C, at pH 7, and with a liquid-to-solid ratio of 6 provided the optimal parameters that allowed a higher yield of oil extraction (68%).

5.2.9 ELECTRICALLY ASSISTED AQUEOUS EXTRACTION OF POLYPHENOLS

Grape pomace (composed of stems, seeds, and skins) is the main by-product of wine making. Further processing of this food by-product can be useful in diminishing its environmental impact. Grape pomace is characterized by a high composition of valuable compounds that are not extracted during wine processing (Jackson et al. 1994). In particular, it contains polyphenols that have attracted great interest as far as they possess antibacterial, antiviral, antioxidative, anti-inflammatory, and anti-carcinogenic properties, and can prevent cardiovascular diseases (Dugand 1980). Industrial extraction of polyphenols is based on a batch or continuous solid–liquid diffusion with a solvent (ethanol or sulfuric acid). To intensify the extraction, pre-treatment by PAED can be applied to the grape pomace before the diffusion step. This technique is discussed in more detail below.

5.2.9.1   Extraction of Polyphenols from Grape Pomace

Boussetta et al. (2009b) proposed the use of PAED to accelerate the aqueous extraction of the total soluble matter and polyphenols from grape pomace. The effects of sulfur dioxide and thermal treatments, which are the usual means of grape pomace preservation, were also studied. The PAED treatment chamber was initially filled with grape pomace (100.0 ± 0.1 g), which was further mixed with distilled water (300.0 ± 0.1 g, 20–60°C). PAED treatment (80 kV/cm, 10 kA, 80 pulses, 0.5 Hz) was applied to fresh, sulfured (0.1% of SO_2), and frozen–thawed grape pomace. Then the diffusion process (20–60°C, 60 min) was studied in a diffusion cell under agitation (160 rpm). The results demonstrated the efficiency of the PAED-assisted extraction at 20°C: it allowed a threefold increase in the total soluble matter content and 12 times acceleration of the extraction rate as compared with diffusion without pretreatment (Figure 5.8).

PAED treatment at 20°C improved extraction of solutes: 2.6 times from the sulfured grape pomace and 1.3 times from the frozen–thawed grape pomaces. PAED also increased the yield of polyphenols (0.44 ± 0.07%) after 1 h of extraction compared with that obtained after 4 h of extraction without PAED (0.26 ± 0.06%). Both yield of solutes and yield of polyphenols increased when the extraction temperature was elevated from 20°C to 40°C and then to 60°C. The electrically induced damages were also more pronounced at 40°C and 60°C. Similar extraction kinetics were observed at 40°C (untreated samples) and at 20°C (PAED-treated samples). These results clearly indicate that the diffusion temperature can be reduced if PAED are applied.

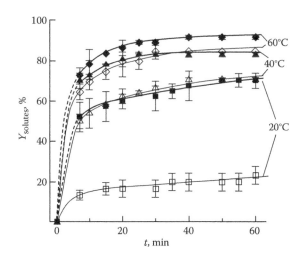

FIGURE 5.8   Effect of extraction temperature on yield of solutes of fresh grape pomace. Dashed lines followed by bold solid lines (solid marks) represent the yield of solutes for grape pomace with PAED extraction (PAED treatment conditions: $d_{electrodes}$ = 5 mm, U = 40 kV, N = 80 pulses, t_i = 10 µs). Nonbold lines (open marks) correspond to experiments without PAED. (From Boussetta, N. et al., *J Food Eng*, 95, 192–198, 2009b. With permission.)

The extraction kinetics of solutes from all experiments were satisfactorily described by Peleg's model (Equation 5.5).

$$Y_{solutes}(t) = Y_0 + t\Big/(K_1 + K_2) \tag{5.5}$$

where $Y_{solutes}(t)$ is the yield of solutes (g/100 g) at the extraction time t (min), Y_0 is the initial yield of solutes at time $t = 0$ (g/100 g) (here $Y_0 = 0$), K_1 is Peleg's rate constant (min·100 g/g), and K_2 is Peleg's capacity constant (g/100 g).

Treatment of samples by PAED, addition of SO_2, temperature elevation, and freezing resulted in increase of initial extraction rate $(1/K_1)$ and equilibrium extraction yield $(1/K_2)$ (Table 5.1).

5.2.9.2 Extraction of Polyphenols from Grape Skins

Boussetta et al. (2009c) compared the effects of electrical discharges and PEF on the acceleration of extraction of polyphenols from grape skins. PEF (1300 V/cm, 38 A, 1000 pulses, 0.5 kHz, 20°C) was applied directly to the grape skins (4.0 ± 0.1 g) without addition of water. The temperature elevation during the treatment was less than 1°C. PAED (40 kV/cm, 10 kA, 60 pulses, 0.5 Hz, 20–60°C) were applied to a mixture of grape skins (40.0 ± 0.1 g) and distilled water (240.0 ± 0.1 g). Diffusion (20–60°C, 180 min) was studied under agitation (160 rpm) for the liquid-to-solid ratio of 6. At 20°C, the yields of polyphenols increased immediately after pulsed electrical treatment (two times after PEF and four times after PAED) compared with the control extraction (simple diffusion). The highest yield of polyphenols (21.4 ± 0.8 μmol gallic acid equivalent [GAE] per gram of dry matter [DM]) was reached after about 60 min of diffusion from PAED-treated samples and after 180 min of diffusion from the PEF-treated skins (Figure 5.9).

TABLE 5.1
Values of the Constants in Peleg's Model

Temperature	Treatment	$1/K_2$ (g/100 g)	$1/K_1$ (g/100 g × min)	R^2
20°C	Fresh	24.4	2.0	0.998
	Fresh + PAED	74.3	23.1	0.997
	SO_2	27.2	2.6	0.987
	SO_2 + PAED	72.7	17.4	1.000
	Frozen	66.5	4.6	0.972
	Frozen + PAED	86.3	35.9	0.991
40°C	Fresh	75.9	18.9	0.998
	Fresh + PAED	87.2	48.4	0.998
60°C	Fresh	90.0	31.3	0.998
	Fresh + PAED	95.9	42.8	1.000

Source: Boussetta, N. et al., *J Food Eng*, 95, 192–198, 2009b. With permission.

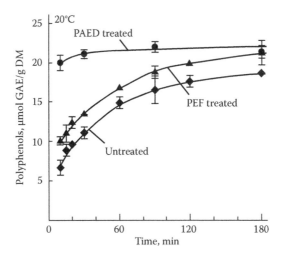

FIGURE 5.9 Total polyphenol content C versus extraction time t for untreated and PEF-treated grape skins at $T = 20°C$. (PEF treatment: $E = 1300$ V/cm, $t_t = 1$ s. PAED treatment: $d_{electrodes} = 10$ mm, $U = 40$ kV, $t_t = 120$ s). (From Boussetta, N. et al., *J Agric Food Chem*, 57, 1491–1497, 2009c. With permission.)

These levels exceeded the maximal yields of polyphenols (19.1 ± 0.5 µmol GAE/g DM) for untreated samples. The high-performance liquid chromatography (HPLC) analysis of treated (by PEF and PAED) and untreated samples after extraction (20°C, 60 min) showed rather similar polyphenolic compositions (Figure 5.10).

Four main polyphenols were identified (catechin, epicatechin, quercetin-3-O-glucoside, and kaempferol-3-O-glucoside). However, PAED-treated samples contained more catechin compared with untreated and PEF-treated samples. Such selectivity may result in PAED-induced supplementary damage of skins. Differences between the extraction kinetics and polyphenolic composition of PEF- and PAED-treated tissues reflected the different modes used for both treatments: PEF was applied to grape skins, while PAED were applied to a suspension of skins turbulized by electrical discharges. Therefore, a rather substantial quantity of polyphenols was extracted from skins directly during PAED. Moreover, PAED treatment leads to tissue fragmentation and the consequent increase of transfer surface area.

Other authors have also studied the effect of PAED on the quality of extracted compounds (Vishkvaztzev et al. 1998). Similar HPLC profiles of soymilk proteins were obtained for PAED-treated and untreated samples. Gros (2005) also compared the HPLC profiles of PAED-treated and untreated extracts. PAED (80 kV/cm) were applied to bovine serum albumin solution. No differences were observed between the HPLC profiles, which indicates the absence of PAED effect on the protein composition.

Extraction kinetics of total solutes was also measured at different extraction temperatures (20°C, 40°C, and 60°C) for estimation of the effective soluble matter diffusion time τ using the diffusion theory formalism (Boussetta et al. 2009c). Grape

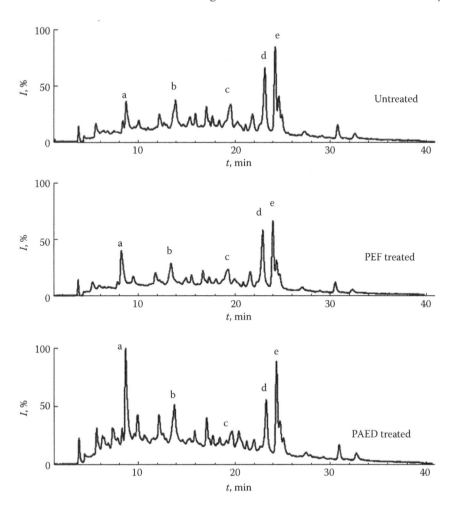

FIGURE 5.10 HPLC profiles from the extracts obtained at 20°C after 60 min of extraction for untreated, PEF-treated, and PAED-treated grape skins. Identified compounds are (a) catechin, (b) epicatechin, (c) quercetin-3-O-glucoside, and (d) kaempferol-3-O-glucoside. (PEF treatment: $E = 1300$ V/cm, $t_t = 1$ s. PAED treatment: $d_{electrodes} = 10$ mm, $U = 40$ kV, $t_t = 120$ s). (From Boussetta, N. et al., *J Agric Food Chem*, 57, 1491–1497, 2009c. With permission.)

skins were assumed to be thin slabs of uniform thickness, and Fick's second law (Equation 5.6) was used for estimation of τ in skins.

$$\frac{\text{Brix} - \text{Brix}_i}{\text{Brix}_f - \text{Brix}_i} = 1 - \frac{8}{\pi^2} \sum_{n=0}^{\infty} \frac{\exp(-(2n+1)^2 t/\tau)}{(2n+1)^2} \tag{5.6}$$

Here $\tau = 4h^2/\pi D_{eff}$, h is the skin thickness (m); D_{eff} is the effective diffusion coefficient (m²/s); and Brix_i and Brix_f are the initial and final values of the soluble matter content (g/100 g), respectively. The value of Brix_i was determined immediately after placing the skins into the diffusion cell, and $Brix_f$ corresponds to the final value after an infinite time of extraction. For large values of time t, this series converged rapidly; thus, only the first five major terms were taken into account for τ estimation. The temperature dependencies of the effective diffusion time τ in the grape skins were satisfactorily described by Arrhenius law, $\tau \propto \exp(W/RT)$, where W is the activation energy (J/mol) and R is the universal gas constant (J/mol/K). Close values of activation energies were obtained for PEF-treated (28.9 ± 5.5 kJ/mol) and untreated samples (31.3 ± 3.7 kJ/mol), which indicates similar mechanisms underlying the thermally activated diffusion of polyphenols.

5.2.10 HYDROETHANOLIC EXTRACTION ASSISTED BY PAED

Boussetta et al. (2010a) studied PAED-assisted extraction in hydroethanolic solvent and considered the potential use of polyphenols as natural antioxidants. Electrical treatment was performed in an aqueous solution, and then diffusion was studied in the presence of ethanol. PAED (40–133 kV/cm, 10 kA, 0.5 Hz, 0–1000 pulses [equivalent to 0–800 kJ/kg of suspension]) were applied to an aqueous grape pomace suspension (liquid-to-solid ratio from 2 to 20, 20–60°C, 0–60 min). The diffusion experiments (0–30% ethanol in water, 20–60°C) were performed under agitation (160 rpm). The optimal conditions of the electrically assisted extraction were proposed: PAED pretreatment at 80 kJ/kg (150 pulses) with 5 mm distance between electrodes (80 kV/cm) in a liquid-to-solid ratio of 5 followed by 30 min diffusion in water with 30% of ethanol at 60°C. Under these conditions, the total polyphenol content reached 2.8 ± 0.4 g GAE/100 g DM with a corresponding antioxidant activity of 66.8 ± 3.1 g Trolox equivalent antioxidant capacity (TEAC) per kilogram of DM. The results clearly indicated that PAED were effective for the enhancement of polyphenol extraction. Addition of ethanol to water after PAED treatment significantly improved extraction of polyphenols (up to three times compared with PAED-assisted aqueous diffusion).

Electrical treatment in the presence of ethanol was also applied for enhancement of the extraction of polyphenols from the grape seeds (Boussetta et al. 2010b). Extraction processes assisted by PAED, PEF, and grinding were compared. In all cases, diffusion experiments (50°C, 60 min) were performed in the presence of ethanol (30%). Ethanol was added to water either after the PEF ($E = 20$ kV/cm, $t_{PEF} = 6$ ms, $T_{PEF} = 50°C$) and PAED ($U = 40$ kV, $t_{PAED} = 1$ ms, $T = 50°C$) treatments, or before the treatment. The results were compared with those obtained for diffusion from untreated seeds and from seeds treated by grinding. In these experiments, the seeds were initially maintained in water or in hydroethanolic solution for the same duration as in experiments with electrical treatment. The results showed that each type of pretreatment of the seeds allowed increasing of the final yield of polyphenols. Grinding appeared to be the most efficient pretreatment method: the yield of polyphenols was 8.6 g/100 g GAE after only 15 min of diffusion. The same result was obtained with PAED after 60 min of diffusion. With PEF treatment, the yield of polyphenols reached 7.4 g/100 g GAE after 60 min of diffusion. When PAED and PEF treatments were applied in the presence of ethanol, it significantly accelerated the kinetics of polyphenol extraction and allowed reaching of the highest yield

of polyphenols (8.6 g/100 g GAE) after shorter diffusion times. The effective diffusion time τ was thus calculated using initial conditions of diffusion (concentration of polyphenols in the extract and seeds) existing after pretreatment of seeds
(Equation 5.7).

$$Y(r,t) = \frac{C - C_e}{C_0 - C_e} = \sum_{n=1}^{\infty} \frac{6\alpha(1+\alpha)}{9 + 9\alpha + q_n^2\alpha^2} \exp\left(\frac{-q_n^2 t}{\tau}\right) \tag{5.7}$$

where C_e is the concentration of polyphenols in the grape seed extract at equilibrium
conditions, C_0 is the initial concentration of polyphenols inside the grape seeds, α is
the ratio of the volume of extracting water to the volume of solution inside the grape
seeds, $\tau = (R^2/D_{eff})$ is the effective diffusion time, R is the particle radius, D_{eff} is
the effective diffusion coefficient, q_n's are nonzero positive roots of the equation tan
$q_n = (3q_n)/(3 + \alpha q_n^2)$. In our case, $\alpha = 7.1$. Only the first term of the series was used and
the effective diffusion time τ was calculated from the slope of ln Y versus time curve.

When the grape seeds were pretreated in water, the effective diffusion time
decreased by a factor of 2 after PEF application, by 4.6 after PAED pretreatment,
and by 8 after grinding, as compared with extraction without pretreatment. When
pretreatment was performed in the presence of ethanol, the effective diffusion time
was reduced to 19 min after PEF treatment and to 15 min after PAED and grinding,
as compared with extraction without pretreatment ($\tau = 72$ min). Thus, any of these
three pretreatments (PEF, PAED, grinding) can decrease the required time of the
diffusion process.

5.2.11 APPLICATION OF PAED AT THE SEMIPILOT SCALE

The application of electrical discharges for the enhancement of extraction of polyphenols from grape pomace has been studied at the semipilot scale (Boussetta et al.
2009d). The amount of product treated at the semipilot scale (1.25 kg/6.25 kg, grape
pomace/water) was 25 times larger than at the laboratory scale (0.05 kg/0.25 kg,
grape pomace/water).

The apparatus used at the semipilot and laboratory scales were of similar shape
but with different size and power (Figure 5.11).

The PAED protocol at the laboratory scale was 80 kV/cm, 10 kA, 0.5 Hz, 0.2 μF,
pulses of 0.16 kJ; the PAED protocol used at the semipilot scale was 80 kV/cm,
10 kA, 0.5 Hz, 5 μF, pulses of 4 kJ. The cylindrical treatment chamber with needle/
plan electrodes was 11 cm × 17 cm (diameter × height) at the laboratory scale and
30 cm × 50 cm (diameter × height) at the semipilot scale. Maintaining a constant
ratio of pulse energy to product mass (0.53 kJ/kg/pulse) was not enough for obtaining similar extraction yields at both scales. Indeed, extraction of polyphenols after
PAED treatment (100 pulses, 20°C) was four times less efficient at the semipilot
scale than at the laboratory scale. This difference may be related to the dimensions
of the treatment cell. PAED induces high-pressure waves and is responsible for the
turbulence and agitation of the liquid inside the treatment cell. However, propagation of the pressure waves and their effects can be diminished in a treatment cell of

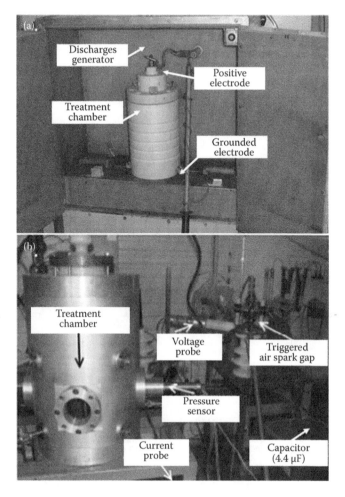

FIGURE 5.11   PAED treatment apparatus used (a) at a laboratory scale and (b) at a semipilot scale.

a large diameter. With the increase in the number of pulses (1–1000) at the semipilot scale, the polyphenol content and antioxidant activity in the extract increased to the maximum values of 350 mg GAE/L and 150 mg TEAC per liter, respectively, at 1000 pulses (Figure 5.12).

These results were similar to those obtained at the laboratory scale after 100 pulses. At the semipilot scale, the minimal number of pulses required for obtaining a significantly higher polyphenol content than in extraction without pretreatment was 300. At these conditions, PAED appeared to be effective for polyphenol extraction.

5.2.12   EFFECT OF THE SECONDARY REACTIONS PRODUCED BY PAED

Electrical discharges generate secondary reactions producing active species such as hydroxyl radicals (during photodissociation of water), atomic oxygen, UV radiation,

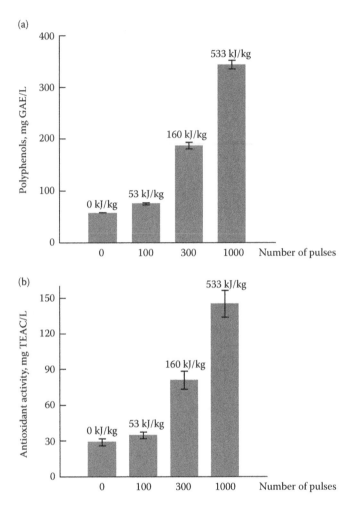

FIGURE 5.12 Effects of the treatment energy on the contents of (a) total polyphenols and (b) antioxidant activity at the semipilot scale.

and ozone (Locke et al. 2005) (Table 5.2). These species can interact with the extracted compounds.

Boussetta et al. (2010a) showed the effect of high-energy treatment on polyphenol degradation. At a rather low energy of treatment (<80 kJ/kg) or small number of pulses (<150 pulses), PAED (80 kV/cm, 100 g of grape pomace per 300 g of water, 20°C) had a positive effect on the extraction of polyphenols and antioxidant activity of the extract. At a higher energy of treatment (>200 kJ/kg) or number of pulses (>350 pulses), both the polyphenol content and antioxidant activity decreased (up to a 30% decrease). These trends were also confirmed by the behavior of individual phenolic compounds (catechin, epicatechin, quercetin-3-*O*-glucoside, and kaempferol-3-*O*-glucoside). It was assumed that the amount of oxidizing species produced under

TABLE 5.2

Typical Reaction Rate Constants for PAED Reactors

Reaction	Reaction Rate Constant ($M^{-1} \times s^{-1}$)
Radical formation	
$H_2O \rightarrow H + OH$	$10^{-8}-10^{-10}$
Radical-molecule reaction	
$OH + organic \rightarrow products$	10^9-10^{10}
Photochemical reactions (tap water)	
$O_2 + organic \rightarrow products$	10^6-10^8 (pH dependent)
Electron molecule	
$e_{aq}^- + H_2O_2 \rightarrow OH + OH^-$	10^{10}
Fenton's reaction	
$Fe^{2+} + H_2O_2 \rightarrow OH + OH^- + Fe^{3+}$	10^2
Ozone molecule	
$O_3 + organic \rightarrow products$	$10^{-2}-10^4$ (pH dependent)
Aqueous electron reactions	
$e_{aq}^- + chloroform \rightarrow products$	10^{10}
$e_{aq}^- + benzene \rightarrow products$	$<10^7$

Source: Locke, B.R. et al., *Ind Eng Chem Res* 45, 882–905, 2005. With permission.

the effect of PAED increased at a high number of discharges, and thus could oxidize polyphenols.

The chemical reactions of oxidation generated by PAED have attracted large interest in relation to the aims of water decontamination (Gao et al. 2003; Chen et al. 2004; He et al. 2005), sterilization (El-Aragi et al. 2008; Chen et al. 2009), and medical treatment (El-Aragi 2009; Kong et al. 2009). Extensive studies of electrical discharges in water applied for destruction of microorganisms were carried out by Russian researchers (Bogomaz et al. 1991). Both the local effects of high-voltage PED in water, arising owing to plasma chemical reactions, and their nonlocal effects exerted through shock wave and UV radiation were taken into account. The authors showed that 1–10 μs pulses with energies within 1–10 kJ created conditions for shock wave formation. The 5 orders of magnitude level of destruction were observed in a suspension of *Escherichia coli* cells (with an initial concentration of 10^9 cm^{-3}) at the 3 J/cm energy input.

During the electrical breakdown, there also arises a high-pressure wave that propagates through the liquid and may affect the product structure. The waveform of this pressure shock wave and the relevant voltage, current, and power (15 kV/cm, 15 kA, 0.5 Hz) in distilled water (20°C) were recorded during the PAED treatment of grape pomace (Boussetta et al. 2009a).

The pressure wave had a biexponential form and was characterized by a rise time of about 2 μs and a bandwidth of 10 μs. The spherical shock wave was created in the interelectrode space and then propagated in radial directions toward the surface of the treatment chamber. The measured pressure peak had a maximum value of 1×10^7 Pa. Coleman et al. (1987) showed that once the shock wave is generated, it is then followed by a rarefaction wave that produces cavitations. The collapsing cavitations create strong secondary shocks with very short duration (about 60 ns that sometimes result in sonoluminescence [excitation of light spikes]). These shocks can interact with the cells and disintegrate them. For example, PAED resulted in complete fragmentation of the grape pomace (Boussetta et al. 2010b).

5.3 FUTURE RESEARCH NEEDS

Despite recent developments in PED applications, several areas need further research to make the technology applicable at the commercial level:

- Confirmation of the mechanisms of discharge establishment in water
- Development and evaluation of the models of extraction kinetics
- Studies aimed at optimization and control of the critical process parameters
- Design uniformity and processing capacity of the treatment chamber
- Identification and application of electrode materials that can provide longer time of operation and lower metal migration
- Development of validation methods ensuring harmlessness of the process
- Process system design (including electric generators), evaluation, and cost reduction.

REFERENCES

Aka-Ngnui, T., Beroual, A. 2001. Bubble dynamics and transition into streamers in liquid dielectrics. *J Phys D: Appl Phys* 34: 1408–1412.

Barskaya, A.V., Kuretz, B.I., Lobanova, G.L. 2000. Extraction of water soluble matters from vegetative raw material by electrical pulsed discharges, in *Proceedings of the 1st Int. Congress on Radiation Physics, High Current Electronics, and Modification of Materials*, G. Mesyats, B. Kovalchuk and G. Remnev (Eds.), pp. 533–535. Tomsk, Russia: Tomsk Polytechnic University.

Bogomaz, A.A., Goryachev, V.L., Remennyi, A.S., Rutberg, F.G. 1991. The effectiveness of a pulsed electrical discharge in decontaminating water. *Sov Technol Phys Lett* 17: 448–450.

Boussetta, N., De Ferron, A., Reess, T., Pecastaing, L., Lanoisellé, J.L., Vorobiev, E. 2009a. Improvement of polyphenols extraction from grape pomace using pulsed arc electrohydraulic discharges, in *Proceedings of IEEE Pulsed Power Conference*, pp. 1085–1090. Washington, United States.

Boussetta, N., Lanoisellé, J.-L., Bedel-Cloutour, C., Vorobiev, E. 2009b. Extraction of soluble matter from grape pomace by high voltage electrical discharges for polyphenol recovery: Effect of sulphur dioxide and thermal treatments. *J Food Eng* 95: 192–198.

Boussetta, N., Lebovka, N., Vorobiev, E., Adenier, H., Bedel-Cloutour, C., Lanoisellé, J.L. 2009c. Electrically assisted extraction of soluble matter from chardonnay grape skins for polyphenol recovery. *J Agric Food Chem* 57: 1491–1497.

Boussetta, N., Reess, T., Pecastaing, L., De Ferron, A., Lanoisellé, J.L., Vorobiev, E. 2009d. Application of high voltage electrical discharges at semi-pilot scale: Extraction of polyphenols from grape pomace. In *Proceedings of the International Conference on Bio and Food Electrotechnologies*, 108–113. Compiègne, France: UTC.

Boussetta, N., Vorobiev, E., Deloison, V., Pochez, F., Falcimaigne-Cordin, A., Lanoisellé, J.L. 2010a. Valorization of grape pomace by the extraction of phenolic antioxidants: Application of high voltage electrical discharges. *Food Chem* 128(2): 364–370.

Boussetta, N., Vorobiev, E., Le, H.L., Cordin-Falcimaigne, A., Lanoisellé, J.L. 2010b. Application of electrical treatments in alcoholic solvent for polyphenols extraction from grape seeds (unpublished data).

Chen, Y.-S., Zhang, X.-S., Dai, Y.-C., Yuan, W.-K. 2004. Pulsed high-voltage discharge plasma for degradation of phenol in aqueous solution. *Sep Purif Tech* 34: 5–12.

Chen, C.-W., Lee, H.-M., Chang, M.-B. 2009. Influence of pH on inactivation of aquatic microorganism with a gas–liquid pulsed electrical discharge. *J Electrost* 67: 703–708.

Coleman, A.J., Saunders, J.E., Crum, L.A., Dyson, M. 1987. Acoustic cavitation generated by an extracorporeal shock wave lithotripter. *Ultrasound Med Biol* 13: 69–75.

Damjanović, B., Skala, D., Patrović-Djakov, D., Baras, J. 2003. A comparison between the oil, hexane extract and supercritical carbon dioxide extract of *Juniperus communis* L. *J Essential Oil Res* 12: 159–162.

Dugand, L.R. 1980. Natural antioxidants, in *Autooxidation in Food and Biological Systems*, M.G. Simic and M. Karel (Eds.), pp. 261–295. New York: Plenum Press.

El-Aragi, G.M. 2009. Pulsed high voltage discharge technology as a new method for killing hepatitis C virus (Hcv) cells. *Environ Eng Manag J* 8: 37–41.

El-Aragi, G.M., Abedel Rahman, Y.M. 2008. Effect of hybrid gas–liquid electrical discharge on liquid foods (milk). *Afr J Food Sci* 2: 011–015.

El-Belghiti, K. 2005. Amélioration de l'extraction aqueuse de solutés des produits végétaux par champs électriques pulsés. PhD dissertation. Compiègne: Université de Technologie de Compiègne.

El-Belghiti, K., Rabhi, Z., Vorobiev, E. 2005. Kinetic model of sugar diffusion from sugar beet tissue treated by pulsed electric field. *J Sci Food Agric* 85: 213–218.

El-Belghiti, K., Moubarik, A., Vorobiev, E. 2008. Aqueous extraction of solutes from fennel (*Foeniculum vulgare*) assisted by pulsed electric field. *J Food Process Eng* 31: 548–563.

Engler, C.R. 1985. Disruption of microbial cells in comprehensive biotechnology, in *Comprehensive Biotechnology*, M. Moo-Young (Ed.), pp. 305–324. Oxford: Pergamon Press.

Gamiz-Gracia, L., Luque de Castro, M.D. 2000. Continuous subcritical water extraction of medicinal plant essential oil: Comparison with conventional techniques. *Talanta* 51: 1179–1185.

Gao, J., Liu, Y., Yang, W., Pu, L., Yu, J., Lu, Q. 2003. Oxidative degradation of phenol in aqueous electrolyte induced by plasma from a direct glow discharge. *Plasma Sources Sci Technol* 12: 533–538.

Gros, C. 2005. Extraction aqueuse et athermique de l'huile de lin assistée par décharges électriques de haute-tension. PhD dissertation. Compiègne: Université de Technologie de Compiègne.

Gros, C., Lanoisellé, J.-L., Vorobiev, E. 2003. Towards an alternative extraction process for linseed oil. *Chem Eng Res Des* 81: 1059–1065.

Harrison, S.T.L., Dennis, J.S., Chase, H.A. 1991. Combined chemical and mechanical processes for the disruption of bacteria. *Bioseparation* 2: 95–105.

He, Z., Liu, J., Cai, W. 2005. The important role of the hydroxy ion in phenol removal using pulsed corona discharge. *J Electrost* 63: 371–386.

Howard, D., Sturtevant, B. 1997. In vitro study of the mechanical effects of shock-wave lithotripsy. *Ultrasound Med Biol* 23: 1107–1114.

Iida, Y., Tuziuti, T., Yasui, K., Kozuka, T., Towata, A. 2008. Protein release from yeast cells as an evaluation method of physical effects in ultrasonic field. *Ultrason Sonochem* 15: 995–1000.

Jackson, R.S. 1994. *Wine Sciences*. New York: Academic Press.

Joshi, A.A., Locke, B.R., Arce, P., Finney, W.C. 1995. Formation of hydroxyl radicals, hydrogen peroxide and aqueous electrons by pulsed streamer corona discharge in aqueous solution. *J Hazard Mater* 41: 3–30.

Katsuki, S., Akiyama, H., Abou-Ghazala, A., Schoenbach, K.H. 2002. Parallel streamers discharges between wire and plane electrodes in water. *IEEE Trans Dielectr Electr Insul* 9: 498–506.

Kleinig, A.R., Middelberg, A. 1998. On the mechanism of microbial cell disruption in high-pressure homogenisation. *Chem Eng Sci* 53: 891–897.

Kob, J.F., Joshi, R.P., Xiao, S., Schoenbach, K.H. 2008. Streamers in water and other dielectric liquids. *J Phys D: Appl Phys* 41: 234–241.

Kong, M.G., Kroesen, G., Morfill, G., Nosenko, T., Shimizu, T., Dijk, J.V., Zimmermann, J.L. 2009. Plasma medicine: An introductory review. *New J Phys* 11: 115–122.

Li, L., Lanoisellé, J.L., Ding, L., Clausse, D. 2009. Aqueous extraction process to recover oil from press-cakes, in *Proceedings of WCCE8*. Montréal, Canada.

Liu, D., Savoire, R., Vorobiev, E., Lanoisellé, J.L. 2010. Effect of disruption methods on the dead-end microfiltration behavior of yeast suspension. *Sep Sci Technol* 45: 1–9.

Locke, B.R., Sato, M., Sunka, P., Hoffmann, M.R., Chang, J.S. 2005. Electrohydraulic discharge and nonthermal plasma for water treatment. *Ind Eng Chem Res* 45: 882–905.

Loginov, M., Lebovka, N., Larue, O. 2009. Effect of high voltage electrical discharges on filtration properties of *Saccharomyces cerevisiae* yeast suspensions. *J Membr Sci* 34: 288–294.

Lovitt, R.W., Jones, M., Collins, S.E., Coss, G.M., Yau, C.P., Attouch, C. 2000. Disruption of baker's yeast using a disrupter of simple and novel geometry. *Process Biochem* 36: 415–421.

Mackersie, J.W., Timoshkin, I.V., Fouracre, R.A., MacGregor, S.J. 2004, in *Proceedings of International Conf. on Gas Discharges and their Applications*, pp. 673–676.

Madhavan, S., Doiphode, P., Kunda, M., Chaturvedi, S. 2000. *IEEE Trans Plasma Science* 28: 1552–1557.

Matsunaga, T., Hieda, K., Nikaido, O. 1991. Wavelength dependent formation of thymine dimers and (6-4) photoproducts in DNA by monochromatic ultraviolet-light ranging from 150 to 365 nm. *Photochem Photobiol* 54: 403–410.

Maurel, O., Reess, T., Matallah, M., De Ferron, A., Chen, W., La Borderie, C., Pijaudier-Cabot, G., Jacques, A., Rey-Bethbeder, F. 2010. Electrohydraulic shock wave generation as a means to increase intrinsic permeability of mortar. *Cement Concrete Res* 40(12): 1631–1637.

Mikula, M., Panak, J., Dvonka, V. 1997. The destruction effect of a pulse discharge in water suspensions. *Plasma Sources Sci Technol* 6: 179–184.

Miraldi, E. 1999. Comparison of the essential oils from ten *Foeniculum vulgare* Miller samples of fruits of different origin. *Flav Fragrance J* 14: 379–382.

Moubarik, A., El-Belghiti, K., Vorobiev, E. 2010. Kinetic model of solute aqueous extraction from fennel (*Foeniculum vulgare*) treated by pulsed electric field, electrical discharges and ultrasonic irradiations. *Food Bioproducts Process* (in press).

Negm, M., Vorobiev, E., Sitohy, M. 2009. Enhancing the aqueous extraction of stevia glycosides from *Stevia rebaudiana* leaves under the action of electric discharge pretreatment, in *Proceedings of BFE*, E. Vorobiev, Lebovka, N., Van Hecke, E., Lanoisellé, J.L. (Eds.). Compiègne: UTC.

Oomah, B.D. 2001. Flaxseed as a functional food source. *J Sci Food Agric* 81: 889–894.

Oomah, B.D., Mazza, G. 1993. Flaxseed proteins—A review. *Food Chem* 48: 109–114.

Saniei, M., Fouracre, R.A., MacGregor, S., Mackersie, J.W. 2004. *Proceedings of International Conference on Gas Discharges and their Applications*, pp. 379–382.

Sato, M., Ishida, N.M., Sugiarto, A.T., Oshima, T., Taniguchi, H. 2001. High efficiency steril-izer by high voltage pulse using concentrated field electrode system. *IEEE Trans Ind Appl* 37: 1646–1651.

Shynkaryk, M.V., Lebovka, N.I., Lanoisellé, J.-L., Nonus, M., Bedel-Clotour, C., Vorobiev, E. 2009. Electrically-assisted extraction of bio-products using high pressure disruption of yeast cells (*Saccharomyces cerevisiae*). *J Food Eng* 92: 189–195.

So, G.C., Macdonald, D.G. 1986. Kinetic of oil extraction from canola (rapeseed). *Can J Chem Eng* 64: 80–86.

Sun, B., Sato, M., Harano, A., Clements, J.S. 1998. Non-uniform pulse discharge-induced radical production in distilled water. *J Electrost* 43: 115–126.

Sunka, P., Babicky, V., Clupek, M., Lukes, P., Simek, M., Schmidt, J., Cernak, M. 1999. Generation of chemical active species by electrical discharges in water. *Plasma Sources Sci Technol* 8: 258–265.

Touya, G. 2003. PhD dissertation. France: University of Pau (http://tel.ccsd.cnrs.fr/tel-00011179).

Touya, G., Reess, T., Pecastaing, L., Gibert, A., Domens, P. 2006. Development of subsonic electrical discharges in water and measurements of the associated pressure waves. *J Phys D: Appl Phys* 39: 5236–5244.

Vishkvaztzev, L.I., Kuretz, B.I., Lobanova, G.L., Filatov, G.P., Barskaya, A.V. 1998. Use of electrical discharges for the treatment of soya beans (in Russian). *Vestnik Rossiyskoy Akademii Selskohozyaistvennih Hauk* 6: 71–72.

Vitkovitsky, I. 1987. *High Power Switching*. New York: Van Nostrand Reinhold Co.

Vorobiev, E., Lebovka, N. 2008. Pulsed electric fields induced effects in plant tissues: Fundamental aspects and perspectives of application, in *Electrotechnologies for Extraction from Food Plants and Biomaterials*, E. Vorobiev and N. Lebovka (Eds.), pp. 39–82. New York: Springer.

Vorobiev, E., Lebovka, N. 2010. Enhanced extraction from solid foods and biosuspensions by pulsed electrical energy. *Food Eng Rev* 2: 95–108.

Zhang, S.Q., Kutowy, O., Kumar, A. 1999. *Stevia rebaudiana* leaves—A low calorie source of sweeteners. *Can Chem News* 51(5): 22–23.

6 Combined Extraction Techniques

Farid Chemat and Giancarlo Cravotto

CONTENTS

6.1 INTRODUCTION

The use of extraction within the food, cosmetics, and pharmaceutical industries is an active subject in research and development. Existing extraction technologies require up to 50% of a company's investment capital in new plants and more than 70% of total energy in food, fine chemical, and pharmaceutical industrial processes. They also present considerable technological and scientific bottlenecks that need to be overcome when attempting to develop new extraction techniques. There is also a need to reduce the energy consumption of extraction processes. This has been caused by increasing energy prices and the drive to reduce CO_2 emissions and eliminate wastewater in order to keep plants in line with environmental restrictions. It is important to meet stringent legal requirements on product safety (hazard analysis and critical control point) and process control (hazard analysis and operability). Another driving force is the need to increase quality as well as functionality to reduce costs and improve competitiveness.

Food and natural products are complex mixtures of vitamins, sugars, proteins, fats and oils, fibers, aromas, pigments, antioxidants, and other organic and mineral compounds. Before such substances can be used as ingredients or additives, they have to be extracted from the food matrix. There is also a constant demand to improve the quality of food ingredients and additives because consumers demand high-quality food, pharmaceutical, or perfumery products. Different methods can be used for this purpose, for example, Soxhlet extraction, maceration, elution, countercurrent solvent extraction, steam and/or hydrodistillation, cold pressing, and simultaneous distillation–extraction. Nevertheless, many food ingredients are well known to be thermally sensitive and vulnerable to chemical changes. Losses of some compounds, low extraction efficiency, and time- and energy-consuming procedures (prolonged heating and stirring in boiling solvents, the use of high solvent volumes, etc.) may be encountered when using these extraction methods. In steam- and hydrodistillation, elevated temperatures over long extraction periods can cause chemical modifications in the oil components and often a loss of the most volatile molecules. When using solvent extraction, it is impossible to obtain solvent-free products, and this process also usually results in the loss of highly volatile components. These shortcomings have led to the use of new sustainable "green" techniques in extraction, which typically involve smaller quantities of solvent and energy, such as ultrasound (US)-assisted extraction (UAE), supercritical fluid extraction (SFE), headspace method, microwave (MW) extraction, controlled pressure drop process, accelerated solvent extraction, and subcritical water extraction (Meireles 2009).

Using these innovative technologies, full reproducible food processes including extraction can now be completed in seconds or minutes with high reproducibility, reducing the processing cost, simplifying manipulation and work-up, giving higher final product purities, eliminating the posttreatment of wastewater, and consuming only a fraction of the time and energy normally needed for conventional processes. The advantages of using these innovative technologies for extraction also include: more effective mixing and micromixing, faster energy and mass transfer, reduced thermal and concentration gradients and extraction temperatures, selective extraction, reduced equipment size, faster response to process extraction control, faster start-up, increased production, and the elimination of process steps (Chemat 2009). Table 6.1 summarizes the most common conventional and innovative extraction processes and presents their advantages and drawbacks.

Extraction under extreme or nonconventional conditions is currently a developing area and a hot topic in applied research and industry. Alternatives to conventional extraction procedures may increase production efficiency and contribute to environmental preservation by reducing the use of solvents, fossil energy, and generation of hazardous substances.

To achieve the development of new extraction technologies or principles, the full potential of conventional and innovative extraction technologies has to be extended by a better use of process conditions, driving forces, and media. The first combinations of conventional innovative technologies could provide solutions such as the combination of MW techniques with the Soxhlet extraction or with the Clevenger distillation. Subsequent combinations of nonconventional processes such as MW and US technologies will permit the development of new operating conditions, driving

TABLE 6.1
Comparison of Innovative Extraction Techniques

Name	UAE	MAE	Pressurized Fluid Extraction (PFE)	Instantaneous Controlled Pressure Drop (DIC)	Supercritical Fluid Extraction (SFE)
Description of system	Sample is immersed in solvent and submitted to US using a US probe or US bath	Sample is placed in vessel with and without solvent and submitted to MW energy	Sample is extracted under high pressure and temperature	Steam is injected into the sample and a sudden drop of pressure is applied toward vacuum	Sample is placed in a high pressure vessel and washed continuously by the supercritical fluid
Investment	Low	Moderate	High	Moderate	High
Ease of operation	Easy to use	Easy to use	Easy to use	Relativity easy to use	Easy to use
Solvent use	50–200 ml per extraction	0–50 ml per extraction	15–60 ml per extraction	50–200 ml per extraction	2–5 ml (solid trap)
Sample size	1–100 g	1–30 g	1–30 g	1–200 g	1–1000 g
Extraction time	10–60 min	3–30 min	10–20 min	1–10 min	10–20 min
Main disadvantages	Large solvent volume; filtration step required	Extraction solvent must absorb MW energy; filtration step required	Possible degradation of thermolabile analytes	Possible degradation of thermolabile analytes	Difficulty of extracting polar molecules without adding modifiers to CO_2
Main advantages	Easy to use; fast extraction; low energy input	Fast extraction; easy to handle; no solvent or moderate consumption	Fast extraction; no filtration necessary; low solvent consumption	Minimal solvent consumption	Fast extraction; no filtration necessary; no solvent consumption

forces, and complex media. Then, combination techniques will permit the development of a kind of extraction technique with high affinity, high process intensification, and allow continuous processes and the possibility to scale up or down.

This chapter presents a summary of all current knowledge on the combination of conventional and innovative techniques (called hybrid technologies) or the combination of nonconventional techniques for the intensification of extractions from food and natural products. It provides the necessary theoretical background and some details (technique, mechanism, applications, and environmental impact) on combined extraction processes that use the most innovative, rapid, and green techniques such as US, MW, the controlled pressure drop process, and enzyme extraction.

6.2 COMBINATION OF CONVENTIONAL AND INNOVATIVE TECHNIQUES: HYBRID TECHNIQUES

6.2.1 ACOUSTICALLY AIDED FILTRATION

The US frequency range can be basically divided into two groups: diagnostic and power US. Diagnostic applications involve high frequency US in the range of 2 to 10 MHz. A typical application is to measure the velocity and absorption coefficient of the acoustic wave in a medium. It is used in medical scanning, chemical analysis, and the study of relaxation phenomena. Power US involves the mechanical and chemical effects of cavitation. The majority of studies use a range of 20–100 kHz. US is used to physically or chemically alter the properties of foods; examples are: pasteurization, sterilization, generation of emulsions, disruption of cells, inhibition of enzymes, tenderization of meat, and modification of crystallization (Mason et al. 2011).

One of the successes of combining US with conventional food processing, and especially separation and extraction, is acoustically aided filtration. The need to remove suspensions of solids from liquids is common in the dairy industry. This separation can be either for the production of solid-free liquid or to produce a solid isolated from its mother liquor. Conventionally, membranes of various sorts are used, ranging from simple semipermeable osmotic-type membranes to size-exclusion principle types for purification. Unfortunately, conventional methodologies often lead to "clogged" filters and, consequently, a need to replace filters on a regular basis.

There are two specific effects of ultrasonic irradiation that can be harnessed to improve the filtration technique. Sonication will cause the agglomeration of fine particles (i.e., more rapid filtration), and at the same time, will supply sufficient vibrational energy to the system to keep the particles partly suspended and therefore leave more free "channels" for solvent elution (Figure 6.1). The combined influence of these effects has been successfully used to enhance the filtration of industrial dairy products, which are particularly time consuming and difficult to process (Muthukumaran et al. 2005). The application of US to filtration, known as "acoustic filtration," suggests that the use of US acts in such a way as to lower the compressibility of both the initial protein deposit and the growing cake. Ultrasonic irradiation significantly enhances the permeate flux with an enhancement factor of 50%. The use of this combination can lead to a doubling in the permeate flux, and the potential for this process is clearly enormous.

Suspension on cross-flow   Suspension on cross-flow

Ultrasound   Ultrasound
Irradiation
Filtrate   Filtrate

FIGURE 6.1   Acoustically aided filtration.

Improvements on the acoustic method use an electrical potential applied across the slurry mixture while acoustic filtration is performed (Muralidhara et al. 1985). The filter itself is made to be the cathode, whereas the anode, on the top of the slurry, functions as a source of attraction for the predominantly negatively charged particulate material. The additional mobility introduced by the electric charge increases the drying efficiency of a 50% coal slurry by a further 10%. When applied to fruit extracts and drinks, this technique has been used to increase the quantity of apple juice extracted from the pulp. Where conventional belt vacuum filtration achieves a reduction in moisture content from an initial value of 85% to 50%, electroacoustic technology achieved the significantly lower rate of 38% (Mason et al. 1996).

6.2.2   MW-ASSISTED SOXHLET EXTRACTION

Fats and oils are traditionally extracted from their matrix using the Soxhlet extraction. Invented in 1879, this apparatus has been widely used in various fields such as environmental applications, foodstuffs, and pharmaceutics. Its principle is relatively easy and proceeds by an iterative percolation of condensed vapors of a boiled solvent, generally n-hexane. Nevertheless, Soxhlet extraction has several disadvantages such as a long operation time (several hours), large solvent volumes, evaporation, and a concentration step needed at the end of extraction. It is also inadequate for thermolabile analytes. There are only few processes in the literature that have reported the combination of Soxhlet extraction with innovative techniques, such as MWs, for the acceleration of fat and oil extraction. The advantages of using MW energy as a noncontact heat source, for the extraction of fats and oils from plant materials include faster energy transfer, reduced thermal gradients, selective heating, and elimination of process steps.

Luque de Castro and García-Ayuso (1998) developed an original focused MW-assisted Soxhlet extraction (FMASE), which uses two energy sources: MWs (applied to the extraction chamber of a modified Soxhlet) and electrical heating (applied to the distillation flask). It was the first device to accelerate the extraction and give the advantages of Soxhlet extraction (repeated leaching with a same solvent in several cycles). However, the performance of the extraction is also dependent on the moisture content of the sample, whereas n-hexane (solvent of choice for lipid extraction) cannot be heated using MWs. FMASE has been used for the determination of the oil content and the fatty acid composition of oleaginous seeds, lipids from sausage products, and fat from cheese and bakery products. This process presents the advantage that the whole extraction process is accelerated.

Recently, another original MW integrated Soxhlet (MIS) has been developed (Virot et al. 2007). Based on a relatively simple principle, this process involves the use of a polytetrafluoroethylene/graphite compound (Weflon, Sorisole, Bergamo, Italy) that allows the diffusion of heat created by the MWs to the surroundings and is particularly useful in the case of solvents that are transparent to MW irradiation, such as hexane (Figure 6.2). MIS extraction combines the advantages of the extraction performed with Soxhlet (extraction repeated by a fresh solvent) and heating by MWs (reduction of extraction time). The process ensures the complete, rapid, and accurate extraction of the samples. No extra heat generator is required. Both dry and wet material may be extracted. Moreover, the apparatus concentrates the final extract and thus eliminates the evaporation step performed under reduced pressure. This system has also been used for the extraction of the oil content and the fatty acid composition of oleaginous seeds, lipids from sausage products, and fat from cheese and bakery products.

Nowadays, the choice of the technique used to perform extractions has to be a result of a compromise between the efficiency and reproducibility of extraction, ease of procedure, together with considerations of cost, time, safety, and degree of automation (Luque de Castro and Priego-Capote 2010). In other words, the green aspect of a new procedure has to be pointed out in order to be recognized. Concerns over global warming are important and a challenge that we cannot disregard. Experiments performed using Soxhlet and MWs together gave savings in the quantity of solvent used (potential recovery of 90%), energy consumed (less than 20%), and CO_2 emitted (less than 20%). Another application developed with this hybrid technique uses green solvents to extract fats and oils. Indeed, despite various attempts to develop

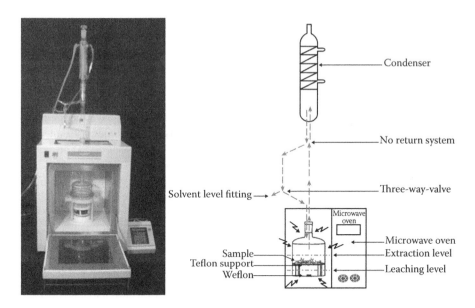

FIGURE 6.2   MWs combination with Soxhlet extraction.

new extraction processes, *n*-hexane is still the solvent of choice for lipid extraction despite concern over health and environment issues. The idea we developed was to use D-limonene, a by-product of the citrus industry, as the solvent and to observe its ability to extract fats and oils. This hybrid technique offers an alternative for fat and oil extraction that is effective, complete, and, most importantly, green.

6.2.3   MW-ASSISTED CLEVENGER DISTILLATION

The traditional method used to isolate volatile compounds as essential oils from plant material (herbs, spices, barks, fruits, etc.) is alembic distillation, which, in chemistry laboratories, is also called Clevenger distillation. This method proceeds by the iterative distillation and boiling of the aromatic matrix in recondensed water vapor, and generally uses large quantities of water and energy. The extraction time can vary from 6 to 24 h. During distillation, fragrant plants exposed to boiling water or steam release their essential oils through evaporation. Recovery of the essential oil is facilitated by distillation of two immiscible liquids: water and the essential oil. This is based on the principle that, at the boiling temperature, the combined vapor pressures equal the ambient pressure. Thus, the essential oil ingredients, for which boiling points normally range from 200 to 300°C, are evaporated at a temperature close to that of water. As steam and essential oil vapors are condensed, both are collected and separated in a vessel traditionally called the "Florentine flask." The essential oil, being lighter than water, floats at the top, whereas water goes to the bottom and is separated.

Historically, there have been three types of distillation: water distillation, water–steam distillation, and steam distillation. In addition, there are numerous other improved methods of producing natural fragrances and essential oils including turbo distillation, hydrodiffusion, vacuum distillation, continuous distillation, and dry distillation. All these conventional extraction techniques have important drawbacks, such as low yields, the formation of by-products, and limited stability. The elevated temperatures and prolonged extraction time can cause chemical modifications in the essential oil components and often a loss of the most volatile molecules.

With growing a flavor and fragrance industry and the increasing demand for more natural products, the need for novel extraction methods has become more intense. The combination of MW energy with Clevenger or alembic distillation has attracted growing interest in the past few years. This has resulted in the development of several techniques such as MW-assisted solvent extraction, vacuum MW hydrodistillation, MW hydrodistillation, MW turbo-hydro-distillation, compressed air MW distillation, MW headspace, MW steam distillation, MW steam diffusion, solvent-free MW hydrodistillation (SFME), and MW hydrodiffusion and gravity (MHG) (Chemat and Lucchesi 2006).

SFME is a new extraction method, patented in 2004, specifically aimed at obtaining essential oils from plant materials (Chemat et al. 2004). The SFME apparatus is an original combination of MW heating and Clevenger distillation at atmospheric or reduced pressure (Figure 6.3). Based on a relatively simple principle, the internal heating of the in situ water within the plant material distends the plant cells and leads to the rupture of the glands and oleiferous receptacles. SFME provides yields comparable with those obtained by traditional hydrodistillation but with extraction times only one-tenth of those required with hydrodistillation. The thermally sensitive

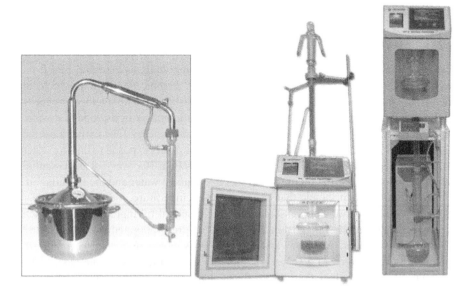

FIGURE 6.3 Conventional Clevenger, combined MW Clevenger, and an "upside-down" MW alembic for extraction of essential oils.

crude materials seem to be preserved with this method, in contrast to conventional Clevenger distillation.

MHG is a new, green technique for the extraction of essential oils (Chemat et al. 2008). This green extraction technique is an original "upside-down" MW alembic method combining MW heating, alembic distillation, and earth's gravity. This physical phenomenon, known as hydrodiffusion, permits the extraction of, in our case, in situ water and essential oil, which had diffused outside the plant material. This then drops, pulled by earth's gravity, out of the MW reactor and falls through the perforated Pyrex disk. A cooling system outside the MW oven continuously cools the extract. Water and the essential oil are collected and separated in a vessel traditionally called the Florentine flask. The essential oil, being lighter than water, floats at the top, whereas water goes to the bottom and can be easily separated. It is important to note that this green method allows essential oils to be extracted without evaporation.

6.2.4 MW- OR US-ASSISTED ENZYME EXTRACTION

US-assisted enzymatic extraction (UAEE) and MW-assisted enzymatic extraction (MAEE) are two efficient, green techniques that generally gave high yields with low energy consumption (Ishimori et al. 1981; Barton et al. 1996; Entezari et al. 2004). These combinations are of paramount importance to select the best operating conditions for scalable, cost-effective extractions of plant material to achieve excellent yields and for minimizing degradation. It is well known that US and MWs can either activate or denature enzymes (Liu et al. 2008). It has been documented that trypsin shows decreased activity with increasing US power from 100 to 500W at 20 kHz (Tian et al. 2004). Similarly, a significant decrease in proteolytic activity is achieved from US treatment at 26.4 kHz and 26 W/cm^2 (Ovsianko et al. 2005).

6.2.4.1 US-Assisted Enzymatic Extraction

The increase in enzyme activity under mild ultrasonic irradiation when the shear force, temperature, pressure, and production of radicals are limited by controlling power and irradiating time, is well described in the literature (Sinisterra 1992; Sakakibara et al. 1996; Yachmenev et al. 2002, 2004, 2007). Matsuura et al. (1994) showed an increase in the fermentation rate of sake, beer, and wine when a relatively low intensity US was applied during fermentation. The proposed mechanism is that the US (a great degassing tool) drives off CO_2 (produced during the fermentation), which normally inhibits the fermentation.

Lieu and Le (2010) determined the optimal conditions of the US-assisted process and combined the US and enzyme process (pectinolytic enzymes) for grape mash treatment in grape juice processing by using response surface methodology. They also compared the efficiency of these treatment methods with that of the traditional enzymatic method. US increased the efficiency of enzymatic treatment with higher extraction yield and shorter treatment time. Grape mash was simultaneously treated by US and enzyme in ultrasonic bath or sequentially with enzymatic treatment after sonication.

As documented below, UAEE and MAEE found a number of successful applications in analytical chemistry, in particular for determination of metals. Because of its great potential as sample treatment for analytical chemistry, the ultrasonic-assisted enzymatic digestion (UAED) for total elemental determination and elemental speciation was described under the most recent achievements published in literature (Vale et al. 2008). Type of enzymes and of ultrasonic system used for the acceleration of the solid–liquid extraction process had to be tailored case by case.

The amount of selenium in crops is an important task for nutritional studies. Selenium enters the food chain through the plants able to transform inorganic Se into a variety of organoselenium species, including bioactive compounds with important implications for human nutrition and health. An efficient method for extraction and analysis of selenium in wheat grain was described in the study of Cubadda et al. (2010), where UAEE was carried out at ambient temperature in a chamber purged with argon in order to prevent species oxidation. This procedure allowed hydrolysis to be completed in 5 min instead of up to 24 h under conventional approaches, with a higher Se recovery. UAED was used for the determination of total Se content in selenium-enriched food supplements. The values obtained were in agreement with those achieved after MW pressurized acid digestion (Vale et al. 2010).

US probe sonication in combination with a mixed enzymic treatment in case of rice and straw samples were applied as sample preparation before arsenic speciation analysis by high-pressure liquid chromatography coupled to inductively coupled plasma mass spectrometric detection (HPLC-ICP-MS) (Sanz et al. 2007).

Li et al. (2009) applied UAEE on the extraction of type I collagen from bovine tendon in the presence of pepsin. The combined US–pepsin treatment achieves higher yield of collagen and shorter processing time than the conventional pepsin isolation method. The activity of pepsin increases distinctly upon ultrasonic irradiation with the experimental ultrasonic intensity. The ultrasonic system alone can have some collagen dissolved although without pepsin. Atomic force microscopy, circular dichroism, and Fourier transform infrared measurements reveal that the fibrillar structure and triple helix conformation of the ultrasound-enzyme soluble collagen are not destroyed by such sonication. The enhanced efficiency of enzymatic extraction of collagen in the presence of US is attributed to the effective agitation caused by ultrasonic cavitation. This improves the dispersal of bulky enzyme aggregates and the open-up of collagen fibrils, thus facilitating the transport of pepsin molecules to collagen substrate surface and subsequent hydrolysis.

UAEE was applied to the extraction of phenolic compounds from the tubers of *Gastrodia elata* Blume (Orchidaceae from China, North and South Korea, and Japan). The authors showed that the yield was increased with the advantage that the typical uncomfortable smell was reduced (Kim et al. 2005).

Sharma and Gupta (2006) described the advantages of ultrasonic preirradiation for extraction of oil by aqueous enzymatic extraction from almond and apricot seeds. A synergic effect was observed by this combination, and the yield was increased from 75% (enzymatic extraction with a mixture of three proteases) up to 95%. Only 2 min of presonication at 70 W reduced the extraction time from 18 to 6 h. Analogously, the sonication as a pretreatment before aqueous enzymatic oil extraction was found to be useful in the case of extraction of oil from the seeds of *Jatropha curcas* L. Ultrasonication lasting 5 to 10 min increased the yield up to 74%, reducing the process time from 18 to 6 h (Shah et al. 2005).

6.2.4.2 MW-Assisted Enzymatic Extraction

Yang et al. (2010) described the feasibility of MAEE for phenolics (corilagin and geraniin) from *Geranium sibiricum* without using any organic solvent. The important parameters involved in the MAEE process were optimized by single-factor experiments, and then the critical parameters were investigated by experimental plan design, to obtain maximum extraction yields. Mass transfer characteristics of MAEE were described and optimal irradiation time of MAEE was obtained through a pseudo first-order equation. These experiments permitted researchers to evaluate whether the cellulase and MAEE could work together to improve extraction yields and total phenolic content and whether the enzymes remain active during the irradiation. The authors used an MW extractor, manufactured by Shanghai Sineo Microwave Products Company (Shanghai, China). One-gram portions of material with cellulase (a few milligrams) was weighed into individual digestion vessels with about 40 ml of acidic water and irradiated in closed vessel under stirring for about 10 min (30–39°C). Results showed that MAEE is a good alternative method for the extraction of natural products from plants because of its higher efficiency and because it is environmentally friendly.

Several important applications of this hybrid technique come from analytical chemistry. A method based on MW-assisted enzymatic digestion and liquid chromatography-tandem mass spectrometry analysis was described for the identification of proteins incorporated within solid matrices. Besides the good recovery rate, a significant decrease in overall analysis time was observed (Stevens et al. 2010).

An efficient extraction of β-glucan from the water-insoluble residue of *Hericium erinaceum* was achieved with proteolytic and chitin degrading enzyme treatments followed by MW irradiation (Ookushi et al. 2008). Two sets of four commercially available enzymes were used in this study (Actinase E, Proteinase K, Chitinase-RS, and Chitinase).

An MAEE method was developed for the simultaneous extraction of arsenic and selenium species in rice products (Reyes et al. 2009). The total arsenic and selenium content in the enzymatic extracts were determined by ICP-MS, whereas speciation analysis was performed by ion chromatography coupled ICP-MS (Guzman et al. 2009). The same MAEE method was also described for the simultaneous extraction of arsenic and selenium species in fish tissues (Reyes et al. 2009).

6.3 COMBINATION OF INNOVATIVE TECHNIQUES

6.3.1 COMBINATION OF MW AND US

The combination of UAE and MW-assisted extraction (MAE) by means of simultaneous irradiation (UMAE) is one of the most promising hybrid techniques for fast, efficient extractions. Since the pioneering work of Chemat et al. (Lagha et al. 1999), several reports have described its success in analytical chemistry, where sample preparation can take much more time than the analysis itself. Typically, UMAE extraction for laboratory-scale applications, can be achieved by inserting a nonmetallic horn in a professional MW oven (Figure 6.4) (Cravotto and Cintas 2007a).

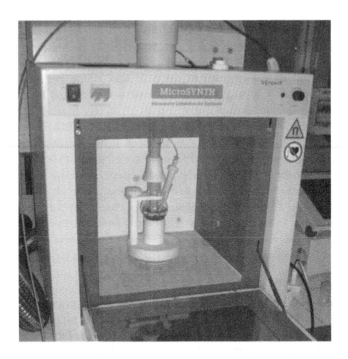

FIGURE 6.4 Simultaneous US–MW-assisted extraction of plant material.

If, on one hand, double simultaneous irradiation can bring additive or even synergic effects to the extraction phenomenon of vegetal matrices, on the other, nonmetallic horns can only be used at moderate power. As described by Cravotto and Cintas (2007b), Pyrex®- quartz- or Peek®-made horns can be safely used up to 90 W; above this point, the intrinsic structure of the material can be irreversibly damaged. This is, however, a minor limitation because UMAE requires lower power levels than the two single energy sources alone. US can dramatically improve the extraction of a target component mainly through the phenomenon of cavitation. The mechanical ultrasonic effect promotes the release of soluble compounds from the plant body by disrupting cell walls, enhancing mass transfer, and facilitating solvent access to cell content. Meanwhile, MW heats the whole sample very quickly, inducing the migration of dissolved molecules. The simultaneous irradiation increases solvent penetration into the matrix, facilitates analyte solvation, and usually increases the solubility of target compounds.

UMAE has been successfully used by Cravotto et al. (2008) as a complementary technique in the extraction of oils from vegetable sources, viz., soybean germ and a cultivated seaweed rich in docosahexaenoic acid. This efficient apparatus consisted of a probe (Danacamerini, Turin, Italy) equipped with an electronic device for frequency/power optimization (Cravotto et al. 2005) and a horn made of PEEK® (polyether ether ketone), inserted into a professional oven (MicroSYNTH; Milestone, Bergamo, Italy). After a comparison of classic and nonconventional techniques, using different types of reactors and setup, it emerged that each technique requires several sets of trials to determine the optimal conditions. Comparing UAE and MAE, it was found that relative yields and extract properties (i.e., antioxidant activity) can often show substantial differences according to which technique was used to extract them. However, UMAE always gave, if not the best results, those surely close to the technique of choice.

Lianfu and Zelong (2008) extracted lycopene from tomato paste applying a US transducer to the bottom of an MW cavity. The combined irradiation, compared with simple UAE, gave a better yield in a shorter extraction time. The UMAE method was applied by Lou et al. (2009, 2010) to extract inulin and polyphenols from burdock roots and leaves (*Arctium lappa* L.), which is a popular vegetable in China and Japan. Apart from a much shorter extraction time, the authors claimed that this extraction technique improved the swelling properties of the extract. The investigation was carried out in a combined extracting apparatus (CW-2000, Shanghai Xintuo MW Instrument Co. Ltd., China), where an ultrasonic transducer is placed on the bottom of the MW cavity. SEM images showed the dramatic destruction of sample microstructure. The results showed that UMAE can be considered the method of choice for the extraction of phenolic components from plant material.

Although performed in separate steps, Hu et al. (2008) also exploited this technique combination to extract and dry isoflavonoids from *Pueraria lobata*. Sonication promoted cell disruption and extraction, then fast MW vacuum drying gave the final product. The authors showed that this sequential treatment increases extraction efficiency while preserving the pharmacological properties of the active compounds.

The extraction of polysaccharides from plants and fungi is an important task because of their many industrial applications. Several authors have described innovative extraction methods in water, assisted by US or MWs. Very recently, Huang and Ning (2010) applied UMAE to extract polysaccharides from a well-known oriental fungus, *Ganoderma lucidum*, which is used in traditional Chinese medicine to promote health and longevity. Using UMAE, the yield of polysaccharides was 115% above that of classical hot water extraction (HWE) and increased by 27.7% over UAE, confirming the great potential of UMAE technology.

Besides all these examples of UMAE, we can also mention examples of UAE plus MAE, namely, two separate nonconventional extraction steps. The best results obtained in the extraction of biologically active compounds (sesquiterpene lactones) from *Achillea millefolium* P. were achieved after 20 min of US treatment and 20 s of MW irradiation (Trendafilova and Todorova 2008).

Although laboratory-scale simultaneous UMAE irradiation for plant extraction seems to be very advantageous, its scaling up is not a trivial task. As already demonstrated for UAE (Cravotto and Binello 2010) (Figure 6.5), the combined US–MW flow reactors that have already been exploited for water decontamination (Cravotto et al. 2007) could be the most promising tools for the scale up of such applications.

Although in recent years several applications of combined US/MW irradiation for plant extraction have appeared in the literature, the great potential of this hybrid technique has not yet been adequately exploited. Because of the high efficiency and the dramatically short extraction time, we believe that UMAE has a great potential for academic and industrial research activity. It is a cost-effective extraction technique for fast sample preparation and a new strategy for process intensification.

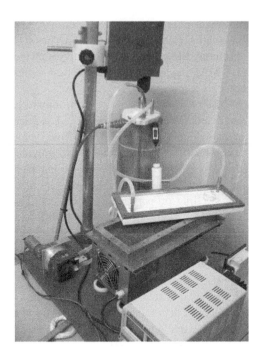

FIGURE 6.5   US-assisted extraction in flow reactor.

6.3.2   COMBINATION OF MW AND DIC PROCESS

The Instantaneous Controlled Pressure Drop process, abbreviated DIC according to the French expression "Détente Instantanée Contrôlée," was developed by Allaf et al. (1993). DIC extraction is based on fundamental studies concerning the thermodynamics of instantaneity. It involves a thermomechanical processing induced by subjecting the product to a fast transition from high steam pressure to vacuum. DIC extraction usually starts by creating a vacuum condition, followed by injecting steam into the material for several seconds, then proceeding to a sudden pressure drop toward vacuum (about 5 kPa with a rate higher than 0.5 MPa/s). By suddenly reducing the pressure, rapid autovaporization of the moisture inside the material will occur. It will swell and lead to texture change that results in higher porosity as well as increased specific surface area and reduced diffusion resistance.

The advantages in heat transfer of applying MW power, which is a noncontact energy source, into the bulk of a material include: faster energy absorption, reduced thermal gradients, selective heating, and a virtually unlimited final temperature. For food production, the resultant advantages could include the following: more effective heating, fast heating of packaged food, reduced equipment size, faster response to process heating control, faster start-up, increased production, and the elimination of process steps.

The combination of these two innovative techniques, electromagnetic (MW) and mechanical (DIC), and their application to physical processes such as extraction and drying appears to be most interesting (Figure 6.6). However, it is not known how

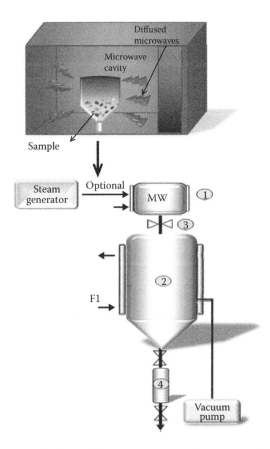

FIGURE 6.6 Combined MWs and instantaneous controlled pressure drop process: (1) MW autoclave; (2) vacuum tank with cooling water jacket; (3) controlled instant pressure-drop valve; (4) extract container; (F1) cooling water flow.

simultaneous MW–DIC action could bring about a physical effect. Therefore, it is expected that the kinetics of the extraction, or drying, processes will improve, but also it is thought that a new effect may occur. For instance, the high pressure level induced by DIC and the difference between internal and external pressure (controlled by Darcy law) could induce particle fragmentation and exudation, and MW polarization could induce dielectric volumetric heating and the selective heating of solid particles and also induce a thermal gradient from the inside to the outside of the treated fruit or matrix. A combination of DIC (pressure phenomena) and MW (dielectric heating) has been successfully applied to drying and extraction with yield enhancements and a reduction in treatment time (Al Haddad et al. 2008).

6.3.3 COMBINATION OF US AND SFE

SFE is a relatively recent extraction technique based on the enhanced solvent power of fluids above their critical point. Its usefulness in extractions is attributed to the combination of gas-like mass transfer properties and liquid-like solvating character-istics with diffusion coefficients that are higher than those of a liquid. The majority of SFE studies have focused on the use of CO_2 because it is nontoxic, nonflammable, cheap, easily eliminated after extraction, and endowed with a high solvating capac-ity for nonpolar molecules. Other possible solvents are Freon, ammonia, and some organic solvents. SFE extraction using carbon dioxide is a suitable method for appli-cation in on-line techniques because the extraction medium is a gas at ambient con-ditions. The major advantages of SFE include pre-concentration effects, cleanness and safety, higher yields, expeditiousness, and simplicity.

On the other hand, US is based on cavitation or collapse phenomena. The tem-perature and the pressure, at the moment of collapse, have been estimated to be up to 4000 K and 1000 atm in an ultrasonic reactor. This creates hotspots that are able to dramatically accelerate chemical reactivity in the medium. When these bubbles

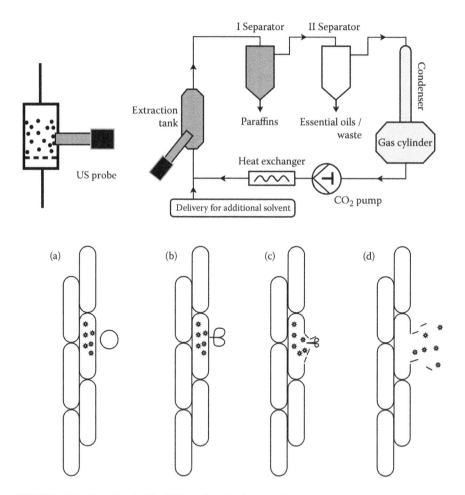

FIGURE 6.7 Combined US–SFE and cavitation bubble collapse near plant material in supercritical CO_2.

collapse onto the surface of a solid material, the high pressure and temperature released generate microjets directed toward the solid surface. These microjets are responsible for the degreasing effect of US on metallic surfaces and are widely used for cleaning materials. Another application of microjets in the food industry is the extraction of vegetal compounds. As shown in Figure 6.7, a cavitation bubble can be generated close to the plant material surface (a), then, during a compression cycle, this bubble collapses (b) and a microjet directed toward the plant matrix is created (b and c). The high pressure and temperature involved in this process will destroy the cell walls of the plant matrix and its content can be released into the medium (d). This is a very interesting tool for the extraction of ingredients from natural products.

Compared with US, SFE processes are more precise in many cases. The use of CO_2 as a supercritical fluid extractor limits the polluting hazards. The drawbacks of SFE versus US are the need for more expensive equipment and the difficulty of extracting polar molecules without adding modifiers to CO_2. Indeed, US permits the extraction of a wide variety of compounds using polar or nonpolar solvents and much simpler equipment. When combined with SFE, US enhances the mass transfer of the species of interest from the solid phase to the solvent used for extraction (Balachandran et al. 2006).

6.4 COMBINED EXTRACTION TECHNIQUES: SAFETY, ENERGY, AND ENVIRONMENTAL IMPACT

Innovative extraction technologies (such as MW, US, DIC, SFE) are simple and can be readily understood in terms of the operating steps to be performed. However, the application of combination techniques can pose serious hazards in inexperienced hands. A high level of safety and attention to detail must be used by individuals using combined or hybrid techniques such as MW/US or DIC/MW when planning and performing experiments. They have to ensure that they seek proper information from knowledgeable sources and that they do not attempt to implement these types of energy and techniques unless proper guidance is provided. Many of these equipments are not commercially available, being laboratory homemade apparatuses. For this reason, only approved equipment and scientifically sound procedures should be used.

Combined extraction techniques are proposed to be "green" extraction processes suitable for high value ingredients contained in food and natural products. The reduced cost of extraction is clearly advantageous for the combined methods in terms of energy, solvent used, and time. Conventional procedures, such as Soxhlet extraction or Clevenger distillation, required an extraction time of 8 h. Hybrid techniques (US Soxhlet or MW Clevenger) or combined methods (MW/US or DIC/MW) generally only required heating and operating for a few minutes. The energy required is respectively 8 kW h for conventional Soxhlet or Clevenger (electrical energy for heating and evaporating) and 0.5 kW h for combined MW with Soxhlet or Clevenger (electrical energy for MW supply). As far as the environmental impact is concerned, the calculated quantity of carbon dioxide emitted into the atmosphere is higher in the case of conventional extraction (600 g CO_2/g product) than for MW extraction (40 g CO_2/g of product). These calculations have been made according to literature sources; to obtain 1 kW h from coal or fuel, 800 g of CO_2 will be emitted to the atmosphere during the combustion of fossil fuel. Further experiments were undertaken with the aim of measuring the ability of each technique to allow solvent recovery. Combined MW technologies allowed the recovery of almost 90% of the solvent used, whereas more than 50% of the solvent was lost during conventional extractions, namely, a Soxhlet extraction followed by vacuum rotary evaporator use. It is estimated that globally, 100,000,000 l of solvent are used per year by analytical laboratories and academia. Combined extraction techniques appear to be green processes that save energy and limit solvent loss, which, at the moment, is a key challenge for the planet.

6.5   FUTURE TRENDS

In this chapter, we have discussed how the concept of combined extraction techniques has already become an important issue in the chemistry of natural products. Detailed analysis of past and present literature explicitly confirms the usefulness of these "original" extraction methods. The understanding, on the molecular scale, of processes relevant to combined extraction techniques has not yet reached the degree of maturity that other topics in chemistry have. Such a challenge is somewhat ambitious and requires a special approach. We hope that this chapter will widen the scope of laboratory and commercial success for the potential applications of combined extraction technologies in food and natural product extraction.

REFERENCES

Allaf, K., Louka, N., Bouvier, J.M., Parent, F., Forget, M. 1993. Drying of products by controlled pressure reduction by heating and pressure increase by heat transfer gas followed by connection to vacuum. French Patent 19920004540.

Al Haddad, M., Mounir, S., Sobolik, V., Allaf, K. 2008. Fruits and vegetables drying using DIC technology and microwaves. *Int J Food Eng*, 4: 1–9.

Balachandran, S., Kentish, S.E., Mawson, R., Ashokkumar, M. 2006. Ultrasonic enhancement of the supercritical extraction from ginger. *Ultrason Sonochem* 13: 471–479.

Barton, S., Bullock, C., Weir, D. 1996. The effects of ultrasound on the activities of some glycosidase enzymes of industrial importance. *Enzym Microb Technol* 18: 190–194.

Chemat, F. 2009. *Essential Oils and Aromas: Green Extractions and Applications*. Dehradun: HKB Publishers.

Chemat, F., Abert-Vian, M., Visinoni, F. 2008. Microwave hydro-diffusion for isolation of natural products. European Patent 1955749.

Chemat, F., Lucchesi, M. 2006. Microwave-assisted extraction of essential oils, in *Microwaves in Organic Synthesis*, A. Loupy (Ed.), pp. 959–983. Weinheim: Wiley.

Chemat, F., Lucchesi, M., Smadja, J. 2004. Solvent-free microwave extraction of volatile natural substances. United States Patent 0187340.

Cravotto, G., Omiccioli, G., Stevanato, L. 2005. An improved sonochemical reactor. *Ultrason Sonochem* 12: 213–217.

Cravotto, G., Cintas, P. 2007a. Extraction of flavourings from natural sources, in *Modifying Flavour in Food*, A.J. Taylor and J. Hort (Eds.), pp. 41–63. Cambridge: Woodhead Publishing Limited.

Cravotto, G., Cintas, P. 2007b. The combined use of microwaves and ultrasound: New tools in process chemistry and organic synthesis. *Chem Eur J* 13: 1902–1909.

Cravotto, G., Di Carlo, S., Curini, M., Tumiatti, V., Roggero, C. 2007. A new flow reactor for the treatment of polluted water with microwave and ultrasound. *J Chem Technol Biot* 82: 205–208.

Cravotto, G., Boffa, L., Mantegna, S. 2008. Improved extraction of natural matrices under high-intensity ultrasound and microwave, alone or combined. *Ultrason Sonochem* 15: 898–902.

Cravotto, G., Binello, A. 2010. Innovative techniques and equipments for flavours extraction. Application and effectiveness of ultrasound and microwaves. *HPC Today* 1: 32–34.

Cubadda, F., Aureli, F., Ciardullo, S., Damato, M., Raggi, A., Acharya, R., Reddy, R.A.V., Prakash, N.T. 2010. Changes in selenium speciation associated with increasing tissue concentrations of selenium in wheat grain. *J Agric Food Chem* 58: 2295–2301.

Entezari, H., Nazary, H., Khodaparast, H. 2004. The direct effect of ultrasound on the extraction of date syrup and its micro-organisms. *Ultrason Sonochem* 11: 379–384.

Guzman Mar, J.L., Hinojosa Reyes, L., Mizanur Rahman, G.M., Kingston, H.M. 2009. Simultaneous extraction of arsenic and selenium species from rice products by microwave-assisted enzymatic extraction and analysis by ion chromatography-inductively coupled plasma-mass spectrometry. *J Agric Food Chem* 57: 3005–3013.

Hu, Y., Wang, T., Wang, M., Han, S., Wan, P. 2008. Extraction of isoflavonoids from Pueraria by combining ultrasound with microwave vacuum. *Chem Eng Process* 47: 2256–2261.

Huang, S., Ning, Z. 2010. Extraction of polysaccharide from *Ganoderma lucidum* and its immune enhancement activity. *Int J Biol Macromol* 47: 336–341.

Ishimori, Y., Karube, I., Suzuki, S. 1981. Acceleration of immobilized [alpha]-chymotrypsin activity with ultrasonic irradiation. *J Mol Cat* 12: 253–259.

Lagha, A., Chemat, S., Bartels, P.V., Chemat, F. 1999. Microwave–ultrasound combined reactor suitable for atmospheric sample preparation procedure of biological and chemical products. *Analusis* 27: 452–457.

Kim, H.J., Kwak, I.S., Lee, B.S., Oh, S.B., Lee, H.C., Lee, E.M., Lim, J.Y., Yun, Y.S., Chun, B.W. 2005. Enhanced yield of extraction from *Gastrodia elata* Blume by ultrasonication and enzyme reaction. *Nat Prod Sci* 11: 123–126.

Li, D., Mu, C., Cai, S., Lin, W. 2009. Ultrasonic irradiation in the enzymatic extraction of collagen. *Ultrason. Sonochem.* 16: 605–609.

Lianfu, Z., Zelong, L. 2008. Optimization and comparison of ultrasound/microwave assisted extraction (UMAE) and ultrasonic assisted extraction (UAE) of lycopene from tomatoes. *Ultrason Sonochem* 15: 731–737.

Lieu, L.N., Le, V.V.M. 2010. Application of ultrasound in grape mash treatment in juice processing. *Ultrason Sonochem* 17: 273–279.

Liu, Y., Jin, Q., Shan, L., Liu, Y., Shen, W., Wang, X. 2008. The effect of ultrasound on lipase-catalyzed hydrolysis of soy oil in solvent-free system. *Ultrason Sonochem* 15: 402–407.

Lou, Z., Wang, H., Wang, D., Zhang, Y. 2009. Preparation of inulin and phenols-rich dietary fibre powder from burdock root. *Carbohydr Polym* 78: 666–671.

Lou, Z., Wang, H., Zhu, S., Zhang, M., Gao, Y., Ma, C., Wang, Z. 2010. Improved extraction and identification by ultra performance liquid chromatography tandem mass spectrometry of phenolic compounds in burdock leaves. *J Chrom A* 1217: 2441–2446.

Luque de Castro M.D., García-Ayuso, L.E. 1998. Soxhlet extraction of solid materials: an outdated technique with a promising innovative future. *Anal Chim Acta* 369: 1–10.

Luque de Castro, M.D., Priego-Capote, F. 2010. Soxhlet extraction: Past and present panacea. *J Chrom A.* 1217: 2383–2389.

Mason, T.J., Paniwnyk, L., Chemat F., Abert-Vian, M. 2011. Ultrasonic processing, in *Alternatives to Conventional Food Processing*, A. Proctor (Ed.), pp. 395–422. London: RSC.

Mason, T.J., Paniwnyk, L., Lorimer, J.P. 1996. The use of ultrasound in food technology. *Ultrason Sonochem* 3: 253–260.

Matsuura, K., Hirotsune, M., Nunokawa, Y., Satoh, M., Honda, K. 1994. Acceleration of cell growth and ester formation by ultrasonic wave irradiation. *J Ferm Bioeng* 77: 36–40.

Meireles, M.A.A. 2009. *Extracting Bioactive Compounds for Food Products: Theory and Applications*. Boca Raton, FL: CRC Press.

Muralidhara, H., Parekh B., Senapati, N. 1985. Solid–liquid separation process for fine particles suspensions by an electric and ultrasonic field. US Patent 4561953.

Muthukumaran, S., Kentish, S.E., Ashokkumar, M., Stevens, G.W. 2005. Mechanisms for the ultrasonic enhancement of dairy whey ultrafiltration. *J Membr Sci* 258: 106–114.

Ookushi, Y., Sakamoto, M., Azuma, J. 2008. Extraction of β-glucan from the water-insoluble residue of *Hericium erinaceum* with combined treatments of enzyme and microwave irradiation. *J Appl Glycosci* 55: 225–229.

Ovsianko, S.L., Chernyavsky, E.A., Minchenya, V.T., Adzerikho, I.E., Shkumatov, V.M. 2005. Effect of ultrasound on activation of serine proteases precursors. *Ultrason Sonochem* 12: 219–223.

Reyes, L.H., Guzman Mar, J.L., Mizanur Rahman, G.M., Seybert, B., Fahrenholz, T., Kingston, H.M. 2009. Simultaneous determination of arsenic and selenium species in fish tissues using microwave-assisted enzymatic extraction and ion chromatography-inductively coupled plasma mass spectrometry. *Talanta* 78: 983–990.

Sakakibara, M., Wang, D., Takahashi, R., Takahashi, K., Mori, S. 1996. Influence of ultrasound irradiation on hydrolysis of sucrose catalyzed by invertase. *Enzyme Microb Technol* 18: 444–448.

Sanz, E., Munoz-Olivas, R., Camara, C., Sengupta, M.K., Ahamed, S. 2007. Arsenic speciation in rice, straw, soil, hair and nails samples from the arsenic-affected areas of Middle and Lower Ganga plain. *J Env Sci Health* 42: 1695–1705.

Sharma, A., Gupta, M.N. 2006. Ultrasonic pre-irradiation effect upon aqueous enzymatic oil extraction from almond and apricot seeds. *Ultrason Sonochem* 13: 529–534.

Shah, S., Sharma, A., Gupta, M.N. 2005. Extraction of oil from *Jatropha curcas* L. seed kernels by combination of ultrasonication and aqueous enzymatic oil extraction. *Bioresource Technol* 96: 121–123.

Sinisterra, J.V. 1992. Application of ultrasound to biotechnology: An overview. *Ultrasonics* 30: 180–185.

Stevens, S., Wolverton, S., Venables, B., Barker, A., Seeley, K.W., Adhikari, P. 2010. Evaluation of microwave-assisted enzymatic digestion and tandem mass spectrometry for the identification of protein residues from an inorganic solid matrix: Implications in archeological research. *Anal Bioanal Chem* 396: 1491–1499.

Tian, Z.M., Wan, M.X., Wang, S.P., Kang, J.Q. 2004. Effects of ultrasound and additives on the function and structure of trypsin. *Ultrason Sonochem* 11: 399–404.

Trendafilova, A., Todorova, M. 2008. Comparison of different techniques for extraction of biologically active compounds from *Achillea millefolium Proa*. *Nat Prod Com* 3: 1515–1518.

Vale, G., Rial Otero, R., Mota, A., Fonseca, L., Capelo, J.L. 2008. Ultrasonic-assisted enzymatic digestion (USAED) for total elemental determination and elemental speciation: A tutorial. *Talanta* 75: 872–884.

Vale, G., Rodrigues, A., Rocha, A., Rial, R., Mota, A.M., Goncalves, M.L., Fonseca, L.P., Capelo, J.L. 2010. Ultrasonic assisted enzymatic digestion (USAED) coupled with HPLC and electrothermal atomic absorption spectrometry as a powerful tool for total selenium and selenium species control in Se-enriched food supplements. *Food Chem* 121: 268–274.

Virot, M., Tomao, V., Colnagui, G., Visinoni, F., Chemat, F. 2007. New microwave-integrated Soxhlet extraction: An advantageous tool for the extraction of lipids from food products. *J Chrom A* 1174: 138–144.

Yachmenev, V.G., Bertoniere, N.R., Blanchard, E.J. 2002. Intensification of the bio-processing of cotton textiles by combined enzyme/ultrasound treatment. *J Chem Technol Biotechnol* 77: 559–567.

Yachmenev, V.G., Blanchard, E.J., Lambert, A.H. 2004. Use of ultrasonic energy for intensification of the bio-preparation of greige cotton. *Ultrasonics* 42: 87–91.

Yachmenev, V.G., Condon, B.D., Lambert, A.H. 2007. Use of ultrasonic energy for intensification of the bioprocessing. In *Proceeding of The 19th International Congress on Acoustics*, Madrid.

Yang, Y.C., Li, J., Zu, Y.G., Fu, Y.J., Luo, M., Liu, X.L. 2010. Optimisation of microwave-assisted enzymatic extraction of corilagin and geraniin from *Geranium sibiricum* L. and evaluation of antioxidant activity. *Food Chem* 122: 373–380.

7 Supercritical Fluid Extraction in Food Processing

Rakesh K. Singh and Ramesh Y. Avula

CONTENTS

7.1 INTRODUCTION

The regulatory legislations on advocating a reduction in the use of organic solvents that are harmful to the environment led to the introduction of alternative methods by the food industry to produce relatively pure ingredients without causing adverse environmental impact (Sahena et al. 2009). Supercritical fluid extraction (SFE) using carbon dioxide (CO_2) as a solvent has provided an excellent alternative to conventional solvent extraction and mechanical pressing since it offers several advantages, including a lack of solvent residue and better retention of aromatic compounds (Zaidul et al. 2007a; Norulaini et al. 2009). Supercritical CO_2 has the unique ability to penetrate through solids like a gas, and dissolve materials like a liquid. Additionally, it can readily change in density upon minor changes in temperature or pressure. A supercritical fluid is any substance at a temperature and pressure above its thermodynamic critical point. Mass transport properties, such as fluid and analyte diffusion coefficients, are greater in supercritical fluid media, resulting in faster extraction fluxes and a substantial reduction in extraction time (Sahena et al. 2009). Supercritical fluids possess both gas-like and liquid-like qualities, and the dual role of such fluids provide ideal conditions for extracting compounds with a high degree of recovery in a short time. The critical temperature (°C) and critical pressure (MPa) of solvents commonly used as supercritical fluids are given in Table 7.1.

TABLE 7.1

Critical Properties of Solvents Used in SFE

Solvent	Temperature (°C)	Pressure (MPa)	Density (g/cm³)
Carbon dioxide	31.1	5.04	0.469
Methanol	239.6	8.09	0.272
Ethanol	240.9	6.14	0.276
Acetone	235.1	4.70	0.278
Hexane	234.5	3.01	0.23
Propane	96.8	4.25	0.217
Ethylene	9.4	5.04	0.215
Acetone	235.1	4.70	0.278
Water	374.1	22.06	0.322

Source: Herrero, M. et al., *Food Chem*, 98, 136–148, 2006. With permission.

Because of its low critical constants, low cost, nontoxicity, chemical inertness, and nonflammability, CO_2 is used as a common solvent in the food industry (Rozzi and Singh 2002). The absence of surface tension allows the rapid penetration of supercritical carbon dioxide (SC-CO_2) into the pores of heterogeneous matrices and helps to enhance extraction efficiencies (Lang and Wai 2001). The nontoxicity and easily separable nature of CO_2 are extremely important for application of SFE technology in the food processing industry.

SFE is an ideal technique to study thermally labile compounds because it performs at low temperatures, which may also lead to extraction of new natural compounds (Khosravi-Darani 2010). Many undesirable reactions (hydrolysis, oxidation, degradation, and rearrangement) were prevented during SFE extraction of ginger. Therefore, the common difficulties encountered with classical water distillation can be avoided in SFE (Sahena et al. 2009). SFE may allow direct coupling with a chromatographic method to extract and directly quantify highly volatile compounds (Cortes et al. 2009). In large-scale SFE processes, CO_2 can be recycled or reused, thus minimizing waste generation. SFE can be applied for analytical scale, to preparative scale, and up to large industrial scale (Sahena et al. 2009). SFE can provide more information pertaining to the extraction processes and mechanisms, which is useful in the quantitative evaluation of the extraction efficiency and appropriate optimization of the process (Lang and Wai 2001). Sample preparation methods based on the use of compressed fluids, such as SFE and pressurized liquid extraction (PLE), provide fast, reliable, clean, and low-cost methods that can be used for routine analysis. However, SFE is useful for extraction of both solid and liquid samples, whereas PLE is used for solid samples. SFE is preferentially used to extract nonpolar compounds, whereas PLE can be used for both polar and nonpolar compounds since all conventional solvents are used in this method. Thermal stability of the compounds is mandatory when using PLE, whereas SFE is good for extraction of thermolabile compounds. Thus, SFE is more selective, allowing users to obtain extracts that can be directly analyzed without additional cleanup steps (Mendiola et al. 2007).

Sahena et al. (2009) also listed the major advantages of SFE over other methods such as microwave extraction, in situ synthesis of fatty acid methyl esters, and Soxhlet extraction.

The major applications of SFE in food processing include total extractions, deodorization, and fractionation. Quantitative recovery of fats/oils, essential oils, volatiles, and flavors are obtained by $SC-CO_2$ alone and with a modifier. Commercial SFE instrumentation has largely been developed in the United States and marketed throughout the world. Isco Inc., USA, Applied Separations Inc., USA, and Leco Corporation, USA, manufacture several types of SFE units that offer considerable flexibility with respect to sample size and experimental design (King 2002). The commercial plants are mainly located in the developed areas of Europe, North America, and Asia. Decaffeination of green coffee beans was the first commercial SFE performed in Germany by Hag AG. Subsequently, the hop extraction process was developed by Carlton and United Breweries in Australia. These commercially successful applications led to improvements for industrial scale up. By manipulating the operating conditions during a supercritical extraction process, the supercritical fluid can effectively and selectively extract specific components such as fats, oils, cholesterol, ketones, aldehydes, and esters while leaving protein and carbohydrates in their intact state (Higuera-Ciapara et al. 2005). Fat extraction in milk and milk products, fat-soluble vitamins from cheese, and removal of cholesterol hold potential for development of novel ingredients with varied functional properties for use in food formulations (Fatouh et al. 2007; Torres et al. 2009; Domagala et al. 2010). The $SC-CO_2$ extracts of plants, algae, and microalgae were found to be effective in antioxidant (Herrero et al. 2006; Chen et al. 2009; Kitada et al. 2009) and antimicrobial activities (Liang et al. 2008), and in some cases, were even found to have similar activity to the antibiotics ampicillin and nystatin (Liu et al. 2009a). Herrero et al. (2006) reviewed the $SC-CO_2$ extraction of functional ingredients from plants (rosemary, coriander, ginger, tamarind) and food industry by-products. Although SFE can perform many types of separation, case-by-case evaluations are always required, and many factors should be evaluated (Andras et al. 2005). Several excellent reviews dealing with $SC-CO_2$ extraction in food processing have been published over the past 10 years (Rozzi and Singh 2002; Sahena et al. 2009; Khosravi-Darani 2010). Recently, a published review indicates that the number of independent patents in the area had reached 8600 (Schutz 2007). This chapter updates these reviews with recent developments in $SC-CO_2$ extraction of food and food products.

7.2 PRINCIPLES AND CURRENT INSTRUMENTATION TRENDS

The densities of the liquid and the gas phases become equal and the distinction between them disappears above the critical temperature, resulting in a single supercritical fluid phase. The critical point of CO_2 is at 31.1°C and 7.38 MPa, and with increasing temperatures, the liquid–vapor density gap decreases up to the critical temperature, at which the discontinuity disappears. Thus, above the critical temperature, a gas cannot be liquefied by pressure increase (Sahena et al. 2009). Carbon dioxide is brought to a specific pressure–temperature combination to allow it to attain supercritical solvent properties for the selective extraction of the sample. The sample is exposed to the supercritical fluid under the controlled conditions to allow dissolution of the analyte. The dissolved analyte will be separated from the supercritical solvent by a significant drop in solution pressure (Wang and Weller 2006). This can also be achieved by changing the temperature, or by washing the solute out of the supercritical fluid using a solvent, and by using packed columns to separate multiple solutes (Rozzi and Singh 2002). Figure 7.1 shows a typical SFE system.

An ideal extraction method should afford total recovery and high purity of the isolated substances. Modification of the extraction conditions depending on the chemical composition of the sample is necessary to maximize extraction efficiency and to allow selective extraction of minute quantities of polar or nonpolar analytes. Modifiers such as methanol, ethanol, water, mixture of water and ethanol, etc., are often used to assist supercritical CO_2 in the extraction of highly polar compounds (Khosravi-Darani 2010). Figure 7.2 shows the schematic diagram of SFE used by Yu et al. (2006) for extraction of limonoids with SC-CO_2 and ethanol cosolvent.

Several researchers have reported that lipid solubility (e.g., lipids in milk) in SC-CO_2 is greatly increased by adding ethanol, and some phospholipids are extracted at levels directly proportional to the added ethanol (Domagala et al. 2010). The presence of water dissolved in the supercritical fluid also increases the solubility of polar compounds, and it has been used successfully to analyze several dairy products (Sahena et al. 2009). The solubility trend in SC-CO_2 for food samples has been studied by several research groups. King (2002) found relatively low weight percent solubilities (5 wt.%) of triglycerides in SC-CO_2 for the 40°C and 50°C isotherms. However, pronounced increase in triglyceride solubility (~15 wt.%) could be observed at above 80 MPa, when the temperature was increased from 50°C to 60°C. These solubility trends for oils in SC-CO_2 have been used by many researchers to perform oil and fat extractions using SC-CO_2 above 60 MPa and temperatures ranging from 80°C to 100°C (Taylor et al. 1997). Rozzi and Singh (2002) described different methods based on solubility parameters and process modeling. It should be emphasized that many lipid solutes have similar solubility parameters, making their separation by SFE difficult. An improvement in the separation of complex lipid mixtures can be achieved by the addition of adsorbent materials such as alumina, silica, celite, florisil, or synthetic resins to fractionate lipids dissolved in SC-CO_2 and SC-CO_2–cosolvent mixtures to the extraction cell along with the sample. Tocopherols, phospholipids,

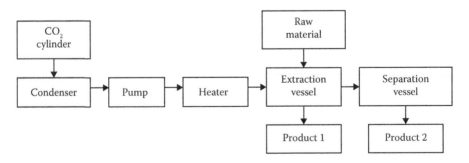

FIGURE 7.1   Flow diagram of a supercritical fluid extraction (SFE) system. (From Rozzi, N.L. et al., *Comp Rev Food Sci Food Saf*, 1, 33–44, 2002. With permission.)

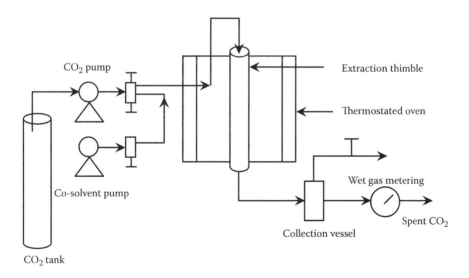

FIGURE 7.2 Schematic diagram of the pilot scale SFE unit. (From Yu, J. et al., *J Agric Food Chem*, 54, 6041–6045, 2006. With permission.)

and phytosterols have been extracted by adding adsorbent materials (Brunner et al. 2009; Sahena et al. 2009). Solubility of β-carotene, α-carotene, and other carotenoids under different extraction conditions has been reviewed by Shi et al. (2007). The solubility of targeted carotenoids in SFE is related to its physical and chemical properties such as polarity, molecular structure, and nature of the material particles.

Particle size affects lipid recovery since it determines the surface area of the sample exposed to SC-CO_2 (Lenucci et al. 2010). The moisture content of samples also affects the extraction efficiency by conditioning their surface structure. High moisture content results in pasty consistency of samples and reduces SC-CO_2 extraction efficiency. Hence, such samples are freeze-dried before SC-CO_2 extraction. Low levels of moisture did not affect extractability, and lipid recovery increased with decreasing moisture content in wet samples of meat and fish (Sahena et al. 2009). For an efficient extraction, characteristics of the sample matrix, preparation of sample, and choice of the extract collection method need to be considered. The level of anticipated extractable and moisture content are key factors affecting the analytical SFE process. Extraction is carried out in high-pressure equipment in batches or continuous modes of operation. In both cases, the supercritical solvent in the extract stream is expanded to atmospheric conditions and the solubilized product is recovered in the separation vessel permitting the recycle of the supercritical solvent for further use.

A continuous system was designed for the processing of anhydrous milk fat (AMF), where a packed column was operated as a stripping column with SC-CO_2 as the continuous phase and AMF as the dispersed phase (Bhaskar et al. 1993). In order to enrich phospholipids in buttermilk powders, regular buttermilk and whey buttermilk (by-product of whey cream after making butter) were microfiltered and then treated with SFE after drying. The total fat, namely, nonpolar lipids, in the powders was reduced by 38–55%, and phospholipids were concentrated by a factor

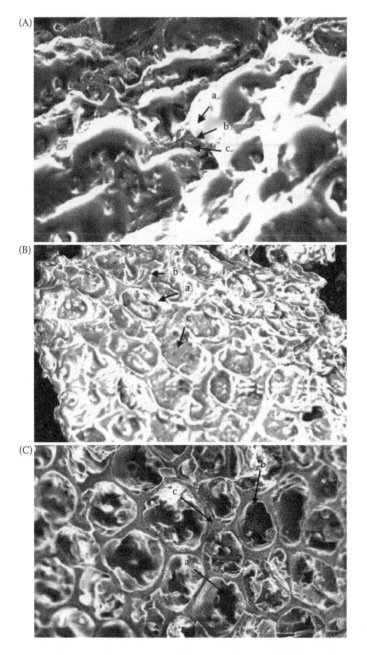

FIGURE 7.3 Scanning electron micrograph (×3000) of ground palm kernel: (A) before extraction, (B) after continuous extraction at 25 MPa, (C) after pressure swing extraction at 25 MPa. [a, oil; b, membrane; c, cell structure]. (From Zaidul, I.S.M. et al., *J Food Eng*, 81, 419–428, 2007b. With permission.)

of 5-fold (Spence et al. 2009). Astaire et al. (2003) also used a two-step method to produce buttermilk derivative ingredients containing increased concentrations of the polar milk fat globule membrane lipids by microfiltration and SFE. The SFE process using SC-CO$_2$ removed exclusively nonpolar lipid material from the microfiltered buttermilk product.

The quality of safflower seed oil obtained by supercritical CO$_2$ extraction is superior to that of oil obtained by conventional methods. A new method of changing flow rate was proposed to improve the process efficiency (Han et al. 2009). A pressurized fluid extraction method for the extraction of pistachio oil was developed mainly as an analytical tool to determine oil content and its quality. Here, the supercritical fluid extractor was modified to be able to pump liquid solvent and CO$_2$ into the extraction vessel alternatively (Sheibani and Ghaziaskar 2008). Palm kernel oil could be recovered completely (47 g/100 g palm kernel) by using combined pressure swing and continuous extraction at 25 MPa as the disruption of the oil glands in palm kernel granules lead to higher yields at 20 MPa and above (Zaidul et al. 2007b). The scanning electron micrographs before and after the extraction process showed that the integrity of membrane structures is very important in the separation (Figure 7.3).

To determine castor oil content and the composition of its individual fatty acid methyl esters, a lipase from *Candida antarctica* was used to catalyze the methanolysis reaction in SC-CO$_2$ (Turner et al. 2004). Grosso et al. (2008) extracted volatile oils from coriander seeds by SC-CO$_2$. The main parts of the SFE apparatus consisted of a diaphragm pump, an extraction vessel with an internal volume of 1 l, and two separators of 0.27 l each, operating in series. A back-pressure regulator is used to control the pressure, which is measured with a Bourdon-type manometer, and the total volume of CO$_2$ is determined with a dry test meter. A preset temperature in the extraction vessel is achieved with the aid of a water jacket. The on-line coupling of SFE equipment with detection/separation instruments paves the way to very interesting analytical possibilities. A robotic interface between SFE and capillary electrophoresis (CE) works as a mimetic human arm for transferring the extracted plug from SFE to the CE vials. It is a unique on-line coupling of SFE and CE through a mechanical interface, which is automatically controlled by a built-in microprocessor using an appropriate electronic interface (Zougagh and Rios 2008). Continuous SC-CO$_2$ extraction technique was demonstrated by Lau et al. (2008) for separation of vitamin E and squalene from carotene. The extraction of fresh palm pressed mesocarp was subjected to SC-CO$_2$ in three stages using continuous extraction technique—that is, extraction of vitamin E and squalene enhanced fraction at 10 MPa, removal of bulk triglycerides at 20 MPa, and production of carotene enriched fraction at 30 MPa. The recovery of vitamin E, squalene, and carotenes was more than 90%.

7.3   FRACTIONATION OF FATS

The major applications of SFE include decaffeination of tea, flavor extraction from herbs, aroma extraction from juices, extraction of colorants, refinement of fats and oils, extraction of antioxidants, deacidification of oil, and inactivation of orange juice pectin esterase (Sahena et al. 2009). Table 7.2 summarizes the extraction studies

TABLE 7.2

Applications of SFE in Food Analysis

Sample	Analyte(s)	Pressure (MPa)/ Temperature (°C)	Reference
Milk fat	Lipid profile	6.9–17.2/40–60	Rozzi and Singh (2002)
	Vitamins		
	Solid fat content		
	Polychlorinated biphenyls		
Butter oil	Cholesterol	10–27.6/40–70	Rozzi and Singh (2002)
Parmigiano cheese	Total fats and vitamins	53.5/100	Mendiola et al. (2007)
Pork	Cholesterol/fat content	7.3–34/50–150	Rozzi and Singh (2002)
Pecan	Fatty acid composition	41.3–66.8/45–75	Rozzi and Singh (2002), Sahena et al. (2009)
Rice bran	Fatty acid composition	17–31/0–60	Sahena et al. (2009),
	α-tocopherol sterols	48.2–62/70–100	Rozzi and Singh (2002)
Sunflower	Fatty acid composition	51.7–62/100	Sahena et al. (2009), Rozzi and Singh (2002)
Sea buckthorn	Fatty acid composition	27.6/34.5	Sahena et al. (2009)
Sardine oil	Fatty acid composition	9.6–19.5/40–80	Sahena et al. (2009)
Annatto	*trans*-Bixin	40–60/40	Rozzi and Singh (2002)
Paprika	Lipids	35/50	Rozzi and Singh (2002)
	Carotenoids	15/40	
	Tocopherol	13.7–41.3/40	
Carrot	Carotenes	60.6/40	Rozzi et al. (2002)
Tomato	Lycopene	33.5–44.5/45–70	Mendiola et al. (2007)
Crustaceans	Astaxanthin	20/60	Mendiola et al. (2007)
Olive products	Tocopherols	35/50	Rozzi and Singh (2002)
	Phenol compounds	33.4/100	
Grape skin	Phenolic compounds	25/60	Mendiola et al. (2007)
Grape seeds	Polyphenolics	45.6/35	Rozzi and Singh (2002)
Rosemary	Antimicrobial and antioxidant compounds	25/60 10–16/37–47	Mendiola et al. (2007); Rozzi and Singh (2002)
	Essential oils	10–40/60	
Citrus maxima fruit	Coumarins	27.6/50	Mendiola et al. (2007)
Orange juice	Antioxidant compounds	16/40	Mendiola et al. (2007)
Red clover and soy bits	Isoflavones	3.5–7.5/10–40	Mendiola et al. (2007)
Black pepper	Essential oil	20–32/45–65	Rozzi and Singh (2002)
Spearmint leaf	Essential oil	27.6/60	Rozzi and Singh (2002)
Onion	Onion oil	10–28/37–50	Mendiola et al. (2007)
Cinnamomum cassia	Cinnamon oil	22.5/50	Mendiola et al. (2007)
Apple, green bean, carrot	Pesticides	32/60	Mendiola et al. (2007)

(continued)

TABLE 7.2 (Continued)
Applications of SFE in Food Analysis

Sample	Analyte(s)	Pressure (MPa)/ Temperature (°C)	Reference
Vegetable soup	Pesticides Organohalogenated and organophosphate	30–50/50–90	Mendiola et al. (2007)
Honey	Pesticide multiresidue	20–60/40–90	Mendiola et al. (2007)

Sources: Sahena, F. et al., *J Food Eng*, 95, 240–253, 2009; Mendiola, J.A. et al., *J Chromatogr*, 1152, 234–246, 2007; Rozzi, N.L., and Singh, R.K., *Comp Rev Food Sci Food Saf*, 1, 33–44, 2002. With permission.

related to food in the past 10 years. Oil extraction using supercritical fluids is an alternative method to replace or to complement conventional industrial processes such as pressing and solvent extraction. The use of SC-CO_2 in fat and oil processing as an extraction, fractionation, concentration, and reaction solvent has been investigated by several research groups (Fatouh et al. 2007; Yu et al. 2007; Zaidul et al. 2007a). Before SC-CO_2 extraction, vegetable substrates are commonly subjected to mechanical pretreatments to improve the rate and yield of the extraction process. The pretreatments may have multiple purposes, including releasing of solutes from cells, facilitating solvent flow through the packed bed, and increasing substrate load onto extraction vessels. One frequently used treatment for high-oil seeds is prepressing to reduce the oil content (Martinez et al. 2008).

Although food industry is a major user of the SC-CO_2 extraction process for extraction and fractionation of fats, oils, essences, pigments, and functional or bioactive compounds, the majority of published reports on SC-CO_2 has focused on oil extraction, and it is not surprising that the technique is becoming a standard method for oil extraction because of its high extraction efficiency, short extraction time, and lack of residue problems (Bhattacharjee et al. 2007; Lang and Wai 2001). Continuous, mechanical screw presses are used to recover oil from oilseeds. Although screw pressing provides a simple and reliable method for processing small batches of seed, its performance depends on the method of preparing the raw material, which consists of a number of unit operations such as cleaning, cracking, cooking, drying, or moistening to optimal moisture content. The application of a thermal treatment before or during pressing generally improves oil recovery, but may adversely influence the oil quality by increasing oxidative parameters (Martinez et al. 2008). Physical (refractive index, color, etc.) and chemical (acid value, iodine value, saponification value, and peroxide value) properties of the oil, and its fatty acid composition are important quality attributes that will determine the oil's acceptance as a food or medicinal supplement. A statistical experimental design based on central composite rotatable design and a response surface methodology are often used to characterize the influence of process variables and arrive at optimal processing conditions (Mitra et al. 2009). Oil extraction by SC-CO_2 has been successfully attempted by several researchers, and several reviews have been published on SFE of fats (Rozzi and Singh 2002; Sahena et al. 2009). In this chapter, we have described a few such examples, highlighting the importance of SC-CO_2 extraction to overcome the problems associated with conventional methods of fat extraction related to the vegetable, seafood, and dairy industries.

Martinez et al. (2008) evaluated the combined effects of temperature and moisture content on oil yield and quality of walnut seeds. The oil extraction process from walnut seeds was carried out by pressing followed by extraction with SC-CO_2. Oil recovery increased significantly as the moisture content was raised, and the highest oil recovery (89.3%) was obtained at a 7.5% moisture content and a temperature of 50°C. The cake resulting from pressing at these conditions was extracted with CO_2 in a high-pressure pilot plant with a single stage separation and solvent recycle. At first, the mass of oil extracted was determined by the oil solubility in CO_2 and a linear relationship was observed. After that, the extraction rate was governed by solubility and diffusion, and continuously decreased with time. The color changed along the extraction from a whitish clear to a yellow product. Tocopherol and carotenoid contents were significantly higher than those obtained by pressing. Among the by-products of the walnut industry, the oil has yet to gain popularity, although it has been demonstrated that consumption of this oil offers a lot of nutritional benefits. The major constituents of walnut oil are triglycerides, in which oleic acid (monounsaturated) and linoleic and α-linolenic acids (polyunsaturated fatty acids) are present in high amounts. The presence of other bioactive minor components, such as tocopherols and phytosterols, has also been documented (Crews et al. 2005; Martinez et al. 2006). Maximum recovery was obtained in the first 2–3 h of extraction at 42 MPa, 50°C, and a flow rate of 30 kg h^{-1} CO_2 during extraction of oil and tocopherols from almond seeds (Leo et al. 2005). Hazelnut particles were extracted at 60 MPa and 60°C, for 180 min with CO_2 flow rate of 2 ml/min to recover a high yield of 59% of oil. The crossover pressure of hazelnut oil was between 15 and 30 MPa. The solubility increased with pressure, but increased with temperature above the crossover pressure (Ozkal et al. 2005a). A maximum yield of oil from hazelnut was obtained for 15 min extraction with 4 g/min solvent flow rate containing 3 wt.% ethanol at 45 MPa and 60°C (Ozkal et al. 2005b).

Pumpkin (*Cucurbita maxima*) seeds have been used as safe deworming and diuretic agents, and the seed oil as a nerve tonic (Mitra et al. 2009). Pumpkin seed oil has strong antioxidant properties (Stevenson et al. 2007) and has been recognized for prevention of the growth and reduction of the size of prostate, retardation of the progression of hypertension, mitigation of hypercholesterolemia and arthritis, alleviation of diabetes by promoting hypoglycemic activity, and lowering levels of gastric, breast, lung, and colorectal cancer (Caili et al. 2006; Mitra et al. 2009). Lang and Wai (2001) reported that conventional extraction methods such as hydrodistillation and organic solvent extraction required the control of several adjustable parameters to improve the extraction processes. The toxic organic residues were considered a serious problem in the solvent extraction of pumpkin seed oil. SC-CO_2 extraction of pumpkin seed yielded 30.7% oil at the optimum conditions of 32.1 MPa and 68.1°C for 94.6 min. The flow rate of CO_2 was maintained at 0.25 l/min (Mitra et al. 2009). Palmitic C16:0 (13.8%), stearic C18:0 (11.2%), oleic C18:1 (29.5%), and linoleic acids

C18:2 (45.5%) were the major fatty acids found in pumpkin seed oil. Oil is a rich source of linoleic acid and has the potential to be used as nutrient-rich food oil. Bernardo-Gil and Lopes (2004) also reported the extraction of pumpkin seed oil at 19 MPa, 35°C, and a superficial velocity of 6.0×10^{-4} m s^{-1}. Monounsaturated fatty acids were predominant in the whole berry oil of sea buckthorn, accounting for more than 62% of the total fatty acids. The optimum extraction conditions for extraction of whole berry oil were predicted to be 27.6 MPa, 34.5°C with a flow rate of 17.0 l/h, and extraction time of 82 min. Under such conditions, the yields of oil, vitamin E, and carotenoids were predicted to be 208.0 g/kg, 288.7 mg/kg, and 620.0 mg/kg dry sea buckthorn berry, respectively (Xu et al. 2008). Sea buckthorn press juice was subjected to percolation with SC-CO$_2$ supercritical carbon dioxide at 13.8 and 27.6 MPa. Percolation extracted about 81% of the oil present in the press juice, lightened the orange color compared with the control juice, and changed the character of the juice as perceived by observers. Control juice was not percolated (Beveridge et al. 2004). Mezzomo et al. (2010) extracted peach kernel oil by using SC-CO$_2$, which showed higher content of flavonoid and presented advantages over conventional methods with respect to high oleic acid content.

The okra seed oil contained linoleic acid (255–297 g/kg), oleic acid (415–419 g/kg), and saturated acids (288–297 g/kg), and can be used for industrial or food purposes. The seed oil also showed higher concentrations of β-sitosterol and tocopherols in the extract compared to those obtained via *n*-hexane Soxhlet extraction. The extracted meal was comparable with other meals in commercial use for feeding livestock (Andras et al. 2005). More than 99% of the total coconut oil could be extracted at pressure and temperature ranges of 20.7–34.5 MPa and 40–80°C, respectively (Norulaini et al. 2009). Vazquez et al. (2009) removed free fatty acids from cold-pressed olive oil by SC-CO$_2$ at 40°C and pressures of 18, 23.4, and 25 MPa. Extractions were performed in a packed column of 3 m high, without external or internal reflux and utilizing 1.8 m of the column as the stripping section. The upper part of the column was used to avoid carryover of the crude oil. Wang et al. (2008) reported a maximum yield of tocols, phytosterols, policosanols, and free fatty acids from sorghum distillers dried grains at 27.5 MPa, and 70°C for 4 h. Higher flow rate of CO$_2$ and shorter extraction achieved lower content of free fatty acids in the lipid extract. Asep et al. (2008) studied the effects of particle size, fermentation, and roasting of cocoa nibs on SC-CO$_2$ extraction of cocoa butter. The extraction yield was increased by reduction in particle size and also by using unfermented cocoa, roasted at 150°C for 35 min. Cocoa butter had similar triglycerides and fatty acid methyl ester composition at 5, 10, and 15 h extraction time.

Shrimp is considered a high-cholesterol product, because it contains more than 150 mg cholesterol/100 g of edible portion. Higuera-Ciapara et al. (2005) reported that at 31.3 MPa, 37°C, and 1875 l of CO$_2$, it is possible to obtain a low-cholesterol shrimp with acceptable organoleptic properties. Polyunsaturated fatty acids, in particular, eicosapentaenoic acid (EPA) and docosahexaenoic acid (DHA), are the ω-3 fatty acids that lower blood cholesterol, thus preventing heart diseases. The main dietary sources of EPA and DHA are fish, such as sardines, mackerels, and anchovies. To concentrate polyunsaturated ω-3 fatty acids in the form of natural fish oil, experimental-phase equilibrium data were measured for a CO$_2$–fish oil system. The best operational conditions to fractionate the oil were 7.8 MPa and 28.1°C, although in all conditions analyzed, EPA could not be fractionated (Correa et al. 2008). These studies showed that triacylglycerol composition is a determinant in predicting the possibility of fractionating the oil with SC-CO$_2$. Dietary ingestion of either lipids or alkoxyglycerols is beneficial because they possess anticarcinogenic and immune stimulating properties. When the endogenous synthesis of alkoxyglycerols is reduced, oral administration is recommended (Mitre et al. 2005). Shark liver oil is a natural source of diesterified alkoxyglycerols, together with squalene and triacylglycerols. However, to obtain a product rich in alkoxyglycerols, squalene needs to be removed from shark liver oil before the saponification or ethanolysis reaction of alkoxyglycerols. Ethanolysis of shark liver oil was carried out to generate a product enriched in nonesterified alkoxyglycerols and fatty acid ethyl esters. Supercritical CO$_2$ extraction was used to fractionate the mixture, achieving a complete elimination of esters and concentrating the alkoxyglycerol compounds in the raffinate product. Extractions were carried out in an isothermal countercurrent 3-m packed column, without reflux under semicontinuous operation and using extraction pressures in the range of 14–18 MPa, temperatures ranging from 45°C to 65°C, and a solvent/feed ratio of 15. An internal reflux induced by thermal gradient was also used to increase nonesterified alkoxyglycerols and monoesterified alkoxyglycerols recovery in the raffinate product (Vazquez et al. 2008).

Milk fat is rich in essential fatty acids and possesses a unique pleasing flavor not found in other fats. Triglycerides represent the major part of milk lipids and form 97% to 98% of the total fat. SC-CO$_2$ extraction of triglycerides carried out over a range of pressures between 10 and 35 MPa resulted in eight fractions (Rizvi and Bhaskar 1995). The first two fractions were liquids, the next three were intermediate in consistency, and the last three were solids. The first fractions were made up of triglycerides with short- and medium-chain fatty acids, the percentage of which decreased in the fractions as their melting point increased. The converse was true for triglycerides with long-chain fatty acids. In a study on distribution of cholesterol in each of the fractions, Arul et al. (1988) found that the liquid fractions were enriched in cholesterol, which decreased as the melting point of the fractions rose. The possibility of producing low-cholesterol milk fat while maintaining the original color and flavor in the extracted product, the fractionation of milk fat, and the concentration of flavor from AMF by using SC-CO$_2$ have been reported (Rizvi and Bhaskar 1995). Milk fat fractions produced with SC-CO$_2$ had distinct and different physical and chemical properties from those obtained by other methods.

In the dairy industry there are several by-products that are rich in phospholipids such as buttermilk, whey, and whey cream. Phospholipids, including sphingolipids, exhibit antioxidative, anticarcinogenic, and antiatherogenic properties, and have essential roles in numerous cell functions (Spence et al. 2009). In countercurrent supercritical fractionation of the fatty acid ethyl esters from butter oil, extracts containing approximately 70% short- and medium-chain fatty acid ethyl esters were obtained at 10.1 MPa and 60°C, and can be used as starting material for the production of highly valuable functional lipids (Torres et al. 2009). Lipids in cheddar cheese were selectively extracted because of the nonpolar properties of SC-CO$_2$, without leaving residual chemicals as is the case with solvent extraction. The SFE parameters were 35 MPa, 35°C, and at a flow rate of 20 g/min for 55 min (Yee et al. 2008). The content of total tocopherols was about 14% higher in the pomegranate seed oil extracted with SC-CO$_2$ than that obtained by Soxhlet extraction (Liu et al. 2009b).

Research has been carried out by several groups with SFE of bovine and sheep milk fat (Bhaskar et al. 1998; Spano et al. 2004). Fatouh et al. (2007) fractionated buffalo butter oil by SFE into four fractions (F1–F4) and determined their physico-chemical and nutritional characteristics. Fractionation was performed at 50°C and 70°C over a pressure range of 10.9–40.1 MPa. SC-CO$_2$ flow rate was maintained at 2 ml/min and extraction time was 2 h. Fractions F1 and F2 were obtained at 50°C and a pressure of 10.9 and 15.0 MPa, respectively. The other two fractions, F3 and F4, were obtained at 70°C and 22.3 and 40.1 MPa, respectively. Short-chain, medium-chain, and saturated fatty acids were decreased from F1 to F4, whereas long-chain and unsaturated fatty acids were increased. Triacylglycerol molecular species exhibited a similar trend as fatty acids. Cholesterol was concentrated in F1, and with increasing fluid density, it decreased by more than 50% in F4. The changes that occurred in the chemical composition of the fractions resulted in distinctive differences in their thermal profile and solid fat content (Figures 7.4 and 7.5). Low melting behavior was exhibited by fractions obtained in the initial stages of frac-tionation, which may be potential candidates in a wide variety of food applications. Low melting fractions may be used in the preparation of cold-spreadable butter and bakery products, and improving the reconstitution of milk powder. High melting fractions may enhance the melting properties of ice cream.

7.4   DETERMINATION OF VITAMINS

Fat-soluble vitamins from dairy products such as UHT milk, milk powder, infant formula, canned baby food, and margarine were extracted by SFE (Singh and Avula 2007). Turner and Mathiasson (2000) developed a method for analysis of vitamins A and E in milk powder, which utilizes SFE, a miniaturized alkaline saponifica-tion procedure and reversed-phase high-performance liquid chromatography with ultraviolet (UV) detection. The extraction efficiency was optimized based on modi-fications of the sample matrix, combinations of static and dynamic modes of extrac-tion, and effects of changes in extraction parameters. The SFE extracted samples were saponified and the vitamins (A, E, and β-carotene) were determined using reversed-phase liquid chromatography with UV or fluorescence detection. Sample throughput was twice the number achievable with a conventional extraction method-ology (Mathiasson et al. 2002). Perretti et al. (2004) determined fat-soluble vitamins in cheese by SFE and compared the results with conventional methods. Since the extraction by SFE was conducted in the absence of light and oxygen and lower tem-peratures, a higher content of α-tocopherol was obtained than that achieved using conventional methods. Extractability of vitamins by SFE was comparable to official methods. Berg et al. (2000) also used SC-CO$_2$ to remove vitamins A and E from pow-dered and fluid milk, and the extraction conditions for both studies were 80°C and 37 MPa. Enzyme catalyzed hydrolysis and alcoholysis of ester bonds in vitamin A and E esters were carried out to facilitate their determination in different dairy food

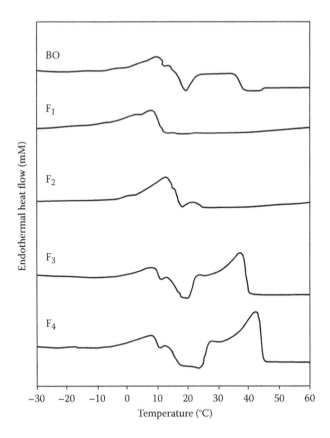

FIGURE 7.4 Differential scanning calorimetric melting thermograms (10°C/min) of buffalo butter oil (BO) and its fractions (F1–F4) obtained at different temperature and pressure conditions of SC-CO_2 extraction. (From Fatouh, A.E. et al., *Lebensm-Wiss Technol*, 40, 1687–1693, 2007. With permission.)

formulas. This method, developed by using immobilized *C. antarctica* preparation, is more beneficial to the oxidation-prone vitamins A and E compared to extraction methodologies based on alkaline saponification, resulting in comparatively higher recoveries (Turner et al. 2001). Vitamin K_1 (phylloquinone) was extracted from commercial soy protein–based and milk-based powdered infant formulas with SC-CO_2 at 55 MPa and 60°C (Schneiderman et al. 1988).

Tocopherol-enriched oil has many potential uses in the food industry. The optimum conditions (20 MPa, 80°C, and CO_2 flow rate at 20 ml/min) yielded a tocopherol concentration of 274.74 mg/100 g oil from Kalahari melon seed (Nyam et al. 2010). SC-CO_2 was found to be a good solvent for extracting both α-tocopherol and γ-oryzanol from rice bran, compared to solvent extraction and Soxhlet extraction because of its higher yields and extraction rate. The best conditions for α-tocopherol extraction from rice bran were 48 MPa and 55°C in the batch and continuous mode. For γ-oryzanol, the best conditions were 48 MPa, 65°C and in the continuous mode (Imsanguan et al. 2008). Although hexane and ethanol could not extract α-tocopherol,

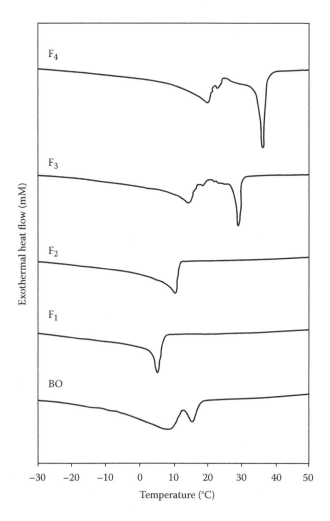

FIGURE 7.5 Differential scanning calorimetric crystallization thermograms (10°C/min) of buffalo butter oil (BO) and its fractions (F1–F4) obtained at different temperature and pressure conditions of SC-CO$_2$ extraction. (From Fatouh, A.E. et al., *Lebensm-Wiss Technol*, 40, 1687–1693, 2007. With permission.)

ethanol at 55–60°C was suitable for γ-oryzanol extraction. The SC-CO$_2$ extraction of milled grape seeds with a particle size range of 300–425 μm contained as much as 265 ppm α-tocopherol, which was higher than that obtained by *n*-hexane, although the overall oil yield was lower (Bravi et al. 2007).

7.5 EXTRACTION OF ANTIOXIDANTS

Naturally occurring chemicals from plants have been consumed by people for many years and have recently begun to receive considerable attention for their medicinal use as safe antioxidants (Herrero et al. 2006). Although about 200 compounds have been identified from Noni (*Morinda citrifolia*) by conventional methods, not much information was available on hydrophobic antioxidants present in the noni plant. Chen et al. (2009) determined the antioxidative properties of crude extracts from leaves, green and brown stems, and fruit of the noni plant obtained by near SC-CO$_2$ extraction. Near SFEs were conducted at pressures of 10.3 and 24.1 MPa, and temperatures of 35°C and 50°C for 3 h of static, followed by 1 h of dynamic extraction. Greater antioxidative capacities and superoxide radical scavenging activity of the extracts were observed, indicating the contribution of nonpolar antioxidants present in the extract in addition to phenolic compounds. The SC-CO$_2$ extraction of emodin (6-methyl-1,3,8-trihydroxyanthraquinone) and picied (stilbenoid glucoside, a major resveratrol derivative) from Japanese knotweed showed 2.5 times higher and 10 times lower yield, respectively, than those obtained using Soxhlet extraction. However, the advantage of SFE over the Soxhlet extraction method is a more than 5 times shorter extraction time (Benova et al. 2010). β-Carotene was the predominant carotenoid in pumpkins. The yields of lutein and lycopene contents were much higher in SFE than those in organic solvent extracts. *cis*-β-Carotene increased by more than 2 times in SC-CO$_2$ extracts, indicating enhanced solubility and isomerization (Shi et al. 2010).

Rosemary extracts are the only commercially available formulations for use as antioxidants in the European Union and the United States (Klancnik et al. 2009). The optimum extraction conditions for rosemary extracts were 38.3 MPa and 120°C for 20 min. Supercritical fluid (SF) rosemary extracts showed a higher concentration of carnosic acid, one of the main compounds responsible for antioxidant activity (Herrero et al. 2006). These extracts did not show any color and had higher activity than synthetic antioxidants and the extracts obtained by hydrodistillation. Evaluation of SC-CO$_2$ extracts from *Ramulus cinnamoni* for scavenging of the free radical DPPH (diphenyl picrylhydrazyl) showed 50% inhibition at a concentration of 2 mg/ml. These extracts also exhibited better antibacterial activity against *Acinetobacter baumannii*, *Pseudomonas aeruginosa*, and *Staphylococcus aureus* than those obtained by ethanol extraction. Extracts obtained from high pressure and low temperature conditions of SFE showed the best activity (Liang et al. 2008). The in vitro antioxidant activity of lotus germ oil extracted by SC-CO$_2$ has been investigated by Li et al. (2009). The high content of phenolic compounds and tocopherol in the lotus germ oil could partially account for the antioxidant activity. Liu et al. (2009c) reported notable antioxidant activity in the seed oil extracted from *Opuntia dillenii* Haw at 46.96 MPa, 46.51°C, for 2.79 h with a CO$_2$ flow rate of 10 kg/h. Grape waste extract obtained by SFE contained bioactive antioxidant molecules and induced antiproliferative effects in human colon adenocarcinoma cells (Lazze et al. 2009). Hu et al. (2005) found higher free radical scavenging activity (33.5%) of SC-CO$_2$ extracts of leaf skin from *Aloe vera* than the ethanol extracts (14.2%) of *A. vera* gel. The SC-CO$_2$ extracts of *A. vera* skin showed stronger antioxidant activities compared to BHT and α-tocopherol. The optimum yield of extracts was obtained at 35 MPa, 50°C, and 36 l/h, and 20% modifier of methanol. Yi et al. (2009) reported the effect of SC-CO$_2$ extraction parameters on the antioxidant activities of lycopene extracts when the extraction time was set to 90 min. For each unit of lycopene extract, the antioxidant activity level was constant below 70°C, but then gradually decreased above 70°C due to isomerization occurring at higher temperature. The ratio of all-*trans*-lycopene to the *cis*-isomers changed from 1.70 to 1.32 when the operating temperature was adjusted from 40 to 100°C, indicating an increased bioavailability due to the generation of *cis*-isomers. No significant effects of pressure or flow rate of SC-CO_2 extraction on the antioxidant activity were observed. The recovery of *trans*-lycopene depended on the content of the compound in the starting material and increased with increases in pressure and solvent flow rate, and with a decrease in the particle size. An increase in the moisture content from 4.6% to 22.8% led to a rise in the extraction yield and to a decrease in the recovery of *trans*-lycopene (Nobre et al. 2009). Eller et al. (2010) found that SC-CO_2 extraction of tomato seed oil resulted in highest total phytosterol as well as individual phytosterol content, compared to pressurized solvent extraction with *n*-hexane and ethanol. Chromatographic analysis indicated that lycopene extracted by SC-CO_2 from tomato skin had negligible degradation at the optimum conditions (40 MPa, 99.8°C, and 2.5 ml of CO_2/min), and the amount extracted represented more than 94% of the total carotenoid content of the sample (Topal et al. 2006).

Lenucci et al. (2010) optimized the biological and physical parameters for SC-CO_2 extraction of lycopene from ordinary and high pigment tomato cultivars. Lycopene matrix also contained β-carotene, lutein, α-tocopherol, and phytosterols, and its extraction resulted in substantial increase in lycopene recovery on addition of oleaginous comatrix. The lycopene extraction from tomato matrix was >80% at 45 MPa bar, 65–70°C with a flow rate of 18–20 kg h^{-1}. Rozzi et al. (2002a) reported the effect of temperature, pressure, flow rate, and CO_2 volume on SFE of lycopene from a by-product of tomato processing. The parameters for extraction were evaluated on the recovery of carotenoids and tocopherols in the tomato product and on the fatty acid composition of the lipids. It was determined that both temperature and pressure had an effect on the extraction of lycopene and that an optimum temperature and pressure combination (86°C and 34.47 MPa for 200 min) resulted in extracting 61.0% of the lycopene present in the sample by using 500 ml of CO_2 at a flow rate of 2.5 ml/min. Huang et al. (2008) reported 93% lycopene recovery at optimal conditions of 57°C, 40 MPa, and 1.8 h. The individual effects of pressure and time, as well as their interactive effects on antioxidant activity of the extract were significant. SC-CO_2 extraction with or without cosolvent of Brazilian red spotted shrimp waste yielded lower carotenoids as compared to organic solvents. However, addition of the entrainer (10% w/w) increased astaxanthin values, which are similar to those observed during extraction with acetone (Sanchez-Camargo et al. 2011). Conjugated linoleic acid (CLA), which is recognized as a growth factor, causes reduction of fat tissue. Milk and dairy products including cheese are the richest sources of CLA. Domagala et al. (2010) compared the yields of CLA obtained by gas chromatography (GC), solvent extraction, and SC-CO_2, and found no clear influence of the extraction method on CLA content in cheese.

SFE has been widely used in bioprocess technologies, especially for disruption of microorganisms, destruction of industrial waste, the gas-antisolvent crystallization, and micron-size particle formation. Enzymatic catalysis in SC-CO_2 is gaining ground as it offers the possibility of integrated synthesis product recovery processes (Khosravi-Darani and Mozafari 2009). One of the most interesting applications of

SFE in bioprocess technology is extraction of food applicable metabolites (fats, vitamins, antioxidants, volatiles, and flavors) from biotechnologically produced biomass such as *Spirulina*, and other microalgae, fungi, and bacteria. SC-CO_2 offers short extraction time for extraction of lipids from microalgae, which are recognized as an important source of bioactive lipids. Natural food ingredients possessing antimicrobial, antiviral, antifungal, and antioxidant activities were extracted by using SC-CO_2 by several research groups. Herrero et al. (2006) reviewed the SC-CO_2 extraction of functional ingredients from algae (species of *Dilophus*, *Hypnea*) and microalgae (species of *Botrycoccus*, *Chlorella*, *Dunaliella*, *Spirulina*).

The cyanobacterium *Spirulina (Arthrospira) platensis* is characterized by high content in γ-linoleneic acid (C18:3ω-6). The kinetics of the SFE process and the effect of operating conditions on lipid extraction and fatty acid composition of extracts from *Spirulina* were described by Andrich et al. (2006). Choudhari and Singhal (2008) extracted the recovery of lycopene from mated cultures of *Blakeslea trispora*. Incorporation of entrainer further enhanced lycopene recovery from 73% to 85%. The optimum conditions were found to be 34.9 MPa, at 52°C for 1.1 h extraction. Lutein is a xanthophyll (family of oxygenated carotenoids) that, together with zeaxanthin, has attracted attention because of its ability to prevent degenerative human diseases. At present, lutein is produced from marigold oleoresin, but certain microorganisms that produce lutein can also become an alternative source (Fernandez-Sevilla and Fernandez 2010). Several researchers have used the SFE technique to extract lutein and carotenoids from microalgal biomass. Kitada et al. (2009) extracted lutein from dry *Chlorella* although the yield was much lower than that obtained when using Soxhlet extraction. SC-CO_2 extraction showed 40% yield of total carotenoids from *Nannochloropsis gadiatana* (Macias-Sanchez et al. 2005) and 50% yield of carotenoids from *Dunaliela salina*, compared to solvent extraction by dimethyl formamide (Macias-Sanchez et al. 2009a,b). Gao et al. (2009) used ultrasound to enhance the extraction yield but found only moderate results. The use of cosolvents cancels out one of the main advantages of SFE: the elimination of use of solvents. SC-CO_2 extraction tends to extract chlorophylls more efficiently than carotenoids, leaving extracts heavily contaminated (Kitada et al. 2009).

7.6   EXTRACTION OF VOLATILES/FLAVORS

Hydrodistillation and steam distillation are the conventional methods used to isolate essential oils from aromatic plants, which may cause hydrolysis and thermal degradation of the product. Supercritical fluids provided a higher solubility of the components of the volatile oil, as well as improved mass transfer rates (Reverchon and De Marco 2006). In addition, the manipulation of temperature and pressure led to the extraction of different components, which can be useful when a particular component is required (Pourmortazavi and Hajimirsadeghi 2007; Vazquez et al. 2008). The influence of SFE on the yield and composition of the volatile oil from Italian coriander seeds was evaluated by Grosso et al. (2008). The best conditions for extractions were found to be 9 MPa and 40°C. The dominant components of the extract found by GC and GC–mass spectrometry were linalool, γ-terpinene, camphor, geranyl acetate, α-pinene, geraniol, and limonene. Aromatic extracts of Roncal cheese were obtained by SFE under different solvent conditions. The extracts retained in an octadecyl silica trap maintained at −5°C during the entire process were reconstituted in different solvents. The reconstitution solvent mixture containing n-hexane/acetone (2:1) represented better Roncal aromatic qualities in sensory evaluation (Larrayoz et al. 2000). The extraction of aroma components from unsmoked ewe's milk Idiazabal cheese was reported by Larrayoz et al. (1999), and the optimized parameters included sample preparation, sample/adsorbent ratio, reconstitution solvent, trap type, and trapping temperature.

Limonoids (limonin and limonin-17-β-D-glucopyranoside) and flavonoids (naringin) were extracted from grapefruit seeds. Less polar limonin was extracted in the first stage by using SC-CO$_2$, whereas in the second stage high polar limonin-17-β-D-glucopyranoside (LG) and naringin could be extracted with SC-CO$_2$ modified by ethanol as cosolvent. The highest yield of limonin (6.3 mg/g seeds) was obtained at 48.3 MPa, 50°C, and 60 min of extraction time, whereas the highest yield of LG (0.62 mg/g seeds) was achieved at 41.4 MPa, 60°C, and 30% ethanol concentration in 40 min. The highest yield of naringin (0.2 mg/g seeds) was achieved at 41.4 MPa, 50°C, and 20% ethanol concentration in 40 min. Mobile-phase flow rate was kept constant at 5.0 l/min (Yu et al. 2007). Isolation of carrot (*Daucus carrota* L.) fruit essential oil was obtained at 40°C and 10 MPa. The supercritical extract was characterized by the presence of heavier molecular weight compounds, whereas some lighter compounds (e.g., pinenes) were not detected. The main component of the supercritical extract was carotol, and it was the most effective compound against Gram-positive bacteria (Glisic et al. 2007). Rozzi et al. (2002b) extracted citral compounds from lemon balm, lemon bergamot, lemon eucalyptus, and lemongrass by using SC-CO$_2$ more efficiently and faster than hydrodistillation. Furthermore, additional citral compounds that are not usually found in hydrodistillates of either plant were extracted by SC-CO$_2$ in the case of lemongrass and lemon eucalyptus. The relative distribution of caryophyllene content from SFE was higher than that found with hydrodistillation (Table 7.3). As the temperature, pressure, CO$_2$ volume, and to a lesser extent flow rate increased, the mass fraction extracted from each plant increased.

Limonoid glucosides are not commercially available, and the extraction and purification of limonoids from juice processing by-products can increase the potential value of citrus crops. It was reported that molasses contains 18% of total limonoid glucosides present in the whole fruit. Miyake et al. (2000) reported that SC-CO$_2$ could be used to remove limonoid aglycones and concentrate limonoid glucosides in citrus juices. Yu et al. (2007) extracted limonoid glucosides from grapefruit molasses by one-step extraction using SC-CO$_2$–ethanol. The conditions were optimized at 48.3 MPa, 50°C, 10% ethanol, and a 5.0 l/min CO$_2$–ethanol flow rate with the extraction yield of 0.61 mg limonoid glucosides/g grapefruit molasses. Hamdan et al. (2008) reported maximum levels of natural pigments and antioxidants in cardamom oil by using SC-CO$_2$ and propane. Propane was more effective at subcritical conditions with a lower ratio of solvent/solid and better quality attributes. Lasekan and Abbas (2010) extracted volatile compounds and acrylamide from roasted almond. Aromatic compounds made up of 27 hydrocarbons, 12 aldehydes, 11 ketones, 7 acids, 4 esters, 3 alcohols, 5 furan derivatives, a pyrazine, and 2 unknown compounds. Acrylamide

TABLE 7.3

Relative Extraction of Caryophyllene in Aromatic Plants (% of Total Distribution of Extract Content)

Extraction Solvent Conditions	Caryophyllene
Lemon balm	
Hydrodistillation	1.23
CO_2: 13.79 MPa, 60°C	8.47
CO_2: 41.37 MPa, 60°C	6.63
Lemongrass	
Hydrodistillation	N/A
CO_2: 27.58 MPa, 40°C	38.63
CO_2: 27.5 MPa, 60°C	13.40
Lemon eucalyptus	
Hydrodistillation	N/A
CO_2: 13.7 MPa, 40°C	2.77
CO_2: 41.3 MPa, 40°C	1.80

Source: Rozzi, N.L. et al., *Lebensm-Wiss Technol*, 35, 319–324, 2002b. With permission.

concentration increased with increase in roasting temperature and time. Thermal degradation of volatiles from oregano, basil, and mint could be avoided in SFE compared to simultaneous distillation–extraction (Rissato et al. 2004). Aroma extracts from Iberian ham showed 41 new volatile compounds that were not extracted by other conventional methods, and higher concentrations of volatile compounds were obtained at 40°C and a density of 0.5 g/ml (Timon et al. 1998). Bhattacharjee et al. (2003) compared the aroma profiles of Basmati rice obtained by SC-CO_2 extraction and Likens–Nickerson extraction. Appreciable extraction of the volatile constituents of the rice could be obtained in SC-CO_2 extraction at 50°C and 12 MPa for 2 h. Catchpole et al. (2003) compared the yields of gingerols from ginger, capsaicins from chilli, and piperines from black pepper obtained by using near-critical CO_2, propane, and dimethyl ether. The yields obtained by SC-CO_2 and dimethyl ether were much higher compared to propane extraction.

The SF extracts of Brazilian cherry contained sesquiterpenes and ketones that strongly contribute to the characteristic flavor of Brazilian cherry. Volatile phenolic compounds responsible for antioxidant activity were also present in high proportion (Malaman et al. 2011). Aroma components with application in blending steps for fast aged rum processes were extracted by SC-CO_2 (Garcia et al. 2009). Concentrated aroma fractions from head alcohol, crude, and aged sugar cane spirit process streams were quantified. The economical profitability of SC-CO_2 extraction was evident as the compounds were concentrated by more than 300 times compared to commercial aged rum.

7.7 DETERMINATION OF PESTICIDE AND DRUG RESIDUES

Cortes et al. (2009) used a method to analyze pesticide residues in the lycopene and carotenoid extracts obtained from tomatoes by SFE and reversed-phase liquid chromatography–GC (RPLC–GC) using the through oven transfer adsorption–desorption interface. Trace amounts of the pesticides were extracted together with the carotenoids. Extractions were performed using an SFE module designed by Iberfluid (Madrid, Spain), which consists of a pump, a flow meter, a 300-ml extraction vessel, and two 100-ml separation vessels. The temperature was set at 40°C and pressure at 15 MPa, respectively, in the first separation vessel and 25°C and 6 MPa in the second separation vessel. The overall extraction time was 3 h. Processes that reduce moisture content of fluid milk may result in a high concentration of sulfamethazine, an animal drug residue not detectable in fluid milk. Sulfamethazine could be extracted quantitatively from dry milk powder at 4.5 MPa and 80°C using SC-CO_2 modified with 10% methanol as the extraction fluid, and it was not detectable in the SC-CO_2 extracted milk powder (Lopez-Avila and Benedicto 1996). Clenbuterol was extracted from diatomaceous earth and lyophilized milk as its 10-camphorsulfonate ion pair using SC-CO_2 for 30 min of dynamic extraction at 38.3 MPa and 40°C (Jimmenz-Carmona et al. 1995). The distribution of polychlorinated biphenyls (PCBs) in the milk fat globule has been studied by sequential extraction of four different lipidic fractions from powdered full-fat milk with SC-CO_2. Extractions were carried out in the dynamic mode in the pressure range 13.6–23.3 MPa at a temperature of 50°C. Extracts obtained at lower pressures were found to be enriched in PCBs, short- and medium-chain triglycerides, and cholesterol (Ramos et al. 2000). A rapid and simple method for the direct screening of paraquat and diquat in olive oil samples involving SFE and liquid chromatography with diode array detection was developed by Zougagh et al. (2008). This arrangement opens up interesting prospects for the direct determination of polar pesticides in complex samples with a good throughput and a high level of automation.

7.8 CONCLUSIONS

SFE using CO_2 as a solvent has provided a viable alternative to the use of chemical solvents. SC-CO_2 is successfully applied for extraction and fractionation of fats, antioxidants, antimicrobials, vitamins, pigments, and flavors from food systems. In order to substitute for the most laborious, time-consuming techniques and classical procedures, SFE methods must be validated so they can be approved as official procedures for food analysis. Future research needs to be focused on the design and development of new/modified SFE instrumentation to minimize use of cosolvents. Continuous SFE units provided with online monitoring systems are essential for commercial units so operators can understand the extraction kinetics and optimize the extraction conditions for various types of foods. Economics of large-scale extractions of metabolites from algae, microalgae, fungi, and bacteria have to be worked out for their application in functional foods and nutraceuticals. Attention may also be drawn to the potential application of SFE in microbial inactivation and destruction of spores to achieve sterilization of processed foods.

REFERENCES

Andras, C.D., Simandi, B., Orsi, F., Lambrou, C., Missopolinou-Tatala, D., Panayiotou, C., Domokos, J., Doleschall, F. 2005. Supercritical carbon dioxide extraction of okra (*Hibiscus esculentus* L.) seeds. *J Sci Food Agric* 85: 1415–1419.

Andrich, G., Zinnai, A., Venturi, F., Florentini, R. 2006. Supercritical fluid extraction of oil from microalga *Spirulina (arthrospira) platensis*. *Acta Aliment* 35: 195–203.

Arul, J., Boudreau, J., Makhlouf, J. 1988. Distribution of cholesterol in milk fat fractions. *J Dairy Res* 55: 361–371.

Asep, E.K., Jinap, T.J., Russy, A.R., Harcharan, S., Nazimah, S.A.H. 2008. The effects of particle size, fermentation and roasting of cocoa nibs on supercritical fluid extraction of cocoa butter. *J Food Eng* 85: 450–458.

Astaire, J.C., Ward, R., German, J.B., Flores, R.J. 2003. Concentration of polar MFGM lipids from buttermilk by microfiltration and supercritical fluid extraction. *J Dairy Sci* 86: 2297–2307.

Benova, B., Adam, M., Pavlikova, P., Fischer, J. 2010. Supercritical fluid extraction of piceid, resveratrol, and emodin from Japanese knotweed. *J Supercrit Fluid* 51: 325–330.

Berg, H., Turner, C., Dahlberg, L., Mathiasson, L. 2000. Determination of food constituent in food based on SFE-application on vitamin A and E in meat and milk. *J Biochem Biophy Met* 43: 391–401.

Bernardo-Gill, M.G., Lopes, L.M.C. 2004. Supercritical fluid extraction of *Cucurbita ficifolia* oil. *Eur Food Res Technol* 219: 593–597.

Beveridge, T.H.J., Cliff, M., Sigmund, P. 2004. Supercritical carbon dioxide percolation of sea buckthorn press juice. *J Food Qual* 27: 41–54.

Bhaskar, A.R., Rizvi, S.S.H., Harriott, P. 1993. Performance of a packed column for continuous supercritical carbon dioxide processing of anhydrous milk fat. *Biotech Progr.* 9: 70–74.

Bhaskar, A.R., Rizvi, S.S.H., Bertoli, C., Fay, L.B., Hug, B. 1998. A comparison of physical and chemical properties of milk fat fractions obtained by two processing technologies. *J Am Oil Chem Soc* 75: 1249–1264.

Bhattacharjee, P., Ranganathan, T.V., Singhal, R.S., Kulkarni, P.R. 2003. Comparative aroma profiles using supercritical carbon dioxide and Likens–Nickerson extraction from a commercial brand of Basmati rice. *J Sci Food Agric* 83: 880–883.

Bhattacharjee, P., Singhal, R.S., Tiwari, S.R. 2007. Supercritical carbon dioxide extraction of cottonseed oil. *J Food Eng* 79: 892–898.

Bravi, M., Spinoglio, F., Verdone, N., Adami, M., Aliboni, A., D'Andrea, A., De Santis, A., Ferri, D. 2007. Improving the extraction of α-tocopherol-enriched oil from grape seeds by supercritical CO_2. Optimisation of the extraction. *J Food Eng* 78: 488–493.

Brunner, G., Gast, K., Chuang, M., Kumar, S., Chan, P., Chan, W.P. 2009. Process for production of highly enriched fractions of natural compounds from palm oil with supercritical and near critical fluids. U.S. Patent 20090155434.

Caili, F., Huan, S., Quanhong, L. 2006. A review on pharmacological activities and utilization technologies of pumpkin. *Plant Foods Hum Nutr* 61: 73–80.

Catchpole, O.J., Grey, J.B., Perry, N.B., Burgess, E.J., Redmond, W.A., Porter, N.G. 2003. Extraction of chilli, black pepper, and ginger with near-critical CO_2, propane, and dimethyl ether: Analysis of the extracts by quantitative nuclear magnetic resonance. *J Agric Food Chem* 51: 4853–4860.

Chen, C.H., Lin, T.P., Chung, Y.L., Lee, C.K., Yeh, D.B., Chen, S.Y. 2009. Determination of antioxidative properties of *Morinda citrifolia* using Near Supercritical Fluid Extraction. *J Food Drug Anal* 17: 333–341.

Correa, A.P.A., Cabral, F.A., Gonclalves, L.A., Ireny, A.G., Peixoto, C.A. 2008. Fractionation of fish oil with supercritical carbon dioxide. *J Food Eng* 88: 381–387.

Cortes, J.M., Vazquez, A., Santa-Maria, G., Blanch, G.P., Villen, J. 2009. Pesticide residue analysis by RPLC-GC in lycopene and other carotenoids obtained from tomatoes by supercritical fluid extraction. *Food Chem* 113: 280–284.

Choudhari, S.M., Singhal, R.S. 2008. Supercritical carbon dioxide extraction of lycopene from mated cultures of *Blakeslea trispora* NRRL 2895 and 2896. *J Food Eng* 89: 349–354.

Crews, C., Hough, P., Godward, J., Brereton, P., Lees, M., Guiet, S., Winkelmann, W. 2005. Study of the main constituents of some authentic walnut oils. *J Agric Food Chem* 53: 4853–4860.

Domagala, J., Sady, M., Grega, T., Pustkowiak, H., Florkiewicz, A. 2010. The influence of cheese type and fat extraction method on the content of conjugated linoleic acid. *J Food Comp Anal* 23: 238–243.

Eller, F.J., Moser, J.K., Kenar, J.A., Taylor, S.L. 2010. Extraction and analysis of tomato seed oil. *J Am Oil Chem Soc* 87: 755–762.

Fatouh, A.E., Mahran, G.A., El-Ghandour, M.A., Singh, R.K. 2007. Fractionation of buffalo butter oil by supercritical carbon dioxide. *Lebensm-Wiss Technol* 40: 1687–1693.

Fernandez-Sevilla, J.M., Fernandez, F.G.A. 2010. Biotechnological production of lutein and its applications. *Appl Microbiol Biotechnol* 86: 27–40.

Gao, Y., Nagy, B., Liu, X., Simandi, B., Wang, Q. 2009. Supercritical CO_2 extraction of lutein esters from marigold (*Tagetes erecta* L.) enhanced by ultrasound. *J Supercrit Fluids* 49: 345–350.

Glisic, S.B., Misic, D.R., Stamenic, M.D., Zizovic, I.T., Asanin, R.M., Skala, D.U. 2007. Supercritical carbon dioxide extraction of carrot fruit essential oil: Chemical composition and antimicrobial activity. *Food Chem* 105: 346–352.

Gracia, I., Garcia, M.T., Rodriguez, J.F., de Lucas, A. 2009. Application of supercritical fluid extraction for the recovery of aroma compounds to be used in fast aged rum production. *Food Sci Technol Res* 15: 353–360.

Grosso, C., Coelho, J.A., Palavra, A.M., Barroso, J.G., Ferraro, V., Figueiredo, A.C. 2008. Supercritical carbon dioxide extraction of volatile oil from Italian coriander seeds. *Food Chem* 111: 197–203.

Hamdan, S., Daood, H.G., Toth-Markus, M., Illes, V. 2008. Extraction of cardamom oil by supercritical carbon dioxide and sub-critical propane. *J Supercrit Fluids* 44: 25–26.

Han, X., Bi, J., Zhang, R., Cheng, L. 2009. Extraction of safflower seed oil by supercritical CO_2. *J Food Eng* 92: 370–376.

Herrero, M., Cifuentes, A., Ibanez, E. 2006. Sub- and supercritical fluid extraction of functional ingredients from different natural sources: Plants, food by-products, algae and microalgae. *Food Chem* 98: 136–148.

Higuera-Ciapara, I., Toledo-Guillen, A.R., Noriega-Orozco, L., Martinez-Robinson, K.G., Esqueda-Valle, M.C. 2005. Production of a low-cholesterol shrimp using supercritical extraction. *J Food Process Eng* 28: 526–538.

Hu, Q.H., Hu,Y., Xu, J. 2005. Free radical-scavenging activity of *Aloe vera* (*Aloe barbadensis* Miller) extracts by supercritical carbon dioxide extraction. *Food Chem* 91: 85–90.

Huang, W., Li, D., Zhang, J., Li, Z., Niu, H. 2008. Optimization of operating parameters for supercritical carbon dioxide extraction of lycopene by response surface methodology. *J Food Eng* 89: 298–302.

Imsanguan, P., Douglas, S., Douglas, P.L., Pongamphai, S., Roaysubtawee, A., Borirak, R. 2008. Extraction of α-tocopherol and γ-oryzanol from rice bran *Lebensm-Wiss Technol Food Sci Technol* 41: 1417–1424.

Jimmenz-Carmona, M.M., Tena, M.T., Luque de Castro, M.D. 1995. Ion-pair-supercritical fluid extraction of clenbuterol from food samples. *J Chromatogr* 711: 269–276.

Khosravi-Darani, K. 2010. Research activities on supercritical fluid science in food biotechnology. *Crit Rev Fod Sci Nutr* 50: 479–488.

Khosravi-Darani, K., Mozafari, M.R. 2009. Supercritical fluids technology in bioprocess industries: A review. *J Biochem Technol* 2: 144–152.

King, J.W. 2002. Supercritical Fluid Extraction: Present status and prospects. *Grasas Aceites.* 53: 8–21.

Kitada, K., Machmudah, S., Sasaki, M., Goto, M., Nakashima, Y., Kumamoto, S., Hasegawa, T. 2009. Supercritical CO_2 extraction of pigment compounds with pharmaceutical importance from *Chlorella vulgaris*. *J Chem Technol Biotechnol* 84: 657–661.

Klancnik, A., Guzej, B., Kolar, M.H., Abramovic, H., Mozina, S.S. 2009. In vitro antimicrobial and antioxidant activity of commercial rosemary extract formulations. *J Food Prot* 72: 1744–1752.

Lang, Q., Wai, C.M. 2001. Supercritical fluid extraction in herbal and natural product studies—A practical review. *Talanta* 53: 771–782.

Larrayoz, P., Carbonell, M., Ibanez, F., Torre, O., Barcina, Y. 1999. Optimization of indirect parameters which affect the extractability of volatile aroma compounds from Idiazabal cheese using analytical supercritical fluids extractions (SFE). *Food Chem* 64: 123–127.

Larrayoz, P., Ibanez, F.C., Ordonez, A.I., Torre, P., Barcina, Y. 2000. Evaluation of supercritical fluid extraction as sample preparation method for the study of Roncal cheese aroma. *Int Dairy J* 10: 755–759.

Lasekan, O., Abbas, K. 2010. Analysis of volatile flavor compounds and acrylamide in roasted Malaysian tropical almond (*Terminalia catappa*) nuts using supercritical fluid extraction. *Food Chem Toxicol* 48: 2212–2216.

Lau, H.L.N., Choo, Y.M., Ma, A.N., Chuah, C.H. (2008). Selective extraction of palm carotene and vitamin E from fresh palm-pressed mesocarp fiber (*Elaeis guineensis*) using supercritical CO_2. *J Food Eng* 84: 289–296.

Lazze, M.C., Pizzala, R., Pecharroman, F.J.G., Garnica, P.G., Rodríguez, J.M.A., Fabris, N., Livia, B. 2009. Grape waste extract obtained by supercritical fluid extraction contains bioactive antioxidant molecules and induces antiproliferative effects in human colon adenocarcinoma cells. *J Med Food* 12: 561–568.

Lenucci, M.L., Cacciopola, A., Durante, M., Serrone, L., Leonardo, R., Piro, G., Dalessandro, G. 2010. Optimization of biological and physical parameters for lycopene supercritical CO_2 extraction from ordinary and high-pigment tomato cultivars. *J Sci Food Agric* 90: 1709–1718.

Leo, L., Rescio, L., Ciurlia, L., Zacheo, G. 2005. Supercritical *carbon dioxide* extraction of oil and α-tocopherol from almond seeds. *J Sci Food Agric* 85: 2167–2174.

Li, J., Zheng, T., Zhang, M. 2009. The in vitro antioxidant activity of lotus germ oil from supercritical fluid carbon dioxide extraction. *Food Chem* 115: 939–944.

Liang, M.T., Yang, C.H., Li, S.T., Yang, C.S., Chang, H.W., Liu, C.S., Cham, T.M., Chuang, L.Y. 2008. Antibacterial and antioxidant properties of *Ramulus cinnamomi* using supercritical CO_2 extraction. *Eur Food Res Technol* 227: 1387–1396.

Liu, X., Zhao, M., Wang, Z., Luo, W. 2009a. Antimicrobial and antioxidant activity of emblica extracts obtained by supercritical carbon dioxide and methanol extraction. *J Biochem* 33: 307–330.

Liu, G., Gao, Y., Hao, Q., Xu, X. 2009b. Supercritical CO_2 extraction optimization of pomegranate (*Punica granatum* L.) seed oil using response surface methodology. *Lebensm-Wiss Technol* 42: 1491–1495.

Liu, W., Wu, N., Liu, X., Zhang, S., Fu, Y., Zu, Y., Tong, M. 2009c. Supercritical carbon dioxide extraction of seed oil from *Opuntia dillenii* Haw. and its antioxidant activity. *Food Chem* 114: 334–339.

Lopez-Avila, V., Benedicto, J. 1996. Determination of veterinary drugs in dry milk powder by supercritical fluid extraction-enzyme linked immunosorbent assay. In *Veterinary Drug Residues: Food Safety*. Washington, D.C.: American Chemical Society.

Macias-Sanchez, M.D., Mantell, C., Rodriguez, M. 2005. Supercritical fluid extraction of carotenoids and chlorophyll a from *Nannochloropsis gadiatana*. *J Food Eng* 66: 245–251.

Macias-Sanchez, M.D., Mantell, C., Rodriguez, M., Martinez, O.E. 2009a. Kinetics of the supercritical fluid extraction of carotenoids from microalgae with CO_2 and ethanol as cosolvent. *Chem Eng J* 150: 104–113.

Macias-Sanchez, M.D., Mantell, C., Rodriguez, M., Martinez, O.E., Lubian, L.M., Montero, O. 2009b. Comparison of supercritical fluid and ultrasound-assisted extraction of carotenoids and chlorophyll a from *Dunaliella salina*. *Talanta* 77: 948–952.

Malaman, F.S., Moraes, L.A.B., West, C., Ferreira, N.J., Oliviera, A.L. 2011. Supercritical fluid extracts from the Brazilian cherry (*Eugenia uniflora* L.): Relationship between the extracted compounds and the characteristic flavor intensity of the fruit. *Food Chem* 124: 85–92.

Martinez, M.L., Mattea, M.A., Maestri, D.M. 2008. Pressing and supercritical carbon dioxide extraction of walnut oil. *J Food Eng* 88: 399–404.

Martinez, M.L., Mattea, M.A., Maestri, D.M. 2006. Varietal and crop year effects on lipid composition of walnut (*Juglans regia* L.) genotypes. *J Am Oil Chem Soc* 68: 322–336.

Mathiasson, L., Turner, C., Berg, H., Dahlberg, L., Theobald, A., Anklam, E., Ginn, R., Sharman, M., Ulberth, F., Gabernig, R. 2002. Development of methods for the determination of vitamins A, E and β-carotene in processed foods based on supercritical fluid extraction: A collaborative study. *Food Addit Contam* 19: 632–646.

Mendiola, J.A., Herrero, H., Alejandro C.A., Elena. I.E. 2007. Use of compressed fluids for sample preparation. Food applications. *J Chromatogr* 1152: 234–246.

Mezzomo, N., Mileo, B.R., Friedrich, M.T., Martinez, J., Ferreira, S.R.S. 2010. Supercritical fluid extraction of peach (*Prunus persica*) almond oil: Process yield and extract composition. *Bioresour Technol.* 101: 5622–5632.

Mitra, P., Ramaswamy, H.S., Chang, K.S. 2009. Pumpkin (*Cucurbita maxima*) seed oil extraction using supercritical carbon dioxide and physicochemical properties of the oil. *J Food Eng* 95: 208–213.

Mitre, R., Etienne, M., Martinais, S., Salmon, H., Allaume, P., Legrands, P., Legrand, A.B. 2005. Humorial defence improvement and haematopoiesis stimulation in sows and off spring by oral supply of shark-liver oil to mothers during gestation and lactation. *Br J Nutr* 94: 753–762.

Miyake, M., Shimoda, M., Osajima, Y., Inaba, N., Ayano, S., Ozaki, Y., Hasegawa, S. 2000. Extraction and recovery of limonoids with the supercritical carbon dioxide microbubble method, in *Citrus Limonoids: Functional Chemicals in Agriculture and Food*, M.A. Berhow, S. Hasegawa, G.D. Manners (Eds.). *Symposium Series* 758: 96–106. Washington, DC: American Chemical Society.

Nobre, B.P., Palavra, A.F., Pessoa, F.L.P., Mendes, R.L. 2009. Supercritical CO_2 extraction of *trans*-lycopene from Portuguese tomato industrial waste. *Food Chem* 116: 680–685.

Norulaini, N.A., Azizi, C.Y.M., Omar, A.K.M., Nawi, A.H., Setianto, W.B., Zaidul, I.S.M. 2009. Effects of supercritical carbon dioxide extraction parameters on virgin coconut oil yield and medium-chain triglyceride content. *Food Chem* 116: 193–197.

Nyam, K.L., Long, K., Man, Y.B.C., Lai, O.M., Tan, C.P., Karim, R. 2010. Extraction of tocopherol-enriched oils from Kalahari melon and roselle seeds by supercritical fluid extraction (SFE-CO_2). *Food Chem* 119: 1278–1283.

Ozkal, S.G., Yener, M.E., Bayindirli, L. 2005a. Response surfaces of apricot kernel oil yield in supercritical carbon dioxide. *Lebensm-Wiss Technol* 38: 611–616.

Ozkal, S.G., Salgin, U., Yener, M.E. 2005b. Supercritical carbon dioxide extraction of hazelnut oil. *J Food Eng* 69: 217–223.

Perretti, G., Marconi, O., Montanari, L., Fantozzi, P. 2004. Rapid determination of total fats and fat-soluble vitamins in Parmigiano cheese and salami by SFE. *Lebensm-Wiss Technol* 37: 87–92.

Pourmortazavi, S.M., Hajimirsadeghi, S.S. 2007. Supercritical fluid extraction in plant essential and volatile oil analysis. *J Chromatogr* 1163: 2–24.

Ramos, L., Hernandez, L.M., Gonzalez, M.J. 2000. Study of the distribution of the polychlorinated biphenyls in the milk fat globule by supercritical fluid extraction. *Chemosphere* 41: 881–888.

Reverchon, E., De Marco, I. 2006. SFE and fractionation of nature matter. *J Supercrit Fluids* 38: 146–166.

Rissato, S.R., Galhiane, M.S., Knoll, F.N., Apon, B.M. 2004. Supercritical fluid extraction for pesticide multiresidue analysis in honey: Determination by gas chromatography with electron-capture and mass spectrometry detection. *J Chromatogr* 1048: 153–159.

Rizvi, S.S.H., Bhaskar, A.R. 1995. Supercritical fluid processing of milk fat: Fractionation, scale-up and economics. *Food Tech* 49: 90–100.

Rozzi, N.L., Singh, R.K. 2002. Supercritical fluids and the food industry. *Comp Rev Food Sci Food Saf* 1: 33–44.

Rozzi, N.L., Singh, R.K., Vierling, R.A., Watkins, B.A. 2002a. Supercritical fluid extraction of lycopene from tomato processing byproducts. *J Agric Food Chem* 50: 2638–2643.

Rozzi, N.L., Phippen, W., Simon, J.E., Singh, R.K. 2002b. Supercritical fluid extraction of essential oil components from Lemon-scented Botanicals. *Lebensm-Wiss Technol* 35: 319–324.

Sahena, F., Zaidul, I.S.M., Jinap, S., Karim, A.A., Abbas, K.A., Norulaini, N.A.N., Omar, A.K.M. 2009. Application of supercritical CO_2 in lipid extraction—A review. *J Food Eng* 95: 240–253.

Sanchez-Camargo, A.P., Meireles, M.A.A., Lopes, B.L.F., Cabral, F.A. 2011. Proximate composition and extraction of carotenoids and lipids from Brazilian redspotted shrimp waste (*Farfantepenaeus paulensis*). *J Food Eng* 102: 87–93.

Sheibani, A., Ghaziaskar, H., 2008. Pressurized fluid extraction of pistachio oil using a modified supercritical fluid extractor and factorial design for optimization. *Lebensm-Wiss Technol* 41: 1472–1477.

Schutz, E. 2007. Supercritical fluids and applications—A patent review. *Chem Eng Technol* 30: 685–688.

Schneiderman, M.A., Sharma, A.K., Mahanama, K.R., Locke, D.C. 1988. Determination of vitamin K1 in powdered infant formulas, using supercritical fluid extraction and liquid chromatography with electrochemical detection. *J Assoc Anal Chem*. 71: 815–817.

Shi, J., Mittal, G., Kim, E., Xue, S.J. 2007. Solubility of carotenoids in supercritical CO_2. *Food Rev Int* 23: 341–371.

Shi, J., Yi, C., Ye, X.Q., Xue, S., Jiang, Y.M., Liu, D.H. 2010. Effects of supercritical CO_2 fluid parameters on chemical composition and yield of carotenoids extracted from pumpkin. *Lebensm-Wiss Technol* 43: 39–44.

Singh, R.K., Avula, R.Y. 2007. Supercritical fluid extraction in dairy and food processing, in *Proceedings of International Conference on Conventional Dairy Foods*, S. Arora, A.K. Singh, R.R.B. Singh, and L. Sabhiki (Eds.), 121–130. Karnal: India.

Spano, V., Salis, A., Mele, S., Madhu, P., Monduzzi, M. 2004. Fractionation of sheep milk fat via supercritical carbon dioxide. *Food Sci Technol Int* 10: 421–425.

Spence, A.J., Goddik, L., Qian, M., Jimenez-Flores, R. 2009. The influence of temperature and pressure factors in supercritical fluid extraction for optimizing nonpolar lipid extraction from buttermilk powder. *J Dairy Sci* 92: 458–468.

Stevenson, D.G., Eller, F.J., Wang, L., Jane, J.L., Wang, T., Inglett, G. 2007. Oil and tocopherol content and composition of pumpkin seed oil in 12 cultivars. *J Agric Food Chem* 55: 4005–4013.

Taylor, S.L., Eller, F.J., King, J.W. 1997. A comparison of oil and fat content in oilseeds and ground beef using supercritical fluid extraction and related analytical techniques. *Food Res Int* 30: 365–370.

Timon, M.L., Ventanas, J., Martin, L., Tejeda, J.F., Garcia, C. 1998. Volatile compounds in supercritical carbon dioxide extracts of Iberian ham. *J Agric Food Chem* 46: 5143–5150.

Topal, U., Sasaki, M., Goto, M., Hayakawa, K. 2006. Extraction of lycopene from tomato skin with supercritical carbon dioxide: Effect of operating conditions and solubility analysis. *J Agric Food Chem* 54: 5604–5610.

Torres, C.F., Torrelo, G., Senorans, F.J., Reglero, G. 2009. Supercritical fluid fractionation of fatty acid ethyl esters from butteroil. *J Dairy Sci* 92: 1840–1845.

Turner, C., Mathiasson, L. 2000. Determination of vitamins A and E in milk powder using supercritical fluid extraction for sample clean-up. *J Chromatogr* 874: 275–283.

Turner, C., Persson, M., Mathiasson, L., Adlercreutz, P., King, J.W. 2001. Lipase catalyzed reactions in organic and supercritical solvents: Application to fat-soluble vitamin determination in milk powder and infant formula. *Enzyme Microb Technol* 29: 111–121.

Turner., C., Linda, C., Whitehand, T.N., McKeon, T. 2004. Optimization of a supercritical fluid extraction/reaction methodology for the analysis of castor oil using experimental design. *J Agric Food Chem* 52: 26–32.

Vazquez, L., Reglero, G., Torres, C.F., Fornari, T., Senorans, F.J. 2008. Supercritical carbon dioxide fractionation of nonesterified alkoxyglycerols obtained from shark liver oil. *J Agric Food Chem* 56: 1078–1083.

Vazquez, L., Hurtado-Benavides, A.M., Reglero, G., Fornari, T., Ibanez, E., Senorans, F.J. 2009. Deacidification of olive oil by countercurrent supercritical carbon dioxide extraction: Experimental and thermodynamic modeling. *J Food Eng* 90: 463–470.

Wang, L., Weller, C.L. 2006. Recent advances in extraction of nutraceuticals from plants. *Trends Food Sci Technol* 17: 300–312.

Wang, L., Weller, C.L., Schlegel, V.L., Carr, T.P. 2008. Supercritical CO_2 extraction of lipids from grain sorghum dried distillers grains with solubles. *Bioresour Technol* 99: 1373–1382.

Xu, X., Wang, Q., Zhao, J., Gao, Y., Liu, G. 2008. Optimization of supercritical carbon dioxide extraction of sea buckthorn (*Hippophaë thamnoides* L.) oil using response surface methodology. *Lebensm-Wiss Technol* 41: 1223–1231.

Yee, J.L., Jimenez-Flores, R., Khalil, H., Walker, J. 2008. Effect of variety and maturation of cheese on supercritical fluid extraction efficiency. *J Agric Food Chem* 56: 5153–5157.

Yi, C., Jiang, Y., Li, D., Shi, J., Xue, S. 2009. Effects of supercritical fluid extraction parameters on lycopene yield and antioxidant activity. *Food Chem* 113: 1088–1094.

Yu, J., Dandekar, D.V., Toledo, R.T., Singh, R.K., Patil, B.S. 2006. Supercritical fluid extraction of limonoid glucosides from grapefruit molasses. *J Agric Food Chem* 54: 6041–6045.

Yu, J., Singh, R.K., Patil, B.S., Dandekar, D.V., Toledo, R.T. 2007. Supercritical fluid extraction of limonoids and naringin from grapefruit (*Citrus paradisi* Macf.) seeds. *Food Chem* 105: 1026–1031.

Zaidul, I.S.M., Norulaini, N.A.N., Mohd Omar, A.K., Smith, R.L. Jr. 2007a. Blending of supercritical carbon dioxide (SC-CO_2) extracted palm kernel oil fractions and palm oil to obtain cocoa butter replacers. *J Food Eng* 78: 1397–1409.

Zaidul, I.S.M., Norulaini, N.A.N., Omar, A.K.M., Sato, Y., Smith, R.L. Jr. 2007b. Separation of palm kernel oil from palm kernel with supercritical carbon dioxide using pressure swing technique. *J Food Eng* 81: 419–428.

Zougagh, M., Rios, A. 2008. Supercritical fluid extraction as on-line clean-up technique for determination of riboflavin vitamins in food samples by capillary electrophoresis with flourimetric detection. *Electrophoresis* 29: 3213–3219.

8 Pressurized Hot Water Extraction and Processing

Charlotta Turner and Elena Ibañez

CONTENTS

8.1 INTRODUCTION

In the food industry and in chemical analysis laboratories worldwide, it is of importance to develop faster, less toxic, and more environmentally sustainable extraction technologies and methods. One of the main factors to consider when proposing to change or replace a current extraction method is the choice of solvent. In most extraction processes, the "hot spot" in terms of environmental impact is the solvent. Depending on the type of solvent used, the degree of environmental impact varies according to the way natural resources are harvested; usage of energy; and emissions to air and water from the production and use of solvents, transportation, and disposal or recycling. Hence, it takes a substantial effort to rank solvents with regard to environmental sustainability. Efforts have been made, using tools such as the Environmental, Health, and Safety (EHS) performance factor (Koller et al. 2000; Capello et al. 2007) and Life Cycle Assessment (LCA) (Pennington et al. 2004; Rebitzer et al. 2004; Capello et al. 2007), to assess the environmental sustainability of a process. Whereas the EHS method is mainly focused on scoring the effects of nine different categories—release potential, fire/explosion, and reaction/decomposition (representing safety hazards); acute toxicity, irritation, and chronic toxicity (representing health hazards); persistency, air hazard, and water hazard (representing environmental hazards)—LCA takes into account the environmental impact of a process or product during its entire life cycle, that is, from cradle (e.g., harvesting of raw material) to grave (e.g., disposal of solvent). EHS and LCA have been combined and applied to a wide range of different organic solvents and acids, demonstrating that simple alcohols such as methanol and ethanol are the most environmentally sustainable organic solvents, whereas tetrahydrofuran, dioxane, and formic acid are among the worst (Capello et al. 2007). However, even though methanol is one of the most eco-friendly solvents, it is largely outcompeted by water. Water has essentially negligible environmental effect, since it is nontoxic to health and the environment, safe to work with, and transportation is accomplished by an already established infrastructure. It is therefore obvious that water is the most environmentally friendly solvent. The only factor that could turn water into a less friendly solvent, is by extracting hazardous compounds, which then would result in a toxic mixture, or if the water is used at extremely high temperatures, which could be dangerous in case of accidents.

When can water be used as an extraction solvent? The answer is in all applications where polar protic solvents are needed. At elevated temperatures, that is, pressurized liquid water at temperatures above 100°C, less polar compounds can also be extracted, since the dielectric constant of water decreases with increasing temperature. It was pointed out by Prof. Jessop (Green Solvent Conference, October 11, 2010, Brechtesgaden, Germany) that pressurized hot water fills an important function as one of the few "green" solvents capable of replacing solvents of different properties, but mainly those of low basicity, high proticity, and high to medium polarity, according to Kamlet–Taft solvatochromic parameters (Marcus 1993). This implies that, in all applications where these kinds of solvents are used, it is worth investigating if water instead can be used as an extraction solvent. One concern is undoubtedly the risk of hydrothermal degradation of the target compounds, which is discussed in this chapter.

What about the sustainability of the extraction process? Is it sufficient to switch to a more "green" solvent such as water? Certainly not. As discussed earlier in relation to LCA of solvents, the entire life cycle of the extraction process needs to be considered with the functional unit being a certain amount of extracted target compound. For instance, perhaps a larger amount of one solvent is needed to extract a certain amount of target compound compared to another solvent. Furthermore, one solvent may require more downstream processing (e.g., cleanup) compared to another. Other aspects of the process that need to be considered are the total amount of sample and other chemicals needed, energy usage, solvent recycling, etc. However, a good start is to select water as the extraction solvent—if the application is suitable.

The water molecule is simple, but still very complex. A search in Web of Science with "water" in the title of the article returns more than 100,000 hits, and a search with "water molecule" in the title garners 1000 hits. Research ranges from modeling of liquid water molecular dynamics (Lindahl et al. 2001), transportation of water across biological membranes (Murata et al. 2000), phase behavior of water (Poole et al. 1992), water pollution (Morel et al. 1998), and wastewater treatment (Scott and Ollis 1995). Approximately 500 articles are about pressurized hot water extraction (PHWE), of which the most cited article (272 hits in ISI Web of Knowledge 2010) is that by Hawthorne et al., which concerns extraction of organic pollutants from sediments and soils using sub- and supercritical water as extracting solvent (Hawthorne et al. 1994).

This chapter describes the extraction process in brief, and the most important parameters used to control the PHWE process: temperature, pressure, extraction time, and flow rate; selectivity and accuracy of the extraction; and removal of water after extraction. The chapter also explains the different parts of the PHWE equipment, differences between commercially available equipment, and several pointers on how to build your own system. Applications reviewed in this chapter range from flavonoids and phenolic acids to pesticides/contaminants, terpenoids, and biofuels/bioethanol production. Industrial applications related to food safety are also discussed, together with new processes based on multiple unit processing extraction and particle formation online. An outline of future trends on PHWE both at industrial and analytical scale summarizes the chapter.

8.2 PRINCIPLES OF PHWE

8.2.1 EXTRACTION STRATEGY

To develop a fast, accurate, and selective extraction method, using the most sustainable solvent as possible, it is pertinent to first understand the actual extraction process. Since this chapter is about PHWE, only solid and semisolid (i.e., "wet") samples are considered. The first question to pose is: Where in the sample are the target analytes? A simple model of a sample particle (Figure 8.1) demonstrates the five main locations where the compounds can be found during an extraction (Waldeback 2005):

1. Deeply embedded in the matrix/inside cell walls or by other means physically hindered to have contact with the extraction solvent. This situation may also include different kinds of chemical bonding to the matrix.
2. Dissolved in the pore water deep inside the sample particle or on the surface.
3. Adsorbed to the surface of the particle by molecular interactions, which can be hydrogen bonding, dipole–dipole, or weaker van der Waals forces.

Upon extraction, analytes are also found in positions 4 and 5:

4. In the bulk of the solvent, in the zone not affected by convection (i.e., only diffusion).
5. In the bulk of the solvent, where convection maintains a higher mass transfer.

The aim of the extraction is to recover the target analytes from these locations out into the bulk solvent where there is preferably some sort of convection with an improved mass transfer rate (marked with 5 in Figure 8.1). The chemical and physical processes that are required to achieve extraction are: (i) desorption from the sample particle surface; (ii) diffusion through stagnant liquid pores and layers; (iii) continuous adsorption and desorption similar to partition chromatography; (iv) solvation into the extraction solvent; and finally, (v) diffusion through the convection-free zone into the final destination in the bulk solvent.

A main obstacle to the extraction is the presence of sample matrices and their interaction with the molecules of interest. Hence, to conduct a successful extraction, it is necessary to understand as much as possible the chemical and physical properties of the matrix and the target compounds. Mass transfer mechanism across chemical and physical interfaces needs to be considered, which for instance differs largely between dried and fresh samples. Unfortunately, very little time is usually spent on studying the sample matrix including effects of different sample pretreatments.

The second important aspect is the choice of solvent and extraction conditions. In this case, since this chapter deals with PHWE, it is mainly an issue to decide whether PHWE is feasible to use in a certain application, and if so, at what conditions (temperature, pressure, extraction time, flow rate). Knowledge about distribution coefficients and solubility parameters is useful in order to select an appropriate solvent as well as extraction condition. For instance, the Hansen solubility parameter (Hansen 1969), which has been used to assess the solubility of flavanoids in pressurized hot water and water/ethanol mixtures (Srinivas et al. 2009), is a very useful tool to predict the solubility of an analyte in subcritical water at different extraction conditions. Equation 8.1 shows the Hansen solubility parameter (δ), also called

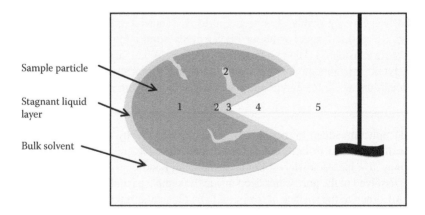

Sample particle

Stagnant liquid layer

Bulk solvent

FIGURE 8.1   Sites on a sample matrix particle where analytes can be found.

the total solubility parameter, which is the sum of the cohesive energy densities originating from hydrogen bonding (H), polar dipole–dipole interactions (P), and dispersive (van der Waals) interactions (D). Figure 8.2 shows the calculated Hansen solubility parameter of water between 20°C and 325°C (Panayiotou 1997; Srinivas et al. 2009). Obviously, hydrogen bonding is the strongest intermolecular interaction between water molecules, and it is also the hydrogen bonding strength that decreases the most with higher temperatures.

$$\delta^2 = \delta_H^2 + \delta_P^2 + \delta_D^2 \tag{8.1}$$

Several models have been proposed to describe the extraction process, using supercritical fluids and pressurized hot liquids as solvents. For instance, the "hot ball" model has been described, originally for supercritical fluid extraction (SFE) by Clifford et al. (1995), and after that by Vandenburg et al. (1998) for the extraction of additives from polymeric samples using pressurized fluid extraction (PFE). By plotting $\ln(m_1/m_0)$, where (m_1) is the mass of analyte remaining in the particle (in the model: sphere) of radius (r) at time (t), (m_0) is the initial amount of analyte, and (D) is the diffusion coefficient of the analyte in the solvent, a linear portion is given in Equation 8.2 (Cotton et al. 1993). However, initially, the plotted line of $\ln(m_1/m_0)$ versus time falls steeply, after which it becomes linear and follows Equation 8.2. See Figure 8.3.

$$\ln(m_1/m_0) = -0.4977 - (\pi^2 Dt/r^2) \tag{8.2}$$

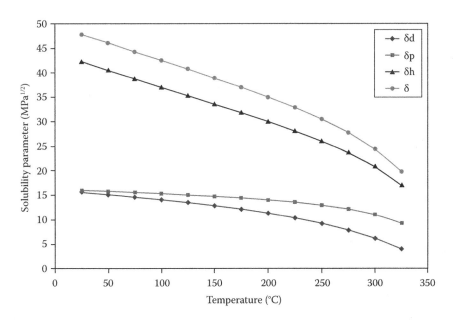

FIGURE 8.2 Solubility parameter of water. (From Srinivas, K. et al., *J Food Sci*, 74, E342–E354, 2009. With permission.)

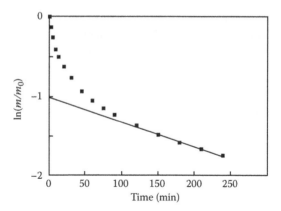

FIGURE 8.3 SFE of chrysene from diesel soot, and hot ball model including rapid fluid entry, reversible release (desorption), transport out of matrix, and removal by fluid. (From Clifford, A.A., *J Chem Soc Faraday Trans*, 91, 1333–1338, 1995. With permission.)

The physical explanation of the shape of the curve is that the analyte near the surface is rapidly extracted until a decreasing concentration gradient is established across the particle. The extraction rate is then completely controlled by the rate at which the analyte diffuses to the surface. By plotting the amount of extracted analyte versus the extraction time for different solvents at different temperatures, the resulting curves showed a good fit to the "hot ball" model (Vandenburg et al. 1998). The "hot ball" model only takes diffusion into account, which is really a limitation. A more accurate kinetic extraction model has been developed by Pawliszyn (2003), which assumes that a matrix particle consists of an organic layer on an impermeable but porous core and the analyte is adsorbed on the pore surface. The extraction process was modeled by considering the following basic steps, which can also be explained using Figure 8.1:

1. Desorption of analyte from the surface
2. Diffusion through the organic part of the particle
3. Solvation into the extraction solvent
4. Diffusion through the stagnant solvent phase out to the solvent affected by convection

Equations describing these steps are given by Pawliszyn (2003). With a good kinetic model of the extraction process, it would be possible to predict experimental parameters and to find out where the extraction is expected to be only diffusion dependent. However, the above-mentioned models can at most be considered to give useful hints when developing new extraction methods based on PHWE.

8.2.2 Temperature Effects

The most important effect of increasing the temperature in PHWE is the weakening of hydrogen bonds, resulting in water with a lower dielectric constant. This effect results in water of a lower polarity, and depending on temperature, polar to medium-polar (100–250°C) and even nonpolar (250–374°C) compounds can be extracted with high solubility in PHWE. Figure 8.4 shows the change in water's dielectric constant as a function of temperature (25–250°C), assuming a large enough pressure is applied to maintain water in its liquid state, as well as the dielectric constant for methanol/water and acetonitrile/water mixtures at ambient conditions (Yang et al. 1998). Figure 8.4 implies that water at 200°C has a dielectric constant similar to either pure methanol or acetonitrile at room temperature (Yang et al. 1998).

An increase in temperature also tends to promote solubility, as the thermal kinetic energy rises. Increasing the temperature will also facilitate analyte diffusion. In fact, the diffusivity of liquid water at 200°C is about 10 times faster than water at ambient temperature. Furthermore, higher temperature will reduce interactions between analytes and the sample matrix by disrupting intermolecular forces such as van der Waals forces, hydrogen bonding, and dipole attractions. Higher temperature also decreases the viscosity of water, thus enabling better penetration of matrix particles. This change in viscosity is especially pronounced during the first 100°C increase in temperature from ambient conditions. Finally, elevating the temperature will decrease the surface tension, allowing the water to better "wet" the sample matrix. An excellent site about the science of water is found at www.lsbu.ac.uk/water.

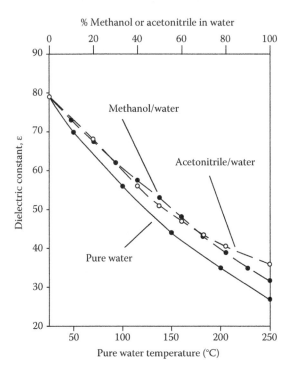

FIGURE 8.4   Dielectric constant of water vs. temperature and methanol/water and aceto-nitrile/water mixtures at room temperature. (From Yang, Y. et al., *J Chromatogr* A, 810, 149–159, 1998. With permission.)

Another, less desirable effect of increasing the temperature of water in PHWE is the higher degree of degradation of thermally labile compounds. An increase in temperature is accompanied by an increase in the reaction rate. Because temperature is a measure of the kinetic energy of a system, a higher temperature implies higher average kinetic energy of all molecules and more collisions per unit time. A common rule of thumb for many chemical reactions is that each 10°C increase in temperature will double the reaction rate.

An example of an unwanted reaction during PHWE is degradation of antioxidants. These oxidation-prone molecules may degrade upon exposure to oxygen, heat, metal, and light. Anthocyanins are extremely labile compounds whose stability depends on pH. An analytical methodology was developed for the determination of anthocyanin species in red cabbage (Arapitsas and Turner 2008) and red onion (Petersson et al. 2008) based on PHWE using water/ethanol/formic acid (94:5:1, vol%) at 99°C and 50 bars as a solvent. However, recent studies have shown that using the same solvent but at 110°C results in the degradation of the anthocyanins soon after extraction from red onion has started (Petersson et al. 2010). Calculated theoretical extraction curves, based on experimental data for extraction plus degradation and degradation only, showed that more anthocyanins, 21–36% depending on the species, could be extracted if no degradation occurred. Recent results have shown that a dynamic extraction with continuous flow of pressurized hot water can be used to extract anthocyanins from vegetables, avoiding chemical degradation.

Another study, concerning PHWE of antioxidants from birch bark, showed that elevated temperature (180°C) gives water extracts with higher antioxidant activity compared to lower temperature (80°C) (Co et al. 2011). Moreover, it was determined that partial degradation of some of the antioxidants occurred at a higher temperature, that is, a higher temperature resulted in both improved extraction efficiency and greater degradation of the extracted antioxidants. However, new antioxidant compounds could be formed and/or partially degraded antioxidants may have larger antioxidant capacity than their precursors, thus increasing the final antioxidant activity of the obtained extract. Hence, it is crucial to optimize the extraction temperature carefully, in order to obtain a fast, but accurate method, even for thermally labile compounds. The above-mentioned degradation studies were performed in static mode, that is, extracted compounds were in contact with the hot extracting solvent for a relatively long time (up to about 45 min). The study conducted by Co et al. (2011) also demonstrated the importance of choosing an appropriate analysis method to determine the extraction yield of antioxidants. A DPPH (2,2-diphenyl-1-picrylhydrazyl) assay showed no difference in antioxidant activity between some of the extracts, whereas high-performance liquid chromatography (HPLC) coupled to electrochemical detection (ECD), diode array detection (DAD), and tandem mass spectrometry (MS/MS), HPLC-ECD-DAD-MS/MS, clearly did. Figure 8.5 shows two overlaid chromatograms of a hot water extract produced at 180°C, and monitored by UV at 280 nm and ECD at +0.4 V vs. Ag/AgCl. The beauty of this hyphenated analysis technique is that individual antioxidant species can be determined as a function of solubility properties (retention time in HPLC), antioxidant capacity (ECD signal strength), and chemical structure (MS/MS and DAD).

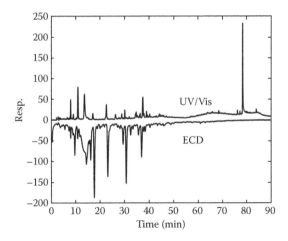

FIGURE 8.5 UV and ECD chromatograms of a birch bark extract produced by a 5-min extraction plus a 5-min hydrothermal treatment in water at 180°C.

Caramelization of sugars is another type of important degradative reaction that may occur with water as a solvent at temperatures of about 160°C and above (Montilla et al. 2006). This reaction may cause largely erratic results as well as problems with plugging filters and tubings of the PHWE equipment. Recent studies have demonstrated the neoformation of antioxidant compounds from Maillard reaction products during PHWE at high temperatures of model systems, involving reducing sugars and amino acids or proteins, and natural sources such as plants and algae (Plaza et al. 2010a,b). These studies showed evidence of the occurrence of Maillard reaction and the contribution of these compounds (Maillard reaction products, Maillard polymers) to the total antioxidant activity, as demonstrated when using water at temperatures higher than 170°C (Howard and Pandjaitan 2008; Plaza et al. 2010a,b). Hence, in food applications, it is advisable to use a temperature of 160°C or lower in order to prevent both caramelization of sugars and Maillard reactions. Naturally, a shorter extraction time or the use of dynamic extraction will minimize the challenges associated with unwanted reactions during extraction. On the other hand, more studies are needed to assess the effectiveness and the safety of the neoformed antioxidants derived from Maillard reaction products since they can have a positive effect as new compounds with bioactive properties.

8.2.3 EXTRACTION TIME AND FLOW RATE

Flow rate determines the extraction yield obtainable per time unit and residence time for extracted analytes in the extracting solvent, if the extraction is not diffusion-limited. Hence, with respect to thermally labile compounds, it is necessary to thoroughly optimize the flow rate in order to assure quantitative extraction yields without causing degradation of the analytes and without excessive dilution of the sample extract. In general, a faster flow rate results in a shorter extraction time as well as shorter residence time for the extracted compounds, but then of course the dilution of the sample extract may lead to difficulties in detection of analytes in chemical analysis and/or more energy demanding downstream processing in industrial applications. Some commercially available PFE equipment do not have the option of dynamic extraction with a flow rate, hence the extraction process is static with several cycles of batch extractions, and each cycle has a certain time length. Typically, the extraction time is 10–15 min, plus a preheating step of 5–10 min depending on the temperature used and instrument capabilities.

Comparison between PHWE and microwave-assisted extraction (MAE) has been reported by Teo et al. (2008, 2009) for the extraction of bioactive compounds from plants. In these studies, both methods were shown to be feasible alternatives for the extraction of thermally labile marker compounds present in medicinal plants although in some cases, MAE provided better results in terms of extraction efficiency with shorter extraction time (Teo et al. 2009). PHWE has also been compared to other conventional techniques, such as hydrodistillation, Soxhlet extraction, SFE (Luque de Castro et al. 1999), and ultrasound-assisted extraction (UAE) (Pongnaravane et al. 2006). In all these cases, PHWE has demonstrated important advantages in terms of extraction time, extraction efficiency, extraction yield, and economic and environmental concerns.

8.2.4   EXTRACTION PRESSURE

Pressure in PHWE has only a marginal effect on the properties of water, because of the low compressibility of liquids in general. However, pressure is a necessary requisite to obtain a liquid at temperature above its atmospheric boiling point. For instance, by considering the vapor pressure curve for water, it can be deduced that in order to maintain water as a liquid at 150°C, a pressure of about 5 atm needs to be applied—or rather, this is the pressure that will be built up in a closed container with water that is being heated to 150°C. In most commercial equipment that can be used for PHWE, a fixed pressure is set to 50 or 100 bars.

Some authors suggest that an extra pressure may increase extraction efficiency since this can permit the fluid to penetrate the sample matrix better, or mechanically disrupt the matrix, resulting in higher extraction efficiency (King 2006); however, it is generally recognized that, for many natural raw materials, as long as they are grinded and pretreated correctly, no significant effects are observed.

8.2.5   SELECTIVITY DURING EXTRACTION

In general, selectivity during the extraction with liquid solvents is usually obtained by varying the solvent power or by replacing the solvent. Another way to achieve selectivity is to include adsorbents in the extraction cell (EC) to provide cleanup of interfering matrix components (Westbom et al. 2008). In PHWE, selectivity is mainly achieved by carefully optimizing the temperature, as discussed above. Basically, a high enough temperature should be used to quantitatively extract the compounds of interest, but not higher than necessary, because this will lead to coextraction of unwanted matrix components and increase the risk of degradation.

For instance, Garcia-Marino et al. (2006) showed that by using water in sequential extraction from 50°C to 200°C, it was possible to extract different procyanidins according to its degree of polymerization and structure. Increasing temperature favored compounds formed by more than one unit of catechin.

On the other hand, Ibañez et al. (2003) studied the selectivity of subcritical water to extract the most active compounds from rosemary via a homemade PHWE device. In this work, HPLC monitoring of the relative amounts of several interesting compounds was carried out. The study showed that the selectivity of subcritical water toward the extraction of antioxidants could be easily tuned considering small changes in the extraction temperature; for instance, low temperatures (50°C) favored the extraction of highly polar compounds, whereas preferential extraction of carnosic acid (a relatively nonpolar antioxidant) was observed at 200°C. Similar results were obtained using a commercial instrument and characterizing the extracts by capillary electrophoresis coupled to mass spectrometry (CE-MS) (Herrero et al. 2005). Analogous results in terms of selectivity were obtained for PHWE of sage (Ollanketo et al. 2002) and oregano (Rodriguez-Meizoso et al. 2006). In the latter study, compound class selectivity was observed at different extraction temperatures providing extracts with different phenolic profiles. Results showed that, at the lowest temperature, the most polar compounds were preferentially extracted (flavanones/dihydroflavonol such as dihydroquercetin, eriodictyol, and dihydrokaempferol), whereas at 200°C, the solubility of less polar compounds was increased by several orders of magnitude.

Kubátová et al. (2001) demonstrated the selectivity of subcritical water toward the extraction of oxygenated flavor compounds from savory and peppermint as compared to hydrodistillation and SFE (Kubatova et al. 2001).

Although one of the main trends in pressurized liquid extraction is the in-cell cleanup to improve the selectivity of the extraction process (Haglund and Spinnel 2010), just a few examples of this approach using PHWE could be found in the literature. For instance, Curren and King (2002) described the in situ sample cleanup using matrix solid-phase dispersion by employing the acrylic polymer XAD-7 HP with further addition of triethylammonium phosphate to the extraction solvent.

Further selectivity before chromatographic analysis can be obtained by on-line coupling the PHWE to trapping methods such as liquid–liquid extraction, solid phase extraction (SPE), flat sheet or hollow fiber microporous membrane liquid–liquid extraction (MMLLE) (Luthje et al. 2004), or solid phase microextraction (SPME) (Dong et al. 2007; Concha-Grana et al. 2010) (see Section 8.4).

8.3   INSTRUMENTATION

8.3.1   Basic Details about Instrumentation

Although a wide range of conditions can be used in PHWE, they all have in common the pressure requirement needed to keep water in its liquid state during heating above its atmospheric boiling point. Temperatures applied usually range from room temperature to 200°C, and pressures are generally set to between 35 and 200 bars. Basic instrumentation may differ depending on whether a commercial or homemade setup is used or whether a static or dynamic process is implemented. Static PHWE is a batch process with one or several extraction cycles with replacement of solvent in between. During extraction, solvent is kept in the EC as the extraction takes place.

In dynamic PHWE, the extraction solvent is continuously pumped at a selected flow rate through the extraction vessel containing the sample.

An example of basic instrumentation for commercial static PHWE equipment is shown in Figure 8.6. First of all, one or several solvent reservoirs can be used to work with the selected solvent (water) or mixture of solvents (e.g., water/ethanol). When different reservoirs are available, a solvent controller is also provided. A high-pressure pump is then used to pump the solvent and fill the EC. A filter paper is inserted into a stainless steel EC followed by the sample, sometimes mixed with a dispersing or drying agent, if needed. However, an advantage of PHWE is that drying of the sample is usually not necessary. The cell is either loaded on a carousel and automatically placed in the oven or—for simpler equipment—manually placed into the oven. Commonly, a static valve (SV) in combination with a pressure valve (PV) and a pressure relief valve (PRV) control the pressure in the sample vessel during static extraction by adding more solvent to the cell or by opening the static valve, whichever is needed to maintain the desired pressure. The operating procedure consists of several steps: (1) Preheating step, to reach thermal equilibrium: during heating, thermal expansion of the solvent occurs and causes an increase in pressure within the cell; the valve keeps the desired pressure by releasing solvent at this step, which can be considered a noncontrolled dynamic extraction period before static extraction. (2) Static extraction: once the target values are achieved, extraction is performed during a selected time, typically 5–30 min. (3) Solvent replacement when extraction cycles are used: after static time, part of the solvent in the EC is replaced with fresh solvent, to start the next extraction cycle. (4) System purge: after the last cycle, the sample cell is purged with an inert gas (nitrogen) to remove the remaining solvent from the cell and the lines to a vial where the extract is collected (collecting vial, CV). (5) Pressure

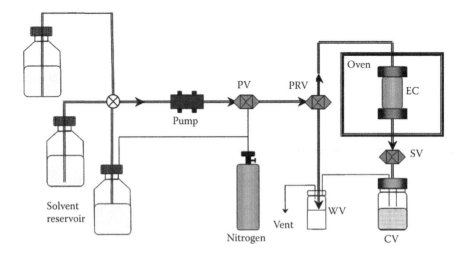

FIGURE 8.6 Schematic of a PHWE system.

release: the system is depressurized at atmospheric conditions (through the vent) to the waste vial (WV). By using this protocol, individual or sequential extractions can easily be performed by repeating the procedure with a new solvent/new temperature conditions and purging in the same or a different collection vial.

Dynamic PHWE is quite similar to static PHWE, but requires a slightly more sophisticated high-pressure or HPLC pump to control the water flow rate as well as a pressure restrictor or a micrometering valve rather than a static open/close valve. At present, there is only one dynamic PHWE equipment available on the market (see next section), whereas different types of commercial static PHWE instruments are available (see Section 8.3.2). Numerous homemade instruments have been described in the literature (Lou et al. 1997; Bautz et al. 1998; Vandenburg et al. 1998; Hawthorne et al. 2000; King 2006). As will be discussed in Section 8.3.3, the advantage of homemade systems is the possibility to perform both dynamic and static extractions with less operating restrictions, the working temperature range, and the possibility of carrying out different processes (reaction, drying, extraction) by adapting the basic setup.

Among other advantages, dynamic operation may prevent, to some extent, thermal degradation of bioactive compounds since water is continuously flowing through the matrix at a certain fluid velocity that improves the efficiency of the extraction and prevents the excessive heating of the sample. King et al. (2003) demonstrated the higher recoveries of anthocyanins with dynamic PHWE extraction compared to static conditions.

8.3.2 COMMERCIALLY AVAILABLE EQUIPMENT

On the market, several equipments are available that perform static PHWE, in either full- or semi-automatic mode. By October 2010, the following vendors have been identified:

Dionex (www.dionex.com) offers products such as ASE® 150 (one extraction vessel at a time; with sizes 1, 5, 10, 22, 34, 66, or 100 ml) and ASE® 350 (carousel with up to 24 extraction vessels of the same sizes as for the 150 system). A solvent controller allows mixing and delivery of up to three solvents. The applicable temperature range is 40–200°C and the fluid delivery pressure is fixed at 1500 lb/in.^{2} (103 bars). pH hardened pathway with Dionium™ components and extraction vessels make the instrument compatible with acid or alkaline pretreated sample matrices. Newly developed technologies such as flow through technology dynamic extraction simplify sample preparation by including in-line filtration and in-cell cleanup.

Applied Separations (www.appliedseparations.com) has two analytical extraction systems, one PSE™ and fast-PSE™, which operate with one and six in-parallel extraction vessels, respectively. Available vessel sizes are 11, 22, and 33 ml. An automated solvent dispenser is optional, and this handles up to four different solvents. Temperature range is 50–200°C and maximum pressure is 150 bars. The company also offers a Process Development Unit Helix™, which, in fact, is a laboratory supercritical fluid process unit that can be configured for PHWE. The main features of the system are flexibility, which allows users to modify the system depending on the particular need, possibility of using temperatures of up to 250°C, and operation in both static and dynamic modes. Volumes of the extraction vessels range from 100 to 1000 ml. The system also offers an optional stirrer.

Fluid Management Systems (www.fmsenvironmental.com) markets PLE™ systems, which can handle from one up to six samples (six modules) at the same time in parallel, with the option of in-line cleanup. Temperature and pressure ranges are 70–200°C and 1500–3000 lb/in.² (103–207 bars), respectively. Extraction vessels come in sizes ranging between 5 and 250 ml.

8.3.3 How to Build Your Own System

The idea behind building your own system comes from the limitations of the available systems, which are mainly based on static PHWE and operate within a strictly limited temperature range. Since more versatile systems are also very expensive, for some users, homemade systems might be a good choice.

Homemade systems allow users to modify and design their own system depending on their requirements and specifications. Thus, if someone wants to conduct research on PHWE, the flexibility provided by home-built systems is an advantage.

Basic components of homemade equipment (Figure 8.7) consists of (1) a water reservoir (sometimes together with an optional heater plate to heat water up to 60–70°C); (2) an ordinary HPLC pump; (3) a stainless steel tubing with the appropriate dimensions enabling the solvent to reach the desired temperature before passing through the extraction vessel (heating coil) that should be placed inside the oven; (4) a thermocouple for measuring the temperature of the solvent just before it enters the extraction vessel; (5) a stainless steel EC (e.g., a preparative or analytical HPLC column can be used, depending on the sample size) wherein the sample is placed before extraction; (6) a simple GC oven (only temperature control is needed; thus, there is no need for injector or detector); (7) a needle valve, micrometering valve, or a back pressure regulator (BPR) to maintain the pressure; (8) a cooling bath, around the collection vial, to bring down the temperature of the extractant to below the boiling point of the solvent and to preserve the extract from thermal degradation; (9) collection vial(s); (10) optional nitrogen gas to flush out the entire system after extraction; and (11) optional gravimetric scale to measure the flow rate during extraction.

The system can also include an additional heating tape to heat the tubing that exits the EC. After passing through the EC, water is cooled down to temperatures in which some of the extracted analytes are not soluble anymore, which allows them to precipitate in the tubing and result in the blockage of the system. Thus, a heating tape is recommended to prevent clogging problems and should be placed in contact with the stainless steel tubing between the exit of the oven and after the pressure regulation valve.

An additional HPLC pump can be used to wash the lines exiting the EC and to recover the extracted analytes. In this case, it should be connected through a tee between the EC and the valve to prevent deposition of analytes when water cools down during collection (Miller and Hawthorne 1998).

It is suggested to install an in-line filter to protect the needle valve/BPR. It is also recommended to install a pressure gauge for control of the pressure as well as a burst

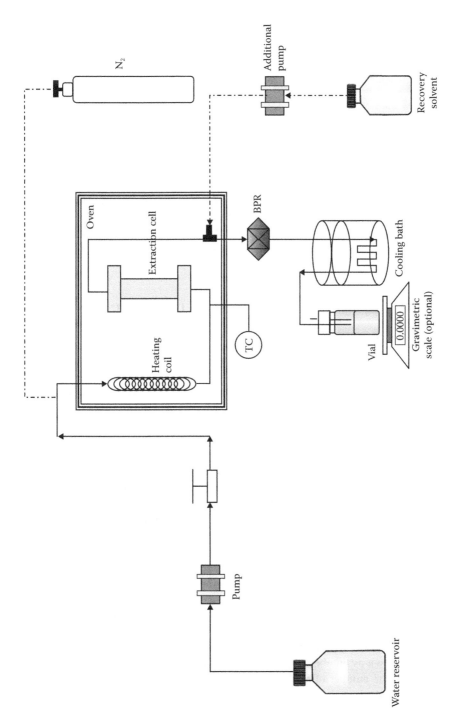

FIGURE 8.7 An example of a home-built dynamic PHWE system.

disk in case the pressure rises too high in the system. There are burst disks available for different pressures and solvents.

Other aspects that should be considered are temperatures and solvents that are allowed in the different parts of the extraction system. For instance, an HPLC pump cannot take a too high temperature, and especially not with some of the organic solvents. If water is used as a solvent, caution has to be taken if temperatures higher than approximately 200°C are used, because water becomes increasingly corrosive at higher temperatures. Hence, higher-quality nickel steel alloy that exhibits high resistance to corrosion should be used, for instance, Hastelloy®. In practical terms, the best solution may be to preheat the water to about 70°C upstream of the HPLC pump, that is, the water reservoir, and the rest of the heating downstream of the pump inside the proposed GC oven.

For water to be used in PHWE, oxygen should be thoroughly removed by using methods such as ultrasound or helium purge. Degassing of water is needed to prevent possible oxidation reactions from occurring during extractions that are promoted at high temperatures and also for safety reasons, to avoid oxidative corrosion of the EC and tubing. Removal of oxygen might also prevent pump cavitation.

8.4 APPLICATIONS IN FOOD AND AGRICULTURAL INDUSTRY

8.4.1 From Analytical to Industrial Scale

Applications of PHWE reported in the literature range from analytical to industrial scale. Target compounds are somehow related to the process conditions used to perform PHWE. In this sense, pressurized hot water is a very versatile medium in both its sub- and supercritical regions covering a wide range of reduced temperatures (T_r) and pressures (P_r). When supercritical or near-critical conditions are considered— that is, T_r (1.02–2.5) and P_r (1.2–1.8) and $T_r = 0.65$–1.05 and $P_r = 0.35$–2.0, respectively—either supercritical water oxidation or molecular transformations such as biomass conversion occur and therefore, applications are developed under a more industrial perspective. For T_r and P_r values in the range of 0.50–0.80 and 0.02–0.35, respectively, extraction devoted to environmental analysis and/or food and agricultural samples is carried out targeting either pollutants or health-beneficial compounds such as different types of antioxidants at analytical scale. Under these conditions, no industrial applications have been found mainly because of the lack of pilot and industrial systems and the need for more economic studies. Moreover, a high value-added product is needed as a target to make the process profitable enough to be developed at an industrial scale, which is quite difficult when talking about bioactive compounds.

Figure 8.8 shows the number of relevant publications from the early 1970s to 2010, based on a search for papers published on PHWE in the Web of Science. The first three bars correspond to the 1970s, 1980s, and 1990s, whereas data from 2000 to 2010 are shown as number of papers/year. As can be seen, there is a clear trend of increase in the number of published articles dealing with extraction of food and agricultural products using subcritical or hot pressurized hot water in recent years. In 2007, this number increased to more than 80 per year, that is, more articles than those published in the entire 1990s.

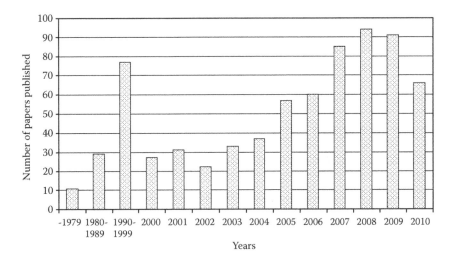

FIGURE 8.8   Record of number of papers published vs. years for PHWE (search done in Web of Science Oct-10 using the keywords "hot water extract* and food* OR agric*"/"subcrit* water extract* and food* OR agric* OR plant*").

Considering the wide range of different applications for PHWE, the reader is referred to a number of review articles that have been published in the field, covering general (Teo et al. 2010); environmental (Giergielewicz-Mozajska et al. 2001; Ramos et al. 2002; Nieto et al. 2008); food and drugs (Carabias-Martinez et al. 2005; Herrero et al. 2006a; Beyer and Biziuk 2008); and medicinal plants, herbs, and agricultural (Huie 2002; Ong et al. 2006; Wang and Weller 2006; Golmohamad et al. 2008) applications. In this chapter, some applications have been selected for their focus in the food and agricultural field, ranging from analytical to industrial applications.

Most of the applications that will be discussed later can be viewed from the perspective of sample preparation or as a first step toward a development of the process at pilot and industrial scale. Even if it is true that some special aspects should be considered for process design (e.g., viability and feasibility studies at a large scale), it is important to consider not only the typical applications, but also the ideas that can revolutionize the field of hot water application (i.e., thinking "out of the box").

8.4.2   Pesticides/Contaminants

Recent studies on pesticide extraction using pressurized hot water refer to the determination of pesticides, contaminants, and insecticides in fruits, sediments, milk, soils, etc. Some articles describe the methodology and the usefulness of PHWE (King 2006; Kronholm et al. 2007) for multiple applications, whereas other specify several interesting applications.

For instance, extraction and analytical conditions were optimized to determine seven pesticides (flutolanil, simazine, haloxyfop, acifluorfen, dinoseb, picloram, and ioxynil) in four Mediterranean summer fruits with high water content (peaches, melon, watermelon, and apricot) by CE-MS. Sample extraction was based on pressurized hot water at 60°C and 1500 lb/in.², followed by an SPE cleanup step (Juan-Garcia et al. 2010). A new method has been developed based on PHWE followed by SPME and determination by gas chromatography-mass spectrometry (GC-MS) for the analysis of organochlorine pesticides in sediment samples at trace levels (Concha-Grana et al. 2010). Obtained results were in excellent agreement with certified materials and provided low detection and quantification limits for organochlorine pesticides in marine sediments. Prof. Riekkola's group, in the University of Helsinki, also developed a new approach for pesticide determination in grapes (Luthje et al. 2005). The authors used the on-line coupling PHWE-MMLLE-GC-MS for pesticide analysis (procymidone and tetradifon) in grape skin. The novelty of introducing MMLLE as a trapping step after PHWE helps to clean and concentrate the extract before on-line transfer to the GC.

More conventional approaches were used for the extraction of organic pesticides in soils (Konda et al. 2002) or triazine pesticides in foods (Curren and King 2001). In Curren and King's work, solubility measurements of atrazine, cyanazine, and simazine were carried out to help optimize water extraction of target analytes. The obtained results indicated that adding a cosolvent to water, in addition to increasing the system temperature, increased the solubilities of triazine pesticides in subcritical water, whereas no degradation/hydrolysis was observed under the tested conditions.

Insecticides were also extracted using PHWE. Examples of carbamate residues in bovine milk (Bogialli et al. 2004a) and fruits and vegetables (Bogialli et al. 2004b) can be found in the literature. The studies conducted by Bogialli et al. were based on the use of matrix solid-phase dispersion technique with heated water as extractant followed by LC-MS.

Another interesting approach was the use of hot phosphate-buffered water as extracting agent for analyzing traces of polar and medium polar contaminants in soils (Crescenzi et al. 2000). A new development of PHWE to extract and quantify antibiotics in animal feeds has been reported (Wang et al. 2008), providing more efficient extraction compared to traditional ultrasonic extraction.

8.4.3 Phenolic Compounds

Food and agricultural applications considering endogenous compounds have been developed in the past few years, mainly involving the use of environmentally friendly solvents such as water and/or ethanol and with the basic idea of isolating bioactive or health-beneficial compounds for later use as food ingredients. Many of these are phenolic compounds (e.g., flavonoids, phenolic acids) present in plants and different types of foods, with antioxidant properties that can be used to improve human health. Some key applications are discussed below.

Flavonoids consist of, for instance, anthocyanins, flavonols, and procyanidins (Andersen and Jordheim 2006). Flavonoids such as quercetin, kaempferol, catechin, and anthocyanidins have been extracted from vegetables and herbal plants using pressurized hot water, ethanol, or water/ethanol mixtures as solvents at temperatures between 50°C and 160°C (Ollanketo et al. 2002; Ibañez et al. 2003; Turner et al. 2006; Arapitsas and Turner 2008; Howard and Pandjaitan 2008). For instance, water and a 70:30 mixture of ethanol and water at temperatures between 50°C and 190°C were used to extract flavonoids from dried spinach (Howard and Pandjaitan 2008). Results showed that the total phenolic content and the antioxidant capacity were maximum in extracts obtained at the highest extraction temperatures (170–190°C) for both water and the water/ethanol mixture, although the solvent mixture was more efficient than pure water in extracting different flavonoids. Howard and Pandjaitan demonstrated that polymeric Maillard reaction products were responsible for the increase in antioxidant capacity of extracts at temperatures above 130°C for water and above 150°C for ethanolic solvents (Howard and Pandjaitan 2008). Naringenin and other major flavonoids (dihydrokaempferol, naringin) have been extracted using PHWE from knotwood of aspen (Hartonen et al. 2007).

Turner et al. (2006) extracted quercetin and its glycosides as well as isorhamnetin and kaempferol from yellow and red onion using water at 120°C and 50 bars, using three consecutive 5-min extraction cycles. Thermostable β-glucosidase was used to convert the polyphenolic glucosides into their respective aglycons within 10 min of reaction, using water at 90°C and pH 5 as a reaction media.

Flavonoids with antioxidant properties (e.g., rosmarinic and carnosic acids, carnosol, and methyl carnosate) have been extracted from sage by PHWE, UAE with methanol, and hydrodistillation (Ollanketo et al. 2002). The highest antioxidant activity was obtained at 100°C. Ibañez et al. (2003) extracted antioxidant compounds such as carnosol, rosmanol, carnosic acid, methyl carnosate, cirsimaritin, and genkwanin from rosemary leaves using subcritical water at temperatures between 25°C and 200°C, showing that it was possible to tune the selectivity of the extraction by varying the temperature. For instance, extractions at 200°C provided higher concentrations of carnosic acid, the most potent antioxidant compound in rosemary leaves. Recent studies demonstrated the ability of a newly developed water extraction and particle formation on-line process (patented process WEPO®; Ibañez et al. 2009) to extract antioxidant compounds from rosemary leaves (Herrero et al. 2010) with bioactivities similar to those obtained with a well-known SFE process.

Other flavonoids have also been extracted from oregano leaves using subcritical water extraction at different temperatures (Rodriguez-Meizoso et al. 2006). Results corroborated that the selectivity of the extraction toward selected compounds was a function of the extraction temperature.

The extraction of anthocyanins from berries using PHWE was also reported (King et al. 2003; Ju and Howard 2005), together with the extraction of catechins and proanthocyanidins from grapes and grape pomace (Pineiro et al. 2004; Garcia-Marino et al. 2006; King and Grabiel 2007). In general, studies involving PHWE of anthocyanins indicate a maximum extraction yield between 120°C and 160°C (King et al. 2003). Studies on the effect of temperature and particle size on subcritical water extraction of anthocyanins from grape pomace showed a maximum yield at about 120°C (King and Grabiel 2007). These studies also indicated the possibility of thermal degradation of the target compounds at higher temperatures.

Petersson et al. (2010) also demonstrated that the anthocyanins extracted from red onions using pressurized hot water in a static batch extractor showed thermal degradation at 110°C and a residence time as short as 8 min (Petersson et al. 2010). The neoformation of antioxidants has been discussed in Section 8.2.2.

Although most research activities are based on the study of the optimal experimental extraction conditions, Srinivas et al. (2010) reported the measurement of the aqueous solubilities of quercetin and its dihydrate at temperatures between 25°C and 140°C using a continuous flow type apparatus.

Recently, a new approach based on the application of the Hansen three-dimensional (3-D) solubility parameter was presented as a tool to optimize both extraction solvent and temperature for, among others, formalvidin-3,5-diglucoside, a component in grape pomace (Srinivas et al. 2009). In a first step, the solvent power of subcritical water above its boiling point was characterized using the Hansen solubility parameter; lately, the 3-D Hansen solubility parameter for the compound was obtained considering that the square of the total solubility parameter (the cohesion energy density [CED]) of a solvent or solute was the sum of the square of the CED components due to dispersion (D), polar (P), and hydrogen bonding (H) intermolecular forces. The solvent conditions with the lowest relative energy difference between the solubility parameters of the solute and that of the solvent was, in principle, the best solvent.

Chen et al. (2007) studied the PHWE of different flavonoids and phenolic acids from Brazilian propolis lumps, obtaining extracts with strong antioxidant activity and growth suppression of leukemia (HL-60, U937), lung cancer (A549, CH27), and liver cancer (Hep G2, Hep 3B) cells in a concentration-dependent manner (Chen et al. 2007).

Other phenolic acids have been extracted from fruits and vegetables using PHWE at temperatures ranging between 20°C and 100°C (Alonso-Salces et al. 2001; Mukhopadhyay et al. 2006; Chen et al. 2007; Waksmundzka-Hainos et al. 2007). In a recent work, Budrat and Shotipruk (2008) studied the effect of extraction temperature, from 130°C to 200°C, on the total phenolic contents of bitter melon and on gallic acid content, that is, the main phenolic acid in bitter melon. Results showed that the concentration of gallic acid increased with the temperature. In general, the main concern for the extraction of phenolic acids as well as many of the flavonoids is the thermal stability, or rather the tendency to degrade during extraction, as discussed above.

8.4.4 TERPENOIDS

Terpenoids (e.g., carotenoids, limonene, pinene, artemisinin, retinol, taxol, squalene, lycopene, carotene) have only limited water solubility even at elevated temperatures, and are commonly extracted with 2-propanol or methanol/ethyl acetate/light petroleum (1:1:1) at temperatures ranging from 40°C to 190°C (Schaneberg and Khan 2002; Breithaupt 2004; Waldeback et al. 2004; Fojtova et al. 2008). Yang et al. (2007) used pressurized hot water, at temperatures between 100°C and 250°C, to extract α-pinene, limonene, camphor, citronellol, and carvacrol terpenoids from oregano and basil leaves; they demonstrated that thermal degradation occurred at all temperatures tested. Carotenoids have been extracted from different types of algae using PHWE at temperatures ranging from 50°C to 200°C (Herrero et al. 2006b; Rodriguez-Meizoso et al. 2008; Cha et al. 2010), although higher concentrations of these compounds were obtained with ethanol as the extracting agent at high temperatures (approximately 160°C).

Other terpenes, for instance, terpene trilactones, have been extracted using PHWE from *Ginkgo biloba* leaves (Lang and Wai 2003). In a more recent study, the stability of five terpenes (α-pinene, limonene, camphor, citronellol, and carvacrol) under subcritical water conditions was investigated (Yang et al. 2007), showing that at temperatures higher than 100°C, terpenes were thermally degraded.

Other applications include alkaloids (Mroczek and Mazurek 2009), oligosaccharides (Bansleben et al. 2008), lignans (Ho et al. 2007; Smeds et al. 2007), and mannitol from olive leaves (Ghoreishi and Shahrestani 2009) and essential oils. The use of subcritical water to selectively extract essential oil components from plants was described for marjoram, eucalyptus, and other plants well known for their characteristic essential oil profile (Jimenez-Carmona and Luque de Castro 1999; Jimenez-Carmona et al. 1999; Luque de Castro et al. 1999; Kubatova et al. 2002; Schaneberg and Khan 2002; Ozel et al. 2003). Lately, new and exotic plants have been extracted to characterize their essential oil (del Valle et al. 2005; Khajenoori et al. 2009), mainly plants used in traditional Chinese medicine. To characterize the compounds associated to a certain bioactivity, PHWE has been used as the sample preparation technique, followed by SPME or liquid-phase microextraction, and coupled to GC-MS to determine and quantify the bioactive compounds (Deng et al. 2005; Dong et al. 2007).

8.4.5 INDUSTRIAL APPLICATIONS RELATED TO FOOD SAFETY

As noted earlier, water under supercritical conditions allows partial or full oxidation of organic matter. Under these conditions, water can be used to eliminate food waste streams, turning into nonpollutant gas, biofuel, and heat, and also to eliminate micropollutants from wastewaters with negligible production of traces of dioxins, furans, or NO_x (Kritzer and Dinjus 2001); these processes prevent a possible further reintroduction of pollutants into the food chain and help to preserve food safety.

Supercritical water is a very reactive medium that can adapt well in many industrial applications. Under supercritical conditions, substantial amounts of gases and organic substances can homogeneously be mixed with water, which can then be fractionated by adjusting the subcritical conditions by forming additional phases. This can also be combined with chemical reactions providing a homogeneous medium leading to integrated processes, for example, homogeneous catalysis (Dinjus and Kruse 2004). Examples involve the use of water to treat organic wastes that are completely miscible in supercritical water and, moreover, in the presence of an oxidizer, can react to form mainly carbon dioxide, water, and other simple molecules that can be removed from the solution through simple operations such as evaporation or distillation (Bellissent-Funel 2001). This process is known as supercritical water oxidation (SCWO) and works under reaction temperatures of 500–700°C and pressures of 24–50 MPa. Reaction times, at which a full oxidation is achieved, seldom exceed several minutes, being usually less than 1 min for most applications. However, by using more moderate conditions ($T = 250$–500°C, $P = 23$–30 MPa), which may be located near the critical point of water, the oxidation reaction behaves less aggressively and lasts longer, with reaction times ranging between 1 and 30 min. In this case, oxidation in pressurized water may be used to obtain intermediate products of the oxidation reaction, usually by hydrolysis.

Another application of supercritical water is production of hydrogen from biomass by hydrothermal gasification; in this process, working at temperatures up to 700°C and pressures of about 30 MPa with an excess of water, the reaction leads to valuable hydrogen instead of synthetic gas, as in conventional gasification processes. Once the reaction is completed, pressurized hydrogen is obtained, together with CO_2 and small amounts of methane and CO; hydrogen can be enriched because of the different partition coefficients of hydrogen and carbon dioxide between the aqueous and gas phases.

One pioneering study demonstrating the subcritical water remediation of polycyclic aromatic hydrocarbons (PAH) and pesticide-contaminated soils at pilot scale was described by Lagadec et al. (2000), using water at 275°C for 35 min. Treatment reduced the levels of PAH and pesticides to below detectable limits in contaminated soils, allowing their reuse as fertile land with no toxicity.

8.4.6 BIOFUELS/BIOETHANOL PRODUCTION

Plant and waste material from the agriculture or food industries represents one of the world's largest resources of lignocellulosic biomass (containing hemicellulose, cellulose, and lignin) and has become the focus of attention for bioethanol production because of their large quantities, low cost, and environmental benefits (Hamelinck et al. 2005).

Lignocellulosic biomass components hemicellulose and cellulose can be converted into reducing sugars by acid or enzymatic hydrolysis, and then into ethanol by fermentation. Among the various types of pretreatment, hot liquid water (HLW) hydrolysis, which uses high pressure and temperature (160–240°C, 1–5 MPa), is recognized as one of the most promising methods, since the process does not need a chemical catalyst, and neither neutralization nor detoxification are required after pretreatment (Nabarlatz et al. 2007). A process for cellulose hydrolysis to ethanol by HLW was developed by Xu (2008), providing about 89% hemicellulose conversion rate for rice straw (Zhuang et al. 2009).

In an interesting review, Schacht et al. (2008) discussed the conversion of plant materials to ethanol by using supercritical fluid technology, including the use of near-critical water as a pretreatment to partly hydrolyze the plant material. Product streams from the hydrolytic treatment were later fermented. The resulting ethanol solution was processed by multistage countercurrent supercritical carbon dioxide extraction to ethanol of 99.8 wt.% concentration. Nonfermentable residues would be subjected to a second hydrolysis or transferred to a biogas production. Solid residues of the biogas reactor, in particular, lignin-containing fractions, could be oxidized with near and supercritical water to gas (Schacht et al. 2008). The authors conclude that only a combination of technologies, that is, SFE and SCWO, would make possible the development of a process for fuel ethanol production competitive enough to substitute for mineral oil–based fuels.

8.4.7 MULTIPLE UNIT PROCESSING—A STEP TO THE FUTURE

One interesting approach suggested by Professor J. W. King in a very recent review (King and Srinivas 2009) is the possibility to develop a platform integrating sub- and supercritical technologies to face some of the challenges in our society: environmental impact, sustainability, energy preservation, and health. The use of pressurized hot water can be exploited with advantages, over a wide range of temperatures and appropriate pressures, to include extraction of bioactive natural products, reaction of targeted substrates, biomass conversion for renewable fuels, and synthesis of chemicals.

The idea of building a multiunit operations system with the possibility of using different fluids can provide unique characteristics and advantages in our aim to develop a "green" processing platform consistent with sustainability. This multiunit operation system should work with environmentally benign solvents such as liquefied or supercritical CO_2 for nonpolar to moderately polar solutes, and with compressed hot water (between its boiling and critical points) for a wider range of polarities, considering also the use of ethanol as cosolvent together with water or carbon dioxide. Multiple critical fluid processing involves the integration of two or more fluids held under pressure applied as either mixtures or in a sequential manner for one or more unit processes.

An example of a multiple unit is the combination of PHWE with supercritical CO_2-assisted dispersion and drying of the extract. One of the concerns about PHWE is that the extract ends up in a diluted water sample, which can be a problem in the analytical determination if the compounds of interest occur at a very low concentration. Furthermore, if the extracted compound will be used as additive in a food or cosmetic formulation, it may not be desirable with dilute water extracts. Hence, after extraction, it may be interesting to remove the water in an efficient and gentle manner to obtain the extracted compounds as a dry powder. This can, for instance, be accomplished by freeze-drying or spray drying. Another way of drying the extract is to combine extraction with drying on the fly. A novel technique has recently been developed, Water Extraction with Particle formation On-line (WEPO®) (Ibañez et al. 2009). Figure 8.9 shows a schematic layout of the WEPO process.

WEPO is a new process that combines PHWE with particle formation on-line using supercritical CO_2 as dispersant and hot nitrogen for drying the produced

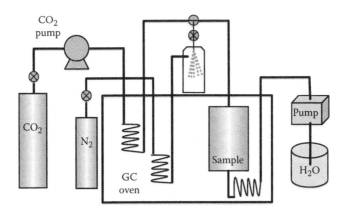

FIGURE 8.9 Schematic layout of WEPO process.

fine droplets. The process is based on the CAN-BD (Carbon Dioxide Assisted Nebulization with a Bubble Dryer®) particle formation process, which was invented by Prof. Sievers (Sievers et al. 2001). Results showed that by using WEPO, antioxidant-rich powders were obtained from oregano and rosemary, giving particle sizes less than 140 µm in diameter, with maintained antioxidant capacity, as compared to the extract without the drying process connected.

This type of platform can be a step into the future of industrial operation in the fields of food and agriculture, toward a greener and more sustainable and efficient processing.

8.5 FUTURE TRENDS

One of the future trends in industrial applications of PHWE is the development of green processing platforms able to perform, using green solvents such as supercritical carbon dioxide and water, multiunit operations consisting of extraction, reaction, raw material pretreatment, biofuel conversion, etc., with the main objective of sustainability. Related to this, another future trend is the use of PHWE coupled with enzymatic catalysis. For instance, thermostable β-glucosidase has been used in hot water to catalyze hydrolysis of quercetin glucosides in onion waste (Turner et al. 2006; Lindahl et al. 2010). Temperatures of up to 85–95°C can be used with maintained activity of the enzyme. The environmental impact of the developed method was compared to a conventional extraction/hydrolysis method based on methanol extraction and hydrochloric acid hydrolysis at 80°C, showing that the new method is preferable in terms of primary energy consumption and global warming potential (Lindahl et al. 2010). In the future, different kinds of enzymes will be developed and used in hot water processes to perform rapid reactions.

In terms of using PHWE as a sample preparation technique, new developments are foreseen directed to the improvement of the extraction process selectivity by combining the selectivity of the solvent and the use of new adsorbents in the EC. By using this approach, in-cell cleanup can be carried out to simplify the extraction of some minor components, leading to a higher sample throughput and a better quality of the analytical determination. The use of more sophisticated materials (such as specially designed polymers, molecular imprinted polymers, immunosorbents, or functionalized particles) can help improve the fractionation of specific analytes while avoiding coextraction of undesired material.

Another likely future trend is that on-line coupling systems will be developed to combine PHWE with separation techniques such as GC, HPLC (including the use of hot water chromatography, fully compatible with PHWE process), CE, and their coupling with MS in different formats. This will be done directly or by introducing a further purification step such as SPE or SPME. This approach will result in the reduction of errors associated with manual labor, lower consumption of solvents, and improved quality of results.

In terms of applications, it is expected that PHWE will find more uses in the food and agricultural field as scientists realize the important advantages of PHWE and the minimal drawbacks associated to its use, which can be solved by an accurate optimization of the extraction conditions.

8.6 CONCLUSIONS

Water at elevated temperature and pressure has the potential to replace environmentally burdensome solvents such as acetonitrile, methanol, dichloromethane, and toluene. Compared to conventional extraction techniques, PHWE is faster, often more automated, and enables a solvent-free work environment. Furthermore, new PHWE methods are fairly easy to develop by only optimizing temperature and extraction time. However, caution has to be taken regarding degradation of thermolabile target compounds during extraction, even though the problems are usually minor if the extraction time and temperature are carefully optimized. Finally, selectivity can be obtained in PHWE by varying the temperature of the water, or by using adsorbents in the EC, or both.

How does PHWE fit to a sustainable development? First of all, PHWE is an organic solvent-free technique, and water has a negligible environmental impact and is low cost in terms of production, transportation, and disposal. Furthermore, heating of water above its atmospheric boiling point consumes significantly less energy than producing hot water steam at the same temperature. By being able to substitute hazardous solvents with the probably most environmentally friendly solvent, water, the toxicity is reduced with positive effects on both the environment and health in general. To assess the environmental impact in specific cases, life cycle assessment should be conducted to calculate energy usage, raw data input, and releases to air and water.

Considering the wide range of applications discussed in this chapter and trying to apply creative thinking, it is advisable that new approaches be introduced to encourage a more rational use of our resources. Thus, new developments are expected, ranging from new methodologies to improve the selectivity/efficiency of a sample preparation technique to new platforms able to, in a sustainable manner, run different sample pretreatments–extractions–reactions–transformations in a more integrated fashion, and always considering the general well-being of the population and the environment.

REFERENCES

Alonso-Salces, R.M., Korta, E., Barranco, A., Berrueta, L.A., Gallo, B., Vicente, F. 2001. Determination of polyphenolic profiles of Basque cider apple varieties using accelerated solvent extraction. *J Agric Food Chem* 49: 3761–3767.

Andersen, O.M., Jordheim, M. 2006. Chemistry, biochemistry and applications, in *Flavonoids*, O. Andersen and K. Markham (Eds.), pp. 471–474. Boca Raton, FL: CRC Press.

Arapitsas, P., Turner, C. 2008. Pressurized solvent extraction and monolithic column-HPLC/DAD analysis of anthocyanins in red cabbage. *Talanta* 74: 1218–1223.

Bansleben, D., Schellenberg, I., Wolff, A.C. 2008. Highly automated and fast determination of raffinose family oligosaccharides in Lupinus seeds using pressurized liquid extraction and high-performance anion-exchange chromatography with pulsed amperometric detection. *J Sci Food Agric* 88: 1949–1953.

Bautz, H., Polzer, J., Stieglitz, L. 1998. Comparison of pressurised liquid extraction with Soxhlet extraction for the analysis of polychlorinated dibenzo-*p*-dioxins and dibenzofurans from fly ash and environmental matrices. *J Chromatogr A* 815: 231–241.

Bellissent-Funel, M.C. 2001. Structure of supercritical water. *J Mol Liq* 90: 313–322.

Beyer, A., Biziuk, M. 2008. Methods for determining pesticides and polychlorinated biphenyls in food samples—Problems and challenges. *Crit Rev Food Sci Nutr* 48: 888–904.

Bogialli, S., Curini, R., Di Corcia, A., Lagana, A., Nazzari, M., Tonci, M. 2004a. Simple and rapid assay for analyzing residues of carbamate insecticides in bovine milk: Hot water extraction followed by liquid chromatography-mass spectrometry. *J Chromatogr A* 1054: 351–357.

Bogialli, S., Curini, R., Di Corcia, A., Nazzari, M., Tamburro, D. 2004b. A simple and rapid assay for analyzing residues of carbamate insecticides in vegetables and fruits: Hot water extraction followed by liquid chromatography-mass spectrometry. *J Agric Food Chem* 52: 665–671.

Breithaupt, D.E. 2004. Simultaneous HPLC determination of carotenoids used as food coloring additives: Applicability of accelerated solvent extraction. *Food Chem* 86: 449–456.

Budrat, P., Shotipruk, A. 2008. Extraction of phenolic compounds from fruits of bitter melon (*Momordica charantia*) with subcritical water extraction and antioxidant activities of these extracts. *Chiang Mai J Sci* 35: 123–130.

Capello, C., Fischer, U., Hungerbuhler, K. 2007. What is a green solvent? A comprehensive framework for the environmental assessment of solvents. *Green Chem* 9: 927–934.

Carabias-Martinez, R., Rodriguez-Gonzalo, E., Revilla-Ruiz, P., Hernandez-Mendez, J. 2005. Pressurized liquid extraction in the analysis of food and biological samples. *J Chromatogr A* 1089: 1–17.

Cha, K., Lee, H., Koo, S., Song, D., Lee, D., Pan, C. 2010. Optimization of pressurized liquid extraction of carotenoids and chlorophylls from *Chlorella vulgaris*. *J Agric Food Chem* 58: 793–797.

Chen, C.R., Lee, Y.N., Chang, C.M.J., Lee, M.R., Wei, I.C. 2007. Hot-pressurized fluid extraction of flavonoids and phenolic acids from Brazilian propolis and their cytotoxic assay in vitro. *J Chin Inst Chem Eng* 38: 191–196.

Clifford, A.A., Burford, M.D., Hawthorne, S.B., Langenfeld, J.J., Miller, D.J. 1995. Effect of the matrix on the kinetics of dynamic supercritical fluid extraction. *J Chem Soc Faraday Trans* 91: 1333–1338.

Co, M., Zettersten, C., Nyholm, L., Sjöberg, P.J.R., Turner, C. 2011. Degradation effects in the extraction of antioxidants from birch bark using water at elevated temperature and pressure. *Anal Chim Acta* doi:10.1016/j.aca.2011.04.038.

Concha-Grana, E., Fernandez-Gonzalez, V., Grueiro-Noche, G., Muniategui-Lorenzo, S., Lopez-Mahia, P., Fernandez-Fernandez, E., Prada-Rodriguez, D. 2010. Development of an environmental friendly method for the analysis of organochlorine pesticides in sediments. *Chemosphere* 79: 698–705.

Cotton, N.J., Bartle, K.D., Clifford, A.A., Dowle, C.J. 1993. Rate and extent of supercritical fluid extraction of additives from polypropylene: Diffusion, solubility, and matrix effects. *J Appl Polym Sci* 48: 1607–1619.

Crescenzi, C., Di Corcia, A., Nazzari, M., Samperi, R. 2000. Hot phosphate-buffered water extraction coupled on line with liquid chromatography/mass spectrometry for analyzing contaminants in soil. *Anal Chem* 72: 3050–3055.

Curren, M.S.S., King, J.W. 2001. Solubility of triazine pesticides in pure and modified subcritical water. *Anal Chem* 73: 740–745.

Curren, M.S.S., King, J.W. 2002. New sample preparation technique for the determination of avoparcin in pressurized hot water extracts from kidney samples. *J Chromatogr A* 954: 41–49.

del Valle, J.M., Rogalinski, T., Zetzl, C., Brunner, G. 2005. Extraction of boldo (*Peumus boldus* M.) leaves with supercritical CO_2 and hot pressurized water. *Food Res Int* 38: 203–213.

Deng, C.H., Yao, N., Wang, A.Q., Zhang, X.M. 2005. Determination of essential oil in a traditional Chinese medicine, *Fructus amomi* by pressurized hot water extraction followed by liquid-phase microextraction and gas chromatography-mass spectrometry. *Anal Chim Acta* 536: 237–244.

Dinjus, E., Kruse, A. 2004. Hot compressed water—A suitable and sustainable solvent and reaction medium? *J Phys Cond Matter* 16: S1161–S1169.

Dong, L., Wang, J.Y., Deng, C.H., Shen, X.Z. 2007. Gas chromatography-mass spectrometry following pressurized hot water extraction and solid-phase microextraction for quantification of eucalyptol, camphor, and borneol in chrysanthemum flowers. *J Sep Sci* 30: 86–89.

Fojtova, J., Lojkova, L., Kuban, V. 2008. GC/MS of terpenes in walnut-tree leaves after accelerated solvent extraction. *J Sep Sci* 31: 162–168.

Garcia-Marino, M., Rivas-Gonzalo, J.C., Ibañez, E., Garcia-Moreno, C. 2006. Recovery of catechins and proanthocyanidins from winery by-products using subcritical water extraction. *Anal Chim Acta* 563: 44–50.

Ghoreishi, S.M., Shahrestani, R.G. 2009. Subcritical water extraction of mannitol from olive leaves. *J Food Eng* 93: 474–481.

Giergielewicz-Mozajska, H., Dabrowski, L., Namiesnik, J. 2001. Accelerated Solvent Extraction (ASE) in the analysis of environmental solid samples—Some aspects of theory and practice. *Crit Rev Anal Chem* 31: 149–165.

Golmohamad, F., Eikani, M.H., Shokrollahzadeh, S. 2008. Review on extraction of medicinal plants constituents by superheated water. *J Med Plants* 7: 1–21, 134.

Haglund, P., Spinnel, E. 2010. A modular approach to pressurized liquid extraction with in-cell clean-up. *LC GC Eur* 23: 292–301.

Hamelinck, C.N., van Hooijdonk, G., Faaij, A.P.C. 2005. Ethanol from lignocellulosic biomass: techno-economic performance in short-, middle- and long-term. *Biomass Bioener* 28: 384–410.

Hansen, C.M. 1969. Universality of solubility parameter. *Ind Eng Chem Prod Res Dev* 8: 2.

Hartonen, K., Parshintsev, J., Sandberg, K., Bergelin, E., Nisula, L., Riekkola, M.L. 2007. Isolation of flavonoids from aspen knotwood by pressurized hot water extraction and comparison with other extraction techniques. *Talanta* 74: 32–38.

Hawthorne, S.B., Yang, Y., Miller, D.J. 1994. Extraction of organic pollutants from environmental solids with subcritical and supercritical water. *Anal Chem* 66: 2912–2920.

Hawthorne, S.B., Grabanski, C.B., Martin, E., Miller, D.J. 2000. Comparisons of Soxhlet extraction, pressurized liquid extraction, supercritical fluid extraction and subcritical water extraction for environmental solids: recovery, selectivity and effects on sample matrix. *J Chromatogr A* 892: 421–433.

Herrero, M., Arraez-Roman, D., Segura, A., Kenndler, E., Gius, B., Raggi, M.A., Ibañez, E., Cifuentes, A. 2005. Pressurized liquid extraction-capillary electrophoresis-mass spectrometry for the analysis of polar antioxidants in rosemary extracts. *J Chromatogr A* 1084: 54–62.

Herrero, M., Cifuentes, A., Ibañez, E. 2006a. Sub- and supercritical fluid extraction of functional ingredients from different natural sources: Plants, food-by-products, algae and microalgae—A review. *Food Chem* 98: 136–148.

Herrero, M., Jaime, L., Martin-Alvarez, P.J., Cifuentes, A., Ibañez, E. 2006b. Optimization of the extraction of antioxidants from *Dunaliella salina* microalga by pressurized liquids. *J Agric Food Chem* 54: 5597–5603.

Herrero, M., Plaza, M., Cifuentes, A., Ibañez, E. 2010. Green processes for the extraction of bioactives from Rosemary: Chemical and functional characterization via ultra-performance liquid chromatography-tandem mass spectrometry and in-vitro assays. *J Chromatogr A* 1217: 2512–2520.

Ho, C.H.L., Cacace, J.E., Mazza, G. 2007. Extraction of lignans, proteins and carbohydrates from flaxseed meal with pressurized low polarity water. *LWT-Food Sci Technol* 40: 1637–1647.

Howard, L., Pandjaitan, N. 2008. Pressurized liquid extraction of flavonoids from spinach. *J Food Sci* 73: C151–C157.

Huie, C.W. 2002. A review of modern sample-preparation techniques for the extraction and analysis of medicinal plants. *Anal Bioanal Chem* 373: 23–30.

Ibañez, E., Kubatova, A., Señorans, F.J., Cavero, S., Reglero, G., Hawthorne, S.B. 2003. Subcritical water extraction of antioxidant compounds from rosemary plants. *J Agric Food Chem* 51: 375–382.

Ibañez, E., Cifuentes, A., Rodríguez-Meizoso, I., Mendiola, J.A., Reglero, G., Señorans, F.J., Turner, C. 2009. Device and procedure for the on-line extraction and drying of complex extracts. Patent: P200900164.

Jimenez-Carmona, M.M., Luque de Castro, M.D. 1999. Isolation of eucalyptus essential oil for GC-MS analysis by extraction with subcritical water. *Chromatographia* 50: 578–582.

Jimenez-Carmona, M.M., Ubera, J.L., de Castro, M.D.L. 1999. Comparison of continuous subcritical water extraction and hydrodistillation of marjoram essential oil. *J Chromatogr A* 855: 625–632.

Ju, Z.Y., Howard, L.R. 2005. Subcritical water and sulfured water extraction of anthocyanins and other phenolics from dried red grape skin. *J Food Sci* 70: S270–S276.

Juan-Garcia, A., Font, G., Juan, C., Pico, Y. 2010. Pressurised liquid extraction and capillary electrophoresis-mass spectrometry for the analysis of pesticide residues in fruits from Valencian markets, Spain. *Food Chem* 120: 1242–1249.

Khajenoori, M., Asl, A.H., Hormozi, F., Eikani, M.H., Bidgoli, H.N. 2009. Subcritical water extraction of essential oils from *Zataria multiflora* Boiss. *J Food Proc Eng* 32: 804–816.

King, J.W., Grabiel, R.D., Wightman, J.D. 2003. Subcritical water extraction of anthocyanins from fruit berry substrates. *Proceedings of the 6th International Symposium on Supercritical Fluids ISSF*, 28–30 April, Versailles (France), pp. 409–418.

King, J.W. 2006. Pressurized water extraction: Resources and techniques for optimizing analytical applications, in *Modern Extraction Techniques: Food and Agricultural Samples*, C. Turner (Ed.), pp. 79–95. Anaheim, CA: ACS Press.

King, J.W., Grabiel, R.D. 2007. Isolation of polyphenolic compounds from fruits or vegetables utilizing subcritical water extraction. United States Department of Agriculture patents. 2007 Apr. 24, no. US 7,208,181 B1, 7 pp.

King, J.W., Srinivas, K. 2009. Multiple unit processing using sub- and supercritical fluids. *J Supercrit. Fluids* 47: 598–610.

Koller, G., Fischer, U., Hungerbuhler, K. 2000. Assessing safety, health, and environmental impact early during process development. *Ind Eng Chem Res* 39: 960–972.

Konda, L.N., Fuleky, G., Morovjan, G. 2002. Subcritical water extraction to evaluate desorption behavior of organic pesticides in soil. *J Agric Food Chem* 50: 2338–2343.

Kritzer, P., Dinjus, E. 2001. An assessment of supercritical water oxidation (SCWO)—Existing problems, possible solutions and new reactor concepts. *Chem Eng J* 83: 207–214.

Kronholm, J., Hartonen, K., Riekkola, M.L. 2007. Analytical extractions with water at elevated temperatures and pressures. *TRAC-Trend Anal Chem* 26: 396–412.

Kubatova, A., Lagadec, A.J.M., Miller, D.J., Hawthorne, S.B. 2001. Selective extraction of oxygenates from savory and peppermint using subcritical water. *Flavour Fragr J* 16: 64–73.

Kubatova, A., Jansen, B., Vaudoisot, J.F., Hawthorne, S.B. 2002. Thermodynamic and kinetic models for the extraction of essential oil from savory and polycyclic aromatic hydrocarbons from soil with hot (subcritical) water and supercritical CO_2. *J Chromatogr A* 975: 175–188.

Lagadec, A.J.M., Miller, D.J., Lilke, A.V., Hawthorne, S.B. 2000. Pilot-scale subcritical water remediation of polycyclic aromatic hydrocarbon- and pesticide-contaminated soil. *Environ Sci Technol* 34: 1542–1548.

Lang, Q.Y., Wai, C.M. 2003. Pressurized water extraction (PWE) of terpene trilactones from *Ginkgo biloba* leaves. *Green Chem* 5: 415–420.

Lindahl, E., Hess, B., van der Spoel, D. 2001. GROMACS 3.0: a package for molecular simulation and trajectory analysis. *J Mol Model* 7: 306–317.

Lindahl, S., Ekman, A., Khan, S., Wennerberg, C., Borjesson, P., Sjoberg, P.J.R., Karlsson, E.N., Turner, C. 2010. Exploring the possibility of using a thermostable mutant of beta-glucosidase for rapid hydrolysis of quercetin glucosides in hot water. *Green Chem* 12: 159–168.

Lou, X.W., Janssen, H.G., Cramers, C.A. 1997. Parameters affecting the accelerated solvent extraction of polymeric samples. *Anal Chem* 69: 1598–1603.

Luque de Castro, M.D., Jimenez-Carmona, M.M., Fernandez-Perez, V. 1999. Towards more rational techniques for the isolation of valuable essential oils from plants. *TRAC-Trend Anal Chem* 18: 708–716.

Luthje, K., Hyotylainen, T., Riekkola, M.L. 2004. Comparison of different trapping methods for pressurised hot water extraction. *J Chromatogr A* 1025: 41–49.

Luthje, K., Hyotylainen, T., Rautiainen-Rama, M., Riekkola, M.L. 2005. Pressurised hot water extraction-microporous membrane liquid–liquid extraction coupled on-line with gas chromatography-mass spectrometry in the analysis of pesticides in grapes. *Analyst* 130: 52–58.

Marcus, Y. 1993. The properties of organic liquids that are relevant to their use as solvating solvents. *Chem Soc Rev* 22: 409–416.

Miller, D.J., Hawthorne, S.B. 1998. Method for determining the solubilities of hydrophobic organics in subcritical water. *Anal Chem* 70: 1618–1621.

Montilla, A., Ruiz-Matute, A.I., Sanz, M.L., Martinez-Castro, I., del Castillo, M.D. 2006. Difructose anhydrides as quality markers of honey and coffe. *Food Res Int* 39: 801–806.

Morel, F.M.M., Kraepiel, A.M.L., Amyot, M. 1998. The chemical cycle and bioaccumulation of mercury. *Annu Rev Ecol Syst* 29: 543–566.

Mroczek, T., Mazurek, J. 2009. Pressurized liquid extraction and anticholinesterase activity-based thin-layer chromatography with bioautography of Amaryllidaceae alkaloids. *Anal Chim Acta* 633: 188–196.

Mukhopadhyay, S., Luthria, D.L., Robbins, R.J. 2006. Optimization of extraction process for phenolic acids from black cohosh (*Cimicifuga racemosa*) by pressurized liquid extraction. *J Sci Food Agric* 86: 156–162.

Murata, K., Mitsuoka, K., Hirai, T., Walz, T., Agre, P., Heymann, J.B., Engel, A., Fujiyoshi, Y. 2000. Structural determinants of water permeation through aquaporin-1. *Nature* 407: 599–605.

Nabarlatz, D., Ebringerova, A., Montane, D. 2007. Autohydrolysis of agricultural by-products for the production of xylo-oligosaccharides. *Carbohydr Polym* 69: 20–28.

Nieto, A., Borrull, F., Marce, R.M., Pocurull, E. 2008. Pressurized liquid extraction of contaminants from environmental samples. *Curr Anal Chem* 4: 157–167.

Ollanketo, M., Peltoketo, A., Hartonen, K., Hiltunen, R., Riekkola, M.L. 2002. Extraction of sage (*Salvia officinalis* L.) by pressurized hot water and conventional methods: antioxidant activity of the extracts. *Eur Food ResTechnol* 215: 158–163.

Ong, E.S., Cheong, J.S.H., Goh, D. 2006. Pressurized hot water extraction of bioactive or marker compounds in botanicals and medicinal plant materials. *J Chromatogr A* 1112: 92–102.

Ozel, M.Z., Gogus, F., Lewis, A.C. 2003. Subcritical water extraction of essential oils from Thymbra spicata. *Food Chem* 82: 381–386.

Panayiotou, C. 1997. Solubility parameter revisited: An equation-of-state approach for its estimation. *Fluid Phase Equilib* 131: 21–35.

Pawliszyn, J. 2003. Sample preparation: Quo Vadis? *Anal Chem* 75: 2543–2558.

Pennington, D.W., Potting, J., Finnveden, G., Lindeijer, E., Jolliet, O., Rydberg, T., Rebitzer, G. 2004. Life cycle assessment: Part 2. Current impact assessment practice. *Environ Int* 30: 721–739.

Petersson, E., Liu, J., Sjöberg, P.J.R., Danielsson, R., Turner, C. 2010. Pressurized hot water extraction of anthocyanins from red onion: A study on extraction and degradation rates. *Anal Chim Acta* 663: 27–32.

Petersson, E.V., Puerta, A., Bergquist, J., Turner, C. 2008. Analysis of anthocyanins in red onion using capillary electrophoresis-time of flight-mass spectrometry. *Electrophoresis* 29: 2723–2730.

Pineiro, Z., Palma, M., Barroso, C.G. 2004. Determination of catechins by means of extraction with pressurized liquids. *J Chromatogr A* 1026: 19–23.

Plaza, M., Amigo-Benavent, M., Castillo, M.D.d., Ibañez, E., Herrero, M. 2010a. Facts about the formation of new antioxidants in natural samples after subcritical water extraction. *Food Res Int* 43: 2341–2348.

Plaza, M., Amigo-Benavent, M., del Castillo, M.D., Ibañez, E., Herrero, M. 2010b. Neoformation of antioxidants in glycation model systems treated under subcritical water extraction conditions. *Food Res Int* 43: 1123–1129.

Pongnaravane, B., Goto, M., Sasaki, M., Anekpankul, T., Pavasant, P., Shotipruk, A. 2006. Extraction of anthraquinones from roots of *Morinda citrifolia* by pressurized hot water: Antioxidant activity of extracts. *J. Supercrit Fluids* 37: 390–396.

Poole, P.H., Sciortino, F., Essmann, U., Stanley, H.E. 1992. Phase behavior of metastable water. *Nature* 360: 324–328.

Ramos, L., Kristenson, E.M., Brinkman, U.A.T. 2002. Current use of pressurised liquid extraction and subcritical water extraction in environmental analysis. *J Chromatogr A* 975: 3–29.

Rebitzer, G., Ekvall, T., Frischknecht, R., Hunkeler, D., Norris, G., Rydberg, T., Schmidt, W.P., Suh, S., Weidema, B.P., Pennington, D.W. 2004. Life cycle assessment Part 1: Framework, goal and scope definition, inventory analysis, and applications. *Environ Int* 30: 701–720.

Rodriguez-Meizoso, I., Marin, F.R., Herrero, M., Señorans, F.J., Reglero, G., Cifuentes, A., Ibañez, E. 2006. Subcritical water extraction of nutraceuticals with antioxidant activity from oregano. Chemical and functional characterization. *J Pharm Biomed Anal* 41: 1560–1565.

Rodriguez-Meizoso, I., Jaime, L., Santoyo, S., Cifuentes, A., Reina, G.G.B., Señorans, F.J., Ibañez, E. 2008. Pressurized fluid extraction of bioactive compounds from *Phormidium* species. *J Agric Food Chem* 56: 3517–3523.

Schacht, C., Zetzl, C., Brunner, G. 2008. From plant materials to ethanol by means of supercritical fluid technology. *J Supercrit Fluids* 46: 299–321.

Schaneberg, B.T., Khan, I.A. 2002. Comparison of extraction methods for marker compounds in the essential oil of lemon grass by GC. *J Agric Food Chem* 50: 1345–1349.

Scott, J.P., Ollis, D.F. 1995. Integration of chemical and biological oxidation processes for water treatment: Review and recommendations. *Environ Prog* 14: 88–103.

Sievers, R.E., Huang, E.T.S., Villa, J.A., Kawamoto, J.K., Evans, M.M., Brauer, P.R. 2001. Low-temperature manufacturing of fine pharmaceutical powders with supercritical fluid aerosolization in a Bubble Dryer (R). *Pure Appl Chem* 73: 1299–1303.

Smeds, A.I., Eklund, P.C., Sjoholm, R.E., Willfor, S.M., Nishibe, S., Deyama, T., Holmbom, B.R. 2007. Quantification of a broad spectrum of lignans in cereals, oilseeds, and nuts. *J Agric Food Chem* 55: 1337–1346.

Srinivas, K., King, J.W., Monrad, J.K., Howard, L.R., Hansen, C.M. 2009. Optimization of subcritical fluid extraction of bioactive compounds using Hansen solubility parameters. *J Food Sci* 74: E342–E354.

Srinivas, K., King, J.W., Howard, L.R., Monrad, J.K. 2010. Solubility and solution thermodynamic properties of quercetin and quercetin dihydrate in subcritical water. *J Food Eng* 100: 208–218.

Teo, C.C., Tan, S.N., Yong, J.W.H., Hew, C.S., Ong, E.S. 2008. Evaluation of the extraction efficiency of thermally labile bioactive compounds in *Gastrodia elata* Blume by pressurized hot water extraction and microwave-assisted extraction. *J Chromatogr A* 1182: 34–40.

Teo, C.C., Tan, S.N., Yong, J.W.H., Hew, C.S., Ong, E.S. 2009. Validation of green-solvent extraction combined with chromatographic chemical fingerprint to evaluate quality of *Stevia rebaudiana* Bertoni. *J Sep Sci* 32: 613–622.

Teo, C.C., Tan, S.N., Yong, J.W.H., Hew, C.S., Ong, E.S. 2010. Pressurized hot water extraction (PHWE). *J Chromatogr A* 1217: 2484–2494.

Turner, C., Turner, P., Jacobson, G., Almgren, K., Waldeback, M., Sjöberg, P., Nordberg-Karlsson, E., Markides, K.E. 2006. Subcritical water extraction and beta-glucosidase-catalyzed hydrolysis of quercetin glycosides in onion waste. *Green Chem* 8: 949–959.

Vandenburg, H.J., Clifford, A.A., Bartle, K.D., Zhu, S.A., Carroll, J., Newton, I.D., Garden, L.M. 1998. Factors affecting high-pressure solvent extraction (accelerated solvent extraction) of additives from polymers. *Anal Chem* 70: 1943–1948.

Waksmundzka-Hainos, M., Oniszczuk, A., Szewczyk, K., Wianowska, D. 2007. Effect of sample-preparation methods on the HPLC quantitation of some phenolic acids in plant materials. *Acta Chromatogr* 19: 227–237.

Waldeback, M., Señorans, F.J., Fridstrom, A., Markides, K.E. 2004. Pressurized fluid extraction of squalene from olive biomass, in *Modern Extraction Techniques for Food and Agricultural Samples*, C. Turner (Ed.), pp. 96–106. Anaheim, CA: ACS Press.

Waldeback, M. 2005. Pressurized fluid extraction: A sustainable technique with added values. PhD thesis, Uppsala University, Department of Physical and Analytical Chemistry, Uppsala, Sweden.

Wang, L.J., Weller, C.L. 2006. Recent advances in extraction of nutraceuticals from plants. *Trends Food Sci Technol* 17: 300–312.

Wang, L.L., Yang, H., Zhang, C.W., Mo, Y.L., Lu, X.H. 2008. Determination of oxytetracycline, tetracycline and chloramphenicol antibiotics in animal feeds using subcritical water extraction and high performance liquid chromatography. *Anal Chim Acta* 619: 54–58.

Westbom, R., Sporring, S., Cederberg, L., Linderoth, L.A., Bjorklund, E. 2008. Selective pressurized liquid extraction of polychlorinated biphenyls in sediment. *Anal Sci* 24: 531–533.

Xu, M.Z. 2008. The study on hot liquid hydrolysis of lignocellulosic biomass. Masteral dissertation, Chinese Academy of Sciences.

Yang, Y., Belghazi, M., Lagadec, A., Miller, D.J., Hawthorne, S.B. 1998. Elution of organic solutes from different polarity sorbents using subcritical water. *J Chromatogr A* 810: 149–159.

Yang, Y., Kayan, B., Bozer, N., Pate, B., Baker, C., Gizir, A.M. 2007. Terpene degradation and extraction from basil and oregano leaves using subcritical water. *J Chromatogr A* 1152: 262–267.

Zhuang, X.S., Yuan, Z.H., Ma, L.L., Wu, C.Z., Xu, M.Z., Xu, J.L., Zhu, S.N., Qi, W. 2009. Kinetic study of hydrolysis of xylan and agricultural wastes with hot liquid water. *Biotechnol Adv* 27: 578–582.

9 Instant Controlled Pressure Drop Technology in Plant Extraction Processes

Karim Salim Allaf, Colette Besombes,
Baya Berka, Magdalena Kristiawan, Vaclav Sobolik,
and Tamara Sabrine Vicenta Allaf

CONTENTS

9.1 INTRODUCTION

9.1.1 Case of Volatile Molecules

Essential oils are generally a volatile mixture of organic compounds derived from a single botanical source. These "oils," sometimes composed of hundreds of chemical aromas and compounds, are primarily responsible for the characteristic smell of the natural plant source. They are well known for their aromatic uses, antimicrobial properties, and even in the field of aromatherapy.

Essential oils are usually extracted using hydrodistillation (HD) or steam extraction. Similar operations are used to separate other volatile molecules. These techniques have changed very little until recently. They provide heat from steam or boiling water by both convection and condensation processes. Steam extraction includes a solid–gas interaction stage, whereas HD involves a solid–liquid interaction. At the operational level, in both techniques, the vapors generated (water and volatile molecules such as essential oils) have to be condensed on an adequate low-temperature condenser, which is followed by a decantation stage in which water is separated from the other condensed liquid. The kinetics of the evaporation stage of all these techniques is known to be particularly slow (a long processing time of up to 24 h and sometimes longer), involving a large consumption of energy and the degradation of sensitive molecules (Bocchio 1985; Chavanne et al. 1991; Capon et al. 1993; Salisova et al. 1997; Gamiz-Gracia et al. 2000; Chemat et al. 2004; Lucchesi et al. 2004; Satrani et al. 2006; Ferhat et al. 2007; Kotnik et al. 2007; Wenqiang et al. 2007; Bendahou et al. 2008; Cassel et al. 2009). Numerous authors have highlighted several other undesirable aspects encountered while using conventional essential oil extraction methods: losses of some volatile compounds, low extraction efficiency (EE), thermal degradation of some compounds in the extract, etc. (Jiménez-Carmona et al. 1999; Gamiz-Gracia et al. 2000).

Allaf and colleagues (1993, 1998) developed an innovative process, instant controlled pressure drop (DIC), which rapidly isolates high-quality essential oils without any requirement for an organic solvent.

9.1.2 Case of Nonvolatile Molecules (Solvent Extraction)

Another effect of DIC treatment is the expansion of the solid matrix, which can act on the solvent extraction. Indeed, this type of treatment increases the porosity and the specific surface area of the treated plant; subsequently, the solvent can easily enter the matter and hence facilitate the extraction of the desired material. DIC texturing can be considered a pretreatment that dramatically decreases the time necessary for the solvent extraction of nonvolatile compounds.

9.2 FUNDAMENTALS

9.2.1 Heat and Mass (Liquid and Vapor) Transfers in Steam

There are special zones involved in the secretion of essential oils located within plant tissues (deep structures) or on the surface of the plant (peltate glandular trichomes). The process of essential oil extraction should depend on the way in which the plant stores oil and the type of secretion zones.

One of the principal aspects of fundamental studies on steam extraction was carried out by Allaf and colleagues (Allaf 2009; Besombes et al. 2010). Aromatic plants are a porous material in which coupled heat and mass transfer processes occur successively. Heat flow is used in part to increase the product temperature, but in the main ensures a phase change (liquid–vapor).

In steam extraction, the external saturated steam ensures heat transfer on the external surface of the solid matrix; it is normally carried out by convection but mostly by condensation. Whatever the "volatile" compound, the partial pressure strictly depends on the temperature. The external mass transport within the surrounding environment takes place from the surface of the grain toward the condenser and concerns water vapor and other volatile molecules. In this case, the essential oil partial pressure gradient is the driving force. The process can be intensified by increasing the difference in temperature and by reducing the distance between the exchange surface of the material and the condenser.

After quickly rising to the highest level, the exchange surface can reach the saturated steam temperature, which ensures a gradual spread of heat within the solid.

Subsequently, an internal heat transfer occurs through a similar conduction phenomenon. The presence of air, water, and other volatile molecules could give rise to vaporization/condensation phenomena in the holes within the porous structure. This generates a heat transfer phenomenon where the temperature gradient is the driving force. This process greatly intensifies the heat transfer as a conduction-type transfer with an effective conductivity value much higher than standard values. The majority of this heat flow is used to evaporate water and essential oils from the holes in the product.

Various types of mass transfer may take place within the plant; usually this is assumed to be the limiting process since the effective heat diffusivity is usually much higher than the effective mass diffusivity. It is assumed that during steam extraction of essential oil, the transfer of liquid essential oil is negligible whereas most mass transfer is a gas diffusion phenomenon, obeying a Fick-type law with the partial pressure gradient of each volatile compound as the driving force.

9.2.2 Paradox of Coupled Heat and Vapor Transfers

It is possible to assume that the partial pressure of each vapor, either water or other volatile molecules, depends only on the temperature. Thermodynamic data on the liquid–gas equilibrium of each molecule at different temperatures can generally be found in the literature. Such data are readily available for water and some well-known molecules but they are lacking for numerous essential oil molecules.

Mass transfer processes, which usually occur with liquids and vapors, depend on the nature and localization of the molecules and the porosity of the medium. These processes may be driven by capillary forces and by diffusion when it is a liquid/solid interaction. The gas phase (gas/solid interaction) mass transfer occurs through diffusion reinforced by a similar DARCY process within the holes. It also depends on several factors including temperature, structure, and the morphological state of the matter (degree of porosity, specific surface area, permeability of the secretion element walls, etc.) (Allaf 2009). In steam extraction, it can be assumed that the transfer of essential oils in the form of a liquid is negligible. The main transfer phenomenon is a gas diffusion phase, which can be shown to be a Fick-type law related to the partial pressure gradient of each volatile compound through an effective diffusivity (D_{eff}). However, the involvement of the structure, the presence of secretion element walls as barriers, and more generally the porosity of the material contribute to a macroscopic mass transfer inside the material.

In similar cases, it is assumed that essential oil extraction necessarily implies an amount of heat capable of transforming the liquid phase into the gas phase within the porous material (Besombes et al. 2010). As the external vapor (partial) pressure is saturated, one can postulate:

$$\vec{\nabla} \cdot \vec{\phi} + (\rho_s c_{ps} + \rho_{eo} c_{pe} + \rho_w c_{pw}) \frac{\partial T}{\partial t} + \frac{\partial \left(p_e \varepsilon \frac{M_e L_e}{RT} + p_w \varepsilon \frac{M_w L_w}{RT} \right)}{\partial t} = 0 \qquad (9.1)$$

As the internal transfer is carried out by conduction:

$$\vec{\nabla}\cdot(-\lambda\vec{\nabla}T)+(\rho_s c_{ps}+\rho_{co}c_{pe}+\rho_w c_{pw})\frac{\partial T}{\partial t}+\frac{\partial}{\partial t}\left[\frac{\varepsilon}{RT}(p_e M_e L_e+p_w M_w L_w)\right]=0 \quad (9.2)$$

The temperature distribution is assumed to be stationary during most of the operation since most of the heat transfer is "only" used to evaporate essential oils and water (Allaf 2009):

$$\vec{\nabla}\cdot(-\lambda\vec{\nabla}T)+\frac{\partial}{\partial t}\left[\frac{\varepsilon}{RT}(p_e M_e L_e+p_w M_w L_w)\right]=0 \quad (9.3)$$

As the mass transfers of vapor of different essential oil compounds and water would be governed by a Fick-type law, the formulations of Allaf (1982) were used separately:

$$\frac{(p_e/T)}{\rho_s}(\vec{v}_{eo}-\vec{v}_s)=-D_{eff_e}\vec{\nabla}\left(\frac{p_e/T}{\rho_s}\right) \quad (9.4)$$

$$\frac{(p_w/T)}{\rho_s}(\vec{v}_v-\vec{v}_s)=-D_{eff_w}\vec{\nabla}\left(\frac{p_w/T}{\rho_s}\right) \quad (9.5)$$

By neglecting possible shrinkage phenomena, one can assume that $\rho_s=$ constant and $\vec{v}_s=0$, 9.4 and 9.5 may be transformed into:

$$(p_e/T)\vec{v}_{eo}=-D_{eff_e}\vec{\nabla}(p_e/T) \quad (9.6)$$

$$(p_w/T)\vec{v}_v=-D_{eff_w}\vec{\nabla}(p_w/T) \quad (9.7)$$

As the external vapor (partial) pressure is saturated, one can assume that the internal vapor partial pressure (p_w) within the porous material is approximately constant, and so 9.7 can be postulated as:

$$-\lambda\vec{\nabla}\cdot(\vec{\nabla}T)+\varepsilon M_e L_e\frac{\partial}{\partial t}\left[\frac{p_e}{RT}\right]=0 \quad (9.8)$$

By assuming that each "particle" of the plant is spherical, homogeneous, and isotropic, Equations 9.6 and 9.8 can be transformed at one dimension (r) respectively into:

$$\left(P_e/T\right)v_{eo} = -D_{eff_e}\left(\frac{\partial\left(P_e/T\right)}{\partial r}\right) \tag{9.9}$$

$$-\lambda\frac{\partial^2 T}{\partial r^2} + \varepsilon M_e L_e\frac{\partial}{\partial t}\left[\frac{P_e}{RT}\right] = 0 \tag{9.10}$$

On the other hand, when (p_e/T) increases, the temperature increases:

$$\frac{\partial\left(P_e/T\right)}{\partial T} > 0 \tag{9.11}$$

The values of essential oil partial pressure (p_e) are thus higher at the exchange surface than in the granule core. This result is a paradox, inducing a motion completely opposite to what is required for an extraction operation (Al Haddad 2007; Al Haddad et al. 2008). Then, in the standard steam extraction of essential oils, the operation is achieved using successive layers with "front progression" kinetics (Figure 9.1).

This may further explain why a pretreatment step of grinding is always required in steam extraction to reduce the granule shape. However, it would be possible to remedy this situation and make a notable improvement in the kinetics by adopting

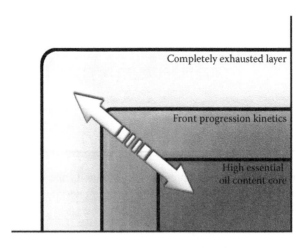

FIGURE 9.1 Paradoxical stage and "front progression" kinetics from coupled transfers of heat and volatile compounds, in standard steam distillation.

other methods for heating and/or moving such molecules by a mass transfer achieved through the total pressure gradient (TPG) process (DIC process).

9.2.3 SPECIFICITY OF DIC EXTRACTION OF ESSENTIAL OILS

9.2.3.1 Equipment

A laboratory-scale DIC reactor (Figure 9.2) was defined and presented in various papers (Kristiawan et al. 2009). It is a 7-l processing vessel with a heating jacket, a 0.7-m^3 vacuum tank (a volume 100 times greater than the treatment chamber) with a cooling water jacket, a water ring vacuum pump, and steam flow valves; a pneumatic valve ensures an "instant" connection (opened in less than 0.1 s) between the vacuum tank and the processing vessel. Both pilot and industrial-scale DIC reactors were used in order to define the scaling up of various treatments.

9.2.3.2 Principle

DIC treatment is based on thermal effects induced by exposing the raw material to a short period of high saturated steam pressure, from about 0.1 up to 0.6 MPa depending on the product, that is, a temperature ranging from 100 to 159°C (Figure 9.3). Hence, the partial pressure of each essential oil compound is higher than that during standard steam distillation (SD).

The high temperature–short time stage is followed by an abrupt pressure drop toward a vacuum at about 5 kPa.

Such a pressure drop, whose rate ($\Delta P/\Delta t$) is higher than 0.5 MPa/s, simultaneously triggers:

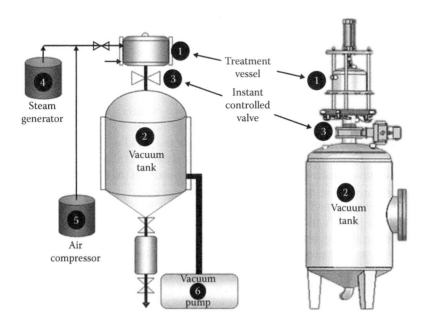

FIGURE 9.2 Instant controlled pressure drop (DIC) apparatus.

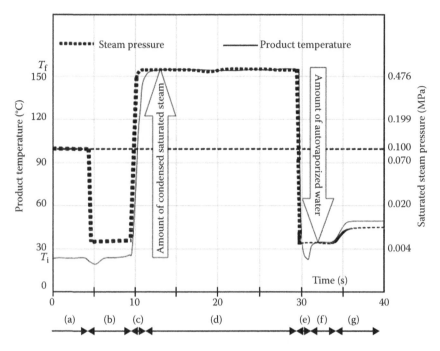

FIGURE 9.3 Temperature and pressure of a DIC processing cycle: T_i is initial temperature of the product, T_f denotes highest temperature of the product: (a) sample at atmospheric pressure; (b) initial vacuum; (c) saturated steam injection to reach the selected pressure; (d) constant temperature corresponding to saturated steam pressure; (e) abrupt pressure drop toward a vacuum; (f) vacuum; (g) release to atmospheric pressure.

- An autovaporization of volatile compounds
- The instantaneous cooling of the product, which stops thermal degradation
- A swelling or even rupture of the cell and the secretion element walls

The porous structure thus created enhances mass transfer, intensifying the effective diffusivity as well as the overall operation kinetics. However, the major impact of the DIC process is that it provides a solution for the paradox of essential oil mass/heat transfers by allowing a mass transfer as a gradient of total pressure (Darcy-type law with a mass transfer by TPG), flowing from the core of the material at high total pressure of vapor to its surface under a vacuum.

The temperature and pressure levels during one DIC cycle are shown in Figure 9.3. After an initial atmospheric pressure stage (a), a vacuum of about 4.5 kPa is established in the autoclave (b) and, just after saturated steam, is injected into the autoclave (c) and maintained at a fixed pressure level for a predetermined time (d). The initial vacuum (a) state allows closer contact between the steam, which is the heating fluid, and the exchange surface of the plant. The heat transfer toward the plant is enhanced, allowing the product to immediately reach saturated steam temperature. After this thermal treatment, the steam is cut off and the spherical valve is opened rapidly (in less than 0.2 s), which results in an abrupt pressure drop within the vessel (e). The vacuum period (f) is followed by a final release in order to attain atmospheric pressure (g), or by stage (c) of a new DIC cycle for a Multi Cycle-DIC treatment (Figure 9.4). Multi Cycle-DIC contains n repetitions of stages (c), (d), (e), and (f), with a total heating time that is the heating time of all these cycles.

By abruptly dropping the pressure, an adiabatic autovaporization of the over-heated water and volatile compounds occurs, inducing an instant cooling of the residual material. The vapor engenders mechanical stresses within the plant, puffing cells, and the structure, depending on the plant's viscoelastic behavior, which can even result in cell wall rupture. The initial water content and temperature of the plant just before the pressure is dropped are very important parameters. The condensates we recover are usually very stable oil-in-water emulsions with a droplet diameter of less than 0.5 μm. Moreover, the residual plant solids underwent structural and chromatography analysis. Structure analyses are mainly carried out with a scanning electron microscopy to compare solid material before and after DIC treatment.

During the heating stage (d), the heat flux from the steam to the product is:

$$\varphi = hS(T_v - T) \tag{9.12}$$

And the quantity of heat during the thermal treatment time of a cycle is then:

$$Q_c = hS \langle T_v - T \rangle_{LN} \cdot t_c \tag{9.13}$$

During this heating stage (d), convection and condensation of saturated steam ensure a high temperature at the exchange surface with a high value of h; the operation is very quick and may be considered as an instantaneous process. The initial vacuum state enables a closer contact between the external product surface and the steam (higher value of the surface S), intensifying this process. During this same

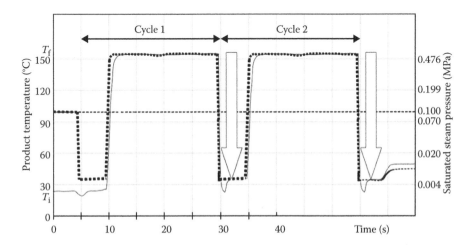

FIGURE 9.4 Pressure change in a multicycle DIC process.

stage, both heat and mass transfers occur within the product. The first occurs through conduction according to a Fourier-type law.

$$\vec{\varphi} = -\lambda_{eff} \cdot \vec{\nabla} T \tag{9.14}$$

The second is a normal Fick-type water diffusion mass transfer (Allaf 1982):

$$\frac{\rho_w}{\rho_s}\left(\vec{v}_w - \vec{v}_s\right) = -D_{eff}\vec{\nabla}\left(\frac{\rho_w}{\rho_s}\right) \tag{9.15}$$

By neglecting the possible expansion phenomena, one can assume ρ_s to be constant and $v_s = 0$, so Equation 9.15 may be transformed into:

$$\rho_w \vec{v}_w = -D_{ws}\vec{\nabla}\rho_w \tag{9.16}$$

Moreover, by assuming a radial transfer, one can write:

$$\rho_w v_w = -D_{ws}\frac{d\rho_w}{dr} \tag{9.17}$$

By using mass conservation and continuity, the second Fick's law becomes:

$$\frac{\partial \rho_w}{\partial t} = \vec{\nabla} \cdot \left[D_{ws}\vec{\nabla}\rho_w\right] \tag{9.18}$$

The diffusivity (D_{ws}) is assumed to be constant only in the hypothesis of both structural and thermal homogeneities. Indeed, D_{ws} varies considerably versus the porosity and temperature of the granule. A general quantification of physical processes and adequate experiments normally confirm this hypothesis, allowing Equation 9.18 to become the second Fick's law:

$$\frac{\partial \rho_w}{\partial t} = D_{ws}\vec{\nabla} \cdot \left[\vec{\nabla}\rho_w\right] = D_{ws}\nabla^2\rho_w \tag{9.19}$$

And, by assuming a unidirectional flow, Equation 9.19 becomes:

$$\frac{\partial \rho_w}{\partial t} = D_{ws}\frac{d^2\rho_w}{dr^2} \tag{9.20}$$

The duration (t_c) of stage (d) must generally be defined so that both the temperature (T) and moisture content (W) are uniform within the product. Usually, the first parameter is much more quickly established; the quantity of added moisture absorbed by the product from the surrounding saturated steam during stage (d) is:

$$m_v = Q/L = m_s(c_{ps} + <Wc_{pw}>) \cdot (T_f - T_i) \tag{9.21}$$

After the pressure-drop stage (e) and during the vacuum stage (f), the internal total pressure is mainly due to water vapor and essential oils in the porous medium. Under DIC conditions, Allaf and colleagues assumed that the total pressure of the volatile mixture vapor in the porous material just after the pressure drop was much higher than the external pressure (Allaf 2009). Therefore, the transfer of essential oils within the expanded granule is assumed to be from the core toward the surrounding medium through the gradient of total pressures. The Darcy-type law describes this transfer:

$$\rho_{mix}\left(\vec{v}_{mix} - \vec{v}_s\right) = -\frac{K}{v_m}\vec{\nabla}P \tag{9.22}$$

Allaf (2009) assumed a negligible expansion phenomenon, possibly due to the autovaporization itself just after the pressure drops (Allaf 2009). They assumed that $\vec{v}_s = 0$. With a radial transfer within a spherical shape material, Equation 9.22 could be written as:

$$\rho_{mix}\vec{v}_{mix} = -\frac{K}{v_m}\frac{\partial P}{\partial r} \tag{9.23}$$

By using mass conservation and continuity, and by integrating between the hole (whose radius is R_o) and the external radius (R_s) of the spherical shape, Equation 9.23 becomes:

$$\dot{m}_m = -\frac{4\pi K}{v_m}\frac{(P_{globule} - P_{ext})}{\left(\dfrac{1}{R_o} - \dfrac{1}{R_s}\right)} \tag{9.24}$$

The value of the total pressure ($P_{globule}$) in the hole decreases versus the time (t), depending on the flow ($\dot{m}_m$). The globule radius (R_o) is assumed to be constant (absence of expansion) as is the temperature, which would appear to remain constant at the value reached just after the pressure drop, defined by the level of water–vapor temperature–pressure equilibrium. Equation 9.24 becomes:

$$\frac{d\dot{m}_m}{dt} = -\frac{4\pi K}{v_m}\frac{\left(-RT/<M>V_{hole}\right)}{\left(\dfrac{1}{R_o} - \dfrac{1}{R_s}\right)}\dot{m}_m \tag{9.25}$$

$$\dot{m}_{\mathrm{m}} = \dot{m}_{\mathrm{om}} \exp\left(-\frac{4\pi KRT}{v_{\mathrm{m}}} \frac{t}{<M>\frac{4}{3}\pi R_{\mathrm{o}}^{3}\left(\frac{1}{R_{\mathrm{o}}} - \frac{1}{R_{\mathrm{s}}}\right)}\right) \tag{9.26}$$

$$\dot{m}_{\mathrm{m}} = \frac{4\pi KRT}{v_{\mathrm{m}}\left(\frac{1}{R_{\mathrm{o}}} - \frac{1}{R_{\mathrm{s}}}\right)}(P_{\mathrm{o}} - P_{\mathrm{ext}})\exp\left(-\frac{4\pi KRT}{v_{\mathrm{m}}} \frac{t}{<M>\frac{4}{3}\pi R_{\mathrm{o}}^{3}\left(\frac{1}{R_{\mathrm{o}}} - \frac{1}{R_{\mathrm{s}}}\right)}\right) \tag{9.27}$$

By integrating between $t = 0$ and $t \to \infty$, the following expression can be obtained:

$$\dot{m}_{\mathrm{m}} = \frac{4}{3}\pi KR_{\mathrm{o}}^{3}\frac{M}{RT}(P_{\mathrm{o}} - P_{\mathrm{ext}}) \tag{9.28}$$

Usually, the time t_{v} of the vacuum stage (f) allows the vapor mixture to be transported toward the surrounding medium and then collected.

One can calculate the amount of heat needed for this autovaporization as:

$$Q = m_{\mathrm{m}}L_{\mathrm{m}} = m_{\mathrm{p}}c_{\mathrm{pp}}(T_{\mathrm{i}} - T_{\mathrm{f}}) \tag{9.29}$$

The DIC process provides an appropriate solution to the disadvantages of the paradoxical phenomenon and greatly intensifies essential oil extraction. It uses autovaporization instead of (or coupled to) an evaporation process, and TPG instead of the normal diffusion phenomenon.

9.3 EXPERIMENTAL WORK

Various raw materials were used, as well as different extraction processes and assessment methods. HD and SD were used as references to compare with the DIC instant autovaporization. The main part of our experimental work was carried out at different scales for analyzing, optimizing, and modeling. The effect of DIC essential oil extraction was studied with saturated steam pressure, total heating time, and the number of cycles as the operative parameters (independent variables). The response parameters we had to consider as dependent variables were the total yield and the variation in extract composition.

9.3.1 RAW MATERIAL

Numerous raw materials, such as angelica seeds, cinnamon bark, chamomile flowers, lemongrass leaves, patchouli leaves, thyme leaves, valerian roots, citrus peels, lavandin, and myrtle, were treated by DIC and compared to conventional extraction methods.

The different materials were studied according to the experimental protocol presented in Figure 9.5. Moreover, the operating parameters were analyzed using the experimental designs.

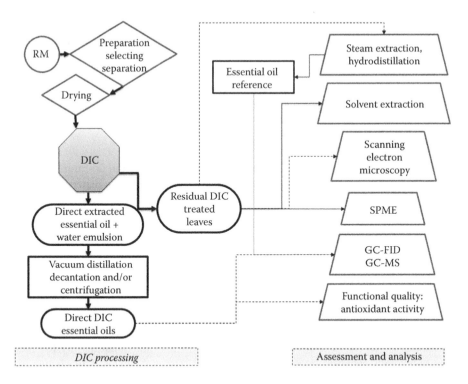

FIGURE 9.5   Experimental protocol for extracting and assessing essential oils.

9.3.2   TREATMENT METHODS

SD and HD were used as a reference for comparison with DIC essential oil extraction.

9.3.2.1   Steam Distillation

A Pignat P3734 laboratory apparatus was used. A packed bed of a variable quantity (approximately 50 g) of aromatic raw plant material was placed on a stainless steel grill fixed above a distillation vessel containing 2 l of water. Electric heating of the water produced 2.5 kg/h of steam, which in turn passed through the plant bed where it evaporated and carried away the essential oil. The vapors were condensed and then collected in a receiver vessel. A cooling tap water system around the receiver vessel cooled the distillate from 100 to 25°C in order to avoid the loss of volatile compounds. During the isolation process, the excess of water in the receiver vessel was recycled into the distillation vessel in order to recover the evaporated water.

9.3.2.2   Hydrodistillation

With most of the products, the pilot plan HD equipment used was a modified Clevenger. Some modifications were made to the equipment in order to carry out various treatments and quantify the amount of essential oils extracted.

Conventionally, an amount of ca. 200 g of raw material was immersed in 2 l of distilled water in a 3-l distillation flask. The extraction of essential oils was carried out over 3 h, from the first drop of distillate until the plant material had been completely consumed. The duration of extraction was first determined by studying the kinetics of the operation using the measurement of the quantity of essential oils versus time.

The oil recovered and dried with anhydrous sodium sulfate was then stored in a refrigerator at 4°C in a tightly closed amber vial, away from sources of degradation (heat or light, etc.), for further analyses. HD was carried out on the raw material.

9.3.2.3 Solvent Extraction

Solvent extraction allowed us to determine the amount of essential oil initially present in the raw materials. This technique consists in placing the raw material in contact with a solvent (the optimized solvent) for a given duration (this is also optimized beforehand) and at a given temperature (which depended on the solvent used).

9.3.3 ANALYSIS AND CALCULATION METHODS

9.3.3.1 Measurement of Antioxidant Activity

The free radical 2,2-diphenyl-1-picrylhydrazyl (DPPH) is a very stable radical in methanolic solution, and is used to evaluate the antioxidant activity of plant essential oils. In solution, this radical has an intense violet color; its reduction is accompanied by fading to yellow. Measuring the effectiveness of an antioxidant (ability to bind free radicals) is achieved by measuring the loss of the violet color. This change in color is attributable to the recombination of the DPPH radical following the oxidation–reduction reaction:

$$DPPH_{ox} + A_{red} \rightarrow DPPH_{red} + A_{ox} \qquad (9.30)$$

We evaluated the antioxidant activity of essential oil using the Brand-Williams method with some modifications (Brand-Williams et al. 1995). This consisted in measuring the absorbance of essential oils using an OMEGA HELIOS Thermoscientifique UV–VIS spectrophotometer (Thermo Fisher Scientific, Saint Herblain, France) at 516 nm as the peak wavelength.

The essential oil was diluted in methanol at different concentrations. From each sample, 1 ml was withdrawn and mixed with 3 ml of a methanol solution of 0.3 mM DPPH. The reaction mixture was vigorously shaken and incubated for 30 min at room temperature away from any source of light. A 4-ml quartz cell was filled carefully with the mixture, and the maximum absorbance was measured at 516 nm against a blank consisting of 1 ml essential oil extract and 3 ml methanol. The antioxidant BHT was submitted to the same procedure to evaluate its antioxidant capacity, which was used as a reference. All measurements were repeated three times. The inhibition percentage was determined using the formula:

$$\%_{inhibition} = \frac{Abs_{control} - Abs_{sample}}{Abs_{control}} \qquad (9.31)$$

9.3.3.2 Yield and EE (Quantitative)

The main response of the response surface methodology (RSM) analysis was chosen as the essential oil availability in the DIC-treated plant. This residual availability denotes the ratio of oil yields in the DIC-treated (residual oil) and nontreated plants (reference oil). By assuming that the sum of the volatiles in residual oil and DIC condensate was equal to the quantity in the reference oil, the amount of oil removed by DIC treatment could be estimated.

The yields of reference oil (y_{of}) and DIC direct and residual oils (y_{od} and y_{os}) are based on the dry matter of plants:

$$y_{of} = \frac{m_{of}}{m_{dm}} \tag{9.32}$$

$$y_{od} = \frac{m_{od}}{m_{dm}} \tag{9.33}$$

$$y_{os} = \frac{m_{os}}{m_{dm}} \tag{9.34}$$

where oil mass was computed from the gas chromatography (GC) peak area of all volatile molecules using the external standard method with methyl nonadecanoate. DIC direct oil is composed of the volatile molecules isolated from the DIC condensate.

Furthermore, we also defined an extraction ratio, using the following equation:

$$\text{Extraction ratio } (\%) = \frac{(\text{raw material EO} - \text{DIC textured sample EO})}{\text{raw material EO}} \tag{9.35}$$

9.3.3.3 Determining Essential Oil Composition (Qualitatively)

The products derived from solvent extraction and the different essential oils were analyzed by GC. The detectors used were either flame ionization (FID) or mass spectrometry (MS).

The identification of each compound was conducted by comparison with spectral data banks (for MS detection) and/or by using a solution of an alkane mixture and Kovats index determination.

9.3.3.4 Effect of DIC Treatment on Compound Availability

By assuming that the DIC treatment only isolates a part of a compound and does not increase its availability (A) in the residual plant solid, the DIC EE could be estimated as:

$$EE = 1 - A = 1 - \frac{y_{os}}{y_{of}} \tag{9.36}$$

It should be noted that the availability of compounds in residual plants was often increased by DIC treatment, which led to an underestimation of EE.

TABLE 9.1

Yield of Essential Oils Obtained Using DIC Extraction in Comparison with Conventional Hydrodistillation Process

Essential Oil	Hydrodistillation (g EO/100 g dry material)	DIC Extraction (g EO/100 g dry material)
Lavandin	(3 h): 4.3%	DIC (0.6 MPa, 2 cycles, 4 min): 4.37%
Oregano	(6 h): 0.67%	DIC (0.6 MPa, 10 cycles, 5 min): 1.97%
Rosemary	(6 h): 0.13%	DIC (0.6 MPa, 10 cycles, 5 min): 0.68%

9.4 RESULTS

9.4.1 KINETICS

In the case of lavandin, oregano, and rosemary, the total yields of total extracted essential oils were measured in DIC extraction in comparison with the conventional HD processes. Optimized DIC extraction systematically provided a higher yield of essential oils compared to HD or SD.

In the case of lavandin, DIC extraction required 4 min instead of 3 h for HD, with a yield that was roughly equivalent. Regarding oregano and rosemary, DIC extraction time was considerably lower than what is required with HD. We can conclude that the extraction by DIC is characterized by much higher kinetics than those obtained by HD.

9.4.2 YIELDS

Regarding the three materials mentioned above, DIC extraction achieves a level of performance equivalent to (in the case of lavender), and often higher (for rosemary and oregano) than, what is obtained by HD (Table 9.1).

In the case of various aromatic plants, the total yields of total extracted essential oils were measured in DIC extraction in comparison with conventional SD and the HD processes. Table 9.2 presents the results obtained with thyme.

Furthermore, more severe DIC conditions were needed for isolating the essential oil from cananga flowers. SD 12 h, whereas Allaf and colleagues found that a oil

TABLE 9.2

Yield of Essential Oils Obtained by DIC Extraction in Comparison with Conventional Steam Distillation and Hydrodistillation Processes

Essential Oil Yields	Steam Distillation	Hydrodistillation	DIC Extraction
	(g EO/100 g dry material)		
Thyme leaves	2.71%	3.68%	4.34%

TABLE 9.3

Response Surface Methodology (RSM) Experimental Design of DIC Treatment of Lavandin at Moisture Content W_t = 20 g H_2O/100 g Dry Material

	RSM Operative Variable Parameters	$-\alpha$	-1	0	$+1$	$+\alpha$
$X_1 = P$	Steam pressure (MPa)	0.14	0.2	0.35	0.5	0.56
$X_2 = t_d$	Thermal treatment time, t_d (s)	60	122	270	418	480

Note: α (axial distance) = $\sqrt[4]{2^N}$, where N is the number of independent variables. In the present case, $N = 2$ and $\alpha = 1.4142$.

yield from cananga of 2.77% of dry matter was obtained with 8 DIC cycles at 0.6 MPa in 4 min compared with 24 h of HD (Kristiawan et al. 2008).

The yield of total DIC extracted essential oils was, in most cases, much higher than with HD and HD.

9.4.3 AUTOVAPORIZATION/EVAPORATION

Two experimental designs (Tables 9.3 and 9.4) had, as variable operative parameters, the total thermal treatment time and the number of cycles; one was characterized by a long treatment time (max. 480 s) and the other was characterized by a short treatment time (max. 240 s). The response parameter was the efficiency defined by Equation 9.36 (EE).

Figure 9.6 presents the Pareto charts of the number of cycles and operating time to identify the impacts of autovaporization and evaporation, respectively, on direct DIC extraction of essential oils from valerian root, lemongrass, chamomile, angelica seeds, thyme, and lavandin essential oils. In all the cases presented, the only operating parameter that is significant is the number of cycles. In contrast, time is an insignificant factor. SD (evaporation of essential oils) is under the direct influence of the processing time. In contrast, the phenomenon of autovaporization is strictly correlated to the number of pressure drops. The obtained results show that when using DIC, evaporation phenomena occur but are insignificant, whereas autovaporization largely predominates.

TABLE 9.4

Independent Variables Used in RSM at a Fixed Steam Pressure Value with Dried Aromatic Plants

	Coded Level	$-\alpha$	-1	0	$+1$	$+\alpha$
$X_1 = C$	Number of cycles	1	2	5	8	9
$X_2 = t_d$	Heating time (s)	80	103	160	217	240

Note: α (axial distance) = $\sqrt[4]{2^N}$, where N is the number of independent variables. In the present case, $N = 2$ and $\alpha = 1.4142$.

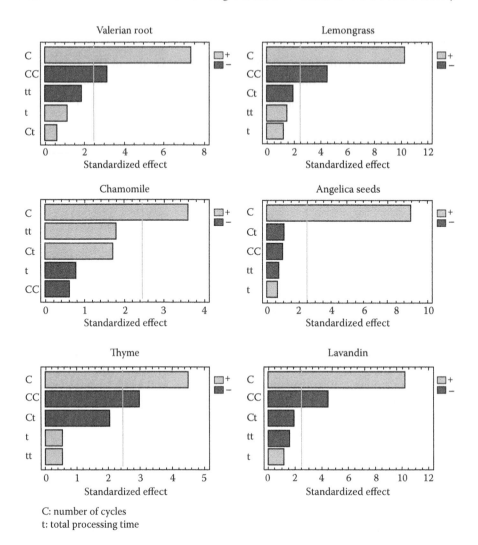

C: number of cycles
t: total processing time

FIGURE 9.6 Pareto charts of the number of cycles and operating time to identify impacts of autovaporization and evaporation, respectively, on direct DIC extraction of essential oils from valerian root, lemongrass, chamomile, angelica seeds, thyme, and lavandin essential oils.

9.4.4 QUALITY

9.4.4.1 Increasing the Proportion of Oxygenated Compounds

This work was carried out using Algerian myrtle leaves. A comparison was made between the compositions of oils obtained by HD with those obtained by DIC extraction (see Table 9.5). Elution order and relative quantification (%) of the different compounds were determined through GC-FID with a nonpolar VF-5ms capillary column, and the retention index was calculated with *n*-alkanes (C5–C28) with the

TABLE 9.5
Chemical Composition of Essential Oil Extracted from Algerian Myrtle Leaves Using Hydrodistillation (HD-EO) and Instant Controlled Pressure Drop (DIC-EO)

No.	Compounds	RI[a]	RI[b]	HD-EO[c] (%)	DIC-EO[d](%)
	Hydrocarbon monoterpenes			55.17	26.15
1	α-Thujene	930	1030	0.19	tr[h]
2	α-Pinene	939	1028	50.81	23.33
3	β-Pinene	979	1116	0.39	0.12
4	δ 3-Carene	1011	1146	0.30	tr[h]
5	p-Cymene	1024	1281	tr[h]	0.18
6	Limonene	1029	1208	2.63	2.20
7	(Z)-β-Ocimene	1043	1236	0.26	tr[h]
8	γ-Terpinene	1059	1263	0.29	0.17
9	α-Terpinolene	1088	1290	0.30	0.15
	Oxygenated monoterpenes			32.9	42.66
10	1,8-Cineole	1032	1216	24.32	21.76
11	cis-Linalool oxide	1072	1448	tr[h]	0.20
12	trans-Linalool oxide	1087	1469	tr[h]	0.23
13	β-Linalool	1098	1552	1.32	2.94
14	Hotrienol	1102	1621	tr[h]	0.19
15	trans-Pinocarveol	1139	1665	0.22	0.28
16	Borneol	1168	1717	tr[h]	0.16
17	Terpinen-4- ol	1177	1610	0.39	0.64
18	p-Cymen-8-ol	1183	1852	tr[h]	0.39
19	α-Terpineol	1192	1711	2.51	5.67
20	Verbenone	1205	1720	tr[h]	0.42
21	trans-Carveol	1219	1835	tr[h]	0.31
22	Nerol	1230	1793	tr[h]	0.18
23	Geraniol	1256	1849	0.51	1.48
24	Linalyl acetate	1257	1556	0.21	0.26
25	Methyl citronellate	1261	1570	tr[h]	0.10
26	Geranial	1268	1725	tr[h]	0.27
27	trans-Pinocarvyl acetate	1298	1626	tr[h]	0.15
28	Myrtenyl acetate	1326	1709	tr[h]	0.64
29	Exo-2-hydroxycineole acetate	1343	1765	0.49	0.11
30	α-Terpinyl acetate	1352	1707	0.85	1.22
31	Neryl acetate	1361	1723	tr[h]	1.23
32	Geranyl acetate	1380	1761	2.08	3.83
	Benzenoid compounds			2.33	10.14
33	Thymol	1292	2192	tr[h]	0.20
34	Carvacrol	1303	2230	tr[h]	0.10
35	Eugenol	1360	2164	tr[h]	4.06
36	Methyl eugenol	1397	2030	2.33	5.38

(continued)

TABLE 9.5 (Continued)
Chemical Composition of Essential Oil Extracted from Algerian Myrtle Leaves Using Hydrodistillation (HD-EO) and Instant Controlled Pressure Drop (DIC-EO)

No.	Compounds	RI[a]	RI[b]	HD-EO[c] (%)	DIC-EO[d](%)
37	(E)-Methyl isoeugenol	1492	2196	tr[h]	0.23
38	Benzyl benzoate	1762	2638	–	0.17
	Hydrocarbon sesquiterpenes			4.76	9.79
39	α-Copaene	1375	1503	tr[h]	0.17
40	β-Elemene	1390	1600	0.28	0.48
41	β-Caryophyllene	1419	1612	0.88	1.86
42	γ-Elemene	1436	1637	0.38	0.47
43	α-Humulene	1454	1676	0.36	1.68
44	β-Chamigrene	1477	1741	0.14	0.40
45	Germacrene D	1485	1722	0.30	0.16
46	β-Selinene	1490	1727	0.76	1.40
47	α-Selinene	1498	1729	0.75	1.80
48	δ-Cadinene	1523	1763	0.91	0.60
49	Cadina-1,4-diene	1534	1789	tr[h]	0.77
	Oxygenated sesquiterpenes			1.68	8.46
50	(E)-Nerolidol	1562	2048	tr[h]	0.42
51	Spathulenol	1578	2146	0.19	0.38
52	Caryophyllene oxide	1583	2008	0.63	4.21
53	Cubenol	1646	2074	0.24	1.60
54	β-Eudesmol	1650	2246	0.20	0.90
55	α-Bisabobol	1685	2228	0.13	tr[h]
56	Juniper camphor	1698	2275	0.29	0.95
	Total aliphatic compounds			0.5	0.34
57	(E)-2-Hexenal	855	1218	0.30	–
58	Isobutyl isobutyrate	896	1086	0.20	0.11
59	(E)-2-Decenal	1263	1632	tr[h]	0.23
	Total non-oxygenated compounds			59.93	35.94
	Total oxygenated compounds			37.41	61.6
	Total compounds			97.34	97.54
	Yield (%)			0.51 ± 0.04 g EO/100 g dry matter	0.56 ± 0.12 g EO/100 g dry matter

Note: tr[h]: trace < 0.05%.

[a] Retention index on apolar column

[b] Retention index on polar column

[c] HD-EO: Essential oils extracted by hydrodistillation from Algerian myrtle leaves dried to 16.28 g H$_2$O/100 g dry material.

[d] DIC-EO: Essential oils directly extracted by DIC treatment from Algerian myrtle leaves dried to 16.28 g H$_2$O/100 g dry material, at 0.6 MPa steam pressure, for 120 s as total thermal treatment time and with 4 cycles.

same column. The identification was carried out by comparing the retention index (RI) with the bibliography, and mass spectra (MS) with MS libraries.

A comparison of the two types of oil showed that the essential oils obtained with DIC were slightly richer in sesquiterpene hydrocarbons than those obtained by HD, but they contain many more oxygen compounds, regardless of their type. This enrichment was at the expense of other compounds, that is, non-oxygenated compounds. Oxygenated compounds are the elements that contribute to the development of aromas.

9.4.4.2 Improvement of Antioxidant Activity

This study was also carried out on Algerian myrtle leaves. We compared antioxidant activities of essential oils obtained by HD and by DIC extraction (Table 9.6).

The amount of direct DIC-EO (essential oils directly extracted by DIC treatment) needed to obtain the same level of antioxidant effect was about 6% less than with HD essential oils. This means that the essential oils obtained through DIC extraction would have slightly superior antioxidant properties compared to those obtained by HD.

9.4.5 Fractioning Extraction

Each compound has specific thermomechanical behavior regarding autovaporization. This normally depends on its own localization and/or the processing temperature (Table 9.7a, b, and c).

In this case of lavandin, 22 main compounds were first identified and then quantified (with or without DIC treatment) by solvent extraction. Some of the most important thermodynamic characteristics of these compounds were not available in the literature (such as partial pressure values at various temperatures).

In the case of rosemary and oregano, 25 compounds were isolated. The change in extraction rate versus time and DIC cycles was monitored.

Whatever the compound, the higher the thermal treatment time and the steam pressure, the higher the yields. However, a steam pressure of 0.6 MPa was used to

---

TABLE 9.6

Antioxidant Activity after Hydrodistillation of Untreated and DIC-Treated Algerian Myrtle Leaves with an Initial Water Content Maintained at 16.2 g H_2O/100 g Dry Matter Obtained by Desorption

	Concentration Required to Inhibit 50% DPPH Free Radicals (EC_{50})	
Type of Essential Oil	**Essential Oil Antioxidant Activity**	**Butylated Hydroxytoluene (BHT) (Antioxidant Reference)**
Direct DIC-EO (0.6 MPa, 120 s, 4 cycles)	1625 ± 15 μg/ml of essential oil	12.03 ± 0.03 μg/ml of BHT
Hydrodistillation HD-EO	1717 ± 15 μg/ml of essential oil	

TABLE 9.7a
Identification of Lavandin Essential Oil Compounds Extracted by DIC

No.	Retention Time	Compounds	Ratio (%)[a]	Boiling Point
1	4.679	Limonene	0.90	174
2	11.599	Eucalyptol	4.22	175
3	16.499	Linalol	36.04	198
4	18.804	Camphor	5.05	204/207
5	20.017	Lavandulol	2.03	229/230
6	20.701	Borneol	4.53	208
7	21.084	Linalol acetate	26.89	220
8	22.384	α-Terpineol	6.57	217
9	25.922	Lavandulyl acetate	3.54	
10	35.658	Caryophyllene	1.15	256/259
11	37.921	β-Farnesene	0.51	206
12	41.519	τ-Cadinene	0.24	271/273
13	49.352	τ-Cadinol	1.09	302/304
14	51.795	α-Bisabolol	1.25	314/315

Sources: Lide, D.R., *Handbook of Chemistry and Physics,* Internet version http://www.hbcpnetbase
.com, 2005; FAO, retrieved from http://www.fao.org/ag/agn/jecfa-flav/search.htm, 2010; The
Good Scents Company, retrieved from http://www.thegoodscentscompany.com, 2010. With
permission.

[a] All compounds with a ratio lower than 0.2% were assumed negligible.

TABLE 9.7b
Identification of Oregano Essential Oil Compounds Extracted by DIC

No.	Retention Time	Compounds	Ratio (%)[a]	Boiling Point
1	9.863	α-Pinene	2.85	155
2	10.631	β-Pinene	2.96	165
3	15.311	Linalol	3.20	198
4	19.949	Borneol	0.60	208
5	20.37	4-Terpineol	1.21	212
6	21.531	α-Terpineol	0.87	217
7	24.842	Thymoquinone	1.39	230/232
8	28.894	Thymol	32.91	231/232
9	29.681	Carvacrol	44.26	236/237

Sources: Lide, D.R., *Handbook of Chemistry and Physics,* Internet version http://www.hbcpnetbase
.com, 2005; FAO, retrieved from http://www.fao.org/ag/agn/jecfa-flav/search.htm, 2010; The
Good Scents Company, retrieved from http://www.thegoodscentscompany.com, 2010. With
permission.

[a] All compounds with ratio inferior to 0.2% were assumed to be negligible.

TABLE 9.7c

Identification of Rosemary Essential Oil Compounds Extracted by DIC

No.	Retention Time	Compounds	Ratio (%)[a]	Boiling Point
1	7.576	α-Pinene	14.06	155
2	8.293	Camphene	4.58	160
3	9.168	β-Pinene	4.39	165
4	9.506	p-Cymène	1.14	176
5	10.894	Limonene	3.17	174
6	11.613	Eucalyptol	20.21	175
7	18.748	Camphor	20.80	204/207
8	19.405	3-Pinanone	2.40	201/213
9	20.552	Borneol	5.48	208
10	20.91	α-Terpineol	0.68	217
11	23.035	Verbenone	5.14	
12	25.756	2,3-Pinanediol	0.68	
13	29.09	Thymol	0.78	231/232
14	29.557	Carvacrol	1.65	236/237
15	31.123	Chrysanthenone	1.44	208/210
16	49.282	τ-Cadinol	0.71	302/304

Sources: Lide, D.R., *Handbook of Chemistry and Physics*, Internet version http://www.hbcpnetbase .com, 2005; FAO, retrieved from http://www.fao.org/ag/agn/jecfa-flav/search.htm, 2010; The Good Scents Company, retrieved from http://www.thegoodscentscompany.com, 2010. With permission.

[a] All compounds with a ratio lower than 0.6% were assumed negligible.

study the effect of the number of cycles and the total thermal treatment time for each compound. The results are shown in Figure 9.7a and b.

It was clear that some compounds (limonene [$C_{10}H_{16}$], eucalyptol [$C_{10}H_{18}O$], ethyl linalool [$C_{12}H_{22}O$], and α-humulene [$C_{15}H_{24}$]) had been completely extracted after the first cycle. After 4 to 6 cycles, most compounds had been extracted. Only some (α-terpineol [$C_{10}H_{18}O$], τ-cadinol [$C_{15}H_{26}O$], and α-bisabolol [$C_{15}H_{26}O$]) seemed to need many more pressure drops. This observation could be very important in a possible strategy involving a fractional extraction by DIC; such an operation would then need to be defined for each case, depending on the raw material and the desired compounds.

Fractioning may be carried out during both extraction (by separating essential oils at each pressure drop) and condensation (by establishing different temperatures).

9.4.6 Saving Energy and Water

Essential oil extraction of lavandin was optimized in terms of pressure, duration, and the number of pressure drops. These *P*, *t*, and *C* optimized values were substituted and used in experimental trials, which were performed at 0.6 MPa saturated steam

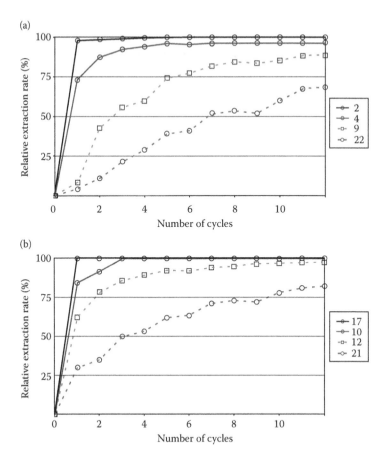

FIGURE 9.7 Relative extraction rate of various compounds of lavandin essential oils versus 20-s cycles. (2, Eucalyptol [$C_{10}H_{18}O$]; 4, camphor [$C_{10}H_{16}O$]; 9, α-terpineol [$C_{10}H_{18}O$]; 10, hexylisovalerate [$C_{11}H_{22}O_2$]; 12, lavandulyl acetate [$C_{12}H_{20}O_2$]; 17, α-humulene [$C_{15}H_{24}$]; 21, τ-cadinol [$C_{15}H_{26}O$]; 22, α-terpineol [$C_{10}H_{18}O$]).

pressure, a total thermal treatment time of 240 s, and 6 cycles, that is, 40 s of thermal treatment time per cycle. Based on these values, it was possible to calculate the quantities of energy and water consumed.

At 0.6 MPa steam pressure, DIC extraction of lavender was achieved in 1 cycle and 20 s for limonene, eucalyptol, α-humulene, and ethyl linalool. Two cycles and 40 s were required for *n*-hexyl butyrate and acetic acid; 4 cycles and 80 s for hexyl isovalerate, 6 cycles and 120 s for camphor, lavendulol, linalool, caryophyllene, linalool acetate, geranyl propionate, borneol, caryophyllene oxide, and 1-terpinen-4-ol; and 10 cycles and 200 s for lavendulyl acetate. Some 70% of α-bisabolol and 80% of α-terpineol and τ-cadinol were extracted after 12 cycles and 240 s. The essential oil extraction by DIC autovaporization has several advantages: very good *kinetics*, *short duration*, and a *fractioning process*, producing essential oils rich in oxygenated compounds.

9.4.7 POSSIBLE RECOVERY OF RESIDUAL MATERIAL

During the DIC process, the raw material is exposed to high temperature–high saturated steam pressure for a short time (few dozens of seconds). The abrupt pressure drop leads to autovaporization and thus an instant cooling of the raw material. The latter undergoes weak thermal degradation, which allows the raw material to be recycled through various uses, especially solvent extraction. Indeed, DIC treatment facilitates solvent extraction.

9.5 DIC AS A PRETREATMENT FOR SOLVENT EXTRACTION

After DIC essential oil extraction, the residual solids maintain a very good aspect. Thanks to this expanded structure, the solvent process becomes easier, quicker, and more effective for the extraction of nonvolatile molecules. In other words, DIC treatment is both a direct extraction of volatile molecules by autovaporization and a pretreatment that makes solvent extraction more attractive in terms of kinetics, yields, and quality.

9.5.1 SOLVENT EXTRACTION

Solvent extraction is usually used to recover compounds from either a solid or a liquid. The sample is placed in contact with a solvent that will dissolve the desired solutes. Solvent extraction is a very important operation in industries such as the chemical and biochemical industries, together with the food, cosmetics, and pharmaceutical industries (Pharmacyebook 2010).

As a unit operation, extraction is used to remove some compounds from plants and certain organs of animals to be used in food, pharmaceuticals, and fragrances, in the form of beverages, drugs, or perfumes. Solvents used in the separation processes of plant products are usually water, alcohols, organic solvents, and/or chlorine. Supercritical fluids are considered specific types of solvents.

Standard extraction techniques include several methods that all consist in the liquid solvent interacting with the solid material in order to dissolve the desired components. They can be listed as follows:

- Percolation—Usually a very hot solvent is flushed onto a bed of finely divided solids. This method is used in the preparation of coffee (Leybros and Frémeaux 1990).
- Decoction—The solid is immersed in a boiling liquid solvent. This is a brutal operation that should be reserved for the extraction of nonthermolabile active ingredients. However, this process is very fast and sometimes necessary (Leybros and Frémeaux 1990).
- Infusion—The solid is immersed in a heated solvent without boiling, followed by cooling of the mixture. The preparation of tea is a typical example (Groubert 1984).
- Maceration—The solid is immersed in a cold solvent. The operation is usually long and with poor results; however, it is the only method used in the

case of a set of fragile molecules. To be effective, maceration can last 4 to 10 days, which may present problems in terms of fermentation or bacterial contamination, particularly if the solvent used is water. These phenomena can cause rapid degradation of active molecules. To avoid or limit these drawbacks, the maceration can be carried out in a covered container and, in some cases, in a refrigerator.

• Digestion—Especially used in perfumery and pharmacy, digestion is a hot maceration. Because it is faster, it generally does not involve any degradation or bacterial contamination problems (Groubert 1984; Leybros and Frémeaux 1990).

• Elution—This consists in removing the solute from the surface of a solid by simple contact with a solvent. It is frequently used in the methods of analysis (Leybros and Frémeaux 1990).

In all these methods, aside from elution, the mass transfer that occurs during solvent extraction is hindered and limited by the structure and the cell wall of the matrix. To avoid this difficulty and to improve such conventional solvent extraction operations, several solutions have been proposed as pretreatments: reducing the size of particles by crushing, breaking cell walls by enzyme treatment, or using ultrasound, microwaves, accelerated solvent, or a special solvent such as a supercritical fluid. These options have not definitively resolved the issue.

9.5.2 Fundamental

9.5.2.1 Principle

The solvent extraction and separation process is, from a technological point of view, an operation based on the diffusion of a carrier fluid (liquid) within a solid in order to bring out specific molecules. It is usually presented as a solid–liquid interaction.

However, a solvent capable of "solubilizing" one or more components from a solid or a liquid generates an extract solution (solvent + solute) (Mafart and Béliard 1993).

The transfer of these active molecules to the surroundings occurs through a diffusion that is mainly the result of a concentration gradient of solute between the inner solution near the solid phase (more concentrated) and the liquid phase. At the end of the operation, the system tends toward equilibrium and the diffusion is near zero. In contrast, if the liquid phase is continuously renewed, the diffusion continues until the complete exhaustion of the solid phase (Dibert 1989).

At the end of the operation, the exhausted solid (residue), inert or insoluble, contains very little or no solute at all. In general, the solution is the noble phase, but it may be that the insoluble solid residue represents the true economic value (Bimbenet et al. 1993).

9.5.2.2 Identification and Intensification of the Limiting Process

The solid/liquid interaction is achieved by close contact between the solid and the solvent. During the extraction, the concentration of solute in the solid varies continuously, which explains the nonstationary mass transfer.

The operation starts with a superficial solute solubilization in solvent. This occurs at the surface (external process) and has to be immediately transported within the surrounding solvent medium.

After this first stage, a series of successive processes takes place, reflecting the interaction between the solid initially containing the solute and the solvent provoking the separation. These successive processes include:

- Solvent diffusion within the solid matrix.
- Internal solute solubilization in the solvent; this occurs within the solvent, which has diffused within the porous solid plant matrix (internal process).
- Solute diffusion in the solvent within the solid matrix toward the surface. This can also be considered as the diffusion of external solvent within the internal mixture of solute/solvent.
- External convection and/or diffusion transfer of the solute from the surface of the solid to the external environment.

The extraction, often studied on a phenomenological basis as a unit operation, can be analyzed in terms of kinetics and total yields.

Kinetics is usually expressed in terms of solute concentration in the solid per unit time (dx/dt). The sequential nature of these four processes allows the operation to take place at the rate of the slowest process, which is then defined as the limiting process.

To identify the limiting process and intensify the whole operation, we propose, first, to analyze the four major processes involved in the operation.

Solubilization of Molecule to Be Extracted in Liquid Solvent

Solubilization is the operation of dissolving a substance such as a plant tissue compound in a solvent. Solubilization tends toward thermodynamic equilibrium in terms of concentration and is described through the distribution, or equilibrium partitioning constant or coefficient m, which is a function of the maximum solubilization (saturation level), thereby leading to an equilibrium concentration between the extract and the dry matter (Schwartzberg and Chao 1982; Gertenbach and Bilkei 2001).

The higher the value of m, the more easily the compound indicated in the solvent dissolves. The value of m depends on the characteristics of the solvent, the compound to be extracted, and the temperature. Usually, solubilization of hydrophobic (nonpolar) molecules is performed using an organic solvent.

Solubilization of the molecules to be extracted (solute) is the limiting process in the extraction only if, for instance, an unsuitable solvent and a low temperature are chosen. In such a situation, characterized by a very low solubility of the solute in the liquid solvent, the solvent would enter the solid matrix with relative ease, and the solute gradient in the solid matrix and in the external environment would eventually tend to zero. The extraction process would hence be entirely driven by the dissolution of the solute in the solvent.

To intensify such a situation, one has to perform the operation by simply using "suitable" solvents with the desired solutes. Often, the appropriate temperature should be as high as possible, although below the boiling temperature and the degradation level of the molecules concerned. The choice of extraction solvent and temperature is usually made according to the desired solutes, and the nature and the varieties of the plant; the aim is to reach high kinetics and large amounts of dissolved molecules.

Thus, it is always easy to intensify the operation to avoid having solubilization as the limiting process. On the contrary, it may often be considered in practice as a very quick and even instantaneous process.

Transport of Dissolved Solute from the Surface toward the Outside

In the external environment, the solute transfer during the extraction can take place both by diffusion and by convection. In the absence of any agitation phenomenon, for instance, it is possible that the limiting process of the extraction operation is the transfer of solute from the solid surface toward the external solvent.

In such a situation, the solvent would quickly penetrate the solid matrix and dissolve the solute. The solute concentration inside the solid would be homogeneous.

In this case, the extraction would occur at the rate of solute diffusion from the external surface of the solid to the surrounding solvent. To overcome this "problem" and to obtain a proper intensification, the nature of the solute transfer in the external environment has to be changed in order to be convective instead of diffusional. It is thus sufficient to achieve a "good" agitation of the outside "solvent medium."

The external transfer resistance becomes negligible. A quantification of the impact of agitation is represented by the Biot number whose value gives an idea of the homogeneity of the system:

$$B_i = \frac{kl}{D_s} \qquad (9.37)$$

In practice, the Biot number is usually greater than 200. One can thus neglect the mass transfer resistance at the interface compared to that inside the solid. It is for this reason that agitation is often included in the extraction process (Schwartzberg and Chao 1982).

Internal Transfer by Diffusion

In the study of solvent extraction kinetics, we have to examine the presence of two types of diffusional transfers: solvent within the plant structure (solid/liquid interaction) and solvent within the solute solution (liquid/liquid interaction).

From a theoretical point of view, the mass transfer of solvent within the porous solid can usually be considered as occurring either by convection or diffusion. However, whatever the porosity of the plant, internal motion by convection in the pores is assumed to be negligible. Therefore, as the main mass transfer, diffusion of the solvent within the solid porous matrix occurs with the densities gradient ratio of the solvent and the solid as the driving force. As the diffusion phenomenon involved in transferring solutes within the solvent appears as a transfer of the solvent toward the solute solution in the holes of the solid matrix (interaction solid/liquid), both diffusion processes can be seen as an effective diffusion of solvent within the solute-soaked solid, depending on the porosity of the material.

Indeed, during this operation, the random movement of molecules guarantees a diffusion transfer with the concentration gradient as the driving force. We can then explain why diffusion (Fick type) appears to be the major transport phenomenon occurring within the solid matrix (Aguillera and Stanley 1999) and the mass transfer process cannot be intensified through any external mechanical or thermal changes. Only a modification of shape (by grinding) and/or an expansion improving the porosity may improve such an operation.

Within the solid matrix, the molecules that the solvent has dissolved owe their "movement" in the solvent to a Fick-type diffusion whose driving force is the concentration gradient of each molecule. This operation may be regarded as a diffusion of solvent at an external solute concentration within an internal solvent assumed to be at saturated solute concentration.

9.5.2.3 Kinetics Analysis

Allaf and colleagues studied the extraction kinetics stages and phenomena. They assumed that a quick initial stage involving the superficial extraction of solute would occur first, allowing an amount Y_A of solute to be extracted just at the starting time (at time $t = 0$) (Mounir and Allaf 2008). Afterward, the internal mass transfer of solute flux within the product would have the highest transfer resistance. Under such conditions, the driving force has been considered to be the gradient of solvent ratio to solid apparent densities and the gradient of solute ratio to solvent apparent densities. The authors thus proposed to systematically consider an overall or effective diffusivity, taking into account the diffusion of solvent through the solid matrix and within the solution of solute present in the solvent.

They then could consider the operation as being similar to Fick's diffusion:

$$\frac{\rho_{sol}}{\rho_m}\left(\vec{v}_{sol} - \vec{v}_m\right) = -D_{eff}\vec{\nabla}\left(\frac{\rho_{sol}}{\rho_m}\right) \tag{9.38}$$

However, if it is possible to assume that damping does not involve any swelling or shrinkage, it would be easy to simplify such a transfer law assuming that it occurs in a presumably motionless solid matrix. This also suggests that the change with time of solid porous medium density is negligible, despite the extraction phenomenon that still occurs. It is then possible to assume that ρ_m is constant, and to have:

$$\rho_{sol}\vec{v}_{sol} = -D_{eff}\vec{\nabla}\rho_{sol} \tag{9.39}$$

Using the balance mass, it is possible to obtain:

$$\frac{\partial \rho_{sol}}{\partial t} = \vec{\nabla} \cdot (D_{eff}\vec{\nabla}\rho_{sol}) \tag{9.40}$$

where t is time.

Although the effective diffusivity (D_{eff}) varies considerably with the system temperature and porosity, Mounir and Allaf (2008) assumed it to be constant in the hypothesis of both structural and thermal homogeneities (see explanation data in the box), which allows the equation to become:

$$\frac{\partial \rho_{sol}}{\partial t} = D_{eff} \cdot \nabla^2 \rho_{sol} \tag{9.41}$$

And, for a unidirectional radial flow, it becomes:

$$\frac{\partial \rho_{sol}}{\partial t} = D_{eff} \cdot \frac{d^2 \rho_{sol}}{dr^2} S \tag{9.42}$$

Allaf and colleagues considered that an amount m_o of the solute is immediately accessible on the surface and can be immediately dissolved in the external solvent that is to be removed from the plant, usually by convection (Ben Amor and Allaf 2006; Ben Amor and Allaf 2007a,b; Ben Amor 2008; Ben Amor et al. 2008). As an initial amount (m_m) of solute was present in the solid plant, the fraction present in the volume must be $(m_m - m_o)$. On the other hand, a certain amount (m_r) of solute can remain closely linked to the plant structure and is thus unavailable for extraction. Conventional operating conditions for solvent extraction carried out with a suitable solvent and with external surrounding agitation, allowing solid–liquid interaction to take place by convection, mean that the complex internal diffusion requiring the solute located within the plant is the limiting process of the operation.

In terms of the internal solute concentration X, we can consider:

X_A: corresponds to the amount of solute m_o initially accessible at the surface; it will be removed very rapidly by convection.
$(X_m - X_A)$: corresponds to the amount of solute $(m_m - m_o)$ initially located (in a manner assumed to be uniform and homogeneous) inside the volume. This quantity changes over time by diffusion. At time t, it is indicated by X.
X_∞: corresponds to the amount m_r of solute unavailable under the operating conditions.

The solutions required for this diffusion equation closely depend on the initial and boundary solvent conditions. The classical Crank's solutions according to the geometry of the spherical expanded granule may be adopted (Crank 1975).

$$\frac{X - X_\infty}{X_m - X_A - X_\infty} = \sum_1^\infty A_i e^{-q_i^2 \tau} \tag{9.43}$$

Where X, X_∞, and X_A are the solute densities on a dry basis in the solid matrix at time t, at $t \to \infty$ (unavailable solute), and at the starting diffusion time, respectively. By using the amount of extract Y, they could get:

$$\frac{Y_\infty - Y}{Y_\infty - Y_A} = \sum_1^\infty A_i e^{-q_i^2 \tau} \tag{9.44}$$

where Y, Y_∞, and Y_A are the amounts of solute extracted at time t, at $t \to \infty$ (total yields), and at time $t = 0$, respectively.

Mass and Heat Diffusivities

A bibliographic analysis showed that the effective diffusivity (D_{eff}) of the entire extraction operation of liquids in solid plant matrices have a magnitude that is generally close to 10^{-11} to 10^{-10} m² s⁻¹, whereas heat diffusivity is usually between 10^{-8} and 10^{-6} m² s⁻¹, depending on water content and porosity.

This confirms, in a way, that the solvent diffusion into a solid matrix must often be the slowest process and appears to be the "main" limiting process throughout the operation. The intensification operation lies in the expansion of this natural structure as a real method for improving the processing ability of the plant regarding solvent extraction.

To study the kinetics and use the experimental data to make a model of the diffusion process, Allaf and colleagues excluded the value $Y = 0$ measured at the starting point (at $t \to 0$) (Mounir and Allaf 2008). The experimental data used for such a diffusion model excludes the points close to $t = 0$; the extrapolation of the model thus obtained showed Y_A to be different to 0 in general. Y_A represents the amount of solute accessible on the surface that is quickly removed and extracted from the surface, independently from the diffusion process. Its value is then calculated by extrapolating the diffusion model toward $t = 0$.

The solvent extraction kinetics are then defined through the values of the yields Y_∞, the starting accessibility Y_A, and the effective diffusivity D_{eff}.

Intensification Procedures

Mounir and Allaf (2008) listed four procedures capable of intensifying the solvent extraction operation:

1. *Grinding* normally reduces the depth of diffusion of a liquid within the granule. It also increases the exchange surface and makes the solute on the surface more accessible. Generally, granules are assumed to be spherical and compact.
2. The aim of *texturing* is to reduce mass transfer resistance. Indeed, the natural plant structure and, more specifically, the cytoplasmic membrane and the cell wall should not normally support the liquid transfer processes. By breaking these membranes using enzymatic treatments, DIC swelling, etc., mass transfer kinetics can be significantly increased.
3. The aim of *expansion* aims to increase porosity and the volume of the holes, which should improve solute transfer in the solvent within the product.
4. *Agitation* of the surrounding solvent allows solute transport from the product surface to the outside to be carried out by convection.

Modifying granule structure, improving porosity, breaking the cell wall, etc., should intensify the solvent extraction operation, as described below.

The kinetics model of solvent extraction leads to the identification of the impacts of the different intensification treatments through the yields (Y_∞), the starting accessibility (δY_s; this is the part of solute normally extracted by solvent convection at the surface, in a very short time), and D_{eff} (effective diffusivity). Y_∞, δY_s, and D_{eff}

TABLE 9.8

Experimental Data of Composite Central Design and Results of Jatropha Oil Yield after a 2-h Extraction

Run No.	DIC Treatment		Crank's Solution R^2	Oil Yields		Diffusivity		Starting Accessibility	
	P (MPa)	t (s)		Y_∞ (g oil/100 g dry material)	Y_∞/Y_{2hRM} (%)	D_{eff} (×10⁻¹² m² s⁻¹)	Improvement $D_{eff}/D_{eff(RM)}$ (%)	δY_s (g oil/100 g dry material)	
1	0.64	63	0.9148	51.5	112	5.44	225	91.30	47
2	0.5	45	0.9082	50.59	110	5.62	232	90.78	46
3	0.5	45	0.9194	50.39	110	5.52	228	91.12	46
4	0.7	45	0.9026	50.56	110	5.32	220	92.04	47
5	0.5	70	0.9082	51.08	111	5.13	212	92.58	47
6	0.5	45	0.9003	50.45	110	5.36	221	91.51	46
7	0.64	27	0.9395	50.06	109	5.29	218	91.60	46
8	0.36	27	0.9007	49.3	107	5.90	244	90.08	44
9	0.36	63	0.9857	49.51	108	5.21	215	90.50	45
10	0.3	45	0.9326	49.1	107	5.36	221	91.29	45
11	0.5	20	0.9521	49.2	107	5.50	227	90.28	44
RM	–	–	0.9045	45.91	100	2.42	100	75.91	35

Note: RM, raw material.

are considered the main response parameters characterizing the modification of the plant "technological abilities" regarding solvent extraction technology.

9.5.3 MAIN TRENDS OF EXPERIMENTAL RESULTS

In all solvent extraction operations carried out with any part of a plant, it was possible to intensify the operation through expansion, which increases the porosity, the specific surface area, and the availability of the desired compounds (Ben Amor et al. 2008). DIC texturing was studied and optimized for the type of operation, the raw material, and the final quality and molecules.

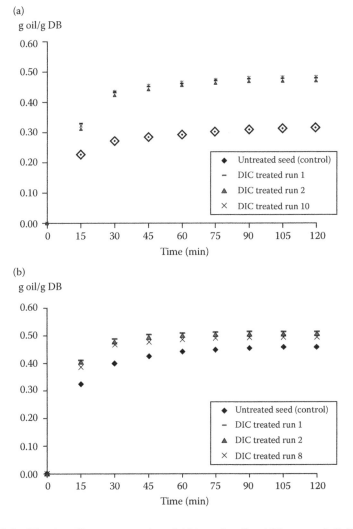

FIGURE 9.8 Kinetics of hexane extraction of: (a) jatropha oil and (b) rapeseed oil. Untreated compared with DIC treated seeds (sample nos. 1, 2, and 10).

TABLE 9.9
Experimental Data of Composite Central Design and Results of Rapeseed Oil Yield after 2 h Extraction

DIC Run No.	DIC Treatment		Oil Yields			Diffusivity		Starting Accessibility	
	P (MPa)	t (s)	Crank's Solution R^2	Y_∞ (g oil/100 g dry material)	Availability Y_∞/Y_{2hRM} (%)	D_{eff} (×10^{-12} m^2 s^{-1})	Improvement $D_{eff}/D_{eff(RM)}$ (%)	δY_s (g oil/100 g dry material)	
1	0.70	70	0.980	48.2	153	7.497	1049	37.3	77.37
2	0.27	105	0.969	47.0	149	7.639	1069	36.2	76.94
3	0.63	35	0.979	48.1	153	7.552	1057	37.1	77.19
4	0.45	70	0.975	48.1	153	7.248	1014	38.7	80.53
5	0.45	20	0.954	46.8	149	7.538	1054	37.0	78.92
6	0.63	105	0.973	47.9	152	7.635	1068	36.8	76.69
7	0.2	70	0.981	46.9	149	7.431	1040	36.6	78.02
8	0.45	120	0.965	48.0	152	7.319	1024	38.4	79.91
9	0.45	70	0.986	48.0	152	7.566	1058	36.7	76.43
10	0.45	70	0.976	48.0	152	7.565	1058	37.0	77.12
11	0.27	35	0.986	46.8	148	8.014	1121	33.3	71.23
RM	–	–	0.951	31.5	100	0.715	100	8.4	26.71

9.5.3.1   Vegetable Oil

Nguyen Van (2010) measured the values of Y_∞, δY_s, and D_{eff} for oil hexane extraction from jatropha seeds and rapeseeds. DIC was able to systematically increase the yields by 112% and 153%, respectively, and decreased the extraction time by 30 and 60 min, respectively, as against the usual time of more than 120 min in both cases (Table 9.8).

This kind of texturing treatment was able to dramatically improve the starting accessibility, with more than 90% of jatropha oil—that is, up to 47 g oil/100 g on a dry basis—extracted during the first stage of the operation. In comparison, with non-DIC-treated jatropha, only 75% of oil—that is, 35 g oil/100 g on a dry basis—was extracted during the first stage of the operation. Only about 25% of oil was extracted through the solvent diffusion within this non-DIC-treated jatropha. The diffusivity of solvent within the solid matrix was twice as high for DIC-treated jatropha, with 5.6×10^{-12} m² s⁻¹ compared to 2.4×10^{-12} m² s⁻¹ for non-DIC-treated jatropha (Figure 9.8a).

Similar results were obtained with rapeseeds (see Table 9.9). Solvent extraction just after the DIC texturing treatment was identified as the factor leading to a higher δY_s, wherein more than 80% of oil—that is, up to 38 g oil/100 g on a dry basis—was extracted during the first stage of the operation, compared to 26.7% (which is 8 g oil/100 g on a dry basis) for non-DIC-treated seeds (Figure 9.8b). The diffusivity of solvent within the solid matrix was 11 times higher for DIC-treated rapeseeds: 8×10^{-12} m² s⁻¹ compared to 0.7×10^{-12} m² s⁻¹ for non-DIC-treated rapeseeds.

Figure 9.9 shows a Pareto chart of the experimental design used with rapeseeds with saturated steam pressure and treatment time as operating parameters and the yield as the response parameter.

Adequate statistical analysis of data from an RSM method allowed us to identify the saturated steam pressure as the most influential processing parameter on both yield and extraction time.

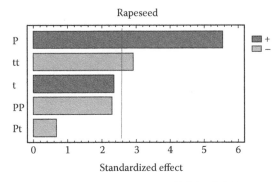

P: Saturated steam pressure
t: Thermal steam time

FIGURE 9.9   Standardized Pareto chart from response surface method experimental design for oil rapeseed oil yields.

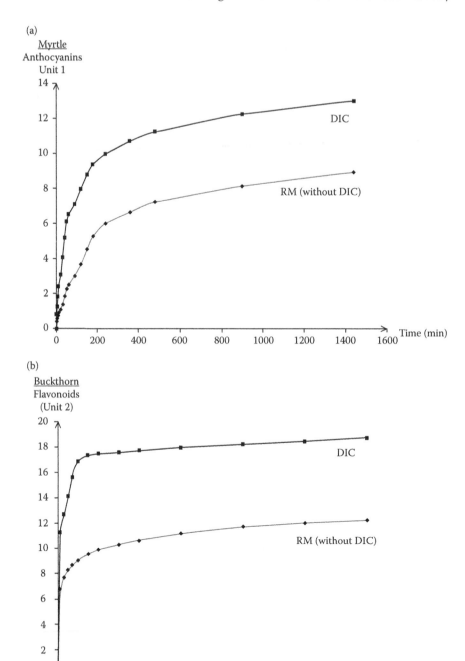

FIGURE 9.10 Solvent extraction of active molecules: (a) anthocyanins from Algerian myrtle with DIC treatment at 0.28 MPa for 9 s; (b) flavonoids from Algerian buckthorn with DIC treatment at 0.5 MPa for 180 s, and 5 cycles;

(c)

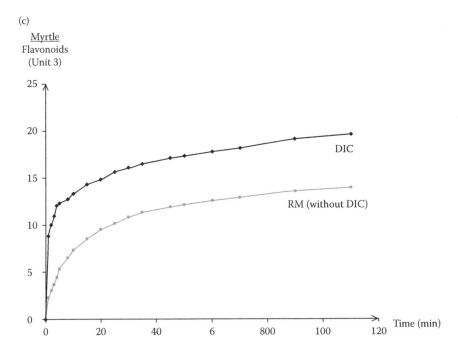

FIGURE 9.10 (Continued)   (c) flavonoids from Algerian myrtle with DIC treatment at 0.5 MPa for 180 s, and 5 cycles. Unit 1: mg of equivalent delphinidine-3-glucocide anthocyanins/g of Algerian myrtle dry matter; unit 2: mg of equivalent kaempferol flavonoids/g of Algerian buckthorn dry matter; unit 3: mg of equivalent myricetin flavonoids/g of Algerian myrtle dry matter.

Finally, it should be pointed out that a possible impact of optimized DIC treatment was to reduce the solvent/material concentration ratio, which may greatly reduce the energy consumption of the total solvent extraction.

9.5.3.2   Nonvolatile Active Molecules

Texturing by DIC treatment is a very important operation when extracting various nonvolatile active molecules. It dramatically increases the yields (through a much higher availability) and decreases the operation time (through higher diffusivity and starting accessibility). Normally, DIC treatment is carried out, defined, and optimized to avoid any discernable thermal degradation, thanks to the substantial cooling effect of the pressure drop toward a vacuum.

The extraction of anthocyanin from Algerian buckthorn (*Rhamnus alaternus*) and from Algerian myrtle was carried out using methanol/0.1% HCl (vol/vol) (Figure 9.10). When the operation followed the DIC texturing treatment, it led to about 50% more overall anthocyanin yields, and a starting accessibility δY_s value that was twice as high as that of the non-DIC-treated samples (Figure 9.10a).

TABLE 9.10

Solvent Extraction of Anthocyanins from Algerian Myrtle Leaves: DIC-Treated and Untreated Samples

	Without DIC	With DIC[a]
Extraction Time (min)	Unit 1: mg of Equivalent Delphinidine-3-glucocide Anthocyanins/g of Algerian Myrtle Dry Matter	
0	0.00	0.00
1	0.42	0.87
3	0.59	1.27
7	0.73	1.87
10	0.93	2.42
20	1.10	3.07
30	1.40	4.08
40	1.86	5.18
50	2.26	6.11
60	2.51	6.54
90	3.01	7.11
120	3.68	7.97
150	4.54	8.80
180	5.28	9.37
240	6.00	9.95
360	6.63	10.69
480	7.23	11.23
900	8.13	12.25
1440	8.93	12.99

[a] Pressure: 0.28 MPa; time: 9 s; number of cycles: 1.

In terms of extraction time, a total quantity of 9 units Eq. mg of Dp-3-Gl/g dm of anthocyanin was extracted in 160 min from Algerian buckthorn previously treated at 0.28 MPa for 9 s and 1 DIC cycle, compared with 1450 min with non-DIC-treated material (Table 9.10).

A similar observation was made with flavonoid extraction. Methanol/water (80:20, vol/vol) and 1.6% HCl was used as a solvent with Algerian buckthorn (Figure 9.10b) and Algerian myrtle (Figure 9.10c). When the DIC texturing treatment was performed before solvent extraction, overall flavonoid yields were about 50% higher and the starting accessibility (δY_s) was twice that of the non-DIC-treated samples.

There was a dramatic decrease in extraction time (Table 9.10) with a total quantity of flavonoid of 12 units Eq. mg of kaempferol/g ms extracted within 2 min from Algerian buckthorn previously treated at 0.5 MPa for 180 s and a 5-cycle DIC, compared to 150 min with non-DIC-treated material (Table 9.11). The same DIC treatment applied in the case of Algerian myrtle achieved a solvent extraction of

TABLE 9.11

Solvent Extraction of Flavonoids from Algerian Buckthorn: DIC-Treated and Untreated Samples

Extraction Time (min)	Without DIC	With DIC[a]
	Unit 2: mg of Equivalent Kaempferol Flavonoids/g of Algerian Buckthorn Dry Matter	
0	0	0
1	6.79	11.24
3	7.71	12.70
5	8.28	14.12
7	8.70	15.62
10	9.05	16.85
15	9.56	17.32
20	9.88	17.46
30	10.31	17.58
40	10.65	17.73
60	11.16	17.94
90	11.70	18.20
120	12.02	18.47
150	12.24	18.72

[a] Pressure: 0.5 MPa; time: 180 s; number of cycles: 5.

14 units Eq. mg of kaempferol/g ms within 13.5 min, as against 112 min with non-DIC-treated Algerian myrtle (Table 9.12).

9.5.3.3 Antioxidant Activities

Antioxidant activity was studied in Algerian myrtle leaves and Algerian buckthorn. We compared the antioxidant activities of untreated and DIC-treated plants.

The amount of BHT (considered as a reference) for the %IC_{50} level of antioxidant effect was 48.13 ± 0.13 µg/ml, with 2.18 ± 0.07 and 8.22 ± 0.01 µg/ml for untreated Algerian myrtle and Algerian buckthorn, respectively. After adequate DIC texturing, this became 0.78 ± 0.05 and 0.12 ± 0.06 µg/ml, respectively. With DIC texturing, the antioxidant activity was 3.57 and 68.62 times higher for Algerian myrtle and Algerian buckthorn, respectively.

9.6   INDUSTRIALIZATION

DIC technology as an innovative technique has been designed and developed on an industrial scale by ABCAR-DIC process for various applications (decontamination of seaweed, herbs, mushroom, drying of fruits and vegetables, steaming of rice, treatment of coffee beans, and extraction) and different industrial sectors. DIC reactors are currently operating in the laboratory, and in semi-industrial and industrial fields. Thus, there are several infrastructure models with different features and capabilities.

TABLE 9.12

Solvent Extraction of Flavonoids from Algerian Myrtle Leaves: DIC-Treated and Untreated Samples

	Without DIC	With DIC[a]
Extraction Time (min)	Unit 3: mg of Equivalent Nyricetin Flavonoids/g of Algerian Myrtle Dry Matter	
0	0.00	0.00
1	2.27	8.79
2	3.07	10.03
3	3.72	10.96
4	4.44	12.06
5	5.34	12.30
8	6.52	12.73
10	7.35	13.30
15	8.54	14.31
20	9.52	14.85
25	10.15	15.62
30	10.81	16.05
35	11.34	16.48
45	11.92	17.11
50	12.14	17.34
60	12.57	17.78
70	12.92	18.16
90	13.59	19.13
110	13.97	19.60

[a] Pressure: 0.5 MPa; time: 180 s; number of cycles: 5.

9.6.1 DIC EQUIPMENT

9.6.1.1 DIC Reactor TMDR0.3

A pilot-scale DIC reactor TMDR0.3 is shown in Figure 9.11. This DIC equipment is located in Queretaro, Mexico. Today, it is mainly used to enhance the drying of fruits and vegetables. The DIC reactor TMDR0.3 is a semiautomated system with a capacity of about 100 kg/h, depending on the product. The treatment chamber has a volume of 30 l, which translates to 60 kg/h with lavandin.

9.6.1.2 DIC Reactor TADR0.25

A pilot-scale DIC reactor TADR0.25 is shown in Figure 9.12. This DIC equipment is located in Alicante, Spain. This pilot-scale equipment has been used in a variety of ways, mainly as a pretreatment for solvent extraction. The DIC reactor TADR0.25 is a semiautomated system with a capacity of about 50 kg/h, depending on the product. The treatment chamber has a volume of 25 l, which represents 48 kg/h with lavandin.

FIGURE 9.11   Pilot-scale DIC reactor TMDR0.3 (located in Queretaro, Mexico).

9.6.1.3   DIC Reactor TLDR0.5

An industrial-scale DIC reactor TLDR0.5 is shown in Figure 9.13. This DIC equipment is located in La Rochelle, France. It is mainly used for bacterial decontamination and enhancing the drying of fruits and vegetables. The DIC reactor TLDR0.5 is an automated system with a capacity of about 2000 kg/h, depending on the product. The treatment chamber has a volume of 100 l, which represents 187 kg/h with lavandin.

FIGURE 9.12   Pilot-scale DIC reactor TADR0.25 (located in Alicante, Spain).

FIGURE 9.13   Industrial-scale DIC reactor TLDR0.5 (located in La Rochelle, France).

9.6.2   Data Concerning Decontamination

DIC technology has been used as a very relevant decontamination process for 20 years to inhibit the growth of spores and vegetative forms, specifically with thermally sensitive dried solids and powders. The three versions of 1-cycle saturated steam (STEAM-DIC), 1 cycle of CO_2, and multicycle hot air are suitable for thermal sensitive products (pepper, mushrooms, herbs, roots, seaweed, microalgae, skim powder, etc.). The combined mechanical and thermal impacts generate high decontamination levels.

9.6.3   Energy Consumption

The calculated energy consumption was 0.110 kW h per kg and per cycle, that is, a total of 662 kW h per ton of raw material. About 42 kg water was used per ton of raw material (Besombes et al. 2010) (Table 9.13).

9.7   CONCLUSION

SD is limited by the diffusion of essential oil vapor within the product. Usually, SD kinetics is much weaker than that imposed in a normal diffusion because of the paradoxical stage. Thus, conventional extraction processes such as HD and SD are tedious and time consuming. Their impact is very prejudicial in terms of yields and quality. To remedy this situation and intensify the essential oil extraction kinetics, DIC treatment is a very relevant technology. DIC extraction is a rapid process in

TABLE 9.13

Energy Consumption of DIC Extraction (Case of Lavandin)

Mass of RM	15 kg
Volume of the processing vessel	100 l
Apparent density of RM	200 kg m^{-3}
Volume ratio in processing vessel	75%
Volume of processed material	133 l
Intrinsic (true) density of RM	1220 kg m^{-3}
Volume of steam in the processing vessel	88 l
Latent heat of water vaporization	2450 kJ kg^{-1}
Treatment temperature	426 K
Saturated steam pressure	0.6 MPa
Mass of steam in the processing vessel	0.27 kg
Specific heat of RM	1.5 kJ kg^{-1} K^{-1}
Heat of steam used for heating the product	2925 kJ
Heat of steam used per cycle	655 kJ
Total heat per cycle	3580 kJ
Number of cycles	2
Total heat for the DIC extraction	2 kW h
Total heat for the DIC extraction per kg of RM	0.13 kW h/kg RM
Heat yield	75%
Heat needed per ton	177 kW h/ton RM
Water needed per ton RM	36 kg
Rate of used water	85%
Water used per ton	42 kg water/ton RM
Electricity power (vacuum pump and compressor)	21 kW
Extraction time	240 S
Hourly treatment capacity	188 kg/h
Total energy consumption/ton of raw material	289 kW h/ton RM
Essential oil yields	4 g EO/100 g dry matter
Hourly essential oil capacity	8 kg EO/h
Total energy consumption/kg of lavandin essential oils	7 kW h/kg EO

comparison with HD and SD. For instance, in the case of thyme oil, a direct oil yield of 4.3% dm was obtained using 9 cycles of DIC in 160 s and at 0.5 MPa compared with 3.5% dm in 7 h for HD and 2.7% in 6 h for SD.

Work performed on a large number of raw materials showed that DIC extraction increases the extraction yield. Conventional extraction treatments, SD or HD, do not achieve the same levels of performance.

DIC extraction increases the amount of essential oil obtained and reduces the duration. It simultaneously improves the kinetics and the yield.

For all aromatic plants, we found that the DIC EE increases with the number of cycles and the heating time. For all plants, the most significant parameter is the number of cycles. All experimental results corroborated this fundamental study: the DIC extraction process is achieved mainly through an *autovaporization* phenomenon; the contribution of evaporation to this type of extraction process was found to be negligible in the standard DIC processing parameter ranges. In this case, we also observed a low EE during initial cycles that increases to an asymptotic value at the end cycles. DIC treatment has a double effect: first, the vapor generated in the flowers during the expansion breaks and alveolates their structure and carries out the oil. Just after the "external" pressure drops, the mass transfer is not controlled by the diffusion of volatile components but by the TPG. These facts prove the hypothesis that the DIC essential oil extraction mechanism is mainly based on the autovaporization of the volatile compounds, which are a complex liquid mixture of water and volatile oils. The DIC isolation process is based on autovaporization of volatile compounds from a modified structure resulting from multicycle pressure drops and not on molecular diffusion in the standard separation methods. The risk of thermal degradation is avoided by using a short heating time (<1 min) in each DIC cycle.

The DIC oil contains more light oxygenated compounds (LOC) (borneol, eucalyptol, linalool, thymol, etc.) and heavy oxygenated compounds (HOC) (caryophyllene oxide, cubenol, eudesmol, etc.) than HD and SD oils. A similar increase in oxygenated compounds was obtained by using solvent-free microwave extraction (Lucchesi et al. 2004). On the other hand, the sesquiterpene hydrocarbons are present at a higher quantity in SD and HD oils. Results from various other raw materials highlighted the same tendency: DIC treatment for essential oil extraction leads to higher yields, a shorter processing time, and "better" aromatic profiles. The oxygenated compounds (LOC and HOC) are highly odoriferous, whereas the sesquiterpene hydrocarbons contribute only little to fragrance and therefore are less valuable. The percentage of oxygenated compounds is higher in DIC treatments, which indicates a higher quality of essential oil.

Solvent extraction of a solute from a plant is normally limited by the diffusion of solvent within the porous solid containing solute solution. Thus, conventional solvent extraction is tedious and time consuming because of a very weak technological capability of the natural raw material due to its structure. Its impact is very prejudicial in terms of performance (yields, kinetics, energy consumption, etc.) and quality. To remedy this situation and intensify the solvent extraction kinetics, DIC texturing was defined, studied, and optimized in various situations (extraction of vegetable oil, active molecules such as anthocyanins, flavonoids). Regardless of the solvent used (hexane, methanol, ethanol, CO_2 supercritical fluid), the operation is highly suitable. Thus, DIC texturing is performed before the conventional solvent extraction itself. This gives much higher yields (between 10% and 50%), the starting accessibility is at least doubled, and diffusivity is increased up to 10-fold, depending on the DIC conditions.

Work performed on a large number of raw materials showed that DIC texturing could dramatically decrease the extraction time, in some cases providing the same yield in minutes compared with hours for conventional extraction treatments.

In terms of quality, no detectable thermal degradation was observed. This is probably attributable to the time–temperature level: DIC treatment is a highly suitable high temperature/low time process, and despite a relatively high temperature range reaching up to 160°C, the high-temperature treatment time is too short to show any detectable degradation. As an example, in all the cases of antioxidant activity plants we studied, DIC treatment resulted in a 10-fold improvement in the antioxidant activity of the plant.

REFERENCES

Aguillera, J.M., Stanley, D.W. 1999. *Microstructural Principles of Food Processing and Engineering*. Gaithersburg, MD: Aspen Publishers.

Al Haddad, M. 2007. Contribution théorique et modélisation des phénomènes instantanés dans les opérations d'autovaporisation et de déshydratation. PhD dissertation, Université de La Rochelle, La Rochelle.

Al Haddad, M., Mounir, S., Sobolik, V., Allaf, K. 2008. Fruits and vegetables drying combining hot air, DIC technology and microwaves. *Int J Food Eng* 4: 1–6.

Allaf, K. 1982. *Thermodynamics and Transfer Phenomena*. Beirut: Lebanese University, Faculty of Science.

Allaf, K. 2009. The new instant controlled pressure–drop DIC technology, in *Essential Oils and Aromas: Green Extraction and Application*, F. Chemat (Ed.), pp. 85–121. New Delhi: Har Krishan Bhalla and Sons.

Allaf, K., Louka, N., Bouvier, J.M., Parent, F., Forget, M. 1993. French Patent F2708419.

Allaf, K., Rezzoug, S.A., Cioffi, F., Contento, M.P. 1998. French Patent 98/11105.

Ben Amor, B. 2008. Maîtrise de l'aptitude technologique de la matière végétale dans les opérations d'extraction de principes actifs: Texturation par détente instantanée contrôlée (DIC). PhD dissertation, Université de La Rochelle, La Rochelle.

Ben Amor, B., Allaf, K. 2006. Improvement of anthocyanins extraction from *Hibiscus sabdariffa* by coupling solvent and DIC process, in *Proceedings of Malta Polyphenols 2006, 3rd International Conference on Polyphenols Application in Nutrition and Health,* St Julian, Malta, October 26–27, 2006.

Ben Amor, B., Allaf, K. 2007a. A new process of extraction of oligosaccharides from *Tephrosia purpurea*: A solvent extraction–DIC coupling process, in *Proceedings of COSMING 2007-4ème Colloque International Ingrédients Cosmétiques et Biotechnologies,* Saint-Malo, France, June 27–29, 2007.

Ben Amor, B., Allaf, K. 2007b. Amélioration de l'extraction des anthocyanes de la Roselle (*Hibiscus sabdariffa*) en couplant procédé DIC et solvant. *Récent Prog Génie Procédés 96.*

Ben Amor, B., Lamy, C., Andre, P., Allaf, K. 2008. Effect of instant controlled pressure drop treatments on the oligosaccharides extractability and microstructure of *Tephrosia purpurea* seeds. *J Chromatogr A* 1213: 118–124.

Bendahou, M., Muselli, A., Grignon-Dubois, M., Benyoucef, M., Desjobert, J.-M., Bernardini, A.-F., Costa, J. 2008. Antimicrobial activity and chemical composition of *Origanum glandulosum* Desf. essential oil and extract obtained by microwave extraction: Comparison with hydrodistillation. *Food Chem* 106: 132–139.

Besombes, C., Berka-Zougali, B., Allaf, K. 2010. Instant controlled pressure drop extraction of lavandin essential oils: Fundamentals and experimental studies. *J Chromatogr A* 1217: 6807–6815.

Bimbenet, J.J. Duquenoy, A., Trystram, G. 1993. *Génie des procédés alimentaires: Des bases aux applications*. Paris: Dunod.

Bocchio, E. 1985. Hydrodistillation des huiles essentielles: Théorie et applications. *Parfums, Cosmét, Arômes* 63: 61–62.

Brand-Williams, W., Cuvelier, ME., Berset, C. 1995. Use of a free radical method to evaluate antioxidant activity. *Lebensm-Wiss U-Technol* 28: 25–30.

Capon, M. Courilleau-Haverlant, V., Valette, C. 1993. *Chimie des couleurs et des odeurs*. Nantes: Cultures et Techniques.

Cassel, E. Vargas, R.M.F. Martinez, N. Lorenzo, D., Dellacassa, E. 2009. Steam distillation modeling for essential oil extraction process. *Ind Crops Prod* 29: 171–176.

Chavanne, M., Beaudouin, G.J., Jullien, A., Flammand, F. 1991. *Chimie organique expérimentale*. Ed. Modulo.

Chemat, S., Lagha, A., Ait Amar, H., Bartels, P.V., Chemat, F. 2004. Comparison of conventional and ultrasound-assisted extraction of carvone and limonene from caraway seeds. *Flavour Frag J* 19: 188–195.

Crank, J. 1975. *The Mathematics of Diffusion*. USA: Oxford University Press.

Dibert, K. 1989. Contribution à l'étude de l'extraction solide–liquide de l'huile et de l'acide du café vert. PhD dissertation, Claude Bernard Lyon I, Lyon.

FAO. 2010. Retrieved from http://www.fao.org/ag/agn/jecfa-flav/search.htm.

Ferhat, M.A., Meklati, B.Y., Chemat, F. 2007. Comparison of different extraction methods: cold pressing, hydrodistillation and solvent free microwave extraction, used for the isolation of essential oil from citrus fruits, in *Proceedings of 38th International Symposium on Essential Oils*, Graz (Autriche), p. 44.

Gamiz-Gracia, L., Luque de Castro, M.D. 2000. Continuous subcritical water extraction of medicinal plant essential oil: Comparison with conventional techniques. *Talanta* 51: 1179–1185.

Gertenbach, W., Bilkei, G. 2001. Der Einfluss von pflanzlichen Futterzusatzstoffen in Kombination mit Linolensäure auf die immuninduzierte Wachstumsverzögerung nach dem Absetzen. *Biol Tiermed* 3: 88–92.

The Good Scents Company. 2010. Retrieved from http://www.thegoodscentscompany.com.

Groubert, A. 1984. Techniques d'extraction végétale. PhD dissertation, Université de Montpellier I. UFR des sciences pharmaceutiques et biologiques, Montpellier.

Jiménez-Carmona, M.M., Ubera, J.L., Luque de Castro, M.D. 1999. Comparison of continuous subcritical water extraction and hydrodistillation of marjoram essential oil. *J Chromatogr A* 855: 625–632.

Kotnik, P., Škerget, M., Knez, Ž. 2007. Supercritical fluid extraction of chamomile flower heads: Comparison with conventional extraction, kinetics and scale-up. *J Supercrit Fluid* 43: 192–198.

Kristiawan, M., Sobolik, V., Allaf, K. 2008. Isolation of Indonesian cananga oil by instantaneous controlled pressure drop. *J Essent Oil Res* 20: 135–146.

Kristiawan, M., Sobolik, V., Allaf, K. 2009. Isolation of Indonesian cananga oil using multicycle pressure drop process. *J Chromatogr A* 1192: 306–318.

Leybros, J., Frémeaux, P. 1990. Extraction solide–liquide: I. Aspects théoriques. *Techniques de l'ingénieur, traité Genie des procédés*, Vol. J1 077 06.

Lide, D.R. 2005. *CRC Handbook of Chemistry and Physics*, Internet version http://www.hbcpnetbase.com.

Lucchesi, M.E., Chemat, F., Smadja, J. 2004. Solvent-free microwave extraction of essential oil from aromatic herbs: comparison with conventional hydro-distillation. *J Chromatogr A* 1043: 323–327.

Mafart, P., Béliard, E. 1993. *Génie Industriel Alimentaire techniaues séparatives*. Paris: Technique et Documentation-Lavoisier.

Mounir, S., Allaf, K. 2008. Three-stage spray drying: New process involving instant controlled pressure drop. *Dry Technol* 26: 452–463.

Nguyen Van, C. 2010. Maîtrise de l'aptitude technologique des oléagineux par modification structurelle; applications aux opérations d'extraction et de transestérification in-situ. PhD dissertation, Université de La Rochelle, La Rochelle.

Pharmacyebook. 2010. Retrieved from http://pharmacyebooks.com/2009/03/vogels-textbook-of-practical-organic-chemistry-5th-edition.html.

Salisova, M., Toma, S., Mason, T.J. 1997. Comparison of conventional and ultrasonically assisted extractions of pharmaceutically active compounds from *Salvia officinalis*. *Ultrason Sonochem* 4: 131–134.

Satrani, B., Aberchane, M., Farah, A., Chaouch, A., Talbi, M. 2006. Composition chimique et activité antimicrobienne des huiles essentielles extraites par hydrodistillation fractionnée du bois de *Cedrus atlantica* Manetti. *Acta Bot Gallica* 153: 97–104.

Schwartzberg, H.G., Chao, R.Y. 1982. Solute diffusivities in leaching processes. *Food Technol* 36: 73–86.

Wenqiang, G., Shufen, L., Ruixiang, Y., Shaokun, T., Can, Q. 2007. Comparison of essential oils of clove buds extracted with supercritical carbon dioxide and other three traditional extraction methods. *Food Chem* 101: 1558–1564.

NOMENCLATURE

A	Availability ratio	
A_i	Crank coefficient depending on the solid geometry	
B_i	Biot number	
c_{pe}	Specific heat of liquid essential oils in the material	(J kg^{-1} K^{-1})
c_{pp}	Specific heat at constant pressure of the product	(J kg^{-1} K^{-1})
c_{ps}	Specific heat of dry material	(J kg^{-1} K^{-1})
c_{pw}	Specific heat at constant pressure of water	(J kg^{-1} K^{-1})
D_{eff}	Effective diffusivity	(m^2 s^{-1})
$D_{eff\,e}$	Effective diffusivity of essential oil within the porous solid	(m^2 s^{-1})
$D_{eff\,w}$	Effective diffusivity of vapor within the porous solid	(m^2 s^{-1})
D_s	Solute diffusivity within the solvent	(m^2 s^{-1})
D_{ws}	Effective diffusivity of liquid water within the solid medium	(m^2 s^{-1})
DIC	Instant controlled pressure drop (Détente Instantanée Contrôlée)	
EE	Extraction efficiency	(%)
EO	Essential oil	
h	Coefficient of heat transfer by condensation and convection	(W m^{-2} K^{-1})
K	Porous medium permeability for essential oil and water vapor mixture	(m^2)
k	Transfer coefficient	(m s^{-1})
L	Mean of the specific latent heat of condensation of water	(J kg^{-1})
l	Characteristic size of particles (thickness/2)	(m)
L_e	Latent heat of essential oil vaporization	(J kg^{-1})
L_m	Mean of the specific latent evaporation heat of essential oil and water mixture	(J kg^{-1})
L_w	Latent heat of water vaporization	(J kg^{-1})
M_e	Molar mass of essential oils	(kg mol^{-1})
m_m	Mass of evaporated essential oil and water mixture of per DIC cycle	(kg)
m_v	Mass of condensate vapor from the saturated surrounded steam	(kg)
M_w	Molar mass of water	(kg mol^{-1})
P	Total pressure of the vapor of the mixture of water and essential oil	(Pa)
p_e	Partial pressure of essential oils in the material	(Pa)
p_w	Partial pressure of vapor in the porous material	(Pa)
Q_c	Quantity of heat furnished by the steam to the surface of the product	(J)
q_i	Coefficient dependent upon geometry of solid	
R	Ideal gas constant	(J mol^{-1} K^{-1})

R_o	Hole radius	
R_s	External radius	
RSM	Response surface methodology	
S	Exchange surface	(m^2)
T	Temperature	(K)
t	Time	(s)
t_c	Duration time of the heating stage (d) of a DIC cycle	(s)
T_v	Temperature of saturated steam	
$\langle T_v - T \rangle_{LN}$	Logarithmic mean of the difference of temperature between the steam (T_v) and the product (T)	
v_{eo}	Absolute velocity of essential oils within the porous solid	(ms^{-1})
V_{hole}	Hole volume	
v_m	Mean velocity of solid matrix	(ms^{-1})
v_{mix}	Velocity of essential oil and water vapor mixture within the porous medium	(ms^{-1})
v_s	Absolute porous solid velocity	(ms^{-1})
v_{sol}	Mean velocity of solvent within the solid matrix	(ms^{-1})
v_v	Absolute velocity of the vapor within the porous solid	(ms^{-1})
v_w	Absolute velocity of the liquid water within the porous solid	(ms^{-1})
$\langle W c_{pw} \rangle$	Mean value of $\langle W c_{pw} \rangle$ during stage (d)	
X	Solute density dry basis in the solid matrix at time t (mg g^{-1} dry material)	
X_∞	Solute density dry basis in the solid matrix at $t \to \infty$ (mg g^{-1} dry material)	
X_A	Solute density dry basis at the starting diffusion time (mg g^{-1} dry material)	
X_m	Solute density dry basis in the solid matrix at $t = 0$ (mg g^{-1} dry material)	
Y	Amount of solute extracted at time t	
Y_∞	Amount of solute extracted at $t \to \infty$ (total yields)	
Y_A	Amount of solute extracted at time $t = 0$	
ε	Porosity	
λ	Global conductivity of the porous wet material	$(J\ m^{-1}\ K^{-1})$
λ_{eff}	Effective conductivity of material	$(W\ m^{-1}\ K^{-1})$
ν_m	Kinematic viscosity of the mixture of essential oils and water	$(m^2\ s^{-1})$
φ	Heat flow within the porous material	$(W\ m^{-2})$
ρ_{eo}	Apparent density of liquid essential oils in the material	$(kg\ m^{-3})$
ρ_m	Apparent density of porous material	$(kg\ m^{-3})$
ρ_{mix}	Apparent density of the essential oil and water vapor mixture	$(kg\ m^{-3})$
ρ_s	Apparent density of dry material	$(kg\ m^{-3})$
ρ_{sol}	Apparent density of solvent	$(kg\ m^{-3})$
ρ_w	Liquid water in the material	$(kg\ m^{-3})$

$$\tau = \frac{D_{eff}\, t}{d_p^2} \quad \text{Fick's number}$$

10 High Pressure–Assisted Extraction: Method, Technique, and Application

Krishna Murthy Nagendra Prasad,
Amin Ismail, John Shi, and Yue Ming Jiang

CONTENTS

10.1 INTRODUCTION

As human population increases, great quantities of waste are generated while more utilization of natural resources is required. With the development of new processing technologies, these waste products can be utilized and recycled better (Butz and Tausher 2002; Corrales et al. 2008). For example, great quantities of by-products in the form of pericarp and seeds are generated during processing of litchi and longan fruits, even though these by-products contain significant amounts of secondary metabolites such as anthocyanins, flavonoids, phenolic acids, hydrolyzable tannins, and polysaccharides (Lee and Wicker 1990; Sarni-Manchado et al. 2000; Zhang et al. 2000; Jiang et al. 2006; Yang et al. 2006; Prasad et al. 2009a, 2010). These by-products are reported to have antibacterial, antiviral, antioxidant, anti-inflammatory, and anticarcinogenic properties (Namiki 1990; Halliwell 2007; Prasad et al. 2009b; Sun et al. 2010), and to exhibit great potential for further utilization in the pharmaceutical, metallurgical, and food industries.

Extraction is an important step in the isolation of bioactive compounds. These compounds exist in plants enclosed in insoluble structures, such as the vacuoles of plant cells and lipoprotein bilayers, which complicates their extraction (Zhang et al. 2004). The extraction processing for bioactive compounds from plant materials can be described as a mass transport phenomenon, where solids contained in a plant matrix are transferred into solvents up to their equilibrium concentration. Mass transport phenomena can be improved by changes in concentration gradients and heating, and with the help of new technologies such as high pressure, ultrasonics, and pulsed electric fields (Butz and Tausher 2002; Knoor 2003; Corrales et al. 2008). Increased extraction yields caused by high pressure are presumably attributable to its ability to deprotonate charged groups and to disrupt the salt bridges and hydrophobic bonds in cellular membranes, which may have led to higher permeability (Dornenburg and Knoor 1993). Based on phase behavior theory, the pressurized cells exhibit increased permeability as pressure increases, which might account for increased extraction yield by high pressure treatment (Zhang et al. 2004, 2006). Thus, application of high pressure can enhance mass transfer rate, increase solvent permeability in plant cells as well as the secondary metabolite diffusion, shorten extraction times, and increase extraction yields. This chapter introduces the method, tools, and techniques of high pressure–assisted extraction (HPE), and then demonstrates its application using litchi and longan fruits as examples.

10.2 HIGH PRESSURE–ASSISTED EXTRACTION

HPE has been recognized as an environmentally friendly technology by the U.S. Food and Drug Administration (U.S. FDA 2007), and has been in use in the chemical, ceramic, and plastic industries for many years, but the food industry recognized its potential application only in the late 1980s (Otero and Sanz 2003; Zhang et al. 2004). HPE as a relatively novel technique is used for extraction of active compounds from plant materials. High pressures ranging from 100 to 1000 MPa are considered an alternative extraction method, which has been proven to be fast and effective. The advantages of HPE are summarized below:

1. High pressure acts immediately and equally through a food mass, independent of its size, shape, or composition.
2. HPE requires minimal heat and can prevent thermal degradation.
3. High pressure does not alter the covalent bonds during pressurization and, thus, retains high bioactivity.
4. High pressure influences the secondary and tertiary structures of proteins and polysaccharides, and can alter the functional properties of these compounds.
5. HPE can help to obtain high extraction yields.
6. HPE can shorten extraction times.

High pressure can effectively enhance the efficiency of mass transfer and improve solvent permeability in cells as well as diffusion of the secondary metabolites, but have no significantly negative effect on the structure and activity of bioactive compounds

(Dornenburg and Knorr 1993; Ahmed and Ramaswamy 2006; Lopes et al. 2010). Based on the phase behavior theory, the dissolution is faster at higher pressure. The differential pressure between the inner part and the exterior of the plant cells is very large under HPE conditions. Under this large differential pressure, the solvent can permeate very fast through the broken membranes into plant cells, and the mass transfer rate of solute or the rate of dissolution is very large, which leads to a very short extraction time using HPE, compared to conventional extraction processes (Zhang et al. 2004). Furthermore, HPE provides the possibility of inactivating degrading enzymes, which may explain the higher extraction yield and antioxidant activity compared to other extraction methods (Ahmed and Ramaswamy 2006). High pressure also has the ability to decrease the pH of the solvent during extraction, whereas the reduction in pH might enhance the extraction of bioactive compounds because most of the compounds are more stable at low pH (Dorenberg and Knoor 1993; Ahmed and Ramaswamy 2006; Corrales et al. 2008; Khosravi-Darani 2010; Lopes et al. 2010).

High pressure may lead to various structural changes in foods, such as damage and deformation of cells and their membranes as well as denaturation of proteins and enzymes. Weak bonds such as the hydrogen bond, the electrostatic bond, the van der Waals bond, and the hydrophobic bond can be broken by high pressure, but small molecules cannot be changed under high pressure conditions (Butz and Tauscher 2002; Knoor 2003; Zhang et al. 2004; Corrales et al. 2008).

HPE has many advantages for extraction of bioactive products. Various researchers have successfully used the HPE technique for the extraction of corilagin from longan fruit pericarp (Prasad et al. 2010), catechins from green tea (Jun et al. 2010), anthocyanins from grapes (Corrales et al. 2008), flavonoids from litchi fruit pericarp (LFP) (Prasad et al. 2009a), and so on. This chapter focuses on the use of HPE from litchi and longan fruits, and then comments on the events pertaining to extraction of bioactive compounds of other fruits and vegetables by HPE since the year 2005.

10.3   OPERATION OF HPE

The main aspect of high-pressure technology is the fluid medium. Water is relatively incompressible under high pressure, and because of its biological compatibility, it is a preferred medium for high pressure application (Lopes et al. 2010). During extraction using high pressure equipment (Figure 10.1a), a sample with the extraction solvent is packed into a flexible plastic bag (Figure 10.1b), and then loaded into a high pressure vessel (Figure 10.1c), surrounded by a pressure transmitting medium. Pressure is applied by a hydraulically driven piston while the pressure chamber is controlled at a constant temperature by a water bath. Pressurization cycles, pressures, and time can be programmed by a computer, which can also control the automatic locking of the safety box and the alarms. After depressurization, the extracted solution is collected, filtered, and used for further study.

10.4   HPE FROM LITCHI FRUIT PERICARP

Litchi (*Litchi chinensis* Sonn.) fruit belongs to the Sapindaceae family and is native to China, India, Vietnam, Indonesia, and the Philippines. The fruit is of high

(a)

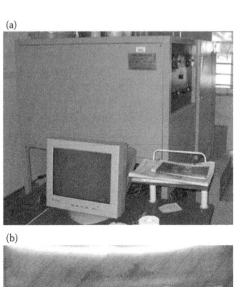

(b)

(c)

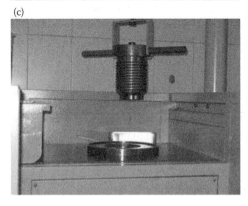

FIGURE 10.1   (a) High pressure extraction equipment, (b) sample with extraction solvent packed in flexible plastic bag, and (c) high pressure vessel.

TABLE 10.1

Comparison of Extraction Yield, Extraction Time, and Individual Flavonoid Content from Litchi Fruit Pericarp Obtained by Conventional, Ultrasonic, and High Pressure (400 MPa) Extractions

Extraction Method	Extraction Time	Extraction Yield (%)	Epicatechin (mg/g dry matter)	Epicatechin Gallate (mg/g dry matter)	Catechin (mg/g dry matter)	Procyanidin B_2 (mg/g dry matter)
Conventional	24 h	19.9 ± 1.12	0.0414 ± 0.001	0.0121 ± 0.003	0.0002	0.0175 ± 0.0003
Ultrasonic	30 min	23 ± 0.5	0.16 ± 0.04	0.06 ± 0.01	0.002 ± 0.0005	0.0731 ± 0.0011
High pressure	2.5 min	29.3 ± 0.19	0.348 ± 0.06	0.2527 ± 0.04	0.016 ± 0.07	0.1346 ± 0.03

Source: Prasad, N.K. et al., *Int J Food Sci Technol*, 44, 960–996, 2009b. With permission.

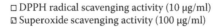

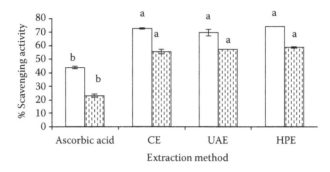

FIGURE 10.2 Comparison of DPPH and superoxide scavenging activity from litchi fruit pericarp using conventional extraction (CE), ultrasonic assisted extraction (UAE), and high pressure–assisted extraction (HPE). For each treatment, means in a row followed by different letters are significantly different at $P < .05$. (From Prasad, N.K. et al., *Int J Food Sci Technol*, 44, 960–966, 2009b. With permission.)

commercial value because of its white translucent pulp and bright red color pericarp, with a sweet taste and pleasant aroma (Jiang et al. 2006; Li and Jiang 2007; Prasad et al. 2009a). The fruit is rich in nutrition and is used as a traditional Chinese medicine to cure neural pain, swelling, and cough (Li 2008). LFP tissues account for approximately 15% by weight of the whole fresh fruit and contain significant amounts of flavonoids (Li and Jiang 2007). Furthermore, the fruit pericarp is rich in anthocyanins such as cyanidin-3-rutinoside, cyanidin-3-glucoside, malvidin-3-glucoside, pelargonidin (Prasad and Jha 1978; Lee and Wicker 1990; Prasad et al. 2009b); phenolic compounds such as gallic acid, catechin, epicatechin, gallocatechin, quercetin, rutin, procyanidin B1, procyanidin B2, epicatechin-3-gallat, and tannins (Roux et al. 1998; Sarni-Manchado et al. 2000; Zhang et al. 2000; Mahattanatawee et al. 2006; Zhao et al. 2006; Li and Jiang 2007; Prasad et al. 2009b); polysaccharides (Yang et al. 2006; Kong et al. 2010); and several volatile compounds (Sivakumar et al. 2008). In recent years, extracts from LFP tissues have exhibited excellent antityrosinase and antioxidant activities, and good anticancer activity (Li and Jiang 2007; Zhao et al. 2007). Thus, LFP tissues can be used as a readily accessible source of natural antioxidants and/or a possible supplement in the food or pharmaceutical industries.

Optimization was carried out to obtain high extraction yields from LFP using varying high pressures (200–500 MPa), time (2.5–30 min), and temperature (30–90°C). Table 10.1 shows that the application of HPE for 2.5 min gave a higher extraction yield than the conventional extraction for 24 h or ultrasonic extraction for 30 min. The results also show that HPE is more effective compared to the other two methods in extracting bioactive compounds from LFP (Prasad et al. 2009a). Additionally, it is noted that the amount of flavonoids obtained by HPE is almost 10 times higher as that obtained when using conventional extraction (Table 10.1).

However, the comparison of LFP extracts obtained from conventional, ultrasonic, and high pressure extractions investigated by Prasad et al. (2009b) exhibited no significant differences in their antioxidant activities by 1,1-diphenyl-2-picryl hydrazyl (DPPH) and superoxide scavenging assay (Figure 10.2), which clearly demonstrates that there is no deleterious effect on bioactivity when using HPE.

10.5   HPE FROM LONGAN FRUIT PERICARP

Longan (*Dimocarpus longan* Lour.) also belongs to the Sapindaceae family and is a highly attractive fruit. Longan fruit pericarp accounts for approximately 20% by weight of the whole fresh fruit and contains significant amounts of phenolic acids, flavonoids, hydrolyzable tannins, and polysaccharides (Rangkadilok et al. 2007; Yang et al. 2008; Prasad et al. 2009c, 2010). These extracts from longan fruit pericarp exhibit antibacterial, antiviral, antioxidant, anti-inflammatory, and anticarcinogenic activities (Rangkadilok et al. 2007; Yang et al. 2008).

The influences of different parameters (solvent concentration, solid/liquid ratio, high pressure, time, and temperature) were determined to obtain high extraction yields and total phenolic contents (Prasad et al. 2009c). The extraction yield and total phenolic content of the high pressure–assisted extract was higher compared to those obtained by conventional extraction from longan fruit pericarp. It was demonstrated that extraction by high pressure was very effective irrespective of different extraction time duration (Prasad et al. 2009c; Figure 10.3). In addition, HPE was successful in extracting a high quantity of corilagin and ellagic acid from longan fruit pericarp, compared to conventional extraction from longan fruit (Prasad et al. 2009d; Figure 10.4).

It was also clearly demonstrated that a high content of corilagin was obtained from longan fruit pericarp when high pressure was used (Prasad et al. 2009e; Figure 10.5).

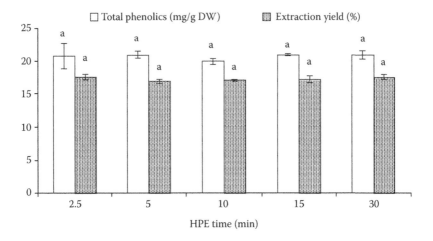

FIGURE 10.3   Total phenolic content and extraction yield from longan fruit pericarp using various extraction times under 500 MPa, with 50% ethanol and 1:50 (w/v) solid/liquid ratio at 30°C. For each treatment, means in a row followed by different letters were significantly different at $P < .05$. (From Prasad, N.K. et al., *Innov Food Sci Emerg Technol*, 10, 155–159, 2009c. With permission.)

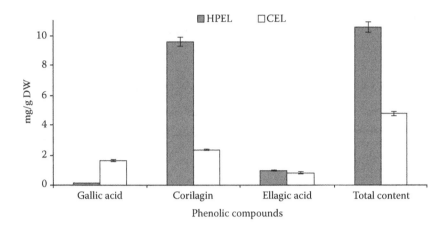

FIGURE 10.4 Phenolic contents of longan fruit obtained by high pressure–assisted extract of longan (HPEL) and conventional assisted extract of longan (CEL). Values are means ± standard deviations of three replicate analyses. (From Prasad, N.K. et al., *Innov Food Sci Emerg Technol*, 10, 413–419, 2009d. With permission.)

In combination with the previous work, the antioxidant activity of the high pressure–assisted extract of longan (HPEL) and the conventional extraction of longan (CEL) were compared by various in vitro assays. The antioxidant activities of HPEL were all higher compared to CEL, tested at various concentrations by total antioxidant activity (Figure 10.6), lipid peroxidation, DPPH radical scavenging activity, and superoxide scavenging activity (Table 10.2).

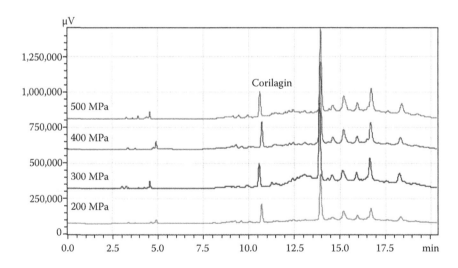

FIGURE 10.5 HPLC profile of corilagin obtained from longan fruit pericarp extracted at different high pressures. (From Prasad, N.K. et al., *Separ Purif Technol*, 70, 41–45, 2009e. With permission.)

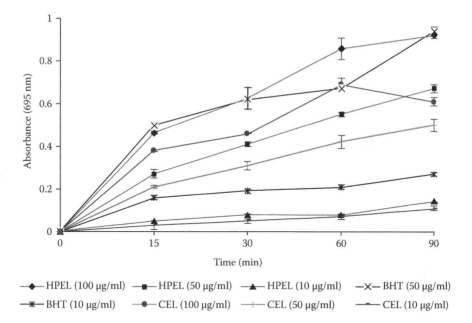

- ◆— HPEL (100 µg/ml) — ■ — HPEL (50 µg/ml) — ▲ — HPEL (10 µg/ml) — ✕ — BHT (50 µg/ml)
- ✱ — BHT (10 µg/ml) — ● — CEL (100 µg/ml) —+— CEL (50 µg/ml) — CEL (10 µg/ml)

FIGURE 10.6 Total antioxidant activity by phosphomolybdenum method of HPEL compared to CEL and BHT. Results are mean ± SD of three parallel measurements. Higher absorbance value indicates higher antioxidant activity. HPEL, high pressure–assisted extract of longan; CEL, convention-assisted extract of longan; BHT, butylated hydroxy toluene. (From Prasad, N.K. et al., *Innov Food Sci Emerg Technol*, 10, 413–419, 2009d. With permission.)

Recently, the use of different high pressures on the extraction yield, and antioxidant and antityrosinase activities of longan fruit pericarp has been demonstrated (Prasad et al. 2010). Ultrahigh pressure–assisted extraction at 500 MPa (UHPE-500) gave the highest extraction yield of 17.6% when compared to other high pressure extractions, significantly ($P > .05$) higher than the conventional extraction (7.2%), which showed that UHPE-500 was more effective in extracting bioactive compounds from longan fruit pericarp compared to other high pressure extraction methods tested (Table 10.3).

The total phenolic content (20.8 mg/g DW) of the extract using UHPE-500 was the highest, whereas the lowest content (11.9 mg/g DW) was obtained with CE, expressed as gallic acid equivalents ($P < .05$). Based on the results, UHPE increased phenolic compound recovery of the extract approximately two times higher than CE.

Furthermore, the antioxidant activity of the extract from longan fruit pericarp using UPHE-500 was the highest, followed by other high pressure extraction and the conventional extraction method, as estimated by DPPH scavenging activity (Figure 10.7), total antioxidant activity (Figure 10.8), and superoxide scavenging activity (Prasad et al. 2010; Figure 10.9). Additionally, antityrosinase (Prasad et al. 2010) and anticancer (Prasad et al. 2009d) activities of the high pressure–assisted extracts from longan fruit pericarp were higher than those obtained from other extractions (Figure 10.10 and Table 10.4).

TABLE 10.2
Antioxidant Activity of Extracts Obtained from Longan Fruit Pericarp Using Different Extraction Methods

Sample	Concentration (µg/ml)	Lipid Peroxidation		DPPH Radical Scavenging Activity		Superoxide Scavenging Activity	
		Inhibition (%)	EC_{50} (µg/ml)	Inhibition (%)	EC_{50} (µg/ml)	Inhibition (%)	EC_{50} (µg/ml)
HPEL	10	27 ± 3	40.23 ± 2.7	28 ± 1.3	39.05 ± 2.1	37 ± 3.5	53.66 ± 2.7
	20	38 ± 1.2		41.7 ± 3.3		42.7 ± 3.3	
	40	51 ± 3.1		53.4 ± 2.3		47.7 ± 4.3	
	80	78 ± 0.6		76.6 ± 1.8		58 ± 1.4	
	100	88 ± 6		82 ± 2		60.2 ± 1.1	
CEL	10	19 ± 4	58.03 ± 2.5	12 ± 1.3	57.46 ± 1.3	18 ± 2.1	84.76 ± 2.4
	20	28.1 ± 0.3		27.7 ± 1		28.5 ± 2.1	
	40	40.2 ± 1.8		41.3 ± 1.8		35.9 ± 2.5	
	80	64.4 ± 2.1		68.5 ± 2		50.7 ± 2.7	
	100	74 ± 4.5		75 ± 0.7		52.6 ± 2.3	
BHT	10	42 ± 4	17.45 ± 4.8	55 ± 0.4	8.31 ± 1.5	13 ± 1.8	470.68 ± 1.0
	20	53.6 ± 3.2		59.9 ± 1.2		14.8 ± 0.7	
	40	64.4 ± 7.2		66.4 ± 2.4		16.3 ± 0.3	
	80	86.1 ± 5.3		79.4 ± 2.1		19.5 ± 1.1	
	100	92 ± 4.7		85 ± 1.5		20.3 ± 1.5	

Source: Prasad, N.K. et al., *Innov Food Sci Emerg Technol*, 10, 413–419, 2009d. With permission.

Note: HPEL, high pressure–assisted extract of longan; CEL, convention extract of longan; BHT, butylated hydroxy toluene; EC_{50}, 50% of radicals scavenged by the test sample.

TABLE 10.3

Comparative Analysis of Extraction Yield and Total Phenolic Content of Longan Fruit Pericarp Using Conventional (CE) and Different Ultrahigh Pressure–Assisted Extractions (UHPE-200, UHPE-300, UHPE-400, and UHPE-500 MPa)

Extraction Method	Extraction Yield (%)	Total Phenolic Content (mg/g DW)
CE	7.2 ± 0.5[c]	11.9 ± 1.2[c]
UHPE-200	15.5 ± 0.4[b]	16.5 ± 0.6[b]
UHPE-300	15.8 ± 0.9[b]	17.2 ± 0.5[b]
UHPE-400	16.3 ± 0.5[b]	18 ± 0.4[b]
UHPE-500	17.6 ± 0.4[a]	20.8 ± 1.9[a]

Source: Prasad, N.K. et al., *J Pharm Biomed Anal*, 51, 471–477, 2010. With permission.

Note: Different letters in the same column indicate significant differences among means of treatments ($P < 0.05$).

In addition, Yang et al. (2009) investigated the effect of high pressure treatments on polysaccharides and lignin contents of longan fruit pericarp. A negative relationship was noted between high pressure and water-soluble polysaccharides, but no significant differences on alkali-soluble polysaccharides and cellulose yield were observed between conventional extraction and high pressure extraction samples.

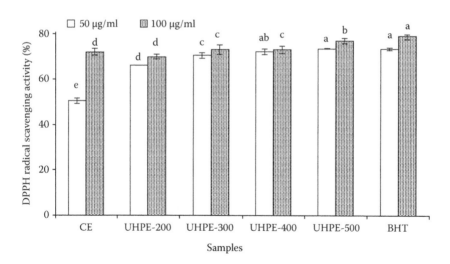

FIGURE 10.7 Comparison of DPPH radical scavenging activity from longan fruit pericarp after application of conventional extraction (CE) and different ultrahigh pressure–assisted extraction conditions (UHPE-200, UHPE-300, UHPE-400, and UHPE-500 MPa). Different letters above bars for the same concentration indicates significant differences among means of treatments ($P < .05$). (From Prasad, N.K. et al., *J Pharm Biomed Anal*, 51, 471–477, 2010. With permission.)

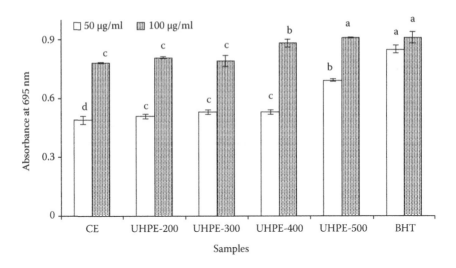

FIGURE 10.8 Comparison of total antioxidant activity from longan fruit pericarp after application of conventional extraction (CE) and different ultrahigh pressure–assisted extraction (UHPE-200, UHPE-300, UHPE-400, and UHPE-500 MPa). Different letters above bars for the same concentration indicates significant differences among means of treatments ($P <$ .05). (From Prasad, N.K. et al., *J Pharm Biomed Anal*, 51, 471–477, 2010. With permission.)

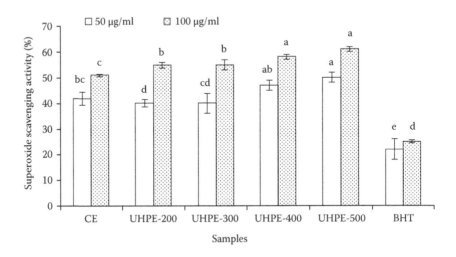

FIGURE 10.9 Comparison of superoxide anion radical scavenging activity from longan fruit pericarp after application of conventional extraction (CE) and different ultrahigh pressure–assisted extraction (UHPE-200, UHPE-300, UHPE-400, and UHPE-500 MPa). Different letters above bars for the same concentration indicates significant differences among means of treatments ($P <$ .05). (From Prasad, N.K. et al., *J Pharm Biomed Anal*, 51, 471–477, 2010. With permission.)

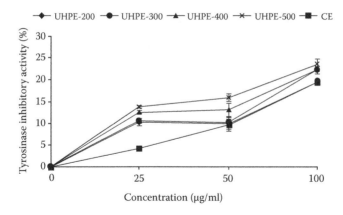

FIGURE 10.10 Comparison of tyrosinase inhibitory activity from longan fruit pericarp after application of CE and different ultrahigh pressure–assisted extraction (UHPE-200, UHPE-300, UHPE-400, and UHPE-500 MPa). (From Prasad, N.K. et al., *J Pharm Biomed Anal*, 51, 471–477, 2010. With permission.)

Several reports have also shown a close relationship between antioxidant activity and total phenolic content (Corrales et al. 2008; Prasad et al. 2009c; He et al. 2010). The phenolic compounds exhibit extensive free radical scavenging activities through their reactivities as hydrogen or electron-donating agents as well as by metal ion chelating properties (Rice-Evans et al. 1996). Longan fruit pericarp contains large amounts of polar compounds such as phenolic acids, flavonoids, and polysaccharides, which have high antioxidant properties. The extract from UHPE contains significantly higher amounts of phenolic compounds such as ellagic acid and corilagin compared to that obtained via conventional extraction (Prasad et al. 2009d,

TABLE 10.4

Anticancer Activity of Extracts Obtained from Longan Fruit Pericarp Using Different Extraction Methods

	Anticancer Activity (%)		
Sample	**SGC 7901[1]**	**HepG[2]**	**A549[2]**
CEL	30.58 ± 1.1[c]	−0.54	NA
HPEL	37.6 ± 2.6[a]	−0.22	11.96 ± 0.80[b]
Cisplatin	34.26 ± 3.2[b]	55.34 ± 2.2	49.57 ± 2.1[a]

Source: Prasad, N.K. et al., *Innov Food Sci Emerg Technol*, 10, 155–159, 2009. With permission.

Note: NA, no activity; SGC 7901, human gastric carcinoma; HepG 2, human hepatocellular liver carcinoma; A 549, human lung adenocarcinoma. CEL, convention extract of longan; HPEL, high pressure–assisted extract of longan and cisplatin (positive control). For each treatment, the means within a row followed by different letters were significantly different at the 5% level.

[1] 50 µg/ml of the samples.

[2] 100 µg/ml of the samples.

TABLE 10.5

Comparison of Extraction Yields of High Pressure–Assisted Extraction with Other Extraction Methods

Essential Component	Extraction	Solvent and Temperature	Duration	Extraction Yield (%)	References
Roots of *Codonopsis lanceolata*	HPE (500 MPa) Conventional	70% Ethanol, 50°C 70% Ethanol, 80°C	30 min 24 h	32.14 21.76	He et al. 2010
Catechins from green tea	HPE (600 MPa) Conventional	50% Ethanol 50% Ethanol, 85°C	15 min 2 h	4956[a] 4962[a]	Jun et al. 2010
Salroside from *Rhodiola sachalinesis*	HPE (500 MPa) Soxhlet extraction Ultrasonic extraction	60% Ethanol, 25°C Methanol, 80°C Methanol, 30°C	3 min 240 min 30 min	0.4 0.48 0.28	Bi et al. 2009
Caffeine from green tea	HPE (500 MPa) Conventional	50% ethanol 50% ethanol	1 min 20 h	4.0 4.2	Jun 2009
Phenolics from *Berberis koreana* bark extract	HPE (500 MPa) Conventional	Water Water, 60°C	5 min 24 h	11.4 8.3	Qadir et al. 2009
Polyphenols from green tea	HPE (500 MPa) Ultrasonic Heat reflux Shaker extraction	50% ethanol, 60°C 50% ethanol, 60°C 50% ethanol, 85°C 50% ethanol, 20°C	1 min 90 min 45 min 20 h	30 29 31 30.5	Jun et al. 2009
Total saponins from roots of ginseng (*Panax ginseng*)	HPE (200 MPa) Ultrasonic Soxhlet	70% ethanol, 60°C 70% ethanol, 60°C 70% ethanol, 90°C	5 min 40 min 6 h	4.39 3.89 4.28	Chen et al. 2009
Phenolics from cherry pomace	HPE (200 MPa) Subcritical fluid extraction	Ethanol, 60°C Ethanol, 60°	25 min 40 min	3.8[b] 0.6[b]	Adil et al. 2008
Anthocyanins from grape byproduct	Pulsed electric field HPE (600 MPa) Ultrasonication Conventional	50% Ethanol, 70°C 50% Ethanol, 70°C 50% Ethanol, 70°C 50% Ethanol, 70°	1 h 1 h 1 h 1 h	370[c] 365[c] 325[c] 210[c]	Corrales et al. 2008
Ginsenosides from roots of *Panax ginseng*	HPE (500 MPa) Reflux Ultrasonic Supercritical CO$_2$	50% ethanol 50% ethanol, 100°C 50% ethanol 3% ethanol	2 min 4 h 30 min 4 h	7.33 4.98 5.89 2.32	Shouquin et al. 2007a
Flavones from roots of *Rhodiola sachalinesis*	HPE (500 MPa) Reflux Ultrasonic Soxhlet	60% ethanol Methanol, 80°C Methanol Methanol, 80°C	3 min 120 min 30 min 240 min	5.23 1.51 2.05 2.1	Shouquin et al. 2007b

(continued)

TABLE 10.5 (Continued)
Comparison of Extraction Yields of High Pressure–Assisted Extraction with Other Extraction Methods

Essential Component	Extraction	Solvent and Temperature	Duration	Extraction Yield (%)	References
Polyphenols from propolis	HPE (500 MPa)	70% ethanol	1 min	290[d]	Xi and
	Leaching	70% ethanol	7 days	296[d]	Shouqin
	Heat reflux	95% ethanol, 80°C	4 h	247[d]	2007
Ginsenosides from roots of *Panax quinquefolium*	HPE (500 MPa)	50% ethanol, 25°C	2 min	0.82	Zhang et al. 2006
	Heat Reflux	50% ethanol, 70°C	6 h	0.76	
	Ultrasonic	70% ethanol, 50°C	40 min	0.71	
	Supercritical CO_2	CO_2 + 3% ethanol, 40°C	4 h	0.32	
	Soxhlet	95% ethanol, 70°C	8 h	0.69	
Flavonoids from propolis	HPE (500 MPa)	70% ethanol	1 min	1.96	Shouquin et al. 2005
	Conventional	70% ethanol	7 days	1.83	
Icarin from *Epimedium*	HPE (600 MPa)	70% ethanol	5 min	10	Zhang et al. 2004
	Reflux	70% ethanol	6 h	6.9	
	Ultrasonics	70% ethanol	1 h	7.2	

[a] Expressed in mg/kg.
[b] Expressed in mg GAE/g.
[c] Expressed in μmol GAE/g.
[d] Expressed in mg/g.

2010). The antioxidant activities of extracts from UHPE-treated longan samples and conventional extraction samples correlate with the amounts of phenolics present in these extracts (Prasad et al. 2009c, 2009d).

10.6 HPE FROM OTHER PLANT MATERIALS

The usage of HPE with a summary of the extraction yields is given in Table 10.5. Roots of *Codonopsis lanceolata* extracted by high pressure exhibited a higher extraction yield compared to that obtained using the conventional extraction process. Additionally, HPE showed the highest antimicrobial activity (MIC < 14 mg/ml, MIC is a minimum inhibitory concentration) against *Listeria monocytogenes*, *Staphylococcus aureus*, *Shigella boydii*, and *Salmonella typhimurium* (He et al. 2010). Moreover, the extract also exhibited the highest antimutagenic activities (82% inhibition).

Bi et al. (2009) extracted salrioside from the roots of *Rhodiola sachalinesis* using high pressure extraction. They compared the extraction efficiency of HPE with other extraction methods including ultrasonication, leaching, and soxhlet extractions. Among all the extraction methods, HPE was found to be the most efficient in extracting salrioside over a very short time. Chen et al. (2009) estimated the content of total saponins and antioxidant activity by the DPPH method, of ginseng root extracts. The DPPH radical scavenging activity and total saponin content using HPE was much higher, compared to Soxhlet and ultrasonic extractions. Furthermore, the antioxidant activity of propolis extracts obtained by HPE was analyzed by Xi and Shouqin (2007).

The antioxidant activity of HPE extracts was all higher compared to other extraction methods. Jeong et al. (2009) reported high anticancer activity of the high pressure–assisted extracts of acer mono against human lung carcinoma (A549), breast adenocarcinoma (MCF-7), colon adrenocarcinoma (Caco-2), and liver adrenicarcinoma (Hep3B).

The influence of different solvents, solvent concentration, solid/liquid ratio, ultra-high pressure, and time were individually determined using mono factor experiments by Shouquin et al. (2007a), when extracting ginsenoside from the roots of *Panax ginseng*. A comparative investigation to extract salidroside and flavonens from the roots of *Rhodiola sachalinesis* between HPE, ultrasonication, soxhlet, and reflux extractions was carried out (Shouquin et al. 2007b). Ginsenoside was extracted using ultrahigh pressure extraction from the roots of American ginseng (*Panax quinqefolium*) by Zhang et al. (2006). The extraction yield of ginsenoside obtained by ultrahigh pressure extraction was 0.821%, which is much higher than that obtained by Soxhlet, ultrasonic, or supercritical carbon dioxide extractions, with extraction yields of 0.697%, 0.761%, and 0.342%, respectively.

Shouquin et al. (2005) extracted flavonoids from propolis using the HPE and conventional methods. The flavanoid content obtained by HPE for 1 min was much higher than for the conventional extraction. These results clearly show the advantage of using HPE as one of the promising extracting techniques. Jun (2009) and Jun et al. (2010) optimized the extraction conditions for caffeine from tea leaves and found that high extraction yields could be achieved when high pressure extraction at 500 MPa was conducted using 50% ethanol as an extraction solvent, with a liquid/solid ratio of 20:1 (ml/g), and extraction time of only 1 min. Various solvents, solvent concentrations, solid/liquid ratios, and different high pressures and times were determined individually for the polyphenolic content of green tea leaves. The extraction yield of polyphenols obtained from HPE for 1 min was comparable to extraction at room temperature for 20 h (Jun et al. 2009).

Qadir et al. (2009) obtained a higher extraction yield (11.4%) from the bark of *Berberis koreana*, compared to that obtained by conventional extraction (8.3%). They also demonstrated high antioxidant and anticancer activities of the HPE extracts. Adil and his coworkers (2008) have optimized and compared the total phenolic content from cherry pomace obtained using various extraction methods. The total phenolic content and antioxidant activity using HPE was higher compared to their counterparts when using other extraction methods.

A comparative investigation of the conventional, ultrasonication, high pressure, or pulsed electric field extraction was carried out by Corrales et al. (2008) using grape by-products. They found out that phenolic content was very high when using pulsed electric field or high pressure. In a further study, Corrales et al. (2009) optimized the extraction conditions of high pressure to extract anthocyanins from grape by-products. In addition, Trejo Araya et al. (2007) reported structural changes in high pressure–treated carrot tissues. Microstructure images of raw carrot showed a more intact, regular, and organized cell distribution and higher cell-to-cell contact throughout the tissues compared with the high pressure–treated cells (Figure 10.11).

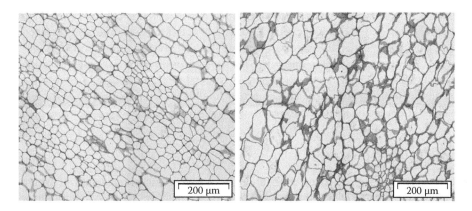

FIGURE 10.11 Microscopy image of raw (left) and high pressure–treated (300 MPa for 2 min, right) carrot tissues. (From Trejo Arya, X.I. et al., *J Food Eng*, 80, 873–884, 2007. With permission.)

10.7 SAFETY ASPECT OF HPE

The vessels of the high-pressure extraction equipment are uniquely designed to safely withstand high pressures (100–800 MPa) over many cycles. During pressure leak, automatic locking of the vessel takes place along with the alarm signal. Since the extraction medium is liquid, HPE is safer than gas medium at high pressure. Electricity is required only when increasing the pressure, whereas no power is required for pressure holding and decompression during the operation of HPE. The main drawback of HPE is the main capital cost of the equipment. With successful design and development in pump and vessel technology, the use of HPE has been adopted by the food industry (Shouquin et al. 2007a). By the end of 2007, about 120 industrial high-pressure processing plants had been installed all over the world, with batch processing volumes ranging from 35 to 420 l and an annual production capacity totaling more than 150,000 tons.

10.8 CONCLUSION AND FUTURE PROSPECTS

The prospect of increasing production of fruits and vegetables and their by-products raises expectations for increased processing, utilization, and extraction opportunities. Most of the fruits and vegetables or their by-products are rich sources of anthocyanins, flavonoids, phenolic compounds, tannins, and polysaccharides with many health benefits such as antioxidants, anticancer activity, and antityrosinase content.

With recent advances in green technologies such as high pressure extraction, bioactive compounds from fruits and vegetables or their by-products can be obtained efficiently. It should be noted that application of HPE can reduce the cost of specific bioactive compounds. For example, the extraction yield of epicatechin from LFP is only 0.04 mg/g DW using CE, whereas it increases nearly eightfold when using HPE (Table 10.1). Thus, taking into the account operation costs, maintenance, and labor charges, the extraction costs of epicatechin obtained by HPE can be reduced up to 40% compared to the conventional extraction technique. It should also be noted that application of high pressure can increase the bioactivities of the extracted compounds. Furthermore, fewer impurities are noticed in the HPE extraction solution as compared to conventional extraction. Additional work is needed to explore the possibility of using HPE to extract bioactive compounds from more plant materials, with an emphasis on their bioactivities.

ACKNOWLEDGMENTS

This work was supported by the National Natural Science Foundation of China (Grant Nos. 30928017, U0631004, and 31071638), the CAS/SAFEA international partnership program for creative research teams, and the Natural Science Foundation of Guangdong Province (Grant No. 06200670).

REFERENCES

Adil, I.H., Yener, M.E., Baymdirh, A. 2008. Extraction of phenolics of sour cherry pomace by high pressure solvent and subcritical fluid and determination of antioxidant activities of the extracts. *Separ Sci Technol* 43: 1091–1110.

Ahmed, J., Ramaswamy, H.S. 2006. High pressure processing of fruits and vegetables. *Stewart Postharvest Rev* 1: 1–10.

Bi, H.M., Zhang, S., Liu, C.J., Wang, C.Z. 2009. High hydrostatic pressure extraction of Saldroside from *Rhodiola sachalinesis*. *J Food Process Eng* 32: 53–63.

Butz, P., Tausher, B. 2002. Emerging technologies: Chemical aspects. *Food Res Int* 35: 279–284.

Chen, R., Meng, F., Zhang, S., Liu, Z. 2009. Effects of ultrahigh pressure extraction conditions on yields and antioxidant activity of ginsenoside from ginseng. *Separ Purif Technol* 66: 340–346.

Corrales, M., Toepfl, S., Butz, P., Knorr, D., Tauscher, B. 2008. Extraction of anthocyanins from grape by-products assisted by ultrasonic, high hydrostatic pressure or pulsed electric fields: A comparison. *Innov Food Sci Emerg Technol* 9: 85–91.

Corrales, M., Garcia A.F., Butz, P., Tauscher, B. 2009. Extraction of anthocyanins from grape skins assisted by high hydrostatic pressure. *J Food Eng* 90: 415–421.

Dornenburg, H., Knoor, D. 1993. Cellular permeabilization of cultured plant tissues by high electric field pulses or ultra high pressure for the recovery of secondary metabolites. *Food Biotechnol* 7: 35–48.

Halliwell, B. 2007. Dietary polyphenols: Good, bad, or indifferent for your health? *Cardiovasc Res* 73: 341–347.

He, X., Kim, S.S., Park, S.J., Seong, D.H., Yoon, W.Y., Lee, H.Y., Park, D.S., Ahn, J. 2010. Combined effect of probiotic fermentation and high pressure extraction on the antioxidant, antimicrobial and antimutagenic activities of deodeok (*Condonopis lanceolata*). *J Agric Food Chem* 58: 1719–1725.

Jeong, M.H., Kim, S.S., Ha, J.H., Jin, L., Lee, H.J., Kang, H.Y., Park, S.J., Lee, H.Y. 2009. Enhancement of anticancer activity of Acer mono by high pressure extraction process. *J Korean Soc Food Sci Nutr* 38: 1243–1252.

Jiang, Y., Wang, Y., Song, L., Liu, H., Lichter, A., Kerdchoechuen, O., Joyce, D.C., Shi, S. 2006. Production and postharvest characteristics and technology of litchi fruit: An overview. *Aust J Exp Agric* 46: 1541–1556.

Jun, X. 2009. Caffeine extraction from green tea leaves assisted by high pressure processing. *J Food Eng* 94: 105–109.

Jun, X., Deji, S., Shou, Z., Bingbing, L., Ye, L., Rui, Z. 2009. Characterization of polyphenols from green tea leaves using a high hydrostatic pressure extraction. *Int J Pharm* 382: 139–143.

Jun, X., Shuo, Z., Bingbing, L., Rui, Z., Ye, L., Deji, S., Guofeng, Z. 2010. Separation of major catechins from green tea by ultra high pressure extraction. *Int J Pharm* 386: 229–231.

Khosravi-Darani, K. 2010. Research activities on supercritical fluid science in food biotechnonlogy. *Crit Rev Food Sci* 50: 479–488.

Knoor, D. 2003. Impact of non thermal processing on plant metabolites. *J Food Eng* 56: 131–134.

Kong, F.L., Zhang, M.W., Kuang, R.B., Yu, S.J., Chi, J.W., Wei, Z.C. 2010. Antioxidant activities of different fractions 1 of polysaccharide purified from pulp tissue of litchi (*Litchi chinensis* Sonn.). *Carbohydr Polym* 81: 612–616.

Lee, H.S., Wicker, L. 1990. Anthocyanin pigments in the skin of lychee fruit. *J Food Sci* 56: 466–468, 483.

Li, J., Jiang, Y.M. 2007. Litchi flavonoids: Isolation, identification and biological activity. *Molecules* 12: 745–758.

Li, J. G. 2008. *In "The Litchi."* Beijing, China: China Agric Press.

Lopes, M.L.M., Mesquita, V.L.V., Chiaradia, A.C.N., Fernendes, A.A.R., Farnandes, P.M.B. 2010. High hydrostatic presure processing of tropical fruits. Importance for maintenance of the natural food properties. *Annu N Y Acad Sci* 1189: 6–15.

Mahattanatawee, K., Manthey, J.A., Luzio, G., Talcott, S.T., Goodner, K., Baldwin, E.A. 2006. Total antioxidant activity and fiber content of select Florida-grown tropical fruits. *J Agric Food Chem* 54: 7355–7363.

Namiki, M. 1990. Antioxidants/antimutagens in foods. *Crit Rev Food Sci* 29: 273–300.

Otero, L. Sanz, P. D. 2003. Modelling heat transfer in high pressure food processing: A review. *Innov Food Sci Emerg Technol* 4: 121–134.

Prasad, U.S., Jha, O.P. 1978. Changes in pigmentation patterns during litchi ripening: Flavonoid. *Plant Biochem J* 5: 44–49.

Prasad, N.K., Yang, B., Zhao, M., Ruenroengklin, N., Jiang, Y. 2009a. Application of ultrasonication or high pressure extraction of flavonoids from litchi fruit pericarp. *J Food Process Eng* 32: 828–843.

Prasad, N.K., Yang, B., Zhao, M., Wang, B., Chen, F., Jiang, Y. 2009b. Effects of high pressure treatment on the extraction yield, phenolic content and antioxidant activity of litchi (*Litchi chinensis* Sonn.) fruit pericarp. *Int J Food Sci Technol* 44: 960–966.

Prasad, N.K., Chun, Y., En, Y., Zhao, M., Jiang, Y. 2009c. Effects of high pressure on the extraction yield, total phenolic content and antioxidant activity of longan fruit pericarp. *Innov Food Sci Emerg Technol* 10: 155–159.

Prasad, N.K., Hao, J., Jhon, S., Liu, T., Jiang, L., Xiaoyi, W., Qiu, S., Xiu, S., Jiang, Y. 2009d. Antioxidant and anticancer activities of high pressure-assisted extract of longan (*Dimpcarpus longan* Lour.) fruit pericarp. *Innov Food Sci Emerg Technol* 10: 413–419.

Prasad, N.K., Yang, B., Zhao, M., Wei, X., Jiang, Y., Chen, F. 2009e. High pressure extraction of corilagin from longan (*Dimocarpus longan* Lour.) fruit pericarp. *Separ Purif Technol* 70: 41–45.

Prasad, N.K., Yang, B., Shi, S., Yi, C., Zhao, M., Xue, S., Jiang, Y. 2010. Enhanced antioxidant and antityrosinase activities of longan fruit pericarp by ultra-high pressure-assisted extraction processing. *J Pharm Biomed Anal* 51: 471–477.

Qadir, S.Y., Kwon, M.C., Han, J.G., Ha, J.H., Chung, H.S., Ahn, J., Lee, H.Y. 2009. Effects of different extraction protocols on anticancer and antioxidant activities of *Berberis koreana* bark extracts. *J Biosci Bioeng* 107: 331–338.

Rangkadilok, N., Sitthimonchai, S., Worasuttayangkurn, L., Mahidol, C., Ruchirawat, M., Satayavivad, J. 2007. Evaluation of free radical scavenging and antityrosinase activities of standardized longan fruit extract. *Food Chem Toxicol* 45: 328–336.

Rice-Evans, C.A., Miller, N.J., Panganga, G. 1996. Structure antioxidant activity relationship of flavonoids and phenolic acids. *Free Radic Biol Med* 20: 933–956.

Roux EL., Doco, T., Sarni-Manchado, P., Lozano, Y., Cheynier, V. 1998. A-type proanthocyanidins from pericarp of *Litchi sinensis*. *Phytochemistry* 48: 1251–1258.

Sarni-Manchado, P., Roux, E.L., Guerneve, C.L., Lozano Y., Cheynier, V. 2000. Phenolic composition of litchi fruit pericarp. *J Agric Food Chem* 48: 5995–6002.

Shouquin, Z., Jun, X., Changzheng, W. 2005. Effect of high hydrostatic pressure on extraction of flavonoids from propolis. *Food Sci Technol Int* 11: 213–216.

Shouquin, Z., Ruizhan C., Changzheng, W. 2007a. Experiment studies on ultra high pressure extraction of ginsenosides. *J Food Eng* 79: 1–5.

Shouquin, Z., Bi, H., Liu, C. 2007b. Extraction of bioactive compounds from *Rhodiola sachalinensis* under ultra high hydrostatic pressure. *Separ Purif Technol* 57: 277–282.

Sivakumar, D., Naude, Y., Rohwer, E., Korsten, L. 2008. Volatile compounds, quality attributes, mineral composition and pericarp structure of South African litchi export cultivars Maurities and MacLean's Red. *J Sci Food Agric* 88: 1074–1081.

Sun, J., Jiang, Y., Shi, S., Wei, X., Xue, S.H., Shi, J., Yi, C. 2010. Antioxidant activities and contents of polyphenol oxidase substrates from pericarp tissues of litchi fruit. *Food Chem* 119: 753–757.

Trejo Araya, X.I., Hendrickx, M., Verlinden, B.E., Buggenhout, S.V., Smale, N.J., Stewart, C., Mawson, A.J. 2007. Understanding texture changes of high pressure processed fresh carrots: A microstructural and biochemical approach. *J Food Eng* 80: 873–884.

U.S. FDA. 2007. Kinetics of microbial inactivation for alternative food processing technologies-high pressure processing, http://www.fda.gov/Food/ResourcesForYou/Students Teachers/ScienceandTheFoodSupply/ucm183567.htm (Accessed on August 12, 2010).

Xi, J., Shouqin, Z. 2007. Antioxidant activity of ethanolic extracts of propolis by high hydrostatic pressure extraction. *Int J Food Sci Technol* 42: 1350–1356.

Yang, B., Wang, J., Zhao, M., Liu, Y., Wang, W. Jiang, Y. 2006. Identification of polysaccharides from pericarp tissues of litchi (*Litchi chinensis* Sonn.) fruit in relation to their antioxidant activities. *Carbohyd Res* 341: 634–638.

Yang, B., Zhao, M., Shi, J., Yang, E., Jiang, Y. 2008. Effect of ultrasonic treatment on the recovery and DPPH radical scavenging activity of polysaccharides from longan fruit pericarp. *Food Chem* 106: 685–690.

Yang B., Jiang, Y., Wang, R., Zhao, M., Jian, S. 2009. Ultra-high pressure treatment effects on polysaccharides and lignins of longan fruit pericarp. *Food Chem* 112: 428–431.

Zhang, D.L., Quantick, P.C., Grigor, J.M. 2000. Changes in phenolic compounds in litchi (*Litchi chinensis* Sonn.) fruit during posthravest storage. *Postharvest Biol Technol* 19: 165–172.

Zhang, S., Junjie, Z., Changzhen, W. 2004. Novel high pressure extraction technology. *Int J Pharm* 78: 471–474.

Zhang S., Chen, R., Wu, H., Wang, C. 2006. Ginsenoside extraction from *Panax quinquefolium* L. (American ginseng) root by using ultra high pressure. *J Pharm Biomed Anal* 41: 57–63.

Zhao, M., Yang, B., Wang, J., Li, B. Jiang, Y. 2006. Identification of the major flavonoids from pericarp tissues of lychee fruit in relation to their antioxidant activities. *Food Chem* 98: 539–544.

Zhao, M., Yang, B., Wang, J., Liu, Y., Yu, L. Jiang, Y. 2007. Immunomodulatory and anticancer activities of flavonoids extracted from litchi (*Litchi chinensis* Sonn.) pericarp. *Int Immunopharmacol* 7: 162–166.

11 Extrusion-Assisted Extraction: Alginate Extraction from Macroalgae by Extrusion Process

Peggy Vauchel, Abdellah Arhaliass, Jack Legrand, Régis Baron, and Raymond Kaas

CONTENTS

11.1 INTRODUCTION

The extrusion process was historically developed for the polymer and food industries; however, lately, specific capacities of this process were exploited for extractive applications. Extrusion-assisted extraction appears to be of great interest for several applications. Hence, this chapter aims at presenting this new development in the extrusion process.

The first part of this chapter is dedicated to extrusion process presentation. Extrusion principle, specificities, and classical applications are described. A review of extrusion-assisted extractions is then presented. The second part of the chapter focuses on a specific application, namely alginate extraction from macroalgae. A comparative study of both classical batch extraction and extrusion extraction is undertaken, so as to highlight the benefits and limits of extrusion in the case of this specific application.

11.2   EXTRUSION PROCESS FOR EXTRACTION

11.2.1   EXTRUSION PROCESS PRINCIPLE

The extrusion process consists in pushing materials through a die, aided by the pressure induced by one or two rotating screw(s) (single- or twin-screw extruder). The classical extrusion process only implies thermomechanical phenomena. When chemical reactions are induced in the extruder, a reactive extrusion process takes place. It is a complex process combining two distinct fields, classical extrusion and chemical reactions, in a unique operation. The extruder then becomes a continuous chemical reactor.

The main elements composing an extruder are presented in Figure 11.1. Solid material is supplied through a hopper equipped with a feed pump. Liquids are generally injected into incompletely filled zones using an external feed pump. All materials are mixed and conveyed along the screws, where they undergo thermomechanical processes (and chemical processes in case of reactive extrusion) and finally pass through the die.

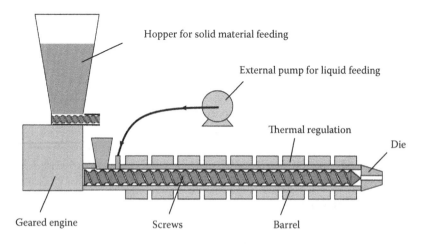

FIGURE 11.1   Main elements of an extruder.

Two types of extruder exist: single- and twin-screw extruders. They mainly differ in terms of transport mechanism and material flow. In a single-screw extruder, material transport is ensured by friction forces, and its efficiency depends on material adhesion to the barrel: if the material sticks to the screw, the material and screw will turn together in the barrel, and the material will not reach the die. In a twin-screw extruder, the first screw channel is constantly cleaned up by the second screw crest, particularly in intermeshing twin-screw extruders (see Figure 11.2). In this case, the extruder behavior can be compared with that of a volumetric pump. This enables working on a larger range of materials, which explains why most reactive and extractive applications are conducted in twin-screw extruders.

The complex geometry of twin-screw extruders, compared to single-screw extruders, leads to difficulties in the comprehension of involved phenomena. However, it also brings many advantages, particularly intense shearing and mixing (axial and radial), which is a key point for reactive and extractive applications. Moreover, the screw profile can be adapted to each application by combining modular screw elements. A screw profile is generally composed of decreasing direct pitch screw elements (from the feeding zone to the die), and few restrictive elements such as reverse screw, mixing, or shearing elements (see Figure 11.2). It is possible to create different zones along the screws, enabling to conduct several operations in the extruder. Each zone corresponds to a specific stage in the process, such as shearing/melting, defibering, chemical reaction, extraction, degassing, and solid–liquid separation. Indeed, reactants or solvents can be introduced at several points along the screws, enabling to conduct chemical or extractive operation in chain. Several degassing points or solid–liquid separation zones can also be created along the screws. Solid–liquid separation equipment for extruders have been specifically developed for extractive applications on bioresources (seeds, vegetables, etc.). Solid–liquid separation is performed through perforated barrel sections: the liquid phase passes through perforations and exits the extruder, while the solid phase is conveyed along the screws. Perforated barrel sections can be inserted anywhere along the barrel, but they are generally used for the final barrel section.

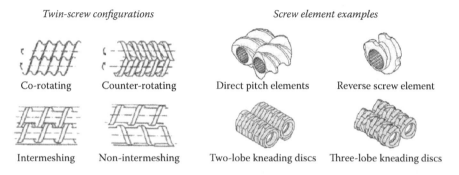

FIGURE 11.2 Twin-screw configurations and screw element examples. (From Vergnes, B. and Chapet M., *Extrusion: Procédés d'extrusion bivis*. Paris: Techniques de l'Ingénieur, AM3653, 2000. With permission.)

11.2.2 Influencing Parameters in Extrusion Process

Two main kinds of parameters have to be taken into account for the extrusion process, namely geometrical and control parameters. The geometrical parameters are screw profile and die design. Restrictive elements can be inserted in the screw profile to enhance mixing and shearing. The main function of die design is shaping of the outcoming product, which is particularly important for food and polymer applications. However, it also has an influence on phenomena taking place within the extruder: the more the die design is restrictive, the more flow is constrained and the higher the pressure becomes within the extruder, again enhancing mixing and shearing. Geometrical parameters cannot be modified during extrusion, contrary to control parameters, namely screw speed, material feed rate, and barrel temperature. Many works have been conducted to understand how these parameters influence extrusion operations. Intrinsic variables can be defined, such as residence time distribution (RTD) or specific mechanical energy (SME), and are much more used to help understand and control extrusion operations. Evolution of these variables depends on the combined effects of several control parameters. Particles incoming in a reactor do not all take the same way to the exit, and consequently do not all stay the same time in the reactor. RTD enables to describe those different residence times in the reactor. It is a classical tool used to characterize flow in a reactor. As regards extrusion, RTD curves are mainly influenced by screw speed, feed rate, screw profile, and die design. Residence time in extruders is generally in the order of a few minutes, and usually never exceeds a dozen minutes. Another important variable for the extrusion process is SME, which corresponds to energy transmitted to extruded material per mass unit. It is defined as the ratio of power supplied by the extruder to material feed rate. It enables to evaluate transformation level of materials, particularly for applications where texturing and shearing are key points. SME values in extrusion are generally in the order of a few hundred kJ kg^{-1} and can reach up to a thousand of kJ kg^{-1} depending on the extruded material. It is also important to know about the physicochemical properties of the extruded material to understand material transformation phenomena and to control the quality of the outcoming product.

11.2.3 Specificities of Extruder as a Reactor

The extrusion process presents several interesting characteristics regarding reactive and extractive applications. One of them is its unique ability to deal with high-viscosity materials, which enables to considerably limit solvent consumption. Solvent saving also induces gains in terms of effluent treatment and process safety (Berzin and Hu 2004).

Another specificity of extruders is their geometry. First, extruders present the advantage of being continuous reactors. Moreover, the screw profile is built with modular screw elements, which enables adaptation to each application. Different zones can be created along the screws with specific functions, such as conveying, mixing, shearing, chemical reactions, and other processes (Xanthos 1992; Berzin and Hu 2004).

Extruders are also interesting considering thermal regulation. Indeed, heat transfer is efficient within material bulk owing to mass transfer from one screw to another, and between material and barrel owing to the high surface to volume ratio

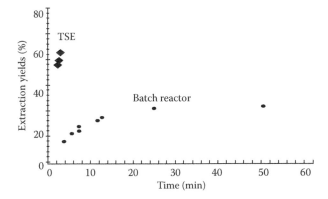

FIGURE 11.3 Comparison of extraction yields and extraction times for hemicellulose extraction from poplar wood in twin-screw extruder (TSE) and in batch reactor. (From Prat, L. et al., *Chem Eng Process*, 41, 743–751, 2002. With permission.)

characterizing this device. Hence, applications requiring thermal regulation can be conducted in extruders. Only highly exothermic reactions and/or important viscous dissipation phenomena may bring problems.

Residence time in an extruder is necessarily short, in the order of a few minutes. Reactions or extractions characterized by too slow kinetics may not be good candidates for extrusion. Conditions within extruders are, however, very specific (pressure, temperature, shearing, mixing, etc.), and operations requiring several hours in a batch reactor may require only a few minutes in an extruder. For example, Prat et al. (2002) compared extrusion and batch processes for hemicellulose extraction from poplar wood: extrusion enabled extraction time reduction and extraction yield enhancement (see Figure 11.3).

Table 11.1 sums up most of the advantages and limits of the extrusion process for reactive and extractive applications.

11.2.4 Extrusion Applications

11.2.4.1 Classical Applications of Extrusion

The extrusion process results from a technology transfer between the polymer and food industries. Both single- and twin-screw extrusion have been developed in parallel in these two industrial fields.

Extrusion is largely used in the food industry to conduct operations combining mixing, shearing, cooking, texturing, and shaping. Three main kinds of applications can be distinguished:

- Simple extrusion, which involves shaping food products (pastas, candies, chewing gums, etc.)
- Extrusion cooking, which is mainly used for cereal products or pet food (breakfast cereals, snacks, bread, etc.)
- Reactive extrusion (modified starch, casein–caseinate conversion, etc.)

TABLE 11.1

Advantages and Limits of Reactive and Extractive Extrusion Process

Advantages	Limits
• High mixing capacity, even with highly viscous materials, enabling reduction of reaction or extraction time	• Corrosive products are banned
	• Highly exothermic phenomena are banned
• Efficient heat transfer between barrel and material and in material bulk owing to mass transfer between screws	• Diffusion is less efficient in a viscous material (extrusion) than in a diluted material (batch)
	• Residence time being reduced, extraction and reaction kinetics have to be fast
• Limitation of solvent consumption	
• Possibilities to inject or to degas products all along the screws	
• High tolerance to temperature and pressure (up to 400–500°C and 500 bars)	
• Continuous process	
• Solvents are confined (process and environment safety)	

Source: Berzin, F., Hu, G.H., Procédés d'extrusion réactive, *Techniques de l'Ingénieur* AM3654, 2004; Ducatel, H., Stadler, T., L'extrusion réactive: Un nouveau procédé adapté à la chimie «verte», in *Proceedings of Congress «50 ans d'innovation et de service—Groupe Clextral»*, pp. 191–205. Firminy: Clextral Group, 2006; Lieto, J., Intensification des procédés. Extrusion réactive, in *Proceedings of Congress «50 ans d'innovation et de service—Groupe Clextral»*, pp. 143–161. Firminy: Clextral Group, 2006.

Extrusion is also largely used in the polymer industry. It can transform polymers, initially under the shape of granules, into a homogeneous melt phase, and shape it (using the die) so as to produce shaped pieces, pipes, sheets, thin films, and others (Colonna and Della Valle 1994). Polymer blendings and alloys are also produced in extruders. Reactive extrusion enables to create new products and to conduct complex operations such as mass polymerization, polycondensation, chemical and rheological modification of polymers, dynamic vulcanization, polymer recycling, and others (Berzin and Hu 2004).

Apart from these two industrial fields, historically linked to extrusion process development, many reactive extrusion applications have been developed in the past few decades. The specificities of the extruder as a chemical reactor have been exploited to get around problems encountered for some applications with classical technologies: low yields, incomplete reactions, long reaction times, important solvent consumption leading to important effluent reject, problems with product quality control, process security, and others. Several published works highlight the benefits brought by extrusion. Some examples are presented in Table 11.2.

11.2.4.2   Extractive Applications of Extrusion

This chapter being dedicated to extrusion-assisted extraction, a special focus is given for extractive applications of extrusion. Extruders were originally not designed for

TABLE 11.2

Examples of Reactive Extrusion Applications and Associated Benefits and Limits

Chemical Reaction Types	Applications	Benefits and Limits
Depolymerization	Oligomer production from starch, guar, cellulose	Continuous process
Sugar hydrolysis	Levulinic acid production	Reaction time reduced, satisfying yield; corrosion problems
	Flavor precursors (Maillard reaction)	Reaction time reduced by eightfold without loss in yield
Specific grafting	Continuous halogenization	Substrate degradation and secondary reactions reduced
	Modified starch transglycosylation	Reactant consumption considerably reduced, high yields, solvent free
Acetylation	Starch acetate production	Gelatinization yield enhanced and secondary reactions reduced
Oxidation	Oxidized starch production	Hydrosoluble compound production, rapid and continuous process
Saponification	Industrial soap production	Continuous process, better work conditions, reaction control enhanced
Cross-linking	Phosphorylated starch production	Phosphorylation yield enhanced, reaction time reduced, no drying
Polymerization	Gluco-oligosaccharides and polydextrose production from glucose	Selective control of polymerization, rapid and continuous reaction
	Oligocondensation of phenolic resins	Continuous process, better work conditions, reaction control enhanced
	Production of high polymers from ε-caprolactone	High molecular weight ("on demand")
	Copolymerization of starch with cationic methacrylate, acrylamide, and acrylonitrile	Reaction time reduced, polymerization yield enhanced, sustainable process
Acid–base reaction	Casein to caseinate conversion	Energy consumption reduced (easier drying, more concentrated solution), investment costs reduced
Chemical treatment and whitening	Natural fibers treatment (cotton, flax, hemp, etc.)	Energy, reactants, and solvent consumption reduced; high yields; no wetting agent use

Source: Ducatel, H., Stadler, T., L'extrusion réactive: Un nouveau procédé adapté à la chimie «verte», in *Proceedings of Congress «50 ans d'innovation et de service—Groupe Clextral»*, pp. 191–205. Firminy: Clextral Group, 2006.

extraction operations. Extrusion process specificities, however, appeared interesting for this kind of application, particularly in the case of extractions from bioresources. Indeed, extruders enable to combine mechanical, thermal, and chemical treatment in the same continuous reactor. Destructuring of biological raw materials and intense mixing provided by screws both facilitate extraction (heat and mass transfers are improved). A few significant examples are given below.

Several works have been conducted on oil extraction from agroresources. In these works, extruders were used to replace classical extraction press, enabling to carry out thermomechanical and chemical treatments (solvent, acid, alkaline) in the same continuous step. Most of these works have been conducted on sunflower oil recovery. Dufaure et al. (1999) and Evon et al. (2007) first worked on sunflower seeds, studying the feasibility of aqueous extraction by twin-screw extrusion and the influence of operating conditions. Then they focused on extraction of residual oil from sunflower press cakes (Evon et al. 2009). Amalia Kartika et al. (2006) also worked on oil extraction from sunflower by twin-screw extrusion. They studied the influence of screw profile and operating conditions on the extraction performance, in terms of extraction yield, oil quality, and SME. In a second work (Amalia Kartika et al. 2010), they proposed to combine thermomechanical treatment of sunflower seeds and solvent extraction, both conducted in an extruder, in a single step. Roque (2009) worked on lesquerella seed oil extraction by extrusion cooking. Freitas et al. (1997) worked on soybean oil extraction. The proposed process involves beans extrusion followed by enzymatic treatment in aqueous phase. The extrusion step before enzymatic treatment enabled to enhance oil recovery and final-product quality and to reduce the environmental impact of the process (no solvent use). Bouzid (1996) worked on aromatic plants (celery fruit and nepeta plants) for essential oils recovery by continuous distillation using the extrusion process. Tandon et al. (2010) also used extrusion to extract and eliminate oil. *Cleome viscosa* seeds have to be defatted before extraction of bioactive molecules (e.g., those with liver-protective activity). Extrusion was used to conduct the defatting pretreatment and enabled to enhance the extraction yield (about 10 times higher) of bioactive molecules. Extrusion is also used as a pretreatment in a work by Jung and Mahfuz (2009). Extrusion before enzyme-assisted extraction of soybean flakes showed some interesting results in terms of oil and protein recovery in specific conditions.

Other kinds of molecules have also been extracted using extrusion. N'Diaye and colleagues worked on hemicellulose alkaline extraction from poplar wood by reactive extrusion (N'Diaye et al. 1996; N'Diaye and Rigal 2000). Extraction yields were similar with extrusion and batch processes, but extraction time and water consumption were sharply reduced with extrusion. Moreover, two operations were conducted simultaneously in the extruder, extraction and liquid–solid separation. Hemicellulose recovery was also conducted by extrusion on wheat straw and bran by Zeitoun et al. (2010). It appeared that twin-screw extrusion resulted in a lower extraction yield compared with stirred reactor, but it has the advantages of a shorter residence time for the material and a lower reactant and water consumption. Lamsal et al. (2010) also worked on lignocellulosic material, namely wheat bran and soybean hull. Extrusion was used as a thermomechanical pretreatment for lignocellulosic ethanol production and showed an interesting potential. In the same field of applications, Karunanithy and Muthukumarappan (2010) used extrusion as a preatreatment on switchgrass before enzymatic hydrolysis to recover fermentable sugars. One can also cite extrusion use for pectin extraction from sunflower culture by-products (Maréchal and Rigal 1999), xylan and arabinoxylan extraction (Rigal et al. 1998), protein extraction (Silvestre et al. 1999), and alginate extraction from macroalgae (Vauchel et al. 2008). This last example is fully detailed in the next section.

11.3 EXAMPLE OF EXTRUSION-ASSISTED EXTRACTION: ALGINATE EXTRACTION FROM MACROALGAE

In this section, an example of an application of extrusion-assisted extraction is presented. The work by Vauchel et al. (2008) on alginate extraction from macroalgae has been chosen as a comparison between the classical batch process used in the industry and the extrusion process as an alternative, enabling to point out advantages and limits of extrusion. Hence, the classical batch process and its limits are first exposed, and then evaluation of extrusion as an alternative process for alginate extraction is presented.

11.3.1 CLASSICAL EXTRACTION OF ALGINATES IN BATCH PROCESS: PRINCIPLE AND LIMITS

Alginates are naturally present in the cell wall of brown seaweeds (Kloareg and Quatrano 1988). These polysaccharides show interesting rheological properties and are widely used in various fields of industry, including textiles, agri-foods, paper, cosmetics, and pharmaceuticals (Pérez et al. 1992). In seaweeds, alginates are present in the form of insoluble salts, mainly sodium alginate (NaAlg), calcium alginate ($CaAlg_2$), and magnesium alginate ($MgAlg_2$). The extraction principle lies in converting these insoluble salts into soluble sodium alginate. The extraction protocol in the alginate industry is divided into five steps: acidification, alkaline extraction, solid–liquid separation, precipitation, and drying (Pérez 1997). These steps are presented in Figure 11.4.

Different conservation modes exist for seaweeds: formalin, freezing, and drying. Formalin treatment is mainly used because it is the cheapest alternative, it enhances final product whiteness (seaweed depigmentation), and it helps eliminating phenols. Seaweeds are rinsed to eliminate formol before lixiviation. During this acid pretreatment, alginate salts are converted into alginic acid (HAlg), according to chemical reactions described in Equations 11.1 through 11.3 (in the case of sulfuric acid use).

$$CaAlg_2 + H_2SO_4 \rightarrow 2\ HAlg + CaSO_4 \tag{11.1}$$

$$2\ NaAlg + H_2SO_4 \rightarrow 2\ HAlg + Na_2SO_4 \tag{11.2}$$

$$MgAlg_2 + H_2SO_4 \rightarrow 2\ HAlg + MgSO_4 \tag{11.3}$$

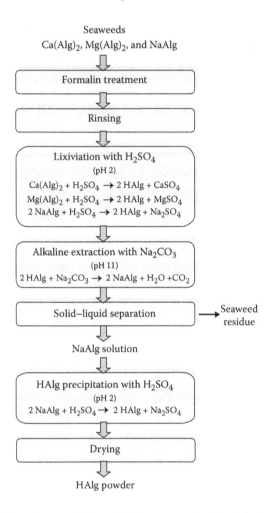

FIGURE 11.4 Schematic presentation of batch process steps for alginate extraction.

Calcium, sodium, and magnesium ions are eliminated in the form of sulfates in rinsing solutions (two or three rinsing water baths). The alkaline extraction step involves immersing acidified seaweeds in a sodium carbonate solution. Insoluble alginic acid is converted into soluble sodium alginate, which passes into the aqueous phase (Equation 11.4).

$$2\,HAlg + Na_2CO_3 \rightarrow 2\,NaAlg + H_2CO_3 \qquad (11.4)$$

During this step, sodium carbonate plays a dual role, disintegrating seaweeds and reacting with alginic acid. At the end of this step, a sodium alginate solution with algal residue particles in suspension is obtained. Solid–liquid separation is then conducted to eliminate algal residues. Sodium alginate solution is finally acidified so as to transform soluble sodium alginate into insoluble alginic acid (Equation 11.5).

$$2\,NaAlg + H_2SO_4 \rightarrow 2\,HAlg + Na_2SO_4 \qquad (11.5)$$

Alginic acid precipitate is recovered and dried. Alginic acid powder can then be treated with suitable bases to produce different commercial alginate salts.

Alkaline extraction is the main step in alginate extraction as it corresponds to the extraction phase itself. This step, which is conducted in a batch reactor, requires high quantities of reactant and water, and several hours are required to attain the optimum extraction yield, depending on the seaweed species concerned (Pérez 1997). The alkaline extraction time is thought to influence the rheological properties of alginates: the reaction conditions seem to favor bacterial development and endogenous alginate lyase activity, the likely cause of alginate degradation (Smidsrød et al. 1963; Moen et al. 1997). Shortening alkaline extraction time while still producing alginate of high rheological quality would undoubtedly be of interest to the alginate industry.

11.3.2 Alginate Extraction by Extrusion Process

This section presents the evaluation of extrusion as a new process for the alkaline extraction of alginates from *Laminaria digitata*. A comparative study of both classical batch extraction and extrusion extraction is undertaken, taking into account several significant points: extraction yield, reactant and water consumption, alginate quality (purity and rheological properties), and the time required.

11.3.2.1 Alginate Extraction Experiments

The extraction protocol was a laboratory adaptation of the industrial process described by Pérez et al. (1992) (see Section 11.3.1). All experiments were conducted on *L. digitata* fronds cut into small pieces. Seaweed pieces were acidified before alkaline extraction. Batch alkaline extractions were conducted in stirred beakers. Extrusion experiments were conducted on a corotative twin-screw extruder (BC21 type provided by Clextral, Firminy, France).

A simple screw profile, composed of decreasing direct pitch screw elements and a small reverse screw element (Figure 11.5), and a 4-mm-diameter and 5-cm-long cylindrical die were used. Algae pieces were introduced to the hopper, and the feed

C2FC12.5/25

T2F100/50   C2F200/33.3   C2F225/25   C2F325/16.6   C2F25/16.6

FIGURE 11.5 Screw profile. Screw elements nomenclature: letters correspond to the screw element type, and numbers correspond to the element length and pitch. T2F, trapezoid groove transfer elements (direct pitch); C2F, U groove transfer elements (direct pitch); C2FC, U groove reverse pitch element; 100/50, 100-mm-long element with a 50 mm pitch.

rate was regulated by means of a feed pump. An external volumetric pump was used to supply the extruder with a 4% (w/w) Na$_2$CO$_3$ solution. As alginate starts to degrade at 40°C, the barrel temperature was maintained at 20°C.

During extrusion, seaweeds and reactant solution were introduced at the barrel opening and carried along the screws. Screws provide an important shearing function, which enables to cut seaweeds in small particles and attain intensive mixing with the reactant. Flowing is constrained owing to the die and restrictive screw elements. At the die exit, a viscous paste is obtained, containing seaweed residues suspended in sodium alginate solution. This paste is then treated, as described by Vauchel et al. (2008), to obtain alginic acid powder: solid–liquid separation by centrifugation, precipitation, drying, and milling.

A specific working point was chosen within the alginate extrusion working field to conduct experiments for comparison of the batch and extrusion processes. This working point, corresponding to the maximum alginate extraction yield, was defined as follows: $Q_a = 1$ kg h^{-1}, $Q_s = 1$ l h^{-1}, and $N = 400$ rpm, where Q_a, Q_s, and N are the algae feed rate, sodium carbonate solution feed rate, and screw speed, respectively.

11.3.2.2 Analysis for Evaluation of Batch and Extrusion Processes

To determine extraction yields, sodium alginate content in solutions (centrifugation supernatants) was quantified according to the method described by Kennedy and Bradshaw (1984). The extraction yield was calculated as the ratio of the dry weight of alginic acid extracted to the dry weight of algae used for the extraction.

The purity of alginic acid powder was analyzed by determining the uronic acid content according to the protocol by Blumenkrantz and Asboe-Hansen (1973), as modified by Tullia et al. (1991). Purity was calculated as the ratio of the dry weight of uronic acids quantified to the dry weight of alginic acid powder used for the quantification.

The intrinsic viscosity of different samples was assessed by means of capillary viscosimetry. The elution time for each sample was measured for different dilutions in a Ubbelohde capillary viscosimeter. Reduced viscosity versus concentration curves were drawn, and intrinsic viscosity was estimated by extrapolating reduced viscosity when the concentration tends toward 0. The model proposed by Mancini et al. (1996) was chosen to estimate the average molecular weight (Equation 11.6).

$$[\eta] = 1228 \cdot 10^{-5} \cdot M_w^{0.963} \tag{11.6}$$

where $[\eta]$ (1 g^{-1}) is the intrinsic viscosity and M_w (Da) is the average molecular weight.

11.3.3 BATCH AND EXTRUSION COMPARISON FOR ALGINATE EXTRACTION

To compare the batch process with the extrusion process for the alkaline extraction of alginate from *L. digitata*, significant criteria had to be identified: extraction yield, water consumption, reactant consumption, and product quality were retained. Product quality is characterized by purity and rheological properties. Both batch and extrusion experiments were conducted with the aim of maximizing extraction yield. Hence, the results presented in Table 11.3 correspond to minimum values for water and reactant consumption, allowing the maximum extraction yield to be attained.

TABLE 11.3

Comparison of Batch and Extrusion Processes for Alkaline Extraction of Alginates from *Laminaria digitata* (Average Values of Triplicates)

Process Type	Batch	Extrusion
Extraction yield (%) (95% confidence bounds)	33.3 ($\pm$1.8)	38.5 ($\pm$1.9)
Na_2CO_3 consumption (kg kg^{-1} dry algae)	0.5	0.2
Water consumption (l kg^{-1} dry algae)	25	10
Time required (min)	60	5
Alginic acid powder purity (%) (95% confidence bounds)	97 ($\pm$1.2)	96 ($\pm$1.4)

An important advantage provided by the extrusion process was large savings in process time: reaction time was decreased from an hour to only a few minutes. Moreover, the extraction yield was more than 15% higher than that obtained using the batch process (relative enhancement). Water and reactant consumption were both significantly reduced to less than half the consumption values recorded for the batch process. It is notable that results presented for the batch process were obtained with a laboratory adaptation of the industrial process. The laboratory batch was an optimized version of the industrial batch. No complete data are available on the industrial process, but differences between the extrusion process and industrial batch process may be even more important, particularly water consumption. Large quantities of water are indeed used in the alginate extraction industry, and savings brought by the extrusion process may be a significant improvement. It also seems that industrial extraction yields may be smaller in the industrial batch process than in the laboratory adaptation, so that extraction yield gain brought by extrusion may be more significant.

Concerning the quality of obtained products, the purity of alginic acid powder was equivalent for both batch and extrusion processes, with values being always higher than, or equal to, 95%. Rheological properties of alginates from both processes were studied by determining their intrinsic viscosity, which is directly linked to their average molecular weights.

Reduced viscosity versus concentration curves for alginates from batch and extrusion processes, and of a high-viscosity commercial alginate, are displayed in Figure 11.6. Alginates extracted by extrusion clearly have superior rheological characteristics, as high-viscosity solutions can be obtained at low concentrations. Besides saving time, this is the key advantage of the alternative process.

These results are confirmed by intrinsic viscosity and average molecular weight values (Table 11.4): extrusion yields alginates with chains that are almost three times longer than those produced using the batch process. This may partly explain the enhanced extraction yield, as precipitation is more efficient with longer molecular chains.

All these enhancements may be due to the high level of shearing and mixing efficiency of the twin-screw extrusion process: algae pieces are destructured and mixed thoroughly with the reactant, therefore considerably reducing the time, water, and

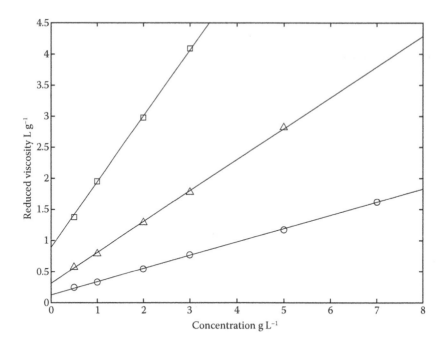

FIGURE 11.6 Reduced viscosity versus concentration curves. (○) High-viscosity commercial sodium alginate, (△) sodium alginate produced by batch extraction, and (☐) sodium alginate produced by extrusion extraction (linear correlations are represented as a solid line).

reactant requirements. Reducing the reaction time helps reduce depolymerization, and consequently an alginate of high rheological quality is obtained.

Globally, results show that the extrusion process is more efficient than the batch process for the alkaline extraction of alginates from *L. digitata* in several key ways: reduced time, water, and reactant requirements; enhanced extraction yield and rheological properties; and high purity of the product. Moreover, extrusion has the advantage of being a continuous process. This makes extrusion an interesting technique

---

TABLE 11.4

Intrinsic Viscosity and Average Molecular Weight of High-Viscosity Commercial Sodium Alginate and Sodium Alginate Produced by Batch and Extrusion Extractions

Sodium Alginate Type	Intrinsic Viscosity ($l\ g^{-1}$) (95% Confidence Bounds)	Average Molecular Weight (Da) (95% Confidence Bounds)
Commercial product (high viscosity)	0.1264 (±0.0248)	14 680 (±2706)
Batch extraction	0.2981 (±0.0529)	35 781 (±5941)
Extrusion extraction	0.8443 (±0.1204)	105 476 (±13,957)

for alginate extraction, both economically and environmentally. The equipment cost may be a drawback for mass production of alginates by extrusion as compared with the batch process, but it could be a promising alternative for the production of high-quality alginates.

11.4 CONCLUSION

In this chapter, recent developments in the use of extrusion process for extractive applications have been presented. The principle and specificities of the extrusion process have been exposed, so as to understand the benefits of this technology in conducting extraction operations. The key point lies in the extruder's geometry. Screws provide intense shearing and mixing. These conditions are favorable to transfer processes (mass and heat), which are key elements in extraction operations. Hence, extruders have obvious abilities to conduct extractions from solid raw material. This is confirmed by a review of works published on extractive applications of extrusion. All of them exploit the extruder's ability to conduct both fractionation of solid raw material and chemical treatment in the same continuous step. In the last section, an example of extrusion-assisted extraction has been detailed: alginate extraction from macroalgae. The extrusion process is compared to the classical batch process in several key ways: time demand, water, and reactant requirements are all reduced, whereas the extraction yield and rheological properties of the product are both enhanced. Hence, extrusion appears as an interesting alternative process in this case. All these elements and examples show that extrusion may be of great interest for many extraction operations from solid raw materials. Extrusion-assisted extraction appears as an interesting alternative to classical extraction processes and may help develop more sustainable processes.

ACKNOWLEDGMENTS

The authors acknowledge financial support from CNRS (National Center for Scientific Research) and IFREMER (French Research Institute for Exploitation of the Sea).

REFERENCES

Amalia Kartika, I., Pontalier, P.Y., Rigal, L. 2006. Extraction of sunflower oil by twin screw extruder: Screw configuration and operating condition effects. *Bioresour Technol* 97: 2302–2310.
Amalia Kartika, I., Pontalier, P.Y., Rigal, L. 2010. Twin-screw extruder for oil processing of sunflower seeds: thermo-mechanical pressing and solvent extraction in a single step. *Ind Crop Prod* doi: 10.1016/j.indcrop.2010.05.005.
Berzin, F., Hu, G.H. 2004. Procédés d'extrusion réactive. *Techniques de l'Ingénieur* AM3654.
Blumenkrantz, N., Asboe-Hansen, G. 1973. New method for quantitative determination of uronic acids. *Anal Biochem* 54: 484–489.
Bouzid, N.E. 1996. Distillation continue de plantes aromatiques en extrudeur bi-vis. PhD dissertation. Toulouse: Institut National Polytechnique de Toulouse.
Colonna, P., Della Valle, G. 1994. *La Cuisson-Extrusion*. Paris: Editions Lavoisier Tec&Doc.

Ducatel, H., Stadler, T. 2006. L'extrusion réactive: Un nouveau procédé adapté à la chimie «verte», in *Proceedings of Congress «50 ans d'innovation et de service—Groupe Clextral»*, pp. 191–205. Firminy: Clextral Group.

Dufaure, C., Mouloungui, Z., Rigal, L. 1999. A twin-screw extruder foir oil extraction: II. Alcohol extraction of oleic sunflower seeds. *J Am Oil Chem Soc* 76: 1081–1086.

Evon, P., Vandenbossche, V., Pontalier, P.Y., Rigal, L. 2007. Direct extraction of oil from sunflower seeds by twin-screw extruder according to an aqueous extraction process: Feasibility study and influence of operating conditions. *Ind Crop Prod* 26: 351–359.

Evon, P., Vandenbossche, V., Pontalier, P.Y., Rigal, L. 2009. Aqueous extraction of residual oil from sunflower press cake using twin-screw extruder: Feasibility study. *Ind Crop Prod* 29: 455–465.

Freitas, S.P., Hartman, L., Couri, S., Jablonka, F.H., de Carvalho, C.W.P. 1997. The combined application of extrusion and enzymatic technology for extraction of soybean oil. *Fett/ Lipid* 99: 333–337.

Jung, S., Mahfuz, A.A. 2009. Low temperature dry extrusion and high-pressure processing prior to enzyme-assisted aqueous extraction of full fat soybean flakes. *Food Chem* 114: 947–954.

Karunanithy, C., Muthukumarappan, K. 2010. Optimisation of switchgrass and extruder parameters for enzymatic hydrolysis using response surface methodology. *Ind Crop Prod*, doi:10.1016/j.ind.crop.2010.10.008.

Kennedy, J.F., Bradshaw, I.J. 1984. A rapid method for the assay of alginates in solution using polyhexamethylenebiguanidium chloride. *Brit Polym J* 16: 95–101.

Kloareg, B., Quatrano, R.S. 1988. Structure of the cell walls of marine algae and ecophysiological functions of the matrix polysaccharides. *Oceanogr Mar Biol* 26: 259–315.

Lamsal, B., Yoo, J., Brijwani, K., Alavi, S. 2010. Extrusion as a thermo-mechanical pre-treatment for lignocellulosic ethanol. *Biomass Bioenergy* 34: 1703–1710.

Lieto, J. 2006. Intensification des procédés. Extrusion réactive, in *Proceedings of Congress «50 ans d'innovation et de service—Groupe Clextral»*, pp. 143–161. Firminy: Clextral Group.

Mancini, M., Moresi, M., Sappino, F. 1996. Rheological behaviour of aqueous dispersions of algal sodium alginates. *J Food Eng* 28: 283–295.

Maréchal, V., Rigal, L. 1999. Characterization of by-products of sunflower culture-commercial applications for stalks and head. *Ind Crop Prod* 10: 185–200.

Moen, E., Larsen, B., Ostgaard, K. 1997. Aerobic microbial degradation of alginate in *Laminaria hyperborea* stipes containing different levels of polyphenols. *J Appl Phycol* 9: 45–54.

N'Diaye, S., Rigal, L., Larocque, P., Vidal, P.F. 1996. Extraction of hemicelluloses from poplar, *Populus tremuloides*, using an extruder-type twin-screw reactor: A feasibility study. *Bioresour Technol* 57: 61–67.

N'Diaye, S., Rigal, L. 2000. Factors influencing the alkaline extraction of poplar hemicelluloses in a twin-screw reactor: Correlation with specific mechanical energy and residence time distribution of the liquid phase. *Bioresour Technol* 75: 13–18.

Pérez, R. 1997. *Ces algues qui nous entourent. Conception actuelle, rôle dans la biosphère, utilisations, culture.* Plouzané: Editions IFREMER.

Pérez, R., Kaas, R., Campello, F., Arbault, S., Barbaroux, O. 1992. *La culture des algues marines dans le monde.* Plouzané: Editions IFREMER.

Prat, L., Guiraud, P., Rigal, L., Gourdon, C. 2002. A one dimensional model for the prediction of extraction yields in a two phases modified twin-screw extruder. *Chem Eng Process* 41: 743–751.

Rigal, L., Ioualalen, R., Gaset, A. 1998. Procédé pour obtenir un extrait de son désamylacé, un raffinat et un matériau obtenu à partir de ce procédé. *Patent* WO9831713.

Roque, L.E. 2009. Oil extraction from lesquerella seeds by dry extrusion and expelling. *Ind Crop Prod* 29: 189–196.

Silvestre, F., Rigal, L., Leyris, J., Gaset, A. 1999. Colle à l'eau à base d'extrait protéique végétal et procédé de préparation. *Patent* EP0997513.

Smidsrød, O., Haug, A., Larsen, B. 1963. The influence of reducing compounds on the rate of degradation of alginates. *Acta Chem Scand* 17: 1473–1474.

Tandon, S., Chatterjee, A., Chattopadhyay, S.K., Kaur, R., Gupta, A.K. 2010. Pilot scale processing technology for extraction of Cliv-92: A combination of three coumarinolignoids clemiscosins A, B and C from *Cleome viscosa*. *Ind Crop Prod* 31: 335–343.

Tullia, M., Filisetti-Cozzi, C.C., Carpita, N.C. 1991. Measurement of uronic acids without interference from neutral sugars. *Anal Biochem* 197: 157–162.

Vauchel, P., Kaas, R., Arhaliass, A., Baron, R., Legrand, J. 2008. A new process for extracting alginates from *Laminaria digitata*: reactive extrusion. *Food Bioprocess Technol* 1: 297–300.

Vergnes, B., Chapet, M. 2000. *Extrusion: Procédés d'extrusion bivis*. Paris: Techniques de l'Ingénieur, AM3653.

Xanthos, M. 1992. *Reactive extrusion: Principles and practice*. Munich: Hanser.

Zeitoun, R., Pontalier, P.Y., Marechal, P., Rigal, L. 2010. Twin-screw extrusion for hemicellulose recovery: Influence on extract purity and purification performance. *Bioresour Technol* 101: 9348–9354.

12 Gas-Assisted Mechanical Expression of Oilseeds

Paul Willems and André B. de Haan

CONTENTS

12.1 INTRODUCTION

Present-day technologies for the production of vegetable oils include hydraulic pressing, screw pressing, and solvent extraction with hexane. Of these methods, solvent extraction has the highest yield, but results in a lower quality of the oil. Pressing methods produce higher quality but lower yields. Today, consumers are becoming more aware of the presence of potentially hazardous chemicals in foodstuff. It is therefore necessary to prevent the use of organic solvents in the production of vegetable oils for consumption. Two potential alternatives for oil production are extraction with supercritical carbon dioxide (SCE) and a novel process called gas-assisted mechanical expression (GAME). In the SCE process, the oil is dissolved in CO_2 and thereby extracted from the plant material. However, the solubility is limited to only a few weight percent at reasonable conditions.

In the GAME process, CO_2 is dissolved in the oil contained in the seed cells. Then expression of the oil–CO_2 mixture is done in a hydraulic press or in a screw press. The resultant press cake has a lower residual oil content than that produced by conventional pressing at the same conditions, because part of the oil has been displaced by CO_2. The oil obtained via this process is unfractionated, of high quality, and free of harmful

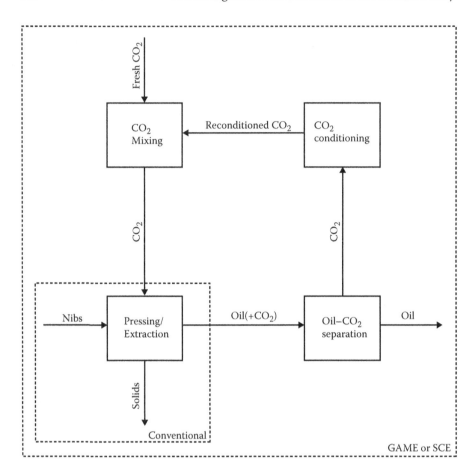

FIGURE 12.1 Unit operations of various oil recovery technologies.

solvents. Compared to SCE, the amount of CO_2 that is needed per kilogram of oil is reduced from roughly 100 kg (SCE) to only a few kilograms (GAME). As one can imagine, this has a significant impact on equipment sizes and economic feasibility of the process. Oil yields can be 30 wt% higher than with conventional presses. A schematic overview of the different technologies is given in Figure 12.1. This chapter will discuss the principle of GAME and show examples of yield improvements and process aspects.

12.2 BACKGROUND

Vegetable oils have been used by mankind for centuries, for both food and non-food applications (Hasenhuettl 1991). Up until the 20th century, the only option to recover oil from seeds has been by conventional mechanical expression. The oil obtained via this method is of a high quality, but the attainable yield is limited to roughly 80 wt% of the oil originally present. The 20th century saw the introduction of solvent extraction. With this method, yields of up to 99% can be obtained, but the oil quality is reduced because of the extensive heat treatment required for the recovery of the

TABLE 12.1

Advantages and Disadvantages of Various Oil Production Processes

Technique	Oil Yield	Oil Quality	Solvent Requirement per Kilogram Oil (kg, Order of Magnitude)	Continuous Process
Mechanical expression	−	+	0	Extrusion: yes Hydraulic pressing: no
Solvent extraction	+	−	2 (organic)	Yes
Supercritical extraction	+	+	100 (CO_2)	Mostly not, but some available
GAME	+	+	1 (CO_2)	Under development

solvent from the product. Undesired components from cell walls are also coextracted with this method. Especially for high value added oils, this quality reduction is unacceptable, limiting the production process to mechanical expression. Maximizing the yield is then limited to optimizing the preconditioning and maximizing the pressure.

A more recent addition to the available methods is extraction with supercritical fluids, mainly CO_2. The principle is the same as with solvent extraction, but a solvent above its critical pressure and temperature is used. In the case of CO_2, this is above 31°C and 7.2 MPa. Solvent removal from the oil is accomplished by reducing the pressure, thus evaporating the solvent. This way solvent residues can be reduced to acceptable levels without high temperature treatment. Also, in the case of CO_2, no organic residues are left in the oil. The disadvantage of this method is the low solubility of vegetable oils in CO_2 at acceptable pressures (roughly below 50 MPa) and the resulting amount of solvent required.

All of the above technologies have some advantages: mechanical expression and supercritical extraction produce high-quality oil, whereas solvent extraction and supercritical extraction give high yields. However, these techniques also have their limitations: mechanical expression has a low yield, solvent extraction results in a reduced oil and meal quality, and supercritical extraction requires a huge amount of CO_2. It is desirable to find a technique that combines the high yields of (supercritical) solvent extraction and the high oil quality of hydraulic pressing and SCE but does not require the large quantities of solvent used in SCE. An overview of the advantages and disadvantages of these techniques is given in Table 12.1, together with the GAME process.

12.2.1 GAS-ASSISTED MECHANICAL EXPRESSION

Some years ago a new technique was described for obtaining butter from cocoa nibs (Clifford 1998; Foidl 1999; Venter 2006). This technique, called gas-assisted mechanical expression (GAME), is a combination of mechanical expression and the use of supercritical CO_2 (hence gas assisted). It was shown that GAME can increase the oil yield from cocoa nibs beyond values obtained in industry at milder conditions than generally used without the need for large quantities of carbon dioxide (Venter 2006).

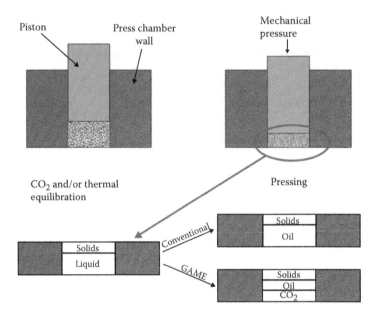

FIGURE 12.2 Principle of GAME. (From Willems, P. et al., *J Supercrit Fluid*, 45, 298–305, 2008. With permission.)

The principle of GAME is illustrated in Figure 12.2. In the GAME process, CO_2 is dissolved in the oil contained in the seeds before pressing. After equilibration, the oil–CO_2 mixture is expressed from the seeds. It was shown for cocoa that the dissolved CO_2 displaces part of the oil during pressing (Venter et al. 2006). It was concluded that at the same effective mechanical pressure (absolute mechanical pressure minus the actual CO_2 pressure, P_{eff}), the *liquid* content is the same in both conventional and GAME press cakes. The liquid in the GAME press cake is saturated with CO_2 (which can be up to 30 wt% CO_2), reducing the oil content compared with the conventional cake by the same amount. The amount of this effect increases with increasing solubility of CO_2 in the oil. Furthermore the dissolved CO_2 reduces the viscosity of oil by an order of magnitude, which increases the rate of pressing. After pressing, the CO_2 content is easily removed from the cake and oil by depressurization. During depressurization of the cake, some additional oil is removed by entrainment in the gas flow, especially when seeds are not dehulled before pressing.

The advantages of GAME include:

- Mechanical pressure can be lower than conventional pressing (50 MPa compared to 100 MPa) while still obtaining higher yields.
- Required CO_2 pressure is lower than for supercritical extraction (10 MPa compared to 45–70 MPa).
- The amount of CO_2 required is much lower than for supercritical extraction (around 1 kg CO_2 per kilogram of oil compared to 100 kg CO_2 per kilogram of oil).

- A totally solvent-free product; accidentally remaining solvent traces are not detrimental to consumer health.
- It is suggested by several authors that CO_2 at the conditions employed has a sterilizing effect (Spilimbergo and Bertucco 2003; White et al. 2006).

12.3 EXPERIMENTAL INVESTIGATIONS

Two extended experimental investigations of GAME are available in literature: one performed at the Technical University of Hamburg, Harburg (Voges 2008) and one from the University of Twente (Venter 2006; Willems 2007). The results from Voges concentrate on uniaxial pressing of flaked rapeseed in batches of 500 g, whereas the experiments from Willems and Venter et al. focused on a variety of seeds, which were generally not flaked, in ~10 g batches. Combined, this provides a wide range of data and conditions. The physical properties of the oil–CO_2 mixtures were also determined to facilitate process design in a later stage.

12.3.1 PHYSICAL PROPERTIES OF OIL–CO_2 MIXTURES

On the basis of available data in literature, a first indication of the magnitude of the effects of CO_2 dissolution can be obtained. As it is assumed that the CO_2 dissolved in the oil is responsible for the increase in oil yield, the solubility of CO_2 in various oils is an important parameter. Figure 12.3 shows the CO_2 content of various vegetable oils as function of pressure. The CO_2 solubility in the different oils is very similar since all the oils consist mainly of triglycerides with similar molecular weight. The

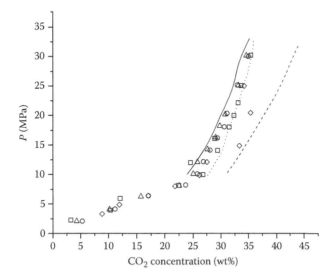

FIGURE 12.3 CO_2 content at 40°C at equilibrium for palm kernel (□), jatropha (○), and linseed (△) oil. Literature values for cocoa butter (Venter et al. 2007) (◇), sesame (Bharath et al. 1992) (-), palm kernel (Bharath et al. 1992) (- -), and rapeseed (Klein and Schulz 1989) (····) oils are included for comparison.

solubility shows a strong increase with pressure up to about 10 MPa, after which the increase levels off. For all the oils, the maximum solubility of CO_2 is around 35 wt% at 30 MPa. The solubility of CO_2 in palm kernel oil is only marginally higher than in jatropha and linseed oils, despite its lower average molecular weight. This is in contrast to results reported in literature, probably due to a difference in triglyceride content. It is expected that all oils will have a similar CO_2 solubility on a mole-per-mole basis and a lower average molecular weight would therefore result in a higher CO_2 solubility on a weight-per-weight basis. Therefore, the triglyceride composition used in this work probably has a higher average molecular weight than the sample used in literature (Bharath et al. 1992). This is supported by the measured density, which is also higher than literature values, as will be shown further on. The solubility in linseed oil is slightly lower because of the high degree of unsaturation of the oil. The CO_2 solubilities in sesame oil (Bharath et al. 1992), rapeseed oil (Klein and Schulz 1989), and cocoa (Venter et al. 2007) from previous works are also very similar to the values presented in Figure 12.3.

Given the solubility of about 30 wt%, a significant impact on especially the viscosity of the mixture can be expected. This is confirmed by the data in Figure 12.4, which shows that at 10 MPa the viscosity is roughly an order of magnitude lower than that of pure oil. As the oils show similar CO_2 contents at a given pressure, the viscosity of the oil–CO_2 mixtures also show comparable behavior. Assuming that oil drainage from the cake is the limiting step, the pressing of oilseeds saturated with CO_2 should be a factor of 10 faster according to the Darcy equation (Voges et al. 2008).

The density of the oil–CO_2 mixtures increased with increasing CO_2 pressure, as shown in Figure 12.5. Density increases almost linearly with pressure above 5 MPa, as was also reported for other oils (Tegetmeier et al. 2000). Compared with the density as function of pressure for pure oils (Acosta et al. 1996), the density of the CO_2-saturated oils shows a stronger increase with pressure. This was also observed by others (Kerst and Schlünder 1998; Dittmar et al. 2002). The majority

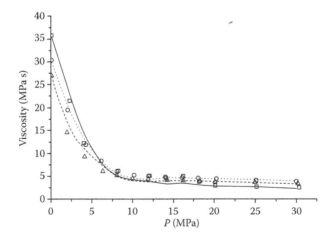

FIGURE 12.4 Viscosity for palm kernel (□,-), jatropha (O,···), and linseed (△,- -) oil at 40°C as function of CO_2–pressure.

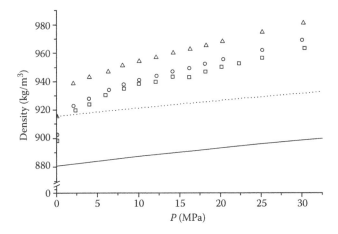

FIGURE 12.5 Density of CO_2 saturated palm kernel ($\square$), jatropha ($\bigcirc$), and linseed ($\triangle$) oil at 40°C as function of CO_2–pressure and density of pure palm kernel (Acosta et al. 1996) (-) and linseed oil (Acosta et al. 1996) ($\cdots$) as function of pressure.

of the density increase can be attributed to the dissolution of the CO_2 and not to the compressibility of the oil (Tegetmeier et al. 2000). The palm kernel oil density measured is much higher than the value reported in literature, which suggests a higher molecular weight of the triglycerides (Bailey et al. 1964). This is consistent with the difference between the measured CO_2 content and literature values.

12.3.2 EXPERIMENTAL SETUP

Up until now, most research into GAME has been performed on uniaxial hydraulic presses, similar to the one shown in Figure 12.6. Several modifications are required to make a conventional hydraulic press suitable for GAME experiments, especially to make the press chamber gas tight. In a typical experiment, seeds are placed on a filter medium in the press chamber and a plunger is lowered onto the seeds. A seal is present between the sides of the plunger and the press chamber walls to prevent gas leakage. The oil collection chamber, which is atmospheric in conventional expression, has to be a closed space. In this case it is achieved by providing two valves in the lines to the chamber: one for CO_2 supply and one for oil drainage after the experiment. During an experiment, the last valve is closed. With the plunger down, a gas-tight volume is achieved and the seeds can be equilibrated with CO_2 before expression. After equilibration, the plunger is lowered with either a preset speed or pressure level. Generally the whole setup is temperature controlled to determine the effect of temperature on the process.

12.3.3 INFLUENCE OF CO_2 PRESSURE ON OIL YIELDS IN GAME

As a significant influence of pressure on CO_2 solubility was found, a significant effect of pressure on oil yield was also expected. Figure 12.7 shows the results of varying

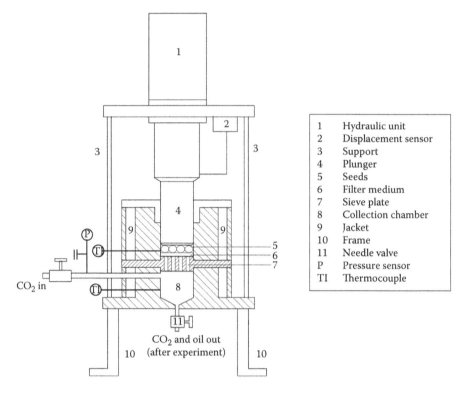

1	Hydraulic unit
2	Displacement sensor
3	Support
4	Plunger
5	Seeds
6	Filter medium
7	Sieve plate
8	Collection chamber
9	Jacket
10	Frame
11	Needle valve
P	Pressure sensor
TI	Thermocouple

FIGURE 12.6   Schematic representation of a hydraulic press. (From Willems, P. et al., *J Supercrit Fluid*, 45, 298–305, 2008. With permission.)

CO_2 pressure on the oil yield for sesame and linseed as well as the yield for conventional expression. Results show that using CO_2 at 8 MPa already increases the oil yield by up to 30 wt% for linseed. Increasing the CO_2 pressure to 10 or 15 MPa shows an additional increase in yield, but it is limited compared with the increase obtained at 8 MPa. This effect can be related to the limited increase in solubility of CO_2 from 8 to 10 and 15 MPa (see Figure 12.3). It was assumed that 10 MPa was a good trade-off between increased yield and increased pressure. Furthermore, the compressibility of CO_2 at 10 MPa is still large, resulting in a limited increase in CO_2 pressure during compression of the seed bed of 1–2 MPa.

12.3.4   INFLUENCE OF PROCESS PARAMETERS ON YIELD

Additional data for rapeseed, palm kernel, and jatropha is also available (see Figure 12.8). In general the conclusions from these seeds are similar to the results presented earlier: a large increase in yield is attained by the addition of CO_2 up to 8 MPa, after which the increase is limited (data not shown). The effect of the effective mechanical pressure (P_{eff}) is the same for both conventional expression and GAME: at a lower pressure, a clear increase with yield is observed with increase in pressure. At

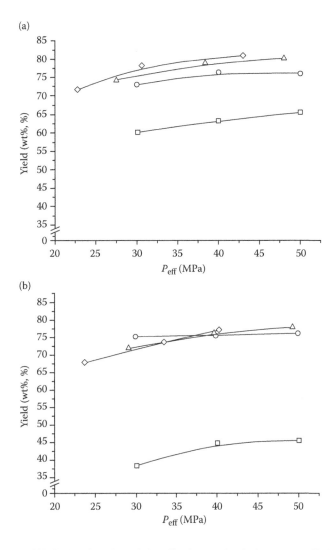

FIGURE 12.7 Oil yield as function of the effective mechanical pressures (P_{eff}) at 40°C for different CO_2 pressures for (a) sesame and (b) linseed: $P_{CO2} = 0$ MPa (□), 8 MPa (○), 10 MPa (△), and 15 MPa (◇). Lines serve as visual aid only. (From Willems, P. et al., *J Supercrit Fluid*, 45, 298–305, 2008. With permission.)

a higher pressure, this increase in yield is less pronounced. Figure 12.8b shows that while in conventional expression the yield of dehulled jatropha is much higher than for hulled jatropha, this difference is reduced for GAME. The oil liberated from the oil cells can be absorbed on the hulls during conventional expression of hulled jatropha. Removing the hulls will prevent absorption and increase the yield. For GAME, either absorption plays a less significant role or the absorbed oil is entrained during depressurization of CO_2.

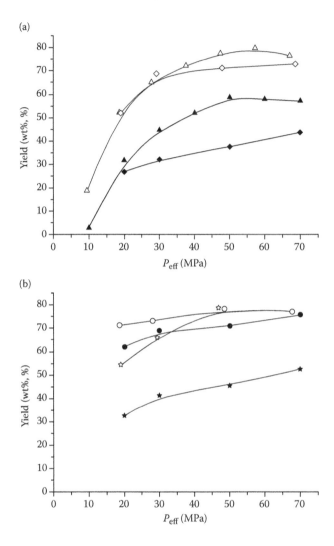

FIGURE 12.8　Oil yield as function of effective mechanical pressure for (a) rapeseed (▲,△) and palm kernel (◊), and (b) jatropha (stars) and jatropha dehulled (pentagrams) at 40°C with 0 MPa CO_2 (closed) and 10 MPa CO_2 (open). Lines serve as visual aid only. (From Willems, P. et al., *J Supercrit Fluid*, 45, 298–305, 2008. With permission.)

In general it can be said that the effect of process parameters such as pressure, temperature, and moisture content have a similar effect on GAME as they have on conventional expression. Since CO_2 only has an effect on the liquid phase and other process parameters generally only influence the solid phase, this was to be expected.

Although the equipment used by Voges et al. has a much larger capacity, it has a similar *l/d* ratio, validating a comparison between results. Owing to the shorter compaction times chosen, the effect of the oil drainage rate is larger in the study by Voges et al. (2008). This results in a clearer influence of temperature and CO_2

pressure (both related to viscosity reduction) and of solid bed height (oil drainage path length). In the experiments performed in our laboratory, the effect of equilibrium CO_2 content on the increased yield is more pronounced.

12.3.5 MULTISTAGE PRESSING

Since continuous pressurization and depressurization is time consuming and put a strain on the equipment, it is expected that industrial implementation of the process will be in a continuous extruder. This process is explained in more detail in

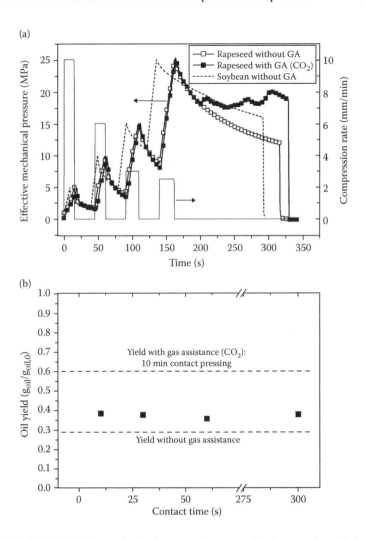

FIGURE 12.9 (a) Effective mechanical pressure for conventional expression and GAME and compression rate as a function of time in four-stage pressing experiments. GA, gas assistance. (b) Oil yield as a function of contact time of CO_2 in four-stage pressing, CO_2 injection after the third stage. (From Voges, S. et al., *Sep Purif Technol*, 63, 1–14, 2008. With permission.)

Chapter 4. In essence this will have to be a two-stage extruder where the solid plug formed at the end of the first and second stage will provide dynamic gas locks (Eggers 1988). This way a closed volume can be created to contain the CO_2. Voges et al. (2008) attempted to get an indication of the performance of such an extruder. To achieve this, they performed four-stage pressure profile in their uniaxial press, with a pressure profile given in Figure 12.9a. After the third pressing stage, they supplied CO_2 to the system at 15 MPa. Yields for various saturation times are given in Figure 12.9b, as well as yields for conventional expression and GAME experiments with saturation before expression. As can be seen from this figure, the oil yield is slightly increased compared with conventional expression for all saturation times. However, the yield for the normal GAME experiments is not reached. This was attributed to the decreased CO_2 permeability of the compressed press cake compared with the uncompressed seed bed. This property enables using the cake as a dynamic gas lock, but also limits the increased yields. In an extruder this can be compensated for by introducing breaking up the plug again or introducing mixing elements in the zone between the two plugs. Therefore, in an extruder design for GAME, an optimum has to be found between compaction to provide the gas lock (and initial oil yield), breakup, and mixing of the cake with CO_2 and additional oil yield in the second stage. The pressure profile and energy requirements of such an extruder need to be considered as well.

12.4 PRINCIPLE

Several effects can contribute to the increased oil yield for the GAME process:

- Decreased viscosity of the oil–CO_2 mixture compared with pure oil, which reduces the energy requirement for oil removal from the cake
- Rupture of the cell walls because of the increased volume of the oil–CO_2 mixture and/or expansion of CO_2 during depressurization
- After liberation from the cells, the oil can be entrained in the CO_2 flow during depressurization
- Increase in volume of the liquid content (oil compared with oil saturated with CO_2), or dilution of the oil content remaining after pressing with CO_2.

These hypotheses have been investigated experimentally by two research groups (Voges et al. 2008; Willems et al. 2008).

A series of experiments in which the order of pressing and CO_2 release was varied was performed to determine the influence of various mechanisms (Willems et al. 2008). Results for sesame and linseed are shown in Figure 12.10. The experiments where CO_2 was dissolved in the seed and released before pressing showed the same yields as the conventional experiments without CO_2. This indicates that cell wall rupture due to increased liquid content does not play a role in increasing the yield. Saturating the press cake with CO_2 after pressing and subsequent release increased the yield for linseed but not for sesame. Entrainment possibly plays a role for linseed by entraining the oil that is absorbed by the hulls after being released from the cells by mechanical expression. This is also confirmed by the entrainment experiments performed by Voges et al. (2008) on rapeseed with a high hull content.

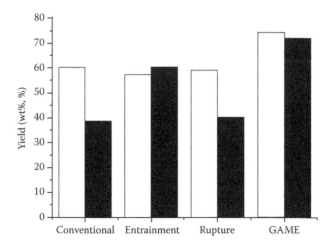

FIGURE 12.10 Yields for conventional, GAME, rupture, and entrainment experiments for sesame (light columns) and linseed (dark columns). (From Willems, P. et al., *J Supercrit Fluid*, 45, 298–305, 2008. With permission.)

As discussed in the previous paragraph, a reduced viscosity of the oil–CO_2 mixture should lead to a faster compression of the press cake assuming liquid drainage is the limiting step in this process. However, data from Willems et al. (2008) show that this is not the case: compression curves of all seeds with and without CO_2 did not significantly differ. This means that solid bed compression is the limiting step in oilseed expression.

The remaining hypothesis for the mechanism is the displacement of the oil by CO_2 in the press cake. For the following discussion it is assumed that for a given effective mechanical pressure, the residual liquid content is the same for both conventional and GAME pressing. In this case the yield for the GAME process can be calculated from the yield for the conventional process and the solubility of CO_2 in the oil for the experimental conditions. Results for this calculation are shown in Figure 12.11. As can be seen from this figure, the oil yield prediction for dehulled seeds (Figure 12.11a) is within experimental error for the dehulled seeds. This confirms the displacement as the major factor for oil yield increase for this type of seeds. For hulled seeds (Figure 12.11b), the prediction underestimates the yields. This is consistent with the additional yield observed for the entrainment experiment shown in Figure 12.10 and data from Voges (2008).

12.5 IMPLEMENTATION

Because of the time required to dissolve the CO_2 in the whole seed, there is a need for a continuous operation to make the process economically feasible. With flaked seeds, on the other hand, maximum yield could already be reached with a dissolution time in the order of minutes (Voges 2008; Voges et al. 2008). A two-stage extruder could be used to flake or crush the seeds and remove part of the oil in the first stage. After this stage, the oilseed paste can be expanded and actively mixed with CO_2 to reduce mass transfer limitations and then pressed for a second time. A schematic of such an extruder is given

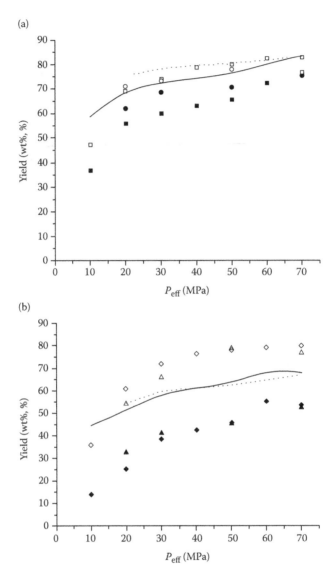

FIGURE 12.11 Prediction of GAME yields (lines) based on conventional yields (closed symbols) and solubility of CO_2 (Bharath et al. 1992; Willems 2007). Experimental GAME yields are also shown. (a) Sesame (■,□,-) and dehulled jatropha (●,○,···), (b) linseed (◆,◇,-) and hulled jatropha (▲,△,···) (P_{eff} = 30 MPa, T = 40°C, P_{CO_2} = 10 MPa). (From Willems, P. et al., *J Supercrit Fluid*, 45, 298–305, 2008. With permission.)

in Figure 12.12. Investigation into the feasibility of such an extruder is ongoing (www .sustoil.org). The major part of the oil recovery is obtained in the two solid plugs, which also serve as seals for the area where supercritical CO_2 is present.

To get an estimate of the performance of such an extruder, a mathematical model was developed (Willems et al. 2009). This is an adaptation of an earlier model by

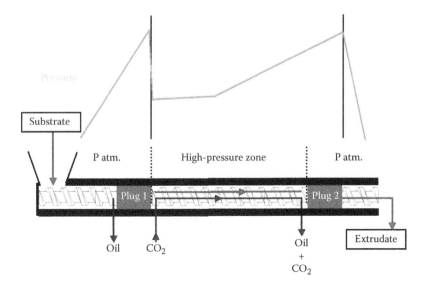

FIGURE 12.12   Schematic of a two-stage extruder for GAME (www.sustoil.org).

Vadke et al. (1988) and describes the oil content and pressure as function of position in an extruder channel. It is originally developed for single-stage expression of crushed oilseed with Darcian filtration of the oil. The description of oil removal was changed to a consolidation-based model to get a more realistic description of the expression of whole seeds. To describe the two-stage extruder, it was assumed that the two stages operate independently and that the model could therefore be run once for the first stage and the outcome of this was used as input for the second stage. The dissolved CO_2 was incorporated by adjusting the calculated liquid volume after the first stage by the appropriate amount. Since oil displacement was identified as the major contributor to the increased yield, this is a valid assumption.

Four cases were studied with the model:

1. Conventional single-stage extrusion
2. GAME extrusion, in which the paste is fully saturated with CO_2 before entering the extruder
3. Two-step conventional extrusion, in which the cake from the first pass is fed to the extruder again
4. Conventional extrusion, after which the cake is saturated with CO_2 and fed to the extruder again.

In all cases the extruder geometry and process conditions were kept constant (Willems et al. 2009). Screw rotational speed was 90 rpm, and feed temperature was 40°C. In both GAME cases, CO_2–pressure was 10 MPa.

Figure 12.13 shows the pressure and *liquid* content profiles for these four cases. In the conventional cases, the liquid content is equal to the residual oil content, whereas in the GAME cases the liquid content is the CO_2-saturated oil content. For

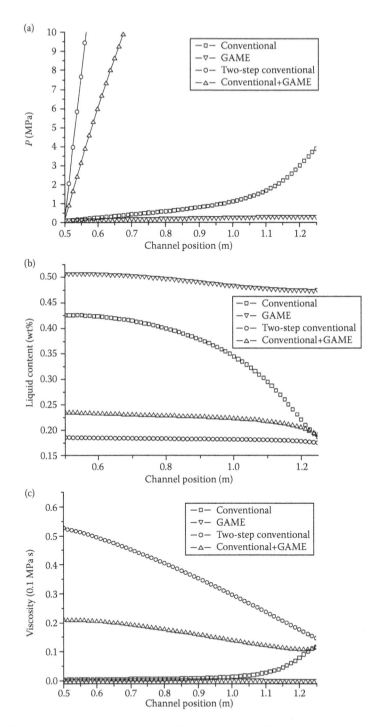

FIGURE 12.13 Predicted (a) pressure, (b) liquid content, and (c) viscosity profiles for the cases studied (90 rpm). (From Willems, P. et al., *J Food Eng*, 90, 238–245, 2009. With permission.)

TABLE 12.2
Final Residual Liquid and Oil Contents for the Four Systems Investigated (90 rpm)

Process	Residual Liquid Content (wt%)	Residual Oil Content (wt%)
Conventional	19	19
GAME	47	40
Two-step conventional	17.5	17.5
Conventional followed by GAME		
• 8 MPa CO_2	19.0	15.2
• 10 MPa CO_2	19.0	14.9
• 15 MPa CO_2	18.9	14.3

comparison, the final *liquid* and *oil* contents are shown in Table 12.2. For the two-step processes, only the second pass is shown. These are preceded by the profile calculated for the conventional case.

From industrial experience, it is known that the majority of the expression and oil removal is done in the last part of the extruder (Carr 1997). This is consistent with the oil content and pressure profiles shown in Figure 12.13: first, pressure increases and liquid (oil) content decreases gradually. The decreased oil content causes an increase in paste viscosity and the pressure increase becomes larger. For this particular extruder design, the model predicts a very small oil yield for the GAME case. Because the liquid content is increased by the dissolved CO_2, the paste viscosity is significantly reduced and pressure build up along the extruder is limited. This removes very little liquid, which in turn limits the viscosity increase and further pressure buildup. If the oilseed passed from the first pass is fed to the extruder again (mimicking a two-stage extruder), pressure buildup is immediately enormous because of the very high viscosity of the press cake. This results in an increased resistance of the cake to oil permeation and therefore the additional oil yield is limited. If, however, this second stage is operated under CO_2 pressure, the liquid content of the paste increases and viscosity of the paste reduces enough to enable more oils to be expelled. The yields corresponding to the residual oil contents for the conventional case and the two-stage GAME case are 69% and 78% oil–oil resp., which corresponds to the increase that can be obtained with hydraulic pressing. The influence of CO_2 pressure on the residual oil content also shows a similar trend to hydraulic pressing. Increasing the pressure to 8 MPa shows a significant increase, but any yield increase at higher pressures is limited. On the basis of these cases, operating GAME in a continuous two-stage extruder seems feasible.

12.6 CONCLUSIONS

GAME is a promising technique to increase the yield of oil from oilseeds. It is capable of producing high oil yields without the use of organic solvents, thus leaving no hazardous traces in the vegetable oil. Experiments have shown that the yield can be increased with up to 30 wt% when using GAME compared with conventional expression. The main mechanisms behind the increase have been shown to be displacement of the oil (both for hulled and dehulled seeds) and entrainment of the oil absorbed on the hulls after expression (for hulled seeds).

Up until now GAME has mainly been studied in uniaxial hydraulic presses. An industrial implementation will require the development of a continuous process, most likely a two-stage extruder. Estimates of required contact time between CO_2 and oilseed paste show that sufficient time is available to saturate the paste with CO_2. This will ensure a maximum effect of the CO_2 content. Case studies with a mathematical model have given an indication of the additional yield that can be obtained. However, the model needs to be significantly improved to provide accurate predictions of reachable yields. This can then be used to design an extruder most suitable for GAME. Research into this subject is still ongoing at several institutes (Agrotechnology and Food Wageningen and Technische Universität Hamburg, Harburg).

ACKNOWLEDGMENTS

The contributions of N.J.M. Kuipers and M.J. Venter to the work described in this chapter are kindly acknowledged.

REFERENCES

Acosta, G.M., Smith, R.L., Arai, K. 1996. High-pressure PVT behaviour of natural fats and oils, trilaurin, triolien and n-tridecane from 303K to 353K from atmospheric pressure to 150 MPa. *J Chem Eng Data* 41: 961–969.

Bailey, A.E., Swern, D., Mattil, K.F. 1964. *Bailey's Industrial Oil and Fat Products*. New York: Interscience Publishers.

Bharath, R., Inomata, H., Adschiri, T., Arai, K. 1992. Phase equilibrium study for the separation and fractionation of fatty oil components using supercritical carbon dioxide. *Fluid Phase Equilibr* 81: 307–320.

Carr, R.A. 1997. Oilseeds processing, in *Technology and Solvents for Extracting Oilseeds and Nonpetroleum Oils*. P.J. Wan and P.J. Wakelyn (Eds.), pp. 101–120. Champaign: AOCS Press.

Clifford, A.A. 1998. Gas assisted press extraction of oil. U.S. Patent Appl. 2 343 898.

Dittmar, D., Eggers, R., Hahl, H., Enders, S. 2002. Measurement and modeling of the interfacial tension or triglyceride mixtures in contact with dense gasses. *Chem Eng Sci* 57: 355–363.

Eggers, R. 1988. Gas tightness of a mechanical pressed rapeseed plug. *Fett Wiss Techn* 90: 184.

Foidl, N. 1999. Device and process for the production of oils and other extractable substances. U.S. Patent 5,939,571.

Hasenhuettl, G.L. 1991. Fats and fatty oils, in *Kirk–Othmer's Encyclopedia of Chemical Technology*, J.I. Kroschwitz and M. Howe-Grant (Eds.), pp. 282–285. New York: Wiley & Sons. 10: 252–287.

Kerst, A.W., Schlünder, E.-U. 1998. Densities of liquid and supercritical mixtures of methyl myristate and carbon dioxide at high pressures. *J Chem Eng Data* 43: 274–279.

Klein, T., Schulz, S. 1989. Measurement and model prediction of vapor–liquid equilibria of mixtures of rapeseed oil and supercritical carbon dioxide. *Ind Eng Chem Res* 28: 1073–1081.

Spilimbergo, S., Bertucco, A. 2003. Non-thermal bacteria inactivation with dense CO_2. *Biotechnol Bioeng* 84: 627.

Tegetmeier, A., Dittmar, D., Fredenhagen, A., Eggers, R. 2000. Density and volume of water and triglyceride mixtures in contact with carbon dioxide. *Chem Eng Proc* 39: 399–405.

Vadke, V.S., Sosulski, F.W., Shook C.A. 1988. Mathematical simulation of an oilseed press. *J Am Oil Chem Soc* 65: 1610–1616.

Venter, M.J. 2006. Gas assisted mechanical expression of cocoa nibs. *Ph.D. thesis.* Enschede: Department of Science and Technology.

Venter, M.J., Willems, P., Kareth, S., Weidner, E., Kuipers, N.J.M., de Haan, A.B. 2007. Phase equilibria and physical properties of CO_2-saturated cocoa butter mixtures at elevated pressures. *J Supercrit Fluid* 41: 195–203.

Venter, M.J., Willems, P., Kuipers, N.J.M., de Haan, A.B. 2006. Gas assisted mechanical expression of cocoa butter from cocoa nibs and edible oils from oilseeds. *J Supercrit Fluid* 37: 350–358.

Voges, S. 2008. Prozessintensivierung durch die einlösung von verdichteten gasen in flüssig-feststoff systeme. Hamburg: Verfahrenstechnik.

Voges, S., Eggers, R., Pietsch, A. 2008. Gas assisted oilseed pressing. *Sep Purif Technol* 63: 1–14.

White, A., Burns, D., Christensen, T.W. 2006. Effective terminal sterilization using supercritical carbon dioxide. *J Biotech* 123: 504.

Willems, P. 2007. *Gas Assisted Mechanical Expression of Oilseeds.* Enschede: Department of Science and Technology.

Willems, P., Kuipers N.J.M., de Haan, A.B. 2008. Gas assisted mechanical expression of oil seeds. *J Supercrit Fluid* 45: 298–305.

Willems, P., Kuipers, N.J.M., de Haan, A.B. 2009. A consolidation based extruder model to explore GAME process configurations. *J Food Eng* 90: 238–245.

13 Mechanochemically Assisted Extraction

Oleg I. Lomovsky and Igor O. Lomovsky

CONTENTS

13.1 BASIC MECHANOCHEMICAL EFFECTS IN EXTRACTION TECHNOLOGY

13.1.1 PREPARATION OF RAW MATERIAL FOR EXTRACTION WITH THE OPTIMAL SIZE OF PARTICLES

Traditionally, the major purpose of the mechanical treatment of raw material before extraction is to reduce the size of its particles to optimize the relationship between internal diffusion inside separate particles and external diffusion in the permeable powder layer. The size of particles is important for internal diffusion resistance. With increase in raw material grinding efficiency, the total of the particle surface area and the contact area between the raw material and the solvent increases, internal diffusion resistance decreases, the number of destroyed cells of the raw material increases, and the extracting agent better penetrates into the cell. However, hydrodynamic conditions of the extracting agent flow through a layer of particles worsen substantially with decreases of the particle size, because the external diffusion resistance increases to a higher extent than the internal diffusion resistance decreases. An optimal particle size making minimal the sum of internal and external diffusion resistances should be determined for the each type of raw material and for each condition of extraction. A usual range of the plant raw material particle sizes optimal for water and water–alcohol mixture extraction is 0.33 to 1.0 mm.

As the particle size decreases, some particles start blocking the surface of others, thus making smaller the size of pores between the particles (where the liquid moves);

it results in the formation of the closed pores. For further intensification of the extraction process, it is necessary to enlarge the surface area of particles participating in interaction with the solvent, improve conditions of mass transfer from the particle surface into solution, and further decrease the external diffusion resistance.

The external diffusion resistance may be affected mechanically by mixing, pressure pulses, low-frequency mechanical oscillations, ultrasound, and electrical pulses, or by arranging the boiling bed conditions. Only 20–25% of the total external surface of particles participate in the process of counterflow extraction, whereas, owing to the low-frequency mechanical oscillations, the solvent flow surrounds almost the entire surface of small particles during extraction if parameters are optimal (Minina and Kauchova 2009).

In the majority of cases, the optimal degree of raw material grinding, temperature, solvent pressure, and the kinetic parameters they determine are found experimentally. In some cases, organoleptic characteristics of the resulting extract may be main indications.

For example, the balanced espresso coffee preparation from the grains of different kinds requires control over the grinding grade and the density of coffee packing. Different coffee kinds have different characteristics, which require different size of particles for achieving the necessary degree of extraction. The optimal fraction is considered to be 50–90 μm. Due to the 50-year-long practice of espresso professionals, it was found that the time of water flow through the coffee layer composed of particles of a certain size during espresso coffee preparation should be 18–22 s. Less than 17 s of extraction leads to a drink not strong enough, with only a small amount of soluble substances and oil gets into the cup. If extraction is insufficient, the espresso taste and odor are not intense and no deep and complicated taste can be felt. Extraction for more than 25 s results in excess concentration of the soluble substances. Such coffee has a bitter taste, is astringent, and has no balance and delicacy.

Thus, grinding of particles of the raw materials to be used for extraction, enhancement of mass transfer, and determination of the optimal extraction conditions are very important in the technology of plant raw material processing. In this chapter, the major attention is paid to the methods of mechanical treatment less frequently applied in extraction of biogenic materials. Mechanical treatment is used to increase the efficiency of extraction by change in concentration of defects and enhancement of the reactivity of solid material and its components.

13.1.2 IMPROVEMENT OF PLANT MATERIAL GRINDING EFFICIENCY USING PRELIMINARY CHEMICAL TREATMENT

Plant raw material grinding was studied most thoroughly in relation to pulp-and-paper industry techniques. To intensity grinding, it was proposed to use preliminary 3-min treatment of the lignocellulose material by 0.1–1% solution of sulfuric acid. The energy consumption for grinding the material treated by sulfuric acid was three to five times lower (Simushkin et al. 1990) than is necessary for grinding a non-treated material. It was assumed that after selective removal of the amorphous part of wood by acid, the remaining material undergoes mechanical grinding more easily.

One of the stages of paper technology is the grinding of lignocellulose materials and their transformation into fibrous wood pulp. This stage consumes a lot of energy, but the major problem here is destruction of the cellulose fibrils and decrease in the degree of cellulose polymerization, thus acid use is impossible in this case.

To increase the process efficiency, it was suggested to use lignolytic enzymes that destroy lignin selectively and simplify the process of obtaining the fibrous wood pulp. In this case, cellulose remains relatively intact. According to different studies, using lignolytic enzymes improves the quality of paper and allows reduction of the energy consumption for grinding the lignocellulose mass (Pandey 2003; Maijala et al. 2008; Bhat 2000). The mechanism of enzymatic activity affecting this process is not quite well understood yet, and the chemical processes underlying these effects are not thoroughly investigated (Ferraza et al. 2008).

Treatment of the wood pulp with cellulase enzymes may also increase efficiency of the process of obtaining the fibrous mass from lignocellulose materials. However, cellulolytic enzymes cause hydrolysis of cellulose, and it substantially worsen the properties of the resulting paper material. The effect of various cellulosolytic enzymes and enzymatic systems on the cellulose properties and rupture strength of material was studied. It was shown that the effect of endoglucanase on the efficiency of the process of fibrous mass obtaining from lignocellulose material was the strongest. On the contrary, the effect of cellobiohydrolase on the pulp properties was not significant, whereas the enzymes with carboxymethylcellulase activity did not affect the process efficiency at all (Ferraza et al. 2008).

In order to achieve fine grinding, special methods of plant raw material embrittlement can be used. Powder plant material with particles less than 50 μm is treated by a small volume of a cellulosolytic enzyme solution. An enzymatic reaction occurs during 24 h of heating at 50°C. As far as the reaction taking place without mixing and running only on the surface of the particles, the degree of cellulose transformation into soluble polysaccharides does not exceed 3–5%, but it is quite sufficient for diminishing the strength of the raw material. Subsequent grinding results in the average size of particles two to three times smaller than the size achieved without preliminary treatment. This allows an increase in the rate of subsequent heterogeneous processes by a factor of 5–10 (Politov et al. 2008).

13.1.3   CHANGES IN THE STRUCTURE OF DEFECTS AND SOLID-PHASE REACTIVITY ENHANCEMENT IN EXTRACTION PROCESSES

Mechanical impact has been efficiently used in recent decades in the extraction of multicomponent mineral raw materials and in hydrometallurgy for governing the reactivity of phases comprising minerals and their mixtures. Using preliminary mechanical treatment of solids, it is possible to increase the concentration of defects in their structure and thus to increase the reactivity of solid reactants in heterogeneous reactions. Under special conditions, the solid-state interaction of the components of raw material with additionally introduced solid reactants during mechanical treatment may cause formation of new chemical compounds that are easily dissolved and extracted. Approaches related to changes in the reactivity of solid reactants and implementation of chemical reactions during mechanical treatment turned out to be useful for mineral raw processing. These approaches helped to increase the yield of extractable components, intensify the processes, and introduce into use new kinds of resources (Balaz 2008; Balaz and Dutkova 2009), including resources of organic compounds (Khrenkova 1993).

Effects connected with changes in the content of defects and the reactivity of components in the solid-state reactions are widely studied and find application in catalysis of organic reactions (Molchanov and Buyanov 2000), in pharmacology (Boldyrev 2004), and in solving many environmental problems (Lomovsky and Boldyrev 2006).

Extraction processes are used both for isolation of the necessary substances from the cell walls and for isolation of substances from the cell content. Obtaining saccharides from the cell walls is a global problem; solving it would allow provision of the food basis for constantly growing population of the earth. However, there also remain other, less global, problems, such as broadening of the food basis for animal breeding and improvement of cellulose and lignocellulose assimilation by microorganisms. The latter is a key problem in production of bioethanol and other kinds of liquid fuel from plant raw materials. For social and political reasons, starchy food and sugary raw materials used in production of liquid biofuel should be replaced by lignocellulose.

Difficulties in chemical and enzymatic isolation of sugar compounds from plant raw materials are caused by the fact that cellulose and lignin form walls that are resistant to the impact of external factors; these walls were constructed by nature for protection of extremely reactive substances of the cell interior. The use of alkalis and sulfur-containing compounds for dissolution of lignin and isolated cellulose is connected with a number of environmental problems, such as high consumption of water and its contamination and the lack of significant need in lignin separated as a by-product. Burning in energy-generating installations still remains the major application of lignin, and it is accompanied by a number of unfavorable environmental consequences. The acid hydrolysis of the hydrocarbon part of the plant raw material contaminates lignin to such a high degree that its combustion causes acid rains. The cellulase-catalyzed enzymatic hydrolysis of lignocellulose is hindered by interaction of the protein enzymes with polyphenols of lignin (tanning reaction) causing formation of very strong complexes and removal of the enzyme from the process. Low rate of plant material hydrolysis in an enzyme solution is an obstacle for development of highly efficient techniques; however, higher environmental safety of enzymatic processes makes them most promising at the present stage.

Mechanochemical approach turned out to be efficient in some cases for making smaller the difficulties related to enzymatic hydrolysis. However, application of the mechanical treatment of enzymes runs across a fundamental problem. The catalytic activity of an enzyme is determined by its very specific spatial structure. The spatial conformational rearrangement of the protein polymer provides macromolecule arrangement in a manner necessary for the reaction. One can hardly expect conservation of this specific structure during mechanical impact on the protein molecule, thus an intense mechanical treatment of enzymes was considered to be undesirable. Even an intense mixing of the reaction medium containing a solution and a solid substrate may cause a sharp decrease in the activity of enzymes. Nevertheless, it turned out that the stability of enzymes may be maintained and even increased if the mixture of the enzyme and the substrate subjected to mechanical treatment has proper mechanical characteristics.

An understanding of the important role of carbohydrates not only in the energy cycle of plants, but also in other specific biological processes, has become the driving force for development of new methods of enzymatic hydrolysis of biopolymers comprising the cell walls. The role of inassimilable (noncellulose) polymeric cell wall carbohydrates of plants, fungi, and microorganisms has changed during the recent decades. There appeared a need in products containing these compounds.

The so-called soluble and insoluble biopolymeric fibers and β-glucans of oat became promising substitutes of fats; β-glucans of fungi turned out to be efficient components of functional antisclerotic nutrition; arabinogalactanes of larch are used as novel excipients in medicinal preparations and nutrition additives, mannanooligo-saccharides of yeast cell walls substitute antibiotics that have been prohibited for use and have lost their efficiency. The use of mechanochemistry in solving the problem connected with the isolation of these compounds will be presented below.

13.1.4 Chemical Reactions of Low-Molecular Intracellular Substances during Mechanical Treatment

The cells contain a broadest range of various biologically active substances (BAS) that are necessary for an adequate diet. Vitamins and microelements traditionally attract much attention from food chemistry specialists. Origin of the functional nutrition caused in the recent decades an abrupt increase in consumption of bioactive compounds as antioxidants: flavonoids, phenolics, carotenoids, some vitamins, and others.

The processes of their isolation, concentration, and transformation into the form suitable for consumption by humans are under development. Mechanochemistry finds efficient application in this area as well.

Efficient extraction of intracellular components requires preliminary destruction of the cell walls. The impact on the cell walls should be chosen so as to minimize denaturation and decomposition of unstable BAS inside the cells. Destruction of cell walls in the biotechnological production of microbial metabolites and callus tissues is a necessary stage of BAS extraction process. The ability to destroy the cell walls with minimum losses of unstable intracellular substances determines efficiency of extraction techniques.

Mechanochemical reaction efficiency depends not only on the chemical properties of reactants but also on their mechanical properties. Mechanochemical processes dealing with soft substances and materials are characterized by the low energy consumption (Avvakumov et al. 2001). The hardness of organic substances is usually much lower than that of inorganic compounds. Mechanochemical reactions occurring in organic systems have 1000 times higher energy yield than reactions in inorganic systems. It was demonstrated that some organic reactions run with higher efficiency in the solid state than in the liquid state (Tanaka and Toda 2000). Thus, mechanochemical processes involving organic substances turn out to be the most promising from the viewpoint of their industrial application.

Mechanochemical approach in the manufacture of BAS preparations allowed proposing of new powder and solid-state techniques. In many cases, mechanical treatment of the powder mixture of biogenic raw material and reactants allows transformation of the different forms of extractable substance into one that is more soluble in a proper extractant.

Eventually, mechanochemical treatment increases the yield of the extractable substance and selectivity of extraction and opens up opportunities for using new kinds of raw material.

This approach turned out to be especially effective in cases when bioassimilable form concentration in the powder produced by mechanical treatment is sufficiently high and no additional concentration is required. The powder product formed directly during mechanical treatment of the powder mixture of a plant food raw material and a suitable food reactant may be used as a food component or an additive for purposes of functional nutrition. Exclusion of a number of stages involving liquid phases and the stage of solid product isolation makes the food technology simpler and decreases concomitant expenses. The absence of liquid phases in the processing is very important and helps prevent oxidation of unstable food components and improve the environmental safety of food manufacture.

13.1.5　Hydrothermal (Quasi-Autoclave) Conditions under Mechanical Treatment of Heterophase Mixtures

Mechanical treatment of a solid–liquid system initiates and accelerates various chemical transformations. These chemical reactions are promoted by phenomena typical for activation of the mixtures of solid components (increase of interfacial area, accumulation of defects, amorphization, and free energy increase). However, in complicated heterophase mixtures, unusual conditions for chemical reactions may occur.

In the case of mechanical activation of a heterophase system containing solid, liquid, and gaseous phases, regions of high pressure and temperature may arise (Boldyrev 2002). The phenomena of adiabatic compression of gas bubbles, viscous flow of the liquid, and friction between the contact surfaces of solid particles were suggested as possible causes of the temperature rise.

Temperature changes caused by compression of gas bubbles may be estimated using the adiabatic equation:

$$T_2 = T_1 \, (p_2/p_1)^{\gamma-1} \tag{13.1}$$

Here, T_1 and T_2 are the initial and final temperatures, p_1 and p_2 are the initial and final pressures, and γ is the ratio of heat capacities C_p/C_v.

In solids and liquids, C_p and C_v differ insignificantly; thus, heating does not exceed $10°$. Intense mechanical treatment involves high-speed displacement of particles, which may destroy hydraulic continuity of the liquid; the uniformity discontinuities (gas bubbles) appear in the system. For the gaseous phase, the difference between heat capacities C_p and C_v is $C_p - C_v = R$, where R is the universal gas constant. It was shown that a 20-fold decrease of the gas bubble volume in the liquid phase of an explosive substance causes local heating up to $500°C$.

During collapse of the gas bubbles caused by hydraulic compression, the velocity of approach of two gas bubbles may reach 100 m/s. Substantial amount of heat is released in collisions of the molecules and particles carried by the flow.

Substantial heat evolution accompanies friction of two solid particles. Temperature change ΔT resulting from friction at the surface of contact between solid particles may be estimated using equation:

$$\Delta T = nmWV/4J\alpha \, (k_1 + k_2), \hspace{2cm} (13.2)$$

where n is the number of contacts, m is the friction coefficient, W is the pressure at contacts, V is the relative velocity of contacting particles, α is the contact area, J is the mechanical equivalent of heat (joules), and k_1 and k_2 are the thermal conductivity coefficients of the contacting particles. This equation shows that, even at low velocities of particles but sufficiently small contact area, the temperature may rise by 500–700°C. Overheating will cause instantaneous boiling of the liquid in the region adjacent to the contact. If the liquid is in the closed volume, vapor formation causes a substantial increase in pressure, which may lead to a chemical reaction.

Mechanical activation of the humid mixtures leads in some cases to the formation of the final products that are characteristic for conditions of hydrothermal reaction. Temuujin et al. (1998) have shown that formation of products typical for hydrothermal synthesis can be observed during mechanical treatment of humid simulating mixtures of inorganic oxides.

The scientific aspect of mechanochemical reactions carried out in a rotor type hydrodynamic flow apparatus is poorly investigated, but their application actively spreads in such areas as processing of cereals, oil-yielding crops, and improvement of the quality of beverages.

13.2 STRUCTURAL FEATURES OF BIOGENIC RAW MATERIALS THAT ARE ESSENTIAL FOR MECHANOCHEMICAL PROCESSES

Plant raw materials, as well as raw materials of microbiological and animal origin, are multicomponent, as with mineral raw materials. A distinguishing feature of biogenic raw materials in comparison to mineral ones is the cellular structure of tissue. Carbohydrate biopolymers cellulose and hemicellulose and polyphenolic polymer lignin, which are the major components of the plant raw materials, jointly accounting for more than 90% of the total mass of dry plant material, are concentrated in the cell walls and intercellular partitions. Some compounds synthesized by plants, e.g., starch, proteins, and lipids, are located in the special plant organs and tissues having specific cellular structure. To isolate these substances and use them for nutrition, it is necessary to destroy the cell walls. Cell content accounting for less than 10% of the plant mass is very diverse in composition, and the major portion of the BAS is concentrated inside the cells.

The overwhelming majority of extracted preparations are made of dried plant raw materials. Due to the loss of protoplasm during drying, the cell walls lose their semipermeability and let the substances pass through in both directions. The intracellular content concentrates in a small region of the intracellular space.

Extraction of intracellular substances from a powdered plant raw material starts from extractant penetration inside the particles of the plant raw material. First percolating through macrocracks, then through microcracks, intercellular channels and vessels, the extracting agent reaches the cells and diffuses through the cell walls. As far as the extracting agent enters the dried cell, its content starts to swell and passes into solution.

From the chemical point of view, the broadest range of biologically active compounds is present in the intracellular substance: acids, alkaline amines, low-molecular carbohydrates and oligosaccharides, alcohols and phenols, glycosides. Many of them occur in biological tissues in different chemical forms possessing different solubility and biological availability. For example, phenolic compounds may be present in the free form, in the form of a salt, in glycosylated form when they are bound with low- and high-molecular carbohydrates, both soluble and insoluble, in the form of polyphenols. For maximum extraction of such compounds, it is necessary to use a series of solvents with different polarity. The usual content of extractable BAS is in the order of some percentage (1–5%); they are highly needed in such substances, and their high cost makes it important to increase the yield even by tenths of a percent.

13.2.1 Structure and Chemical Composition of Some Plant Tissues

A part of the cross section of a herbaceous plant (rye) culm is presented in Figure 13.1.

The cells of the vascular plants, including cereals, are incorporated into tissues that are classified in accordance to the functions they perform as ground tissues,

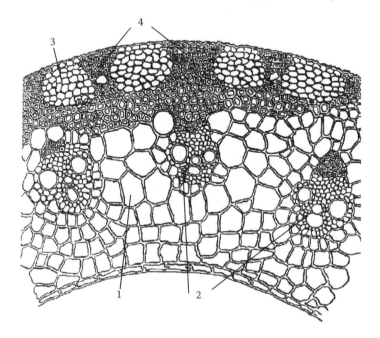

FIGURE 13.1 Cross section along the culm of a herbaceous plant (rye).

vascular tissues, and epidermis. The largest tissue cells reach up to 40 μm in the cross section, the smallest have the size of 5–8 μm.

Parenchyma (1), a precursor of all the other tissues, is represented by a continuous mass of large cells in the middle of the stem. The shape of cells is usually polyhedral, and the cell walls are weakly lignified. Conducting bundles (2) are immersed into an internal layer having a total thickness of about 200 μm and composed of the largest cells of the parenchyma. The conducting bundles have complicated structure and are composed of cells with different diameter (ranging from approximately 1 to 40 μm) highly extended along the vertical direction. The size of the conducting bundle itself is 100–120 μm. Among the tissues under consideration, conducting bundles have the most complicated structure. They are composed of extended cells located closely to each other and forming the bundles. The conducting bundles have higher density but lower cellulose content in comparison to parenchyma and chlorenchyma. The chlorenchyma (3), which is a chlorophyll-bearing parenchyma, is located directly beneath the thin protective layer of transparent epidermis. The cells of the chlorenchyma contain chloroplasts, which enable photosynthesis. The cells of the chlorenchyma are loosely arranged, and large areas of free space between them make easy circulation of gases necessary for photosynthesis. The chlorenchyma appears to be the weakest plant tissue from the viewpoint of mechanical destruction. The parenchyma, conducting bundles, and chlorenchyma contain mainly cellulose.

Mechanical tissue sclerenchyma (4) contains most of the plant lignins and is composed of cells with uniformly thickened and lignified walls. The walls of sclerenchyma cells possess the hardness of steel. They surpass steel in their ability to resist dynamic load without residual deformation.

The external layer of sclerenchyma is 80–120 μm thick and is composed of thick-walled cells having 5–8 μm in size. It also includes chlorenchyma regions up to 80 μm in size composed of 10 μm thin-walled cells (Raven et al. 1986).

13.2.2 DISINTEGRATION OF PLANT RAW MATERIAL TO PARTICLES OF ABOUT 80 μM IN SIZE

The supracellular organization in the plant raw materials is presented by differentiated tissue regions with characteristic size about 80–100 μm.

During destruction of a composite plant material consisting of regions with different mechanical properties, the residues of conducting bundles in the powder are represented by elongated particles several hundred μm long; their thickness is several times smaller than their length. Particles less than 80 μm in size may have a uniform structure and differ in composition and other properties. The structure and chemical composition of the particles of powdered material correspond to the structure and composition of the separate tissues of straw. For example, one may expect that the particles sized less than 80 μm will consist of separate tissues: chlorenchyma, parenchyma and conducting bundles, and sclerenchyma.

Because sclerenchyma is a lignified tissue with high additional lignin content, its particles possess a decreased density. The density of lignin is 1.25–1.45 g/cm³, which is somewhat lower than the density of cellulose 1.52–1.54 g/cm³. Lignification of cells occurs with preservation of their geometric size, so the density of particles increases in the row of chlorenchyma–parenchyma–sclerenchyma particles, as the cellulose content decreases.

13.2.3 Structure and Chemical Composition of Cell Walls

The walls of plant cells are composed mainly of polysaccharides. In addition to polysaccharides, cell walls may include lignin, proteins, mineral salts, pigments, and lipids. Only the lignin content may be comparable with the polysaccharide content, whereas the content of other components is usually not higher than several mass percent.

Cellulose, or β-1,4-D-glucan, is a polymer which may have significantly different number of glucose residues. In the cell wall, the polymer molecules with chain length from several tens to several thousand units are arranged parallel to each other and with molecules together, so that a characteristic lattice forms. Covalent interactions and hydrogen bonds result in formation of microfibrils—the thinnest fibers creating the structural framework of the cell wall.

The diameter of microfibrils is 10–30 nm; their length may reach several μm. Amorphous regions occasionally occur in microfibrils, which affect the mechanical and chemical properties of cellulose. Microfibrils are arranged into 0.4- to 0.5-μm-thick macrofibrils; the latter are visible in an optical microscope. Microfibrils of cellulose are elastic and strong against breaking.

Microfibrils are immersed into an amorphous matrix of the cell wall, which is a plastic gel saturated with water. The matrix is a complicated mixture of polymers, where dominate polysaccharides with different molecular masses, such as hemicelluloses and pectins, which can be treated as oxidized sugars. The polysaccharides of the matrix do not have any crystalline structure.

Lignin is a mixed amorphous polymer of phenol series. Its content in cell walls may reach 30%. Lignin deposits at the end of cell wall growth during the process of lignification. Lignification causes changes in mechanical properties of the cell wall: loss of plasticity and sharp increase in hardness and strength. Cell wall permeability for water decreases.

The walls of cells in the above-ground parts of some plants contain large amounts of inorganic substances, most frequently silica, oxalate, and calcium carbonate. These substances render hardness and fragility to the cell walls.

13.2.4 Disintegration of Plant Raw Material to Particles of about 1 μm in Size

If a plant raw material is ground to particles smaller than the thickness of the cell walls, particles of the product become homogeneous by their chemical composition.

In lignified cell walls, three concentric layers differing in thickness, chemical composition, and mechanical characteristics can be distinguished: the external layer, the medium layer, which is the thickest one (thickness of up to 10 μm), and the narrow internal layer, which is adjacent to the cell cavity. The internal layer is very thin (down to 0.1 μm) and contains a lot of hemicellulose. The medium layer is the richest in cellulose.

Fine particles of the plant raw material sized less than 5 μm may significantly differ by their origin and density. The particles that are residues of the most ligni-fied part of cell wall will have higher density than the particles originating from the internal part of the cell wall. The particles differing from each other in density may be separated using different methods including aerodynamic classifying.

13.2.5 STRUCTURAL FEATURES OF BIOGENIC UNICELLULAR RAW MATERIAL (EXAMPLE: YEAST BIOMASS)

Yeast cell walls and plant cell walls differ by their structural components. The major component of the yeast cell wall responsible for sustaining its strength is β-glucan that forms the medium layer of the cell wall; it is coated with mannanoproteins from the outer and inner sides (Biryuzova 1993; Kalebina and Kulaev 2001; Harthmann and Delgado 2004). The yeast glucan comprises several types of polysaccharide molecules formed by glucose residues connected with β-1,3- and β-1,6-bonds. The rigidity and strength of the cell wall and its stability against environmental effects are determined by β-1,3-glucan rather than by other components (Pereira 2000; Harthmann and Delgado 2004). The structure of the yeast glucan is similar to the structure of cellulose; similar to cellulose, this glucan molecules form micro-fibril and macrofibril containing crystalline and amorphous parts. The proteins of the yeast cell wall make covalent bonds with mannanooligosaccharides (Biryuzova 1993; Kalebina and Kulaev 2001), thus forming mannanoproteins and, in interacting with glucan, forms the supramolecular arrangement of the cell wall.

Mannanooligosaccharides of yeast cell walls are oligosaccharides consisting of mannose residues; they are able to block bacterial lectins (proteins responsible for binding bacteria to the surface of mucous membrane cells of gastrointestinal tract in mammals) and bind mycotoxins (Wold et al. 1990; Emmerik et al. 1994; Dawson and Sefton 2006). In this connection, they attract attention of the specialists in food and forage chemistry.

Mannanooligosaccharides are usually isolated from the cell walls using prolonged autoclave heating and hydrolysis by acids and alkalis followed by product separation (Kanegae et al. 1989; Belousova et al. 1990; Lazzari 2000). Most of these procedures have disadvantages, such as the requirement to maintain high temperatures for a long time and to use aggressive, volatile, and flammable reagents.

The mechanochemical approach is quite competitive with chemical methods. It was demonstrated that the rupture of cell walls is not the only process that occurs during mechanical treatment of yeast biomass. The breakage of cell walls is accom-panied by changes in the supramolecular structure of the cell wall polymers and breakage of covalent bonds in the polymers. The structural layers of the cell walls get disordered. The product of mechanical activation shows increased reactivity. Mechanical treatment of yeast biomass makes higher the biological availability of mannanooligosaccharides for alkaline extraction.

In the reactions of enzymatic hydrolysis, the most reactive compound is the alkali-soluble β-1,3-glucan. However, mechanical treatment of the least active alkali-soluble β-glucan allows increasing of its reactivity.

13.2.6 DESTRUCTION OF BIOGENIC UNICELLULAR RAW MATERIALS DURING MECHANICAL TREATMENT

The amino acid composition of yeast is close to the amino acid composition of milk; extraction of intracellular yeast proteins is used for obtaining milk substitutes. Equipment most frequently used for processing large amounts of biomass includes ball or bead mills, flow-type disintegrators, and jet-type mills. The advantages of the mechanical method over chemical and enzymatic techniques are short time of treatment, negligible degradation of cell components, and low cost. A significant disadvantage of the mechanical method is the difficulty in scaling the processes from laboratory installations to their industrial analogues.

At present, due to the availability of high-strain devices, such as jet-type mills, only several-minute-long treatment is sufficient for destruction of 90–95% of cells in most of crops (Lazzari 2000).

The number of intact cells decreases during mechanical disintegration according to the law close to exponential (Kasatkina and Levagina 1999; Heim et al. 2007)

$$N = N_0 \exp(-t/\tau), \tag{13.3}$$

where N is the current number of intact cells within the volume of disintegrator, N_0 is the initial number of cells, t is the disintegration time, and τ is a constant.

Cell destruction is not the final goal of the process; more frequently, disintegration is carried out for separation of intracellular components. Usually, the bonds of extractable intracellular compounds with the cell wall are mechanically weak and they easily come out after the wall is destroyed. The excessive treatment of the cell content may cause its additional inactivation. Therefore, the process modes providing onefold impact of milling bodies on the cell should be used for destruction. The destructed cell is to be removed from the reactor volume. Efficient destruction of the cells of yeast biomass occurs in the jet mills. The yeast cell biomass gets accelerated up to a high speed and then collides with the immobile wall of the jet chamber. The degree of cell destruction reaches 90% during a cycle of treatment, which makes the method suitable for industrial applications (Lazzari 2000).

13.3 OPTICAL AND ELECTRON MICROSCOPIC STUDIES OF THE ACTIVATION OF PLANT RAW MATERIALS DURING INTENSIVE MECHANICAL TREATMENT

Direct electron microscopic observations of the results of intensive mechanical treatment of certain plant tissues were carried out. In particular, the destruction of the oil palm (*Elaeis guineensis*) bunches was studied. This is a promising raw material for obtaining soluble sugar compounds (Bychkov et al. 2010).

Micrographs in Figure 13.2a–c show a cell of the raw material tissue before its mechanical treatment. The 80-nm-thick layers can be distinguished in the structure of the cell walls. The vascular channels are stained; the distance between them is usually 2–4 μm (Figure 13.2a). Micrographs in Figure 13.2d–f show the raw material

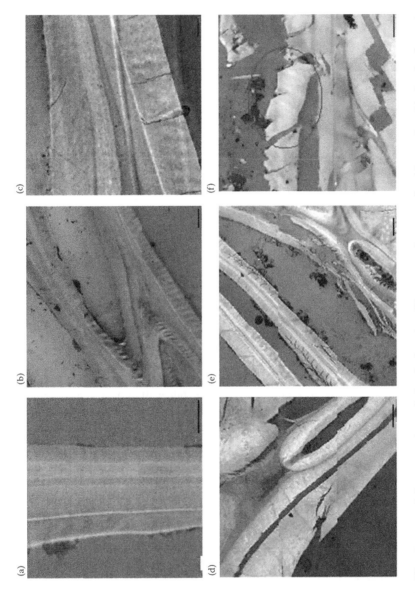

FIGURE 13.2 Ultrafine cuts of the preparations of oil palm bunches. The preparations were fixed by osmic acid; the ultrafine cuts were stained by uranyl acetate. Notations are explained in the text. Scale bars correspond to (a,c) 1 μm and (b,d–f) 2 μm. (From Bychkov, A.L. et al., *Chem Plant Raw Mater*, 1, 49–56, 2010. With permission.)

after intensive mechanical treatment. Mechanical destruction is observed in the sites of "channels" (Figure 13.2d,e); an ultrafine structure of the cell wall becomes distinguishable at a level of 20–30 μm. The lamination of the cell wall is observed (Figure 13.2e). The micrograph in Figure 13.2f shows formation of the separate particles sized 2–7 μm. It demonstrates that formation of particles occurs mainly due to propagation of cracks along the structural vascular channels, followed by separation of blocks from the elongated elements of the cell wall due to cross-propagation of the cracks.

The content of defects in the carbohydrate component of the cell walls was studied using specific reaction with silver ammonia complex solution, and in the study of defect content in the cell wall proteins, the osmic acid solutions were applied. Fixation of the preparations of mechanically activated tissues by osmic acid allows the observation of approximately 30-nm-thick layers. This effect may be caused by two factors: displacement (sliding) of the layers with respect to each other, which make observable thinner elements of the ultrastructure, an increase of the ability to interact with reagents used for revealing defects and for staining.

13.4 MECHANOCHEMICAL REACTIONS USED FOR INCREASING EFFICIENCY OF THE EXTRACTION PROCESS

13.4.1 DEPOLYMERIZATION OF CELL WALL BIOPOLYMERS

Mechanical strains of polymers may cause destruction and formation of intermolecular bonds, distortion of the valence angles, and breakage of chemical bonds in the molecules. Weak intermolecular bonds are first to break. These processes cause various mechanochemically induced transformations of the substance, including amorphization and structural disordering, conformational rearrangements and polymorphous transitions, mechanochemical activation of the chemical reactions, formation of free radicals and their recombination. Grinding leads to formation of new free functional groups, which increase the chemical activity of the polymer and rates of heterogeneous reactions (Dubinskaya 1999; Prut and Zelensky 2001).

Different theories were proposed for explanation of the mechanisms of mechanochemical reactions. According to these theories, chemical reactions can be initiated due to heat evolution caused by collisions of solid particles with each other in the mills with free collisions or with the milling bodies and cylinder walls or caused by friction and plastic flux in the mills with constrained shock. Solid-phase reactions may be accelerated owing to mass transfer under plastic deformation and formation of defects and broken and deformed chemical bonds in the surface layers of solids during their destruction. Displacement of separate segments of the polymer chains during mechanical impact leads to disordering of the initial packing of chains and supramolecular formations. Some of the crystalline regions tend to amorphize and previously amorphous regions become less dense.

Mechanical treatment of cellulose is accompanied by disordering and amorphization of the crystalline regions, rupture of intermolecular hydrogen bonds, breakage of glycoside bonds in the chains, and changes in the molecular mass distribution. All these changes modify the chemical properties of cellulose. Amorphization increases the availability of samples to various reactants and causes an increase in the reactivity of polymers in heterogeneous reactions. Preliminary treatment of pure cellulose and lignocellulose materials in a jar roller allowed increase of the glucose yield in a subsequent enzymatic reaction by a factor of 12 for pure cellulose and by 17 for lignocellulose materials (Tassinary and Macy 1977).

As a result of intensive mechanical impact, separate regions of cellulose molecules displace, which leads to formation of loose and disordered macromolecular structures. Thus, the initial crystalline structure of cellulose transforms into a less ordered structure. Cellulose grinding results in the disappearance of all the peaks and the origin of a halo characteristic to the amorphous state in the diffraction patterns of cellulose. Similar diffraction patterns of amorphous cellulose are observed after the mechanical treatment of different cellulose preparations.

13.4.1.1 Acid Hydrolysis of Polysaccharides in Lignocellulose Raw Materials

The oldest method of polysaccharide conversion to monosaccharides is based on acid hydrolysis (Grethlein 1978). Both concentrated and diluted acids may be used for this process. The concentrated acid technique implies application of 72% sulfuric acid or 42% hydrochloric acid for cellulose dissolution at room temperature, followed by dilution of the mixture to 1% acid concentration and heating to 100–120°C during 3 h for hydrolysis of cellulose oligomers to glucose. This method is suitable for obtaining high yields of glucose, but formation of toxic side products does not allow the acid hydrolysis to take a significant place in the production of alimentary glucose and alcohol. Similar difficulties may also appear when concentrated organic acids are used for conversion of cellulose into glucose.

The process involving exposure to 0.5–2% diluted sulfuric acid at 180–240°C takes from several minutes to several hours (Brink 1993, 1996) and is performed in two stages. The first stage is carried out under the soft conditions and involves depolymerization of hemicellulose to xylose and other monosaccharides. The second stage is depolymerization of cellulose to glucose. The amount of acid used in the procedure is small, and acid regeneration is not required. The maximal yield of glucose is only 55% of the cellulose content; moreover, the side products of hydrolysis by diluted acid also suppress the fermentation process, which prevents the industrial use of diluted acids.

The problem related to regeneration of sulfuric acid in the acid hydrolysis technique remains unsolved; the efficient regeneration of the acid is very expensive. Another unsolved problem of the acid technique is using lignocellulose as a raw material for obtaining lignin free from acid contaminations. Only such lignin may be used as an environmentally safe fuel and as a component in forage mixtures.

13.4.1.2 Enzymatic Hydrolysis of the Polysaccharides of Lignocellulose Raw Materials

Treatment by enzymes usually starts by obtaining a 5–12% suspension of lignocellulose material and then addition of enzymes. Hydrolysis is carried out 24–150 h at 37–50°C and pH 4.5–5.0. After hydrolysis, the substances present in the liquid are soluble monosaccharides, whereas the unhydrolyzed portion of cellulose, lignin, and other insoluble components of the substrate remain in the solid part. The glucose solution is separated from the solid part of suspension, and the solid residue is washed to obtain a higher yield of glucose.

The availability of the substrate for enzymes is the basic factor affecting efficiency of the enzymatic hydrolysis of cellulose (Nazhad et al. 1995). The efficiency of the enzymatic hydrolysis depends on the specificity and mechanism of enzyme activity. For example, cellulase of *T. longibrachiatum* binds tightly to cellulose, and reversible inactivation of the enzyme is observed (Brooks and Ingram 1995). The degree of binding is determined by intensity of mixing (Kaya et al. 1994). This problem may be solved by application of conditions that provide intensive mass transfer. A very high hydrolysis rate was achieved in a reactor with intensive mass transfer (Gusakov et al. 1996).

13.4.1.3 Factors Determining the Efficiency of Using Enzymes

Unfortunately, the enzymatic method of treatment of the cellulose-containing raw materials does not allow obtaining glucose and other fermentation sugar species at a relatively low cost. Even most efficient methods of preliminary treatment known at present allow attaining of the transformation degree not exceeding 77–84% of soluble carbohydrates from the total of polysaccharides contained by the lignocellulose raw material (Cullingford et al. 1993), whereas the amount of enzymes required for conversion of lignocellulose polysaccharides into biologically available carbohydrates is too high.

The necessity in development of a cost-effective technique of sugar production from lignocellulose stimulates an increasing number of works studying the effects of preliminary treatment. The efficiency of preliminary treatment is characterized by the further yield of soluble sugar species, and by the amount of enzyme required for conversion of a specified amount of cellulose into glucose.

In many cases, preliminary treatment involves mechanical impact. Mechanical treatment causes dispersion of material and substantial increase of the interfacial area. The subsequent heterogeneous hydrolysis of cellulose in an aqueous solution proceeds on the interfaces. Amorphization of the crystalline cellulose occurs, which leads to substantial acceleration of the subsequent enzymatic hydrolysis of cellulose. Newly implemented approaches include direct introduction of enzymes into the lignocellulose substrate during mechanical activation of the substrate and enzyme mixture, which is impossible if aqueous enzyme solutions are applied.

13.4.1.4 Preliminary Treatment of the Raw Material before Hydrolysis

Preliminary treatment involving one or several techniques is used for increasing the rate and degree of hydrolysis. The material is subjected to mechanical impact using methods of grinding, ultrasonic treatment, or steam explosion (vigorous steam extrusion, when raw material is fed under the steam pressure and is destroyed by an abrupt pressure drop when it passes through the outlet).

Mechanical impact is combined with material dissolution using chemical reactants, such as alkalis, ammonia, chlorite, sulfur dioxide, amides, diluted and concentrated acids, and, also, hydrolysis by high-temperature steam (220–270°C), microwaves, electron, and gamma irradiations. Examples of commonly used preliminary treatment methods are presented in the review (Sinitsyn et al. 1995).

If additional chemical reactants are used at the stage of preliminary treatment, they should be removed from the final product.

Mechanical grinding and amorphization. Mechanical treatment usually includes shock, shear, pressure, crushing, mixing, compression/expansion, and other kinds of mechanical impact.

An increase in the surface area of the substrate was considered as a reason of the influence of preliminary treatment. The size of cellulosolytic enzyme molecules is about 50 Å. Thus, the surface area accessible for cellulosolytic enzyme is an important factor. The specific surface, measured on the basis of sorption of such small molecules as nitrogen, does not correlate with the rate of enzymatic hydrolysis of the substrate. Surface area determination by means of gas adsorption takes into account small pores that are inaccessible for enzymes and thus do not participate in the enzymatic hydrolysis (Grethlein and Converse 1991).

Coarse particles of the raw material are subjected to soft hydrolysis, during which hemicellulose splits without any essential hydrolysis of cellulose. After the second grinding, the substrate is subjected to enzymatic hydrolysis with cellulosolytic enzymes resulting in formation of the aqueous solution of glucose (Brink 1997). An increase in the efficiency of enzymatic processes after preliminary treatment is explained by formation of micropores during hemicellulose removal and changes in the crystallinity of the substrate, as well as by the decrease in the degree of cellulose polymerization.

Steam explosion treatment. One of the major methods of preliminary treatment of lignocellulose raw materials is steam treatment (Foody 1984). In this method, the biomass is placed into a chamber, the so-called steam gun. An acid solution (up to 1%) is added into the chamber with the biomass. Then the vessel is rapidly filled by steam and kept the required time under high pressure. After that, the pressure drops quickly and the treated biomass is taken out of the chamber. This is why the method is called steam explosion. The action of preliminary treatment depends on the time of pressure impact, temperature, acid concentration, and the size of particles in the raw material. The range of steam pressure is 17–72 atm and the temperature range is 208–285°C.

A flow reactor of continuous type resembling an extruder may be used for preliminary treatment of lignocellulose raw materials (Grethlein 1980). The rotating screws push out the raw material suspension through a small orifice; mechanical impact at the outlet of reactor destroys the structure of the raw material.

There exists a particle size limitation at the first stage of steam explosion methods. Particles that are too small cannot be subjected to steam explosion treatment; the optimal size of particles is 200 μm.

Ultrasonic treatment. Ultrasonic oscillations are elastic mechanical oscillations in a liquid medium within the frequency range of 10^5–10^8 Hz. The ultrasonic wave propagation inside a liquid is accompanied by compression and rarefaction of the medium. Pressure decrease to a certain value below the cohesion strength of the medium results in formation of a cavity that gets filled by the liquid vapor and gases dissolved in the liquid. The cavity collapses with increase of pressure. A rapid adiabatic compression of vapor and gases inside the cavity causes local temperature rise. The size of bubbles is small in comparison to the liquid volume, so heat dissipates and only locally exerts significant effect (de Castro and Capote 2007).

Ultrasonic treatment of lignocellulose raw materials in the presence of enzymes accelerates subsequent hydrolysis of polysaccharides (Yachmenev et al. 2002, 2004). Duration and conditions of the ultrasonic treatment should be chosen to avoid mixture heating above the temperature of enzyme denaturation. The ultrasonic treatment leads to a noticeable destruction of the crystalline structure of cellulose. In comparison to conventional methods, this procedure helps to reduce consumption of enzymes by a factor of 2–3 (Ingram and Wood 2001). According to Ingram and Wood (2001), the ultrasonic activation of the process may be due to disaggregation of the enzyme molecules, an increase in the rate of the enzyme molecule diffusion to the substrate surface through the liquid layer and removal of the products of the enzymatic hydrolysis from the reaction zone. A special chapter dealing with the effect of ultrasound on extraction processes is provided in the monograph, so we will only mention an unexpected feature, which is similar to the features of other methods of mechanical impact. The ultrasonic treatment increases activity of cellulases in the presence of a substrate and decreases activity of free cellulase enzymes without a substrate (Gao et al. 1997).

Due to numerous investigations of the methods for preliminary treatment of lignocellulose raw materials, a better understanding of mechanisms underlying acceleration of the enzymatic hydrolysis was achieved; thus, the basis of the process of sugar isolation from the biomass was formed. However, at present, there are no profitable environmentally safe and industrially applicable methods combining preliminary treatment of lignocellulose and enzymatic hydrolysis of polysaccharides. The most promising approaches are considered in the next section.

13.4.1.5 Mechanoenzymatic Isolation of Saccharides from Lignocellulose Raw Materials

Mechanoenzymatic treatment implies mechanochemical processing of a mixture of reagents and enzymes. This kind of treatment may be applied to plant raw materials of complex structure, to natural polymers with supramolecular structure, and to individual organic polymers. Mechanoenzymatic treatment is aimed to increase the reactivity of substrate; in some cases, chemical reactions catalyzed by enzymes may proceed directly during treatment.

The components of the reactive system get mixed under the effect of intense mechanical loads and arrange in close to each other so that their diffusion paths substantially shorten.

Mechanical treatment of a mixture of solid phases increases mobility of the components. Mobility grows due to disordering of the crystalline structure of solid phase. The solid components may accumulate defects and amorphize; as a result, reactivity of the components and of the system as a whole increases.

From the economical point of view, mechanical energy is an expensive kind of energy; therefore, it is important to use it efficiently. In some cases, mechanical treatment of a solid mixture can be interrupted at the early stage of reactant transformation, whereas complete chemical transformation takes place as a result of other less energy-consuming processes, which are usually carried out in a liquid phase.

In the case of such an approach, we achieve at the stage of mechanochemical treatment:

- Introduction of defects into the crystalline structure of the reactants
- Decrease in the degree of crystallinity and amorphization of reactants
- Formation of mechanocomposites

A mechanocomposite is a product of mechanochemical treatment of a solid heterogeneous mixture. It is a system with physicochemical properties substantially differing from those of the initial mixture, which are determined by significant morphology changes of the components. One of the system components, enzyme, for example, can be mechanically introduced in molecular form into the mass of another biopolymer. Mechanocomposite can be also composed of the particles of different phases; for example, the particles of inorganic alkali can be distributed in the bulk of a biopolymer. Biocomposite differs from conventional mechanical biphase mixture of particles by the presence of developed interface with apparent interfacial interaction. The boundary between phases possesses physicochemical characteristics that are not inherent to any of the initial components and individual phases. It is assumed that the interfacial area in a biocomposite may be increased by a factor of 100 in comparison to a simple mixture of particles.

Figure 13.3 illustrates processing importance of mechanoenzymatic treatment. Mechanoenzymatic treatment of a solid mixture of substrate and enzyme not only affects the enzyme structure and activity but also provides due distribution of the enzyme molecules over the substrate bulk. In the case of the traditional method of substrate introduction into an aqueous enzyme solution (presented schematically to

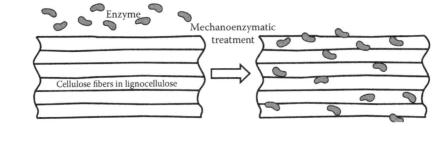

FIGURE 13.3 Mechanochemical treatment results in enzyme introduction into the bulk of lignocellulose material and into reaction zone (to the right); for comparison, addition of the substrate to the aqueous enzyme solution (to the left).

the left of Figure 13.3 for purposes of comparison), the major portion of enzyme molecules turn out to be outside the substrate and fails to be efficient. Mechanical treatment results in distribution of the enzyme molecules over the substrate bulk and their localization in the sites where they act as a catalyst of hydrolysis.

It is better to add the enzyme into the system by mechanical mixing rather than by diffusion through lignocelluloses so that the enzyme would not lose its reactivity while passing from the bulk of solution inside the cellulose particle of the composite.

Thus, the mechanoenzymatic method is based on application of preliminary and/ or intermediate mechanical treatment that enhances the rate of hydrolysis of polysaccharides, increases the yield of carbohydrates, and allows reduction of material and energy consumption. Preliminary treatment involves the mechanochemical treatment of lignocellulose and enzyme mixture with certain intensity and duration in a mechanochemical reactor.

Using soft lignocellulose as a raw material and optimal intensity and duration of mechanical impact provide mechanocomposite formation and enzyme activity preservation, which are necessary for efficient isolation of soluble sugar compounds from lignocellulose and other biopolymers.

Application of preliminary mechanoenzymatic treatment enlarges the degree of lignocellulose conversion up to 90%, whereas the consumption of cellulosolytic enzymes necessary for hydrolysis of polysaccharides decreases substantially in comparison with other known methods.

13.4.1.6 Mechanochemical Extraction of the Yeast Cell Wall Components

The yeast biomass is a multicomponent system containing such important substances as proteins, mannanoproteins, and glucans. In this regard, the majority of physicochemical methods of treating the yeast biomass are aimed at isolation of one or several target components from the biomass; more rarely they can be aimed at simple physical destruction of cells. The most widely spread methods for treatment of microorganisms can be divided into the following groups (Middelberg 1995):

- Mechanical methods involving ball and bead mills (Heim et al. 2007), homogenizing devices and disintegrators (Kim and Yun 2006), and ultrasonic devices (Feliu and Vilaverde 1994)
- Physical methods based on decompression, osmotic shock, and thermolysis
- Chemical methods using antibiotics, chelating agents, organic solvents, detergents, and chaotropic agents (Duda et al. 2004; Danilevich et al. 2006)
- Enzymatic methods involving hydrolytic enzymes (Bozhkov et al. 2004; Milic et al. 2007), or autolysis (Babayan and Latov 1992; Lazzari 2000)
- Irradiation methods (Choi et al. 2007)

The majority of the listed methods has few applications in laboratory practice. For example, treatment of yeast cultures by chaotropic reactants (guanidinium chloride, guanidinium thiocyanate, etc.) provides control of the porosity and permeability of cell walls and isolation of pure nucleic acids from the cells for purposes of genetic investigations. However, commercial application of this method is not advantageous due to high costs of the reactants.

Combined methods are also under development. For example, β-1,3-glucan is obtained using alkaline extraction and enzymatic hydrolysis (Toneeva-Davidova 2002).

Mechanically activated enzymatic hydrolysis may be used for obtaining manna-nooligosaccharide components of functional nutrition and forage. The first procedure of the mechanoenzymatic process is mechanical treatment; it provides supramolecular disordering of the cell wall polymers and increases β-glucan reactivity; the second stage—enzymatic hydrolysis—ensures efficient hydrolysis of glucans and leads to increase in the availability of mannanooligosaccharides (Bychkov et al. 2010).

13.4.2 MECHANOCHEMICAL REACTIONS OF INTRACELLULAR BIOLOGICALLY ACTIVE COMPONENTS RESULTING IN SOLUBLE COMPOUNDS

In relation to chemical composition of the intracellular contents, the vacuole and the cytoplasm should be distinguished in the cell. A typical plant cell has a large vacuole filled with a liquid. The vacuole often occupies almost the whole of the cell volume, and the cytoplasm is a thin layer adjacent to the cell shell. The vacuole contains the cell sap. It is an aqueous solution of many compounds: carbohydrates, organic acids including amino acids, pigments, vitamins, tanning agents, alkaloids, glycosides, and inorganic salts.

Cytoplasm is the inner cell component; it is composed of hyaloplasm and various intracellular structures. Hyaloplasm, the basic substance of the cell, is a transparent gel-like or fine-grained substance composed of water, mineral salts, proteins, carbohydrates, fatty acids, amino acids, and other compounds. Hyaloplasm is permeated with thin protein microtubules forming the cytoskeleton of the cell, which sustains its steady shape. The intracellular structures play the most important role in the vital activity of the cell, but they account only for a very small part of its chemical composition. Extraction of their chemical components requires development of the special biochemical methods.

Drying of plant materials turns the intracellular content into a thin layer lining the cell from inside or gathering in a small region of the intracellular space.

Mechanochemical extraction involves chemical reactions between the components of plant raw material and especially introduced reactants resulting in formation of the soluble compounds.

Discovery of the chemical reactions running during mechanical treatment of the mixtures of low-molecular organic compounds and solid reactants was an achievement of mechanochemistry. In particular, there were demonstrated possibilities of solid-state reactions of neutralization and exchange, esterification and reesterification, numerous reactions of nucleophilic substitution, acylation, etc. It was a surprise for researchers to discover the fact that many reactions of organic compounds in the solid state proceed at a higher rate and with a higher selectivity than in the liquid medium (Boldyreva and Boldyrev 1999; Tanaka and Toda 2000; Dushkin 2004).

Figure 13.4 presents for comparison the schemes of mechanochemically assisted extraction and conventional method of extraction using organic solvents. The scheme on the left side of the figure corresponds to the most widespread method of extraction. According to this method, the raw materials are subjected to grinding and

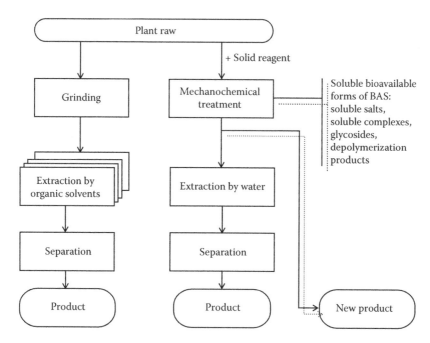

FIGURE 13.4 Comparison of the known methods of extraction by organic solvents and mechanochemically assisted extraction.

multiple extraction using organic solvents of different polarity. The last stage of the process is the separation of the product, e.g., by evaporation of the extracting solvent. The scheme on the right side of the figure is for mechanochemical extraction, when a mixture of a plant raw material and a solid reactant is mechanically treated in the especially designed mills (mechanochemical reactors). During mechanical treatment, the extractable substance transforms into a chemical form most soluble in water. Water is used for separation of the extractable substance.

Mechanochemical treatment yields a powdered product releasing a soluble form of the targeted BAS into the added water. This new product can be used as an individual food additive or a component of functional nutrition. The yield of the targeted product can be higher than that in conventional extraction processes, and the composition of the product of mechanochemically assisted extraction may differ from that of the extract derived using organic solvents. From the practical viewpoint, the key aspect in the mechanochemical technology is the choice of the solid reactant. A solid reactant should be selected that interacts with the raw material producing soluble and biologically available forms of the targeted substance, such as soluble salts, soluble complexes, glycosides, or low molecular products of depolymerization. If selection of a proper reagent is succeeded, the process of new product manufacture appears rather efficient economically, since no liquid extractors are used and no solid or liquid waste is formed. It was shown for a number of alkaloids, acids, glycosides, and phytosterols that this technique allows to increase the yield of the extractable products and selectivity of extraction.

Unfortunately, currently, no general theory can sufficiently describe the mechanochemical reactions occurring during mechanical treatment of the mixtures of reactants (Beyer and Clausen-Schaumann 2005). Only some aspects and possible phenomena, e.g., formation of active surface radicals and the role of interfacial processes (Butyagin 1997), are considered. As far as even dry plant material usually contains substantial amounts of water (up to 15%), the possibility of application of the hydrothermal chemical processes was considered (Boldyrev 2002).

13.4.2.1 Stability of Biologically Active Compounds and Enzymes during Mechanical Treatment

Many organic compounds, especially those having a complex structure, are unstable during intensive mechanical treatment. We have studied the degradation of hypericin, a compound with a complex dianthrone structure, during its mechanical treatment under different conditions and during its storage, because instability and low extraction yield prevent usage of this compound in the food industry (Ang et al. 2004). Different modes of intensive mechanical treatment were applied to hypericum herb. The temperature of the samples immediately after treatment did not exceed 50°C. After activation, the samples were analyzed for the level of hypericin in the product.

Figure 13.5 shows that hypericin starts to decompose almost instantly at very high intensity of mechanical impact (when calculated acceleration of the milling bodies (balls) is 40–60 g). Hypericin degradation may be caused directly by its destruction under mechanical impact, by thermal decomposition of hypericin in the regions of

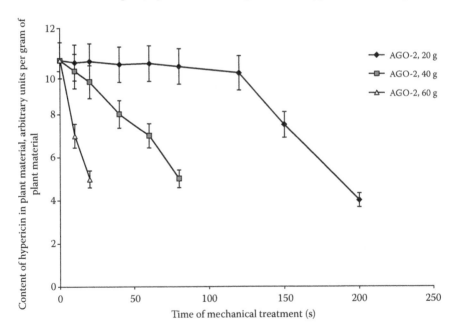

FIGURE 13.5 Hypericin content in the plant raw material versus time and intensity of mechanical treatment in mill activator AGO-2. Acceleration of milling bodies are 20, 40, and 60 g.

local overheating, or by hypericin interaction with free radicals produced by breakage of bonds in the matrix of the plant raw material.

The most interesting is the behavior of samples treated with specific energy input of 5 W/g, which corresponds to the calculated balls acceleration of 20 g. Under such conditions, the hypericin content remains constant until the threshold time equals 120 s and then starts to decrease. Thus, it was shown experimentally that even for a very reactive biologically active compound, such as hypericin, there exist conditions of mechanochemical activation without decomposition of the active substance.

The stability of enzymes under the conditions of mechanical treatment is even more problematic than the stability of other complex organic molecules, because their activity is connected with a certain conformational structure. For estimation of the individual mechanical stability of enzymes, experiments were carried out in which a weighed portion of cellulase (Cellolux 2000, celloveridin) was subjected to mechanical treatment. It was shown (Figure 13.6, curve 1) that a gradual decrease in the cellulolytic activity was observed during mechanical treatment of the enzymatic preparation without the substrate. Activity of the preparation decreases by 10% after 20 s of its mechanical treatment and by 80% after 2 min of treatment (Bychkov and Lomovsky 2010).

It is reasonable to assume that the portion of preparation experiencing the impact of the ball load loses its catalytic activity and does not take part in the reaction.

Quite different is the situation occurring during mechanical treatment in the presence of a substrate (microcrystalline cellulose). In this case, the cellulolytic activity of the preparation increases first and then starts to decrease. Decrease of the cellulolytic activity observed in this case is slower than in the case of enzyme treatment in

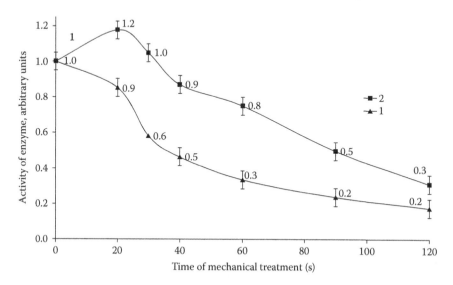

FIGURE 13.6 Changes in activity of the cellulase under mechanical treatment: (1) treatment of the preparation without a substrate and (2) concurrent treatment of the preparation and microcrystalline cellulose. (From Bychkov, A.L., and Lomovsky, O.I., *Russ J Appl Chem* 83, 1106, 2010. With permission.)

the absence of cellulose. A possible cause of initial activity elevation is amorphiza-tion of the substrate and enzyme distribution in the bulk of substrate due to mechani-cal impact.

13.4.2.2 Formation of Soluble Salts

It is characteristic for compounds with pronounced acid–base properties that their salt forms are usually more soluble in water than the relevant molecular acid or base forms. Neutralization of molecular forms in the solid phase during the process of mechanical activation allows to increase the solubility of the target substance and thus to decrease the volumes of solvents used for its subsequent extraction. Mechanochemical reaction of neutralization was observed during mechanical acti-vation of a mixture of ascorbic acid with sodium (or potassium) carbonate and dicar-bonate. Mechanochemical interaction of ascorbic acid with potassium hydroxide also results in formation of ascorbinates; in contrast to reaction running in a solution, it does not cause the opening of the lactone cycle (Boldyrev 1997).

Trace amounts of solvents, mainly water, facilitate mass transfer processes and accelerate the heterogeneous reactions. This effect is especially easy to observe in the systems where pronounced donor–acceptor interaction between the acid and the base components is possible.

For example, mechanochemical interaction of 1,3,5-cyclohexane tricarboxylic acid with weak bases in the presence of microamounts of methanol or cyclohexane results in a crystalline complex of 1:1 composition (Shan et al. 2002); this reaction does not proceed under completely anhydrous conditions. Mechanochemical treat-ment with solid bases under certain conditions results in formation of the salts of triterpenic acids directly in the matrix of a plant raw material (Korolev et al. 2003). The yield of extraction of the target components increases by a factor of 1.5–2.

For instance, mechanical activation of eleutherococcum without additional reac-tants results in increase of the yield of isoflaxocin extraction by a factor of 1.5; note that addition of 2% sodium carbonate causes a fourfold increase in extraction yield due to the formation of isoflaxocin salt (Liu et al. 2007).

The solid phase reaction of neutralization has rather high activation energy and requires substantial energy consumption. The limiting factor of the solid-state reac-tion is mass transfer. Quite often, great efforts are required for efficient transfer of large organic molecules; it results in degradation of substances under the effect of mechanical strain, substantial heating of material and thermal decomposition of the target components. This is the reason why mechanochemical reaction of neutraliza-tion is often terminated before formation of the final salt, at the stage of a composite, which quickly transforms into the salt with water addition or under heating.

13.4.2.3 Formation of Soluble Complexes

For substances without acid–base properties, there exists a possibility of mechano-chemical reactions yielding soluble complexes with organic ligands. In this case, a complex rather than an individual substance passes into solution. If the solubility of this complex is higher than that of the initial compound, it will be possible to attain higher levels of the active substance in solution. Such an approach is traditionally used for increasing the solubility of poorly soluble pharmaceutical substances.

Selective preparation of molecular complexes with characteristics more attractive from the practical standpoint is used in pharmacology. In some cases, the purposed change in solubility or biological activity of a substance, or reduction of its toxic effect, appears to be possible. Mechanochemical synthesis of the complexes of triterpenic glycosides with acetyl salicylic acid and orthofen (Sorokina et al. 2002), ursodeoxycholic acid with phenanthrene (Oguchi et. al 2000), cholic acid with ibuprofen (Oguchi et al. 2002), and D-glucose with carbamide (Kaupp 2003) can be referred to as examples.

The components of the plant raw materials may be used as complexing agents increasing the solubility of low-soluble inorganic substances. Gallocatechins, the polyphenolic compounds of green tea, as components of a plant matrix, enter the mechanochemical reaction with silicon dioxide, producing soluble silicon dioxide compounds.

Figure 13.7 demonstrates that mechanical activation of silicon dioxide in the presence of gallocatechins causes a twofold to threefold increase in silicon dioxide dissolution at neutral pH in comparison with the samples of silicon dioxide mechanical activated under the same conditions but without any additives. An increase in the rate of dissolution is achieved due to the formation of chelate complex compounds of catechins and silicon dioxide (Shapolova et al. 2010). The medicines on the basis of chelate complexes of silicon dioxide are promising for prevention of osteoporosis (Calomme et al. 1997).

13.4.2.4   Reactions of Glycosylation and Reglycosylation

Reactions of glycosylation are the least studied reactions of mechanochemistry despite the fact that the role of glycosides in the transport of BAS, especially poorly soluble ones, in liquids of living organisms can hardly be overestimated. Mechanochemical

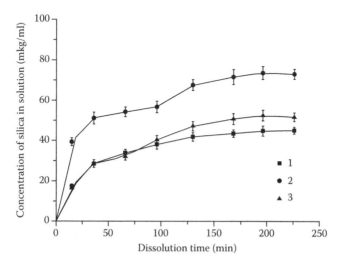

FIGURE 13.7   Dynamics of silicon dioxide dissolution for silicon dioxide from mechanocomposite on the basis of green tea and silica gel: (1) activated silica gel, (2) concurrently activated silica gel, and tea powder (10:1), and (3) activated silica gel in tea solution.

treatment is applied to the crowned serratula and leusea for further isolation of phytoecdysteroids.

A lot of attention is paid to natural biologically active additives with an adaptogenic activity in relation to development of a healthy nutrition concept, as well as prevention and treatment of many diseases. The necessity in improvement of the quality of life is especially acute in the north, where humans are burdened by a number of ecologically unfavorable factors. Plant ecdysteroids have proved to be good food additives in the past two decades; they possess a complex of useful characteristics and exert no side effects on humans and animals (Kholodova 2001).

Methods traditionally used for isolation of phytoecdysteroids include defatting of the raw material, extraction with water, re-extraction with ethyl acetate, chromatographic separation, and crystallization. The major products are 20-hydroxyecdysone and inocosterone (Dinan 2001). Some types of raw materials contain high concentrations of phytoecdysteroids (up to 1.5–2.0%); however, even in this case, the cost of the extracted substances is too high for use as components of functional nutrition.

An alternative method for preparation of available phytoecdysteroids using raw materials with low content of target components is mechanochemical activation of the plant raw material with saccharose.

Mechanochemical treatment of initial raw materials containing a number of ecdysteroids changes the ratio between 20-hydroxyecdysone and inocosterone in comparison to water-extracted product from 1:1 to 2:1. Mechanochemical treatment is likely to result in reglycosylation: destruction of the glycoside bonds between 20-hydroxyecdysone and insoluble cellulose and formation of new bonds with soluble saccharose. The total yield of water-extracted phytoecdysteroids increases by more than 1.5 times (Lomovsky and Korolev 2004).

13.4.2.5 Formation of Surface Complexes with Soluble Compounds

Solid dispersions composed of active poorly soluble substance distributed in a neutral soluble matrix are traditionally used in pharmacology (Shakhtshneider et al. 1997; Boldyrev 2004). Mechanochemical treatment is an alternative method for preparation of dispersions, which can replace procedures with concurrent melting of the active substance and the carrier. The advantage of the mechanochemical approach lies in the absence of the need in heating the substance, which reduces probability of the processes of partial decomposition and, moreover, excludes the use of solvents.

Compounds of different nature can be used as a carrier. The most commonly used in practice are soluble polymers containing carboxylic or hydroxy groups (polyvinyl alcohol, polyethylene glycol, polylactic acid, carboxymethylcellulose), cyclodextrins and their acyclic analogues, as well as low-molecular carriers capable of holding the substances through donor–acceptor interactions (saccharose, lactose, mannitol). The mechanism of carrier activity may be connected with formation of the surface molecular complexes, micellar solubilization, formation, and stabilization of a metastable polymorphous modification (Krishna and Flanagan 1989; Boldyrev et al. 1994; Watanabe et al. 2002).

Determination of the mechanism of interaction is a separate problem that sometimes turns out to be very complicated. The indirect evidence of the surface compound formation are changes in the IR spectra. For example, water-soluble polyethylene glycol is a very efficient tool of mechanochemical solubilization. It was shown that mechanical treatment of the mixture of polyethylene glycol and ibuprofen results in formation of stable surface complexes. The solubility of polyethylene glycol based molecular dispersions is much higher than solubility of the active substance, and it makes possible the several-fold increase of active substance concentration in solution (Shakhtshneider et al. 1996). The polylactic acid tightly holds 20-hydroxyecdysone phytoecdysteroid (Ditrich et al. 2000). Mechanical activation of the sulfathiazol–polyvinylpyrrolidone system causes amorphization of sulfathiazol and formation of hydrogen bonds with the matrix, which makes possible formation of surface molecular complexes. The solubility of the preparation changes along with changes in the mass content of the polymer (Kaupp et al. 2001). The low-molecular hydrocarbons and their analogues also may be used as carriers.

Phytosterols occur in many natural sources, however, in a form unavailable for living organisms. The development of the methods allowing transfer of these substances into biologically available forms opens up new possibilities for production of anticholesterol components of functional nutrition.

Mechanochemical treatment of the sterol-containing plant raw material and soluble saccharides results in formation of a powdered product. Figure 13.8 illustrates the difference between the powder obtained by mechanical treatment of the raw material (millet husk) and the product of treatment of the mixture of raw material and a soluble saccharide obtained under the same conditions. After addition of water the mechanochemical product releases soluble phytosterols into water (Lomovsky and Salenko 2004).

The mechanochemical treatment with polymers can be also applied to dry extracts prepared in advance. Interaction of dry extracts of flavonolignans with carboxymethylcellulose, polyvinylpyrrolidone, and β-dextrin allows 2- to 31-fold enhancement of the solubility of flavonolignans (Voinovich et al. 2009a,b). The product is characterized by a high solubility and high rate of dissolution, which is much greater than that of the preparations obtained by other methods.

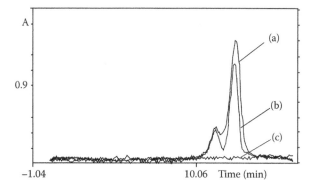

FIGURE 13.8 HPLC chromatogram with electrochemical detection: (a) reference sample prepared from the soluble forms of campesterol and β-sitosterol, (b) product of the mechanochemical treatment of the raw material in the presence of a soluble saccharide, and (c) product of the mechanical treatment of the raw material without a saccharide.

Mechanical activation with insoluble matrices helps in solving the reverse problem of diminishing the rates of dissolution. For example, mechanical treatment of the mixtures of benzimidazole with microcrystalline cellulose and other polymers (Khalikov et al. 1995) makes variation of benzimidazole solubility by an order of magnitude possible.

13.4.3 FORMATION OF BIOCOMPOSITES

Formation of composites contained uniformly distributed particles of the plant raw material and reagent and characterized by increased interfacial area occurs at early stages of mechanical treatment and enhances the rate of chemical processes that take place either during mechanical activation or during dissolution of the composite.

The size of the plant cells is about 3–10 μm, the thickness of cell walls reaches 100–500 nm. Mechanoenzymatic treatment results in a composite in which the initial cell structure is completely destroyed and 200- to 600-nm fragments of the cells are components of the composite.

Chemical interactions in the raw material–solid reagent system accelerate due to destruction of the cell walls and increase of surfaces available for interaction in subsequent heterogeneous reactions. The term *nanobiocomposites* reflects the biological origin of the composite and the level of its spatial organization.

We consider below as an example the preparation of nanobiocomposites containing triterpene compounds. Triterpene compounds are among the most important classes of organic compounds of plant origin. The triterpenic acids are components of popular food products: onion and cranberry fruits. It is believed that active components of these products affect composition of the intestinal flora depressing the pathogenic microorganisms. The flour of dried hawthorn fruits is used as an additive for baking bread and confectionery and also as a component of functional nutrition—enhancing blood circulation in the brain vessels and relieving heart discomfort. The triterpenic acids are components of panax saponins panaxosides.

They play a role of hormones in the vital activity of plants and thus are actively used for control of the growth and development of agricultural plants. Almost all triterpene compounds have an irregular pentacyclic carbon skeleton; but some of them possess a nucleus composed of four rings, which is characteristic of steroids. The most widespread triterpene acids are represented by the lanostane-type terpenoids, such as 3-hydroxy-23-oxolanostanic acid, 3,23-dioxo acid, and other related compounds.

Parameters of mechanochemical treatment for biocomposite preparation, the ratio of components, period of exposure to the impact of milling bodies, etc. should be chosen depending on characteristics of the raw material, mainly its humidity and the content of triterpenic acids (Korolev and Lomovsky 2007). The reactant (alkali) concentration in the mixture subjected to treatment or, otherwise, time of mechanochemical treatment required for efficient solubilization of the acids increases to thechologically unacceptable values. The efficiency of neutralization of the acidic components sharply falls down if water content in the plant raw material is less than 10%.

Additional introduction into the system of the chemically neutral abrasives allowed for use in the food industry, such as zeolites, increases the ease of biocomposite manufacture. The presence of abrasive material in the mixtures subjected to treatment promotes destruction of the cell walls and provides fine grinding of the raw material.

Also, inclusion of the small abrasive particles into the particles of plant raw material takes place at the stage of mechanical activation; it results in increase of the apparent density of the product, which simplifies its settling and centrifugation of water extracts.

When composite is kept in water for 20 min, more than 75% of triterpenic acids contained by the raw material transform into the salts and pass into solution. Thus, mechanochemical treatment of the plant raw material mixed with solid alkalis and auxiliary substances can be used for obtaining mechanocomposite powders containing triterpenic acids in a water-soluble form. Mechanochemically obtained powders may be extracted by water for preparation of triterpenate solutions. Mechanochemical treatment enlarges the yield of triterpene compounds from 2.5% to 4.5% by weight of the dry raw material and improves the qualitative composition of triterpenic acids. Such powders may be used directly as a food additive.

13.5 PROSPECTS OF MECHANOCHEMICAL APPROACH IN EXTRACTION TECHNOLOGY

13.5.1 INCREASE OF THE YIELD OF EXTRACTABLE SUBSTANCES

The yield of extractable substances from biogenic raw materials is determined by a number of factors, e.g., a possibility of optimal conditions for grinding the particles of raw material; the degree of destruction of the plant tissue cells, or improved conditions of diffusion processes. However, the main factor that is distinct in kind from the factors of liquid phase processes is the possibility of conversing the variety of chemical forms of the extractable compound into one most soluble in the extractant. As the concentration of the reactant in the particle formed by the mechanochemical route is much higher than the reactant concentration that can be delivered into the particle from solution, the efficiency of extraction increases. Studies show that the most efficient mechanochemical extraction is the extraction from stronger tissues of roots and stems. Mechanochemical approaches are poorly efficient in the cases when tissues contain high concentrations of soluble compounds in the initial state, as, for example, flowers and young leaves. The scale of extraction yield increase may be illustrated by specific examples, either those mentioned above or those presented below. In some of the studied cases, the ratio of the amount of mechanochemically extractable substance to the amount of the substance isolated by liquid-phase extraction, in grams per 100 grams of dry raw material, is as follows: 2.0/0.8 for isoquinoline alkaloid berberin; 1.3/0.5 for purine alkaloid theobromine; 5.0/2.0 for alantolactones and isoalantolactones; 0.040/0.014 for cardiac glycoside erychroside.

13.5.2 INCREASE IN THE SELECTIVITY OF ISOLATION

Mechanochemical techniques can not only increase in the solubility of required components but also decrease the yield of ballast substances. Mechanochemical reaction

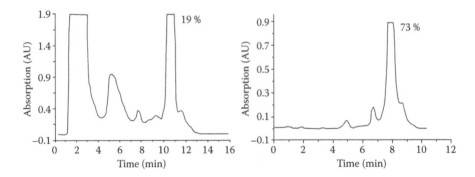

FIGURE 13.9 Chromatograms of the lappaconitine-containing product of extraction for the cases of extraction by solvents (left) and mechanochemically assisted extraction (right).

of ballast substances with an especially introduced insoluble compound can fix them on the surface preventing their release into solution. For example, the alkaloid lappaconitine is a substance extractable quite easily from northern aconitum. The major problem of its isolation from the plant is the high total content of water-soluble ballast substances. Mechanochemical activation of the plant raw material with an insoluble Al_2O_3 matrix, which does not interact with lappaconitine, allows achievement of a substantial decrease of the total yield of water-soluble ballast substances. Figure 13.9 presents chromatograms of the products isolated using two raw materials. They show that the lappaconitine content in the extract of the mechanochemically activated sample is 73%, whereas in the extract of the initial sample, it is 19%.

It simplifies substantially subsequent stages of product purification (Pankrushina et al. 2007).

13.5.3 CHANGES IN THE QUALITATIVE COMPOSITION OF EXTRACTABLE SUBSTANCES

The BAS of plant origin often can be arranged into the groups of related compounds differing by size and number of functional groups. The study of triterpenates mechanochemically extracted through their interaction with solid alkalis has shown that not only monocarboxylic triterpenic acids but also polycarboxylic triterpenic compounds efficiently react with an alkali under the effect of mechanical impact, which increases substantially the yield of triterpenic acids.

As a result, the ratio of components in solution changes in favor of more polar polycarboxylic acids, which cannot be achieved by traditional methods (Korolev et al. 2003).

13.5.4 POSSIBILITIES OF UNSTABLE BAS EXTRACTION

Mechanochemical treatment not only increases extraction yields but also makes it possible to obtain substances that are extremely difficult to extract using conventional procedure. The kinetics of solid state reactions is usually several orders of magnitude slower than that of reactions in solution. In particular, the rates of hypericin degradation in the air in the absence and in the presence of liquid phase are $(1.2 \pm 0.4) \times 10^{-8}$ and $(2.3 \pm 0.7) \times 10^{-5}$, respectively.

The active substance can remain in the raw plant material without any change for a year or more but will rapidly degrade if extracted once they enter oxidation reactions or polymerization. To achieve a reasonable yield, it is necessary to carry out extraction under oxygen-free conditions.

Mechanochemical treatment of a mixture of powdered raw material and solid alkali is carried out in the solid state and finally results in formation of the salts of hypericin, and their solubility in water is times higher than that of initial hypericin. Mechanochemically treated material releases hypericin into solution tens of times quicker, which allows a 5- to 10-fold increase in extraction yield.

13.6 CONCLUSIONS

The application of mechanochemically assisted extraction of plant raw materials allows the enhancement of yield of biologically active and nutritious substances. In the case of unstable substances, their extraction yields can be increased by one order of magnitude. The selectivity of extraction can be also enhanced and qualitative composition of the extractable substances can be improved due to formation of water soluble components in mechanochemical reactions.

REFERENCES

Ang, C.Y., Hu, L., Heinze, T.M., Cui, Y., Freeman, J.P., Kozak, K., Luo, W., Liu, F.F., Mattia, A., DiNovi, M. 2004. Instability of St. John's wort (*Hypericum perforatum L.*) and degradation of hyperforin in aqueous solutions and functional beverage. *J Agric Food Chem* 52: 6156–6164.

Avvakumov, E.G., Senna, M., Kosova, N.V. 2001. *Soft Mechanochemical Synthesis: A Basis for New Chemical Technologies.* Boston: Kluwer Academic Publishers.

Babayan, T.L., Latov, V.K. 1992. Extraction of physiologically active mannan and other polysaccharides from cooking yeast autolisates. *Biotechnol* 2: 23–26.

Balaz, P. 2008. *Mechanochemistry in nanoscience and minerals engineering.* Springer.

Balaz, P., Dutkova, E. 2009. Fine milling in applied mechanochemistry. *Miner Eng* 22: 681–694.

Belousova, N.I., Gordienko, S.V., Eroshin, V.K., Ilchenko, V.A. 1990. Production of amino acid mixtures from autolisates of yeast *Saccharomyces* grown on ethanol or sugars. *Biotechnol* 3: 6–9.

Beyer, M.K., Clausen-Schaumann, H. 2005. Mechanochemistry: The mechanical activation of covalent bonds. *Chem Rev* 105: 2921–2944.

Bhat, M.K. 2000. Cellulases and related enzymes in biotechnology. *Biotechnol Adv* 18: 355–383.

Biryuzova, V.I. 1993. *Ultrastructural Organization of Yeast Cell.* Moscow: Nauka.

Boldyrev, V.V. 1986. Mechanochemistry of inorganic solids. *Proc Natl Ind Sci Acad A* 52: 400–417.

Boldyrev, V.V. 1997. Application of mechanochemistry in development of "dry" technological processes, *Soros Educ J* 12: 48–52.

Boldyrev, V. 2002. Hydrothermal reactions under mechanochemical action. *Powder Technol* 122: 247–254.

Boldyrev, V.V. 2004. Mechanochemical modification and synthesis of drugs. *J Mater Sci* 39: 5117–5120.

Boldyrev, V.V., Shakhtshneider, T.P., Burleva, L.P., Vasilchenko, M.A., Severtsev, V.A. 1994. Technological processes based on solid state chemical reactions. 1. Production of fast soluble dispersed forms of medicines via method of mechanical activation, *Chem Sustain Dev* 2: 455–459.

Boldyreva, E., Boldyrev, V. (Eds.). 1999. *Reactivity of Molecular Solids*. London: John Wiley & Sons.

Bozhkov, A.I., Leonova, I.S., Oblak, V.I. 2004. Ablation of *Saccharomyces cerevisiae* cell walls by *Chaetomium globossum* enzyme complex. *Biotechnol* 6: 46–53.

Brink, D. L. 1993. Method of treating biomass material. U.S. Patent 5221357.

Brink, D. L. 1996. Method of treating biomass material. U.S. Patent 5536325.

Brink, D.L. 1997. Enzymatic hydrolysis of biomass material. U.S. Patent 5628830.

Brooks, T.A., Ingram, L.O. 1995. Conversion of mixed office paper to ethanol by genetically engineered *Klebsiella oxytoca* strain P2. *Biotechnol Progr* 11: 619–625.

Butyagin, P. Yu. 1997. The role of interphases in low temperature reactions of mechanochemical synthesis. *Colloid J Russ Acad* 59: 460–467.

Bychkov, A.L., Korolev, K.G., Lomovsky, O.I. 2010. Obtaining mannanooligisaccharide preparations by means of the mechanoenzymatic hydrolysis of yeast biomass. *Appl Biochem Biotech* 162: 2008–2014.

Bychkov, A.L., Korolev, K.G., Ryabchikova, E.I., Lomovsky, O.I. 2010. Changes of cell walls under mechanical activation of plant and yeast biomass. *Chem Plant Raw Mater* 1: 49–56.

Bychkov, A.L., Lomovsky, O.I. 2010. Mechanical treatment effect on the cellulolytic specimen activity. *Russ J Appl Chem* 83: 1106–1108.

Calomme, M. Berghe, D. 1997. Supplementation of calves with stabilized orthosilicic asid *Biol. Trace Elem.* 56: 153–156.

Choi, S.K., Kim, J.H., Park, Y.S., Kim, Y.J., Chang, H.I. 2007. An efficient method for the extraction of astaxanthin from the red yeast *Xanthophyllomyces dendrorhous*. *J Microbiol Biotechn* 17: 847–852.

Cullingford, H.S., George, C.E., Lightsey, G.R. 1993. Apparatus and method for cellulose processing using microwave pretreatment. U.S. Patent 5196069.

Danilevich, V.N., Petrovskaya, L.E., Grishin, E.V. 2006. Fast and effective extraction of soluble proteins from gram negative micro organisms without breakage of cell walls. *Bioorg Khim* 32: 579–588.

Dawson, K.A., Sefton, A.E. 2006. Method and compositions for control of coccidiosis. U.S. Patent 7048937.

de Castro, L.M.D., Capote, P.F. 2007. Techniques and instrumentation in analytical chemistry—V. 26: Analytical applications of ultrasound. Amsterdam: Elsevier.

Dinan, L. 2001. Phytoecdysteroids: Biological aspects. *Phytochemistry* 57: 325–339.

Ditrich, M., Solich, P., Opletal, L., Hunt, A.J., Smart, J.D. 2000. 20-Hydroxyecdysone release from biodegradable device: The effect of size and shape. *Drug Dev Ind Pharm* 26: 1285–1291.

Dubinskaya, A.M. 1999. Transformation of organic substances under action of mechanical stresses. *Usp Khim* 68: 708–724.

Duda, V.I., Danilevich, V.N., Suzina, N.E., Shorokhova, A.P., Dmitriev, V.V., Mokhova, O.N., Akimov, V.N. 2004. Changes of fine structure of micro organisms cell walls under action of chaotrope salts. *Microbiology* 73: 406–415.

Dushkin, A.V. 2004. Possibilities of mechanochemical technology in organic synthesis and material chemistry. *Chem Sustain Dev* 12: 251–274.

Emmerik, L.C.V., Kuijper, E.J., Fijen, C.A.P. 1994. Binding of mannan-binding protein to various bacterial pathogens of meningitis. *Clin Exp Immunol* 97: 411–416.

Feliu, J.X., Vilaverde, A. 1994. An optimized ultrasonic protocol for bacterial cell disruption and recovery of beta-galactosidase fusion proteins. *Biotechnol Tech* 8: 509–514.

Ferraza, A., Guerra, A., Mendonc, R., Masarin, F., Vicentim, M.P., Aguiar, A., Pavan, P.C. 2008. Technological advances and mechanistic basis for fungal biopulping. *Enzyme Microb Technol* 43: 178–185.

Foody, P. 1984. Method for increasing the accessibility of cellulose in lignocellulosic materials, particularly hardwoods agricultural residues and the like. U.S. Patent 4461648.

Gao, D., Yu, S., Min, Y. 1997. Effect of ultrasonic waves on cellulase vitality. *J S China Univ Technol (Nat Sci)* 11: 22–26.

Grethlein, H.E. 1978. Chemical breakdown of cellulosic materials. *J Appl Chem Biotechn* 28: 296–308.

Grethlein, H.E. 1980. Process for pretreating cellulosic substrates and for producing sugar therefrom. U.S. Patent 4237226.

Grethlein, H.E., Converse, A.O. 1991. Common aspects of acid prehydrolysis and steam explosion for preheating wood. *Bioresource Technol* 36: 77–82.

Gusakov, A.V., Sinitsyn, A.P., Davydkin, I.Y., Davydkin, V.Y., Protas, O.V. 1996. Enhancement of enzymatic cellulose hydrolysis using a novel type of bioreactor with intensive stirring induced by electromagnetic field. *Appl Biochem Biotech* 56: 141–153.

Harthmann, C., Delgado, A. 2004. Numerical simulation of the mechanics of a yeast cell under high hydrostatic pressure. *J Biomech* 37: 977–987.

Heim, A., Kamionowska, U., Solecki, M. 2007. The effect of microorganism concentration on yeast cell disruption in a bead mill. *J Food Eng* 83: 121–128.

Ingram, L.O., Wood, B.E. 2001. Ethanol production from lignocelluloses. U.S. Patent 6333181.

Kalebina, T.S., Kulaev, I.C. 2001. Role of proteins in formation of structure of yeast cell walls. *Usp Biol Khim* 41: 105–130.

Kanegae, Y., Sugiyama, Y., Minami, K. 1989. Method for producing yeast extract. U.S. Patent 4810509.

Kasatkina, N.S., Levagina, G.M. 1999. Disintegration of micro organisms in tornado mill. *Obrab Dispers Mater Sred* 9: 66–68.

Kaupp, G. 2003. Solid-state molecular syntheses: Complete reactions without auxiliaries based on the new solid-state mechanism. *CrystEngComm* 5: 117–133.

Kaupp, G., Schmeyers, J., Boy, J. 2001. Waste-free solid-state syntheses with quantitative yield. *Chemosphere* 43: 55–61.

Kaya, F., Heitmann, J.A., Jr., Joyce, T.W. 1994. Cellulase binding to cellulose fibers in high shear fields. *J Biotechnol* 36: 1–10.

Khalikov, S.S., Zhuravleva, G.P., Larin, P.P., Kogan, G.P., Arkhipov, K.N. 1995. Study of micro crystal cellulose properties after mechanical activation with benzimidazyl medicines. *Chem Nat Compd* 2: 303–307.

Kholodova, Y.D. 2001. Phytoecdysteroids: Biological effects, application in agriculture and complementary medicine. *Ukr Biokhim Zh* 73 (3): 21–26.

Khrenkova, T.M. 1993. *Mechanochemical Activation of Coals.* Moscow: Nedra.

Kim, K.S., Yun, H.S. 2006. Production of soluble β-glucan from the cell wall of *Saccharomyces cerevisiae*. *Enzyme Microb Technol* 39: 496–500.

Korolev, K.G., Lomovskii, O.I., Rozhanskaya, O.A., Vasil'ev, V.G. 2003. Mechanochemical preparation of water-soluble forms of triterpene acids. *Chem Nat Compd* 39: 366–372.

Korolev, K.G., Lomovsky, O.I. 2007. Method of extraction of biologically active mix of triterpenic acids. RU Patent 2303589.

Krishna, A.K., Flanagan, D.R. 1989. Micellar solubilization of a new antimalarial drug, beta-arteether. *J Pharm Sci* 78: 574–576.

Lazzari, F. 2000. Product based on polysaccharides from bakers' yeast and use as a technological coadjuvant for bakery products. U.S. Patent 6060089.

Liu, Y., Jin, L.-J., Li, X.-Y., Xu, Y.P. 2007. Application of mechanochemical pretreatment to aqueous extraction of isofraxidin from *Eleutherococcus Senticosus*. *Ind Eng Chem Res* 46: 6584–6589.

Lomovsky O.I., Korolev, K.G. 2004. Method of ecdisteroids extraction from plant raw. RU Patent 2230749.

Lomovsky, O.I., Salenko, V.L. 2004. Premix. RU Patent 2223662.

Lomovsky, O.I., Boldyrev, V.V. 2006. *Mechanochemistry for Solving Environmental Problems.* Novosibirsk: GPNTB SO RAN.

Maijala, P., Kleen, M., Westin, C., Poppius-Levlin, K., Herranen, K., Lehto, J.H., Reponen, P., Maentausta, O., Mettal, A., Hatakka, A. 2008. Biomechanical pulping of softwood with enzymes and white-rot fungus *Physisporinus rivulosu. Enzyme Microb Technol* 43: 169–177.

Middelberg, A.P.J. 1995. Process-scale disruption of microorganisms. *Biotechnol Adv* 13: 491–551.

Milic, T.V., Rakin, M., Silver-Marinkovic, S. 2007. Utilization of baker's yeast (*Saccharomyces cerevisiae*) for production of yeast extract: Effects of different enzymatic treatments on solid, protein and carbohydrate recovery. *J Serb Chem Soc* 72: 451–457.

Minina, S.A., Kauchova, I.E. 2009. *Chemistry and Technology of Plant Preparations.* Moscow: GEOTAR Publishing Group.

Molchanov, V.V., Buyanov, R.A. 2000. Mechanochemistry of catalysts. *Russ Chem Rev* 69: 476–493.

Nazhad, M.M., Ramos, L.P., Paszner, L., Sadler, J.N. 1995. Structural constraints affecting the initial enzymatic hydrolysis of recycled paper. *Enzyme Microb Technol* 17: 68–74.

Oguchi, T., Kazama, K., Yonemochi, E., et al. 2000. Specific complexation of ursodeoxycholic acid with guest compounds induced by co-grinding. *Phys Chem Chem Phys* 2: 2815–2820.

Oguchi, T., Tozuka, Y., Hanawa, T., et al. 2002. Elucidation of solid-state complexation in ground mixtures of cholic acid and guest compounds. *Chem Pharm Bull* 50: 887–891.

Pandey, A. 2003. Solid-state fermentation. *Biochem Eng J* 13: 81–84.

Pankrushina, N., Lomovsky, O., Boldyrev, V. 2007. The new "green" methodology for isolation of natural products from medicinal plants utilizing a mechanochemical approach. *Planta Med* 73: 931.

Pereira, R.S. 2000. Detection of the absorbtion of glucose molecules by living cell using atomic force microscopy. *FEBS Lett* 475: 43–46.

Politov, A.A., Bershak, O.V., Lomovsky, O.I. 2008. Production method of wood powder. RU Patent 2318655.

Prut, E.V., Zelensky A.N. 2001. Chemical modification and mixing of polymers in extruder-reactor. *Usp Khim* 70: 72–87.

Raven, P.H., Evert, R.F., Eichhorn, S.E. 1986. *Biology of Plants.* London: Worth Publishers.

Shakhtshneider, T.P., Vasilchenko, M.A., Politov, A.A., Boldyrev, V.V. 1997. Mechanochemical preparation of drug carrier solid dispersions. *J Therm Anal Calorim* 48: 491–501.

Shakhtshneider, T.P., Vasiltchenko, M.A., Politov, A.A., Boldyrev, V.V. 1996. The mechanochemical preparation of solid disperse systems of ibuprofen–polyethylene glycol. *Int J Pharm* 130: 25–32.

Shan, N., Toda, F., Jones, W. 2002. Mechanochemistry and co-crystal formation: Effect of solvent on reaction kinetics. *Chem Commun* 20: 2372–2373.

Shapolova, E.G., Korolev, K.G., Lomovsky, O.I. 2010. Mechanochemical interaction of silica with chelating polyphenol substances and production of molecular silicon forms. *Chem Sustain Dev* 18: 663–668.

Simushkin, S.D., Marone, I.Y., Zysin, L.B. 1990. Production method of wood powder. U.S. Patent 1591924.

Sinitsyn, A.P., Gusakov, A.V., Chernoglazov, V.V. 1995. *Bioconversion of lignocellulose materials.* Moscow: Moscow University Publishing.

Sorokina, I.V., Tolstikova, T.G., Dolgikh, M.P., et al. 2002. Pharmacological activity of complexes of nonsteroidal anti-inflammatory drugs with glycyrrhizic acid obtained by liquid and solid-state synthesis. *Pharm Chem J* 36: 11–13.

Tanaka, K., Toda, F. 2000. Solvent-free organic synthesis. *Chem Rev* 100: 1025–1074.

Tassinary, T., Macy, Ch. 1977. Differential speed two roll mill pretreatment of cellulose materials for enzymatic hydrolyses. *Biotechnol Bioeng* 19: 1321–1330.

Temuujin, K.O., Okada, K., MacKenzie, K.J.D. 1998. Role of water in the mechanochemical reactions of MgO–SiO$_2$ systems. *J Solid State Chem* 138: 169–177.

Toneeva-Davidova, E.G. 2002. Method of production of beta-glucanes of yeast call walls. RU Patent 2216595.

Voinovich, D., Perissutti, B., Grassi, M., Passerini, N., Bigotto, A. 2009a. Solid state mechanochemical activation of *Silybum marianum* dry extract with betacyclodextrins: Characterization and bioavailability of the coground systems. *J Pharm Sci* 98: 4119–4129.

Voinovich, D., Perissutti, B., Magarotto, L., Ceshia, D., Guiotto, P., Billia, A.R. 2009b. Solid state mechanochemical simultaneous activation of the constituents of the *Silybum marianum* phytocomplex with crosslinked polymers. *J Pharm Sci* 98: 215–228.

Watanabe, T., Ohno, I., Wakiyama, N., Kusai, A., Senna, M. 2002. Stabilization of amorphous indomethacin by co-grinding in a ternary mixture. *Int J Pharm* 241: 103–111.

Wold, A.E., Mestecky, J., Tomana, M., Kobata, A., Ohbayashi, H., Endo, T., Edén, C.S. 1990. Secretory immunoglobulin a carries oligosaccharide receptors for *Escherichia coli* type 1 fimbrial lectin. *Infect Immun* 58: 3073–3077.

Yachmenev, V.G., Bertoniere, N.R., Blanchard, E.J. 2002. Intensification of the bio-processing of cotton textiles by combined enzyme/ultrasound treatment. *J Chem Technol Biot* 77: 559–567.

Yachmenev, V.G., Blanchard, E.J., Lambert, A.H. 2004. Use of ultrasonic energy for intensification of the bio-preparation of greige cotton. *Ultrasonics* 42: 87–91.

14 Reverse Micellar Extraction of Bioactive Compounds for Food Products

A. B. Hemavathi, H. Umesh Hebbar, and Karumanchi S. M. S. Raghavarao

CONTENTS

14.1 INTRODUCTION

Downstream processing (DSP) is an integral part of the bioprocess industry. It usually consists of a series of separation and purification steps, which finally aim to obtain the product at a desired level of purification. The separation of biomolecules is often performed in batch mode by small-scale processes such as column chromatography, electrophoresis, salt, or solvent precipitation, for which scale-up poses a considerable problem, making them uneconomical unless the product is of high value. Liquid–liquid extraction (LLE) is a traditional chemical engineering unit operation for which the design and scale-up are already established. LLE is well known to operate continuously on a large scale with high throughput; unlike conventional chromatography, electrophoresis and precipitation techniques, ease of operation, and high flexibility in its mode of operation are additional advantages (Treybal 1980). In recent years, LLE methods like aqueous two-phase extraction (ATPE) and reverse micellar extraction (RME) have been recognized as potential DSP techniques for bioactive components. Selective extraction of a target biomolecule from mixture/ crude extract into reverse micelles (RMs) can be achieved by varying extraction parameters (Harikrishna et al. 2002).

RME is an attractive LLE method for DSP of biological products, as many biochemicals including amino acids, proteins, enzymes, and nucleic acids can be solubilized within reverse micelles (RMs) and recovered easily without loss of native function/activity. Other biomaterials such as large biopolymers like DNA (Goto et al. 1999, 2004), organelles, and even entire cells have also been solubilized into the RMs (Cinelli et al. 2006). Some trends currently developed are taking advantage of the semipermeable membranes/membrane reactors, temperature phase behavior of the microemulsions or microemulsion-based organogels (Nagayama et al. 2008) for carrying out enzymatic reactions. Partitioning selectivity of proteins can be achieved in RME, based on the hydrophobic nature of proteins due to the fact that RMs provide both hydrophobic and hydrophilic environment to biomolecules simultaneously. Recovery of biomolecules from the reverse micellar phase can be easily facilitated by exploiting the deassembling nature of RMs. Although a number of research articles and review papers have been published in the past relating to various aspects of RME, engineering aspects have received scant attention when compared with the physicochemical and biological aspects. For example, mathematical modeling, which is very important for the prediction of distribution/solubilization of biomolecules in RMs, has not been given its due importance. Several reports on the application of RME for separation and purification of biomolecules are available for model systems (commercial samples of proteins and enzymes) (Zhang et al. 2002; Shin et al. 2004; Dovyap et al. 2006; Mutalik and Gaikar 2006; Raikar et al. 2007; Hebbar and Raghavarao 2007). When compared with model systems, only a few studies on RME of biomolecules from natural systems were reported (Gaikar and Kulkarni 2001; Mutalik and Gaikar 2003; Liu et al. 2004; Noh et al. 2005; Chen et al. 2006; Hasmann et al. 2007; Hebbar et al. 2008; Nandini et al. 2009, 2010b; Hemavathi et al. 2007, 2008, 2010).

14.2 REVERSE MICELLAR SYSTEM

Reverse micelles are thermodynamically stable, optically transparent, nanometer size droplets of an aqueous solution stabilized in an apolar environment by the surfactant present at the interface. In these systems, the biomolecules are solubilized inside the water core of surfactant shell that protects the biomolecule from denaturation by organic solvent. Figure 14.1 shows the solubilization of different biomolecules in reverse micelles. RMs are generally smaller than their hydrophilic counterparts (micelles), their aggregation number being mostly lower than 50. The solubilization properties of surfactants are often expressed by a three- or four-component phase diagram, after determination of regions of optical transparency.

Figure 14.2 depicts the phase diagram of AOT (sodium bis(2-ethyl-1-hexyl) sulfo-succinate)/isooctane/water system (De and Maitra 1995). The phase diagram will be very useful in determining the composition of new reverse micellar system (RMS) and also in obtaining a transparent organic phase containing RMs in the case of injection method of RME. Characteristic properties of RMSs are (i) thermodynamic stability, (ii) spontaneous formation, (iii) low interfacial tension ($<10^{-2}$ mN/m), (iv) transparent nature (nanometer size, <100 nm), (v) large surface area (10–100 m^2/cm^3), (vi) viscosity comparable with the pure organic solvents, and (vii) capability to dissolve polar substances. The following section briefly summarizes the role of basic phase components of RMS.

14.2.1 SURFACTANTS

Surfactants are special group of lipids that possess both lipophilic and hydrophilic parts. They adsorb at surfaces or interfaces and change the interfacial free energy.

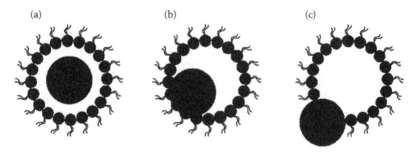

(a)   (b)   (c)

FIGURE 14.1 Solubilization of different biomolecules in reverse micelles: (a) hydrophilic, (b) surface active, and (c) hydrophobic. (From Martinek, K. et al., *Eur J Biochem*, 55, 453–468, 1986. With permission.)

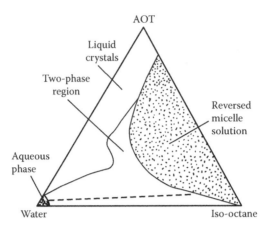

FIGURE 14.2   Equilibrium phase diagram of AOT/isooctane/water. (From De, T.K., and Maitra, A., *Adv Colloid Interface Sci*, 59, 95–193, 1995. With permission.)

A surfactant molecule consists of two distinct chemical groups: the head, which is hydrophilic, and the tail, which is hydrophobic. The surfactants are classified based on their charge as anionic, cationic, nonionic, and zwitterionic. The details of commonly used surfactants for RME are given by Harikirshna et al. (2002). The RMs of AOT, a commonly used anionic surfactant, despite their popularity as a research model system, have some serious limitations such as poor ability to release the proteins into aqueous medium during back-extraction (BE) and slow phase separation. These problems could be avoided using some new/novel surfactants. Goto et al. (1995) synthesized di(tridecyl) phosphoric acid (DTDPA) and used it for the formation of RMs that could extract many proteins.

Nonionic surfactants have many unique properties that are superior to those of ionic surfactants for the synthesis of nanoparticles, such as remarkably low CMC (critical micellar concentration), high efficiency in reducing surface tension (Chen and Cheng 2005), and better solubilizing properties, which make them potentially useful in a wide variety of industrial applications. A few reports on the application of nonionic surfactants such as sugar ester DK-F-110, a mixture of sucrose esters of fatty acids for *Rhizopus delemar* lipase (Naoe et al. 1999), Triton X-100 for BSA (Hebbar and Raghavarao 2007), Tween 85 for chicken egg white (Sun et al. 2000), $C_{12}E_3$ system for horseradish peroxidase (HRP) (Motlekar and Bhagwat 2001), non-ylphenol polyethoxylate (NP) for the synthesis of nanoparticles (Morrison et al. 2005), Triton X-100, SDS (sodium dodecyl sulfate), CTAB (cetyltrimethyl ammonium bromide), Brij 58 and lipophilic Brij 30 mixed surfactants system for DNA polymerases (Anarbaev et al. 2005), and DK-F-110 for cytochrome c (Noritomi et al. 2006a) are available. Mixed reverse micellar (MRM) systems of AOT and nonionic surfactants (Tween 20, Tween 80, Tween 85, and Triton X-100) in isooctane were reported by Hemavathi et al. (2010) for the extraction and primary purification of β-glucosidase from the aqueous extract of barley.

14.2.2 SOLVENTS

Organic phase occupies about 80–90% of the RMS. Water solubilization capacity of the RMs for a given surfactant is strongly dependent on the type of solvent used. A decrease in micellar size with an increase in the molecular size of the solvent was reported. As the molecular size of the solvent increases, their ability to penetrate into the interfacial surfactant layer decreases, thereby intermicellar attractions between the surfactant tails increases, making the micelles sticky. Organic solvents commonly used in RME are isooctane, hexane, and *n*-octane. Goklen and Hatton (1985) studied the effect of solvent structure on maximum RM size and water content ($W_{0, max}$). Sivasamy et al. (2005) reported an increase in pepsin solubilization by AOT reverse micelles with the increase of polarity and molar volume of the solvent employed. The physical properties of the organic solvent have a much more pronounced effect on the enzyme catalysis than on the protein/enzyme solubilization. For example, in the cationic RMs, the substitution of *n*-octane with isooctane does not influence the reversed micelle size and structure but there was a great difference in the initial rate of conversion for the lipase-catalyzed esterification (Tonova et al. 2006). The branched solvents have been reported to permit the highest rates in reactions catalyzed by *Candida rugosa* (Wu et al. 2002) or *R. delemar* lipase (Naoe et al. 2004b).

14.2.3 COSOLVENTS

Cosolvent/cosurfactant is a type of solvent that help surfactants to dissolve in the organic solvent and form RMs thereafter. The cosurfactant seems to buffer the strong repulsive ion–ion interaction between the surfactant head groups, thereby allowing their close packing in order to form the inner core of a RM. Only those alcohols having low solubility in water are used as cosolvents. The cosurfactant weakens the interactions between the enzyme and surfactant headgroups, therefore, the enzyme activity and structure are less influenced during the extraction cycle or in the course of the bioconversions in RMS (Goncalves et al. 2000). Some of the general cosolvents employed are long-chain alcohols, acetates, butyrates, etc. The role of cosurfactant in various aspects of RME are discussed by Krei and Hustedt (1992), Lazarova and Tonova (1999), Zhang et al. (2002), and Tonova and Lazarova (2005).

14.3 EXTRACTION USING REVERSE MICELLES

For the extraction and purification of target biomolecules, RMs should exhibit two characteristic features. First, they should be capable of solubilizing target biomolecules selectively, and they should be able to release these biomolecules into an aqueous stripping phase so that a quantitative recovery of the purified biomolecule can be achieved. In a number of recent publications, extraction and purification of enzymes has been demonstrated using various RMS (Hemavathi et al. 2007, 2008, 2010; Hebbar et al. 2008, 2010; Nandini and Rastogi 2009, 2010a, 2010b; Lakshmi and Raghavarao 2010). The studies revealed that the extraction process is generally controlled by various factors such as concentration and type of surfactant, pH and ionic strength of the aqueous phase, concentration and type of cosurfactant, salt type, charge of the biomolecule, water content, size and shape of RMs, temperature, etc. By manipulating these parameters, selective separation of the target biomolecule from mixtures can be achieved. These parameters are discussed below.

14.3.1 SURFACTANT CONCENTRATION

The concentration of surfactant required to form RMs in an organic solvent is more than the CMC. The concentration of surfactant has been shown to have little effect on the structure and size or aggregation number of the RMs. However, it changes the number of RMs formed, which in turn increases the protein solubilization capacity of the RMs. At higher surfactant concentration the reverse micellar interactions may occur, which leads to interfacial deformation of RMs and percolation of solute. The reverse micellar clustering at higher concentrations of surfactant decreases the interfacial area available to host the biomolecules causing a decrease in the solubilization capacity of the RMs (Nishii et al. 2002).

14.3.2 SURFACTANT TYPE

The uptake of biomolecule is mainly dependent on the charge difference between the biomolecule and the surfactant head groups. In addition to the charge, other surfactant-dependent parameters such as the size of RMs, the energy required to enlarge the RMs, and the charge density on the inner surface of the RMs may also influence the protein solubilization. Most of the work on RME reported to date has been with AOT (Nishiki et al. 2000; Okada et al. 2001; Mathew and Juang 2005; Chen et al. 2006; Zhao et al. 2008; Hemavathi et al. 2008), which is an anionic surfactant. Cationic surfactants such as quaternary ammonium salts have also been used for protein solubilization (Dekker et al. 1991; Zhang et al. 2000, 2002; Shin et al. 2003b; Shen et al. 2005; Hemavathi et al. 2007; Hebbar et al. 2008; Lakshmi and Raghavarao 2010). The studies with ionic surfactants in many cases have shown a rapid degradation of the protein activity after their solubilization (Aires-Barros and Cabral 1991). Recently, many researchers have studied protein transfer using non-ionic surfactants (Liu et al. 2006; Noritomi et al. 2006a,b; Hebbar and Raghavarao 2007). It was demonstrated that a nonionic surfactant has an apparent advantage over ionic surfactants due to the absence of strong charges at the aqueous/organic interface (Ayala et al. 1992; Hemavathi et al. 2010). These advantages are of great importance in the purification of diagnostic or therapeutic proteins and in the bioconversions. Despite the fact that alkyl sorbitan esters (series of Tweens or Spans) as well as sugar esters are nonionic surfactants, they still exhibit weak electrostatic interactions sufficient enough to solubilize significant amounts of some small proteins as shown in the case of cytochrome c (Vasudevan and Wiencek 1996; Naoe et al. 1999).

14.3.3 REVERSE MICELLES SIZE

The size of reverse micelle is dependent on the molar ratio of water to surfactant, $W_0 =$ [H_2O]/[surfactant]. The RMs of particular dimension can accommodate only proteins of certain size. Hence, micelle size may be conveniently used to include or exclude certain proteins. However, it should be noted that several micelles can regroup to form larger micelles when certain operating conditions are altered. It was also hypothesized that a protein can create around itself a new larger micelle of a requisite size to facilitate solubilization. As the ionic strength increases, the reverse micellar size decreases due to decrease in the electrostatic repulsion between the head groups of the surfactant. Besides the ionic strength, the type of the solvent and surfactant used also influence the reverse micelle size. The location of the solubilized protein and the reversed micelle size govern the interactions at protein–micelle interface that dominantly influence the protein structure and are responsible for refolding and superior activity (Wu et al. 2006). The water pool radius can range between 15 and 100 Å and is the main parameter determining the potential application of the RMs for the protein selectivity and enzyme activity, as well as for the preparation of nanosized particles. Naoe et al. (2004a) reported that the interaction between protein and micellar interface was a dominant feature influencing protein structure in RMs and is goverened by the location of solubilized protein.

14.3.4 WATER CONTENT

The water content (W_0) of the RMs is defined as the molar ratio of the water to that of the surfactant per RM. W_0 strongly depends on hydrophilic–lipophilic balance (HLB) of the surfactant, and it increases with HLB. W_0 has a major role in protein solubilization and stability. Krei and Hustedt (1992) have observed a significant influence of W_0 or size of the RM on the partitioning behavior of proteins in various RMS containing cationic surfactants. The nature of the water in the core of the RM is of great importance since biomolecules reside in this water pool. The water pool is generally regarded to be a composite of two different types, the bound water (lining the interior wall of the RMs) and the (remaining) free water. The relative amounts depend on the W_0 and also on the solvent nature. The effect of water content on extraction of the biomolecule and its activity have been reported by many researchers (Motlekar and Bhagwat 2001; Michizoe et al. 2003; Tonova and Lazarova 2005, 2008; Anarbaev et al. 2005).

14.3.5 AQUEOUS PHASE pH

The aqueous phase pH determines the ionization state of the surface-charged groups on the protein molecule. Solubilization of protein in RMs is found to be dominated by electrostatic interaction between the charged protein and the inner layer of the surfactant head groups. Solubilization of protein is favored at pH values above pI of protein in case of cationic surfactants, while the opposite is true for anionic surfactants. For small molecular weight proteins such as cytochrome C, lysozyme, and ribonuclease (MW range 12,000–14,500 Da), the pH–pI value required for optimum solubilization is much lower (<2) when compared with that of larger proteins such as α-amylase (MW 48000 Da). An empirical relation among pI, molecular weight of the solute, and the pH value at which maximum transfer efficiency could be achieved was developed by Wolbert et al. (1989) on the basis of experimental data of Goklen and Hatton (1987) for AOT reverse micelles.

14.3.6 Ionic Strength and Type

The influence of ionic strength on the solubilization of proteins in RMs is explained purely as an electrostatic effect. The electrostatic potential of a protein molecule is inversely proportional to the ionic strength of the solution and is characterized by Debye length (Harikrishna et al. 2002). In general, it was observed that as the ionic strength of the aqueous solution increases, the protein intake capacity of the RMs decreases (Aires-Barros and Cabral 1991). Two reasons were given to explain this phenomenon. First, increasing the ionic strength decreases the Debye length, thereby reducing the electrostatic interaction between the protein molecule and the surfactant head group of the RMs. Second, increasing the ionic strength reduces the electrostatic repulsion between the charged head group of the surfactants in a RM, thereby decreasing the size of the RM. The smaller RMs will have larger curvature, which increases the density of the surfactant monolayer near the surfactant head groups, resulting in a gradual expulsion of protein molecules residing inside the RMs, which is termed as a "squeezing-out effect" (Leodidis and Hatton 1990). Although the lower side of the ionic strength favors the protein transfer, experiments could not be performed at very low ionic strength as the solution becomes cloudy under these conditions.

Apart from ionic strength, the type of the ions also plays an important role in determining the partition behavior of proteins in RMs. The effect of ionic strength and type on extraction efficiency was reported by many researchers (Okada et al. 2001; Kinugasa et al. 2003).

14.3.7 Temperature

The effect of temperature on the water uptake capacity of anionic and cationic surfactants has been reported, and the variation in water uptake was attributed to a change in the aggregation number of the surfactants. Increased temperature is reported to favor the BE more than the forward extraction (FE). The disruption of RMs and higher rate of migration of solute containing RMs toward the interface are reported to be the reasons for enhanced recovery during BE at higher temperatures. Temperature plays an important role also in reverse micellar assisted bioconversion. The catalytic activity of laccase hosted in AOT reverse micelles displayed highest activity at 60°C and then decreased above 65°C due to instability of RMs (Michizoe et al. 2001). The enzyme polyphenoloxidase in reverse micelles displayed 15°C higher optimum temperature than in aqueous medium. This allows the use of the enzyme at higher temperatures with a gain in its stability (Rojo et al. 2001).

Desolubilization at higher temperature is a much faster process compared with the conventional liquid–liquid transfer because of the reduced interfacial resistance (enhanced fluidity of the interface and hydrophobicity of the surfactant). An increase in the temperature would improve the BE yields if the enzyme was solubilized inside the water core (electrostatic driven solubilization), as shown for the systems α-chymotrypsin/AOT and glucoamylase/CTAB (Forney and Glatz 1995). When driving force for the extraction is hydrophobic in nature, an increase in the temperature reduces the recovery as it favors the hydrophobic protein–surfactant interactions. Thus, lysozyme solubilized in sucrose fatty acid ester RMS was recovered at 35°C with 95% of activity recovery without the BE aqueous phase (Noritomi et al. 2006b).

14.3.8 MISCELLANEOUS FACTORS

In addition to the chemical and biochemical factors, protein transfer also depends on physical aspects such as duration of phase contact and the area of contact between two phases, mass transfer efficiency, protein charge, electrostatic potential of the RMs, phase volume ratio, and presence of other ions such as Ca^{+2}, Mg^{+2}, Ba^{+2}, etc. Higher contact area favors the extraction/release of solutes during RME. However, prolonged contact time may lead to the reduction in enzyme activity. Also, the equipment used for contacting or the nature of phase contact influences the extraction efficiency. Incorporation of the enzyme into micelles is greatly dependent on the presence of divalent cations. Leodidis and Hatton (1989) observed that RMs are not formed when Li^+, Be^{2+}, and Mg^{2+} are present in the aqueous phase. This was due to strong hydration of these cations, which precludes stabilization of the microemulsion. Regalado et al. (1996) also observed a poor solubility of HRP in AOT reverse micelles mainly due to the presence of high mineral content in the crude extract, which has reduced the W_0 value of RMs significantly. Addition of calcium scavengers (citric acid–citrate and EDTA–disodium salt) improved the solubilization and recovery of the enzyme. Phase volume ratio (V_{org}/V_{aqu}) is a critical parameter in the extraction and concentration of enzymes. Ideally, this ratio should be low for FE and high for BE to achieve concentration of the solutes. However, the change in the volume ratio could also adversely affect the extraction efficiency of the system. Yu et al. (2003) reported an increase in extraction yield of yeast lipase with the phase volume ratio when the aqueous phase volume is fixed. This is due to an increase in number of RMs. Shin et al. (2003a) observed protein–surfactant complex formation at aqueous–organic interface and a decrease in extraction efficiency at volume ratio higher than 0.2.

14.4 MECHANISM AND METHODS OF BIOMOLECULE SOLUBILIZATION DURING REVERSE MICELLAR EXTRACTION

Although many studies have been performed on the RME of proteins and the catalytic properties of enzymes in RMs, very little is known about the mechanism of protein solubilization in RMs. There are three commonly used methods (Matzke et al. 1992) to incorporate biomolecules into RMs: (i) injection of a concentrated aqueous solution, (ii) addition of dry lyophilized protein to a reverse micellar solution, and (iii) phase transfer between bulk aqueous and surfactant-containing organic phases. The three enzyme solubilization methods are shown schematically in Figure 14.3. The injection and dry addition techniques are commonly used in biocatalytic applications, the latter being well suited for hydrophobic proteins. The phase transfer technique is widely used for extraction of proteins from dilute aqueous solutions.

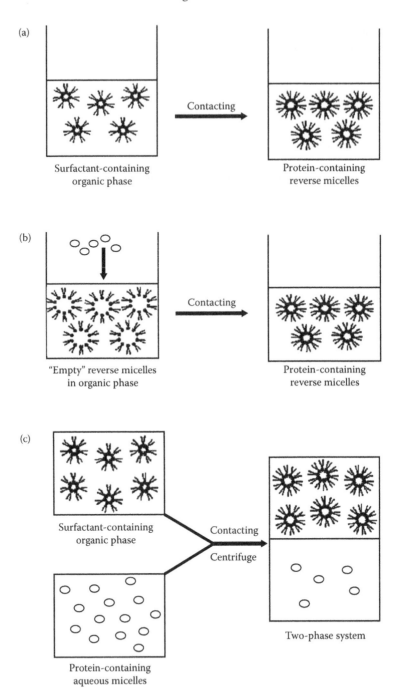

FIGURE 14.3 Methods of carrying out RME: (a) injection of aqueous phase containing solute, (b) addition of dry powder, and (c) phase transfer. (From Matzke, S.F. et al., *Biotechnol Bioeng*, 40, 91–102, 1992. With permission.)

14.4.1 INJECTION METHOD

In the injection method, the protein already solubilized in a concentrated stock aqueous solution (typically, 7% by volume) is added to the surfactant containing organic solvent. The resulting mixture is vigorously shaken until an optically transparent solution is obtained. This procedure, which is commonly used in biocatalytic applications, is simple, faster, and most effective. In this method, RMs are forced to form with the protein already inside. Hence, micelle sizes do not significantly affect the protein solubilization. For small micelle sizes, the injection method solubilized more protein than the dry addition method. The injection mode takes less time for extraction and ensures a high enzyme activity. However, a very few studies on extraction have been made using the injection mode for the solubilization and separation of enzymes. Hemavathi et al. (2008) reported RMS of AOT/isooctane for the extraction and primary purification of β-galactosidase from the aqueous extract of barley (*Hordeum vulgare*) using the injection method. β-Galactosidase was extracted with an activity recovery of 98.74% and purification of 7.2-fold. The same group recently reported a mixed reverse micellar system of AOT/Tween 20 for extraction and purification of β-glucosidase using the injection method (Hemavathi et al. 2010).

14.4.2 DRY ADDITION METHOD

It consists of the initial introduction of the required amount of water into the surfactant solution in order to attain the required surfactant hydration degree. A dry (lyophilized) protein can be dissolved in the resulting solution under vigorous shaking. This method is commonly used in biocatalytic applications and is well suited for hydrophobic proteins. One of the drawbacks of this method is prolonged contact between the enzyme/protein molecule and the organic solvent/surfactant that may lead to a partial denaturation of the former. In this method, protein solubilization is strongly dependent on micelle size. In fact, the protein is appreciably solubilized only when the diameter of the RM is either similar or larger than that of protein. The reason is that the energy barrier for solubilization of a large protein in small micelle is too large to overcome. However, in contrast, for a larger micelle, since the micelle is not required to rearrange its contents to incorporate a protein, the energy barrier is lower and the protein is easily solubilized.

14.4.3 PHASE TRANSFER METHOD

In this method, there are two bulk phases (aqueous and organic), which are brought to equilibrium. Although the enzyme solubilization is slow in this method, a large amount of solute solubilization is possible with minimum values of W_0. The pH of the aqueous phase, size and isoelectric point of the protein, and the surfactant type were shown to have significant effect on the protein solubilization by this method. The LLE process by RMs can be performed in one or two stages. The one-stage procedure consists only of the FE step, by which contaminants are transferred from an aqueous phase to a reverse micellar phase, thus leading to purification of the target biomolecule. The two-stage procedure of phase transfer method of RME consists of two fundamental steps, namely, FE and BE. During FE, the biomolecule is transferred from an aqueous solution into a reverse micellar organic phase, and during BE, the biomolecule is released from the RMs and transferred into a fresh aqueous phase (stripping phase) (Brandani et al. 1996). Recently, Nandini and Rastogi (2010b) reported extraction of lactoperoxidase (LPO) from cheese whey using single-stage RME and compared with two-stage RME. The single stage resulted in extraction of contaminating proteins and recovery of LPO in the aqueous phase leading to its purification. At optimized condition, two-stage RME resulted in an activity recovery of 86.60% and purification of 3.25-fold, whereas single-stage RME resulted in higher activity recovery (127.35%) and purification (3.39-fold).

14.4.3.1   Forward Extraction

FE involves diffusion of biomolecules such as protein from bulk aqueous solution to interface, formation of a biomolecule containing micelle at the interface and diffusion of biomolecule containing micelle into the bulk organic phase. One of the key factors that determine protein transfer into a reverse micellar phase during FE is the size of the protein. Wolbert et al. (1989) reported that as protein size increases, the partition of protein into RMs becomes much more difficult because larger protein requires a RM larger than what would be thermodynamically stable. This can be solved by the addition of cosurfactants or by manipulating organic continuous phase. FE efficiency is controlled by various process parameters such as concentration and type of surfactants, pH, and ionic strength of the aqueous phase, concentration and type of cosurfactants, salts, charge of the protein, temperature, water content, size and shape of the RMs, etc. By manipulating these parameters, selective separation of the targeted biomolecule can be achieved (Wolbert et al. 1989).

14.4.3.2   Back-Extraction

BE involves coalescence of biomolecule filled RMs with the interface to transfer the biomolecule to aqueous phase (also termed as stripping phase). In general, the FE of proteins into RMs has a high efficiency, and the process step that is low yielding, and thus requiring attention, is the back-transfer step. Kinetics of the BE was reported to be much slower as compared with that of FE (Dungan et al. 1991), and BE rate could be enhanced (more than 100 times) with the addition of counterionic surfactant (Jarudilokkul et al. 1999). The counterionic surfactant was reported to interact with the oppositely charged surfactant molecules and facilitate the release of solutes into the stripping phase due to collapse of RMs (Mathew and Juang 2005). During BE, pH value of the stripping solution should be adjusted to prevent the protein–protein interaction (similar to the isoelectric point of the protein to be purified) and salt at high concentration was added to reduce electrostatic interaction between the surfactant and protein (Kinugasa et al. 1992).

Most of the earlier studies tactically assume that conditions, which normally prevent protein uptake in the FE, would promote their release in the BE. The assumption, however, was later found not true and resulted in low protein recovery. Hence, some alternative approaches for enhanced recovery of the proteins from RMs were investigated. These are use of silica particles for the sorption of the proteins as well as surfactants and water directly from the protein-filled RMs and use of ion exchange columns, addition of dewatering agents such as isopropyl alcohol, and dehydration of RMs with molecular sieves to recover the protein, addition of large amount of second organic solvent such as ethyl acetate to destabilize the RMs and hence to release the protein, formation of clathrate hydrates via pressurization, use of temperature to dewater the RMs, and hence to release the protein (Dekker et al. 1991), addition of sucrose to enhance the protein recovery by reducing the protein–surfactant interactions and through the gas hydrate formation. The alcohols with chain >C5 suppress the micellar cluster formation (i.e., the intermicellar attractions decrease), thus enabling an easier release especially of core-solubilized proteins (Mathew and Juang 2005). The branched short-chain alcohols are more preferred as they probably partition into the micelle core and destabilize the hydrophobic protein–surfactant complex (Hemavathi et al. 2007). The role of alcohols in the BE of various proteins/enzymes from the RMS has been recently reviewed by Mathew and Juang (2007). Accordingly, the addition of long-chain alcohols (*n*-octanol, *n*-hexanol) to AOT reverse micelles were found to have little effect on the release rates, the recovery of active protein was not improved (Mathew and Juang 2005). Best results were obtained with short-chain alcohols such as ethanol (Aires-Barros and Cabral 1991; Yu et al. 2003; Chen and Cao 2007), *iso*-propyl alcohol, *iso*-butanol (Carlson and Nagarajan 1992). However, the short-chain alcohols are usually added in large amounts, 10–15%, to the recovery phase. Their presence may change the quality of the aqueous solution and also some risk of denaturation exists. Using gas hydrate, which can be easily formed in the RMs by different gases at moderate pressures (3–5 atm), the protein can be precipitated out from the RMs in a solid state. The formation of clathrate hydrate destabilizes the RMs and decreases W_0. More hydrophilic and polar gases, such as chlorodifluoromethane and tetrafluoroethane were recommended to form gas hydrate, but recently, CO_2 hydrates have been also used (Aydogan et al. 2007). Pressurized CO_2 can precipitate completely lysozyme as a solid without activity loss (Zhang et al. 2001; Chun et al. 2005). Shin et al. (2003a) used acetone for recovering lysozyme as solid precipitate from RMs, which was unable by conventional method of pH and salt concentration adjustment. Lee et al. (2004a) used thiols and nonionic surfactants such as span 85 for the improved BE rate of C.C. lipase from AOT reverse micelles. Lee et al. (2004b) studied the effect of the addition of alcohol and carboxylic acid on the BE of small globular proteins, wherein suppression of RM clusters formation enhanced the BE of proteins.

14.5 MASS TRANSFER KINETICS AND MATHEMATICAL MODELING

Rate of protein transfer to or from a reverse micellar phase and factors affecting the rate are important for the practical applications of RME of proteins/enzymes and for scale-up. The overall mass transfer rate during an extraction process will depend on the rate limiting step. A number of studies (Dungan et al. 1991; Sun et al. 1999; Nishiki et al. 2000; Dovyap et al. 2006; Liu et al. 2006) have been reported on mass transfer of proteins and enzymes during RME.

Mathematical modeling of the solubilization of biomolecules in RMs is essential for an in-depth understanding and for effective use of RME in DSP of biomolecules.

However, the quantitative modeling of protein solubilization in RMs is a complex problem, as many parameters such as hydrophobic interactions of ions with the protein and surfactant, the free energy changes associated with the change in size of the RMs on protein uptake, and the distribution of charged groups on the protein molecule are unknown or difficult to quantify (Dekker et al. 1989). Models presented in the literature for protein solubilization range from simple geometric models (Levashov et al. 1982; Sheu et al. 1986; Krei and Hustedt 1992; Jolivalt et al. 1993) to more rigorous molecular thermodynamic models (Bratko et al. 1988; Caselli et al. 1988; Haghtalab and Osfouri 2003). Models for interfacial mass transfer were proposed by Dungan et al. (1991) and Okada et al. (2001). In all the approaches, an equal distribution of surface charge on a globular protein molecule was assumed. Padalkar and Gaikar (2003) proposed a two dimensional mathematical model to elucidate the mechanism of protein transfer during FE. Here a protein molecule was treated as a polyelectrolyte with multiple charges on the surface in contrast to protein as a single charge species by Dungan et al. (1991). They reported that the complete encapsulation forming a protein-loaded RM as the controlling step during FE. The interface was found to increasingly deform as the protein approaches it. The deformation of the interface closer to the protein was the highest. Haghtalab and Osfouri (2003) developed a thermodynamic model using vacancy solution theory (based on a surface pressure) to predict the partitioning of proteins in equilibrium RMS. In this model, one vacancy solution represents the bulk aqueous phase and the other the reverse micellar phase. Protein molecules are considered as adsorbate, which can be laid into the vacancy of RMs. The results of the prediction were in good agreement with the experiment. This model is also suitable for prediction of protein partitioning in multicomponent mixtures (solution containing more than one protein).

Okada et al. (2001) measured overall mass transfer coefficients of lysozyme using a flat interface cell. The overall mass transfer coefficient of lysozyme increased as the interfacial tension of the system reduced. According to this study, protein destabilizes the bulk interface and enhances the reverse micelle formation, but Dungan et al. (1991) suggested lysozyme to be in the aqueous phase close to the interface that deforms the surfactant layer. Liu et al. (2006) studied the partitioning equilibria and the kinetics of lysozyme and bovine serum albumin. Interfacial mass transfer coefficients were estimated by Dovyap et al. (2006) using two film models for the extraction of L-isoleucine.

Hecht and Peled (2006) studied structural modification of AOT reverse micelles upon incorporation of model proteins (lysozyme and BSA) at various W_0 and protein concentration using SAXS and found that the presence of proteins induced changes in the micelle size distribution. Feng et al. (2006) proposed a multicomplex model to explain the effect of compressed CO_2 on the stability of RMs and the relationship between pressure and protein solubilization. At present, however, the models are yet to be shaped to perfection to give accurate predictions of the distribution coefficient of a protein. The recent studies explore a variety of ultrafast laser techniques to uncover details about structure and dynamics of RMS. Using ultrafast vibrational spectroscopy, researchers have probed hydrogen bond dynamics and vibrational energy relaxation in RMs. These studies have developed an understanding of reverse micellar structure, identifying varying water environments in the RMs and their impact on a range of chemical reactions (Levinger and Swafford 2009).

14.6   APPLICATION OF REVERSE MICELLAR EXTRACTION AS A BIOMOLECULE EXTRACTION AND PURIFICATION METHOD

Commercial interest in cost-effective methods that can separate, concentrate, and purify valuable bioproducts continuously and can be easily scaled up is still the topic of the day. A number of publications in recent years have dealt with various RMS for purification of enzymes from their natural sources aiming production of food, pharmaceutical ingredients, or biochemical reagents. In RME, the possibility of fine tuning the capacity of the RMS through the physicochemical properties of either phase allows protein purification to be achieved either by the target protein solubilization or through the impurities solubilization. Table 14.1 shows the biomolecules of food application studied using various RMSs. A RMS of ionic surfactants is reported by Hemavathi et al. (2007) for the extraction and primary purification of fruit bromelain from the aqueous extract of pineapple (*Ananas comosus* L. Merryl). Studies carried out with ionic surfactants AOT and CTAB confirmed electrostatic interaction as the main driving force for the extraction of fruit bromelain (activity recovery 97.56% and purification 4.5-fold). Reverse micellar phase components were recovered and efficiently reused for fresh or subsequent extraction, which contributes favorably to the process economics and environmental issues. The same group (Hebbar et al. 2008) reported the extraction and primary purification of bromelain from crude aqueous extract of pineapple waste, namely core, peel, crown, and extended stem. A fairly good activity recovery (106%) and purification (5.2-fold) of bromelain was reported.

Hemavathi et al. (2008, 2010) reported the extraction and purification of β-galactosidase and β-glucosidase, respectively, from barley using RME. Lakshmi and Raghavarao (2010) carried out solubilization of soy hull peroxidase (SHP) in CTAB reverse micelles. The active SHP was recovered after a complete cycle of FE and BE. FE efficiency of 100%, BE efficiency of 36%, overall activity recovery of 90%, and purification of 4.72-fold were reported, which is higher compared with purification achieved by ammonium sulfate and acetone precipitation. The phase transfer of soy hull peroxidase was found to be controlled by electrostatic and hydrophobic interactions during FE and BE, respectively.

Lipase, a key enzyme of food and biotechnological industries, was purified using RME (Nandini and Rastogi 2009). Nandini and Rastogi (2010a) also reported optimization of RME of lipase using response surface methodology. Exploration of the response surfaces indicated a complex interaction between the process variables. Global increase in the production of cheese and casein has led to the disposal of large volumes of whey as a waste product. LPO, alkaline phosphatase, lysozyme, xanthine oxidase, xanthine dehydrogenase, etc., are the several enzymes present in the whey. Single-stage RME was used by Nandini and Rastogi (2010b) to purify LPO from milk whey with higher activity recovery (127%). RME is used as an alternative to a traditional oil extraction process that permits simultaneous extraction of oil, proteins, and glucosinolates from cruciferous oilseed meals in three main steps:

TABLE 14.1
Biomolecules of Food Application Studied Using RMS

Biomolecule	Source	RMS	Reference
α-Amylase	Bacillus licheniformis	CTAB/isooctane/isobutanol/n-hexanol	Krei and Hustedt 1992; Tonova and Lazarova 2005
	Bacillus amyloliquefaciens	BDBAC/isoocatane/n-hexanol	Krei et al. 1995
	Bacillus subtilis	TOMAC/isooctane/n-octanol	Brandani et al. 1996
	Aspergillus oryzae	CTAB/isooctane/isobutanol/n-hexanol	Tonova and Lazarova 2008
	Aspergillus niger	AOT/isooctane	Bera et al. 2008
Glucoamylase	Aspergillus awamori	TOMAC/Revopal HV5/n-octanol	Forney and Glatz 1994
		CTAB/ isooctane/butanol/hexanol	Forney and Glatz 1995
		Triton X-100/xylene/hexanol	Shah et al. 2000
Trypsin	Porcine pancreas	Tetraoxyethllene monodecylether/hexane	Adachi et al. 1998
	Pig pancreas	AOT/isooctane	Chen and Cao 2007
α-Chymotrypsin	Bovine pancreas	AOT/isooctane	Dungan et al. 1991
		DTDPA/isooctane	Goto et al. 1995
		DODMAC/isooctane/decanol	Rabie and Vera 1998
		CK-2,13/isooctane	Rairkar et al. 2007
Pepsin	Porcine, bovine	AOT/isooctane	Carlson and Nagarajan 1992
		AOT/isooctane and CTAB/isooctane/butanol/hexanol	Sivasamy et al. 2005
Lysozyme	Chicken egg white	AOT/isooctane	Okada et al. 2001; Zhang et al. 2001; Chun et al. 2005; Hecht and Peled 2006
		DODMAC/isooctane	Shin et al. 2003b
		Span 85/n-hexane	Liu et al. 2007
		CDAB/isooctane/butanol/hexanol	Noh and Imm 2005
		CK-2,13/isooctane	Rairkar et al. 2007

Enzyme	Source	System	Reference
Lipase	Chromobacterium viscosum	AOT/isooctane	Aires-Barros and Cabral 1991; Lee et al. 2004a
		AOT/Tween 85/isooctane	Hossain et al. 1999
		AOT/isooctane/DMSO	Moniruzzaman et al. 2006
		CTAB/isooctane/hexanol	Debnath et al. 2007
		CTAB/Brij 30/isooctane/hexanol	Shome et al. 2007
	R. delemar	DK-F-100/IPA/hexane	Naoe et al. 2001
		AOT/isooctane and AOT/cyclohexane	Naoe et al. 2004b
	Aspergillus niger	CTAB/isooctane/n-butanol/n-hexanol	Nandini and Rastogi 2009, 2010a
	Yeast	AOT/isooctane	Yu et al. 2003
	Mucor javanicus	AOT/isooctane	Talukder et al. 2007
Phospholipase D	Cabbage	Triton X-100/phosphotidyl choline/diethyl ether	Subramani et al. 1996
Lipoxygenase	Soybean	AOT/isooctane	Shkarina et al. 1992; Nowak et al. 1996
Glucose oxidase	Aspergillus niger	AOT/n-octane	Kamyshny et al. 2002; Shipovskov et al. 2005
		CTAB/isooctane/butanol/hexanol	Ferreira et al. 2005
Invertase	Bakers' yeast	AOT/isooctane	Mutalik and Gaikar 2006
Bromelain	Pineapple fruit	CTAB/isooctane/n-butanol/n-hexanol	Hemavathi et al. 2007
	Pineapple waste	CTAB/isooctane/n-butanol/n-hexanol	Hebbar et al. 2008, 2010
β-Galactosidase	Escherichia coli	AOT/isooctane	Shiomori et al. 2000; Chen et al. 2001
	Penicillium canescences	AOT/isooctane	Kouptsova et al. 2001
	Barley	AOT/isooctane	Hemavathi et al. 2008
Glucosidase	Trichoderma reesei	AOT/isooctane	Zamarro et al. 1996
	Sweet almond	AOT/isooctane	Kouptsova et al. 2001
	Barley	AOT/isooctane/Tween 20	Hemavathi et al. 2010
LPO	Milk whey	CTAB/isooctane/n-butanol/n-hexanol	Nandini and Rastogi et al. 2010b

(i) seed conditioning, (ii) solid–liquid extraction by RMS, and (iii) stripping of the extracted compounds from RMS. The extract was enriched in more than 90% of soluble proteins and glucosinolates. Moreover, the seed–meal component fractionation is possible by varying W_0, which affects glucosinolates and protein extractability differently (Ugolini et al. 2008).

Imm and coworkers showed a one-step separation of the target protein by adjusting system properties to recover the lysozyme (Noh and Imm 2005) or lactoferrin (Noh et al. 2005) in the residual aqueous phase, while the major egg white or whey proteins were solubilized in cetyldimethylammonium bromide (CDAB)–RMs. Lysozyme was efficiently purified more than 30-fold with a single FE step. Of the initial lactoferrin, 96% remained in the aqueous phase after FE and fully maintained its bacteriostatic activity against *E. coli*. The one-step separation is advantageous in simplicity, time, cost, and yield compared with classical separation technology (direct crystallization or chromatography) or traditional FE and BE procedures. Separation of immunoglobulin G from the other colostral whey proteins was also achieved by solubilization of non-IgG proteins inside the RMs and recovery of the target immunomodulatory protein in the aqueous phase (Su and Chiang 2003). Residual quantities of surfactant (AOT) were not detected in the aqueous phase, which implies food application of the purified IgG. The studies on lysozyme RME from chicken egg white (Sun et al. 2000; Liu et al. 2007) could also be addressed to the field of food technology. Particularly, due to its antimicrobial activity, lysozyme is considered as a natural food preservative. The most researched applications of RME as a protein purification method concerning hydrolytic enzymes, which have the widest variety of uses in food, detergent, textile, pharmaceutical, diagnostics, and fine chemical industries. Extraction of industrial lipase with CTAB-based RMs resulted in 70% of activity recovery (Shen et al. 2005). However, using anionic (AOT)-based RMS, no purification was achieved when neither lipase (Yu et al. 2003) nor alkaline protease was the target (Monteiro et al. 2005). Leser et al. (1989) proposed an approach to simultaneously extract oil and protein from vegetable meals using RMs. The use of RMs to simultaneously extract oil and protein from soybeans is attractive, since soybeans represent one of the major oil seeds for producing edible oils.

Zhao et al. (2008) used AOT based RMS for soy protein and isoflavones extraction from soy flour. Comparison of amounts of isoflavones recovered using conventional ACN/HCl and AOT/H_2O reverse micellar solution showed that the amounts of isoflavones extracted using reverse micellar solution were many fold higher than those extracted using ACN/HCl. RMs are able to take up both hydrophobic and hydrophilic isoflavone conjugates from soy flour during FE. By modifying the extraction temperature, contact time, and ionic strength, soy protein enriched with daidzin, genistin, daidzein, and genistein was produced from soy flour. Kim et al. (2003) reported extraction of water soluble anthocyanins using the RMS of AOT/hexane. Anthocyanins are frequently used as a natural colorant in various food products. Despite attractive color and pharmaceutical potential, low stability of anthocyanins limited their application as a food colorant. Anthocyanins solubilized in RMs showed about four times greater color intensity and the overall stability was better than that of buffered aqueous anthocyanin (control) in the same storage condition. This study provides insight to control anthocyanin stability in a lipid soluble environment such as oil and salad dressing. Recently, Zhu et al. (2010) evaluated functional properties and secondary structures of defatted wheat germ proteins extracted by RMs, alkaline extraction, and isoelectric precipitation. The results showed that RME samples had a higher fat absorption capacity and emulsifying stability, and they also contained high levels of threonine, histidine, alanine, arginine, glycine, serine, cysteine, proline, and lysine compared with other samples. The denaturation temperature of RME samples was higher than those of other samples and RME samples retained its ordered secondary structure, whereas other samples did not.

The approach of contaminants removal by RMS has also been reported for purification of some intracellular enzymes of worth in medical and analytical practices. The possibility of using RME in a continuous countercurrent mode (a packed column with Raschig's rings) for glucose-6-phosphate dehydrogenase purification was investigated by Hasmann et al. (2007). Reverse micellar-aided permeabilization of *Aspergillus niger* cells provided simultaneous extraction of protein and purification of intracellular catalase. The rate of release of catalase from RM-treated cells is much faster compared with that from untreated cells (Manocha and Gaikar 2006). Separation of intracellular enzymes xylose reductase and xylitol dehydrogenase was achieved by CTAB based RMS (Cortez et al. 2004). Purity of xylose reductase was 5.6-fold in the BE phase, while xylitol dehydrogenase was about 80% recovered in the remaining aqueous phase with 1.8-fold purification. The enzymatic xylose/xylitol conversion could become an alternative to the conventional production of xylitol by inorganic catalyst. Nowadays, xylitol is in great demand in food and pharmaceutical industries. The separated xylitol dehydrogenase could be used as analytical reagent. CTAB was used for RME of large molecular weight alcohol dehydrogenase (141 kDa) from baker's yeast cell extract (Zhang et al. 2000), both FE and BE completed within 15 min.

RME can be applied as a simple, scalable, and inexpensive primary purification step in pharmaceutical plasmid DNA production (Streitner et al. 2007). The separation of RNA and plasmid DNA was possible in the BE, which would allow DNA concentration by changing the phase ratio. The RMs could take nucleic acids up to a concentration of 2 g/dm^3, thus exceeding the capacity of classical anion exchange materials for plasmid DNA (Goto et al. 1999). Complete BE of DNA from cationic surfactant was achieved by adding an alcohol without any conformational change (Goto et al. 2004). Recently, the separation of structurally related impurities from pharmaceutical plasmid DNA was attempted using RME by Tschapalda et al. (2009). RME is capable of complete separation of chrDNA from plasmid DNA independent on the initial chrDNA concentration. RME has already been applied to small fragments of genomic DNA (Goto et al. 1999), and their capability to separate plasmid DNA from RNA has also been demonstrated (Streitner et al. 2007, 2008), thus making them an interesting alternative to common purification strategies. A novel fibrinolytic enzyme, nattokinase, which is considered as a promising agent for thrombosis therapy, was purified by RME (Liu et al. 2006). Nearly 80% of nattokinase activity in the fermentation broth was recovered with purification of 3.9-fold. Papain and bromelain enzymes, widely used in medicine and cosmetics, were purified from their natural sources (Mathew and Juang 2005; Hemavathi et al. 2007; Hebbar et al. 2008). Gaikar and Kulkarni (2001) have shown RME for direct solubilization and purification of shear sensitive penicillin acylase from *E. coli*. Instead of energy intensive ultrasonication where whole cells are broken down, making subsequent processing difficult and expensive, reverse micellar treatment gave enzyme in a pure form. Pandit and Basu (2004) used RME for the removal of ionic dyes from effluent wastewater wherein recovery of solvent and reuse of dye is possible. They proposed an ion exchange reaction model for the recovery of dye using RMs. Reta et al. (2006) reported AOT/heptane RMS for separation and concentration of biological and pharmacological significant polyhydroxy compounds such as shikimic acid, gallic acid, gallotannic acid, rutin, and quercetin. Extraction was reported to be fast and reproducible.

14.7 APPLICATION OF REVERSE MICELLES AS BIOCONVERSION MEDIUM

Reverse micelles offer new possibilities of studying bioconversions of nonpolar compounds. Furthermore, because of the ability of RMs to stabilize hydrophobic molecules, they can be advantageously used to study the catalysis of water-insoluble substrates, as a system that mimics a cell environment. The possibilities to refold proteins and achieve "superactivity" inside the structured RMs caught the researchers' attention and a branch of micellar biocatalysis in RMs was emerged along with the extraction applications. The structure of RMs allows a heterogeneous biocatalysis to proceed in one macro phase wherein the water/oil interface is distributed over nano-sized dynamic entities. Thus, the inherent biphasic problems caused by different solubility and constrained contact between the enzyme and the substrates can be solved on a molecular scale, leading to the higher bioconversion efficiency. Applications of RMs/microemulsions/surfactant emulsions were recognized as a simple and highly effective method for enzyme immobilization for carrying out several enzymatic transformations. Bioconversions that occur in the reversed micelle aqueous core are reported to have better performance compared with usual aqueous media in terms of several fold greater activity and/or stability (Shah et al. 2000). RMS are actually considered in the biocatalysis as common organic media for microencapsulation of enzymes (Carvalho and Cabral 2000). Various enzymatic reactions in RMs have been reported in the literature, which includes the production of triglycerides, steroid conversions, peptide synthesis, and amino acid synthesis (Harikrishna et al. 2002).

Use of membrane extractors in combination with RMs is more attractive for bioconversion. The principle is that the membranes can be used to retain the enzymes and their hosting RMs in the reactors while the products can be recovered in a stripping solution on the other side of the membrane. An electro-ultrafiltration bioreactor for separation of RMs containing enzyme from the product stream (Hossain et al. 1999), ultrafiltration with a tubular ceramic membrane (Carvalho and Cabral 2000; Melo et al. 2001), and hollow fiber membranes were reported. However, the low rejection of the RMs and the permeation of the substrate might be drawbacks for a commercial operation (He et al. 2001). The mixed reversed micelles contribute to the activity improvement by decreasing the strong interactions (electrostatic and/or hydrophobic) with the surfactant and by increasing the microviscosity of the micellar water, which limits the fluctuations of the enzyme structure. A pronounced increase in the papain activity was obtained when a moderate amount of Tween 80 was added into the RMs consisted of either AOT or quaternary ammonium bromide (Fan et al. 2001). A significant amount of work was done on searching the appropriate RMS as bioconversion media.

RMs were explored for studying enzyme catalytic activity and enzyme kinetic behavior (Michizoe et al. 2003). Das et al. (2005) investigated the activity of the interfacially located enzymes (lipase and peroxidase) in cationic RMS and found that the increase in head-group size allowed the enzyme to attain a flexible conformation and to perform better catalysis. Inhibition of biocatalyst with increasing unsaturation at the polar head of the surfactant is observed in the case of HRP (Debnath et al. 2007). The lipase activity was also found to depend on surfactant tail length (Dasgupta et al. 2005). Moniruzzaman et al. (2006) have reported improved activity and stability of *C. viscosum* lipase-catalyzed hydrolysis of olive oil in AOT/water/isooctane RMs. This increased stability of lipase suggested that AOT/ dimethyl sulfoxide (DMSO) reverse micelles provided a better microenvironment for enzymes than that of simple AOT reverse micelles. Recently, an improved catalytic activity of surface-active enzymes (lipase and HRP) upon addition of nonionic surfactants to CTAB/*n*-hexanol/*i*-octane has been shown (Shome et al. 2007). The activity enhancement was attributed to the resulted lower charge density and subsequently higher interfacial area, which could be smoothly occupied by the surface-active enzyme. Particularly, in the case of lipase, the higher activity was also due to the presence of *n*-hexanol (competitive inhibitor of lipase) content at the interfacial region of the cationic microemulsion. AOT microemulsions can be also modified by addition of a nonionic surfactant (Lan et al. 2008), solvents (DMSO; Moniruzzaman et al. 2007), or short-chain PEG 400 (Talukder et al. 2007), or AOT can be chemically modified (He et al. 2001). All these modifications aim to increase in size of the interfacial area and thus to decrease its rigidity, which leads to an enhanced contact between the substrate and the enzyme active site. The use of RMs as reaction media could be an effective method for increasing the enzymatic hydrolysis of cellulose (Chen et al. 2006).

14.8   OTHER APPLICATIONS OF RMS

Other important applications of RMS include tertiary oil recovery, extraction of metals from raw ores, and in drug delivery. RMs have wide applications in enzymology, protein chemistry, bioorganic synthesis, and variety of biotransformations. RMS enable the compartmentalization of water and organic soluble reactants in the respective phases and allow control over the position of equilibrium to favor synthesis (Klyachko and Levashov 2003). Many other approaches such as affinity based RME and separations for high selectivity of protein extraction (Senstad and Mattiasson 1989; Paradkar and Dordick 1993; Choe et al. 1998, 2000; Liu et al. 2006), RME in hollow fibers for higher mass transfer (Dahuron and Cussler 1988), micellar-enhanced ultrafiltration to separate dissolved organic compounds from aqueous streams (Tzeng et al. 1999), micellar electrokinetic capillary chromatography for the separation of neutral and partially charged species, and reverse-micellar–assisted supercritical fluid extraction (SCFE) (Ayala et al. 1992) have been reported.

Yonker et al. (2003) reported use of RMs coupled with UF membranes for the separation of polar macromolecules in near-critical and supercritical solvents, whereas conventional SCFE with CO_2 is restricted to low molecular weight nonpolar solutes. The use of membrane eliminates the need for depressurization of the fluid to recover the target compound.

RMs were used for the controlled synthesis of precursor dipeptides. The studies on the structural and catalytic properties of enzymes, esterification reaction catalyzed by enzyme, and protein refolding (Goto et al. 2000; Sakono et al. 2000) in RMs have been reported. Jimenez-Carmona and Castro (1998) used reverse micelle formation as a strategy for improving the extraction of polar species with supercritical CO_2. Singh et al. (1999) used RMs to entrap photosynthetic bacterium and enhance H_2 production rate, as compared with cells suspended in normal aqueous medium.

Recently, the controlled growth of microporous crystals in RMs has attracted many researchers. Microporous crystals with sodalite structures (zincophosphate) were prepared in RMs by Reddy et al. (1996) by introducing zinc and phosphate ions into separate micelles to control the crystallization by collision and exchange kinetics. The reverse micellar interface provided the site for crystal nucleation. This approach provided a means of controlling the morphology as well as the size of growing crystals. The formation of several metal nanoparticles reported using this technique includes: copper metallic particles employing functionalized surfactants, silver sulfide semiconductors, composite CdS–ZnS nanoparticles, and titania nanoparticles. Rojo et al. (2001) carried out a thermostability study of polyphenoloxidase in AOT/cyclohexane RMs and found improved thermal stability in RMs.

Kouptsova et al. (2001) used RMS for synthesis of alkyl glycosides catalyzed by β-glycosidases, which is advantageous compared with multistage traditional chemical synthesis. Chen et al. (2003) reported synthesis of galactooligosaccharides (GOS) and transgalactosylation modeling in RMs. Feng et al. (2006) reported enhanced solubilization of BSA in AOT/isooctane RMs by compressed CO_2 and recovered completely at high pressures while AOT remains in the solution. It is well known that the use of enzyme catalysts in organic media facilitates the enantioselectivity as well as the possibility of transforming hydrophobic substrates. The RMS solubilize the biomolecules homogeneously and possess an enormous intrinsic interfacial area of contact ($10–100$ m^2/cm^3); the dynamic character and flexibility of the reversed micelles are profitable to enzyme reactivity (Carvalho and Cabral 2000; Melo et al. 2001).

Recently, Pandey and Pandey (2008) has shown a novel method to produce biohydrogen from photosynthetic bacteria entrapped inside RMs. Choi et al. (2010) studied inhibition of citral degradation in beverage emulsions using RMs. Citral is a flavor component that is widely used in the beverage, food, and fragrance industries. It chemically degrades over time in aqueous solutions due to acid catalysis and oxidative reactions, leading to loss of desirable flavor and the formation of off-flavors. The study showed improved chemical stability of citral in beverage emulsions after reverse micelle encapsulation. Yoksan and Chirachanchai (2010) reported silver nanoparticle synthesis using RMs and further fabrication of silver nanoparticle-loaded chitosan–starch-based films, which exhibited enhanced antimicrobial activity suitable for food packaging and/or biomedical applications. Lopez-Jimenez et al. (2010) reported supramolecular solvent made up of RMs of decanoic acid for the simple and rapid extraction of Sudan I, II, III, and IV from chilli-containing foodstuffs before their liquid chromatography–photodiode array determination. Raghavendra et al. (2010) reported synthesis of ethyl valerate (also known as the green apple flavor, which finds wide applications in food, pharmaceuticals, and cosmetics industries) in organic solvents by *C. rugosa* lipase immobilized on AOT microemulsion-based organogels (MBGs).

Vaidya et al. (2010) reported a reverse micellar route for synthesis of core–shell nanocomposites using titanium hydroxyacylate has a shell-forming agent for the first time. Recently, Chen and Dong (2010) reported pH-sensitive polypeptide-based RMs for *in vitro* controlled drug release (anticancer drug delivery). Aragon et al. (2010) reported RMs technique for the preparation of different electrode materials for lithium-ion batteries. RMs are widely used today in the synthesis of many types of nanoparticles. Excellent control of the final powder stoichiometries with possibilities of obtaining homogeneity and mixing on the atomic scale, narrow particle size distributions, negligible contamination of the product during the homogenization of the starting compounds, low energy consumption, low aging times, simple equipment, improved control of the particle sizes, shapes, uniformity, and dispersity are the advantages of reverse micellar synthesis compared with high-temperature traditional routes for the synthesis of common ceramic and metallic materials (Uskokovi and Drofenik 2005). The synthesis of nanoparticles as a combination of reversed micelles and sol-gel technique has been reported for different purposes (antibodies encapsulation, biosensors construction) (Tsagkogeorgas et al. 2006; Morales et al. 2005).

Another important application of RMs is in refolding of proteins produced via genetic engineering. Hagen et al. (1990) were the first to demonstrate the feasibility of this process using RNase as a model protein. An attractive feature of this method is that protein concentration in the range of 1–10 g/l can be processed, which is at least 1000-fold higher than the concentrations employed in the conventional refolding processes. Sakono et al. (2000) reported cytochrome c refolding in AOT based RMS at a high protein concentration, which is advantageous compared with the conventional dilution method. Thus, RME is a useful tool for the renaturation of denatured recombinant proteins. When using RMS, it is advisable to refold proteins at high concentrations, which compares favorably to the dilution refolding in the bulk aqueous phase at higher activity (Sakono et al. 2000; Wu et al. 2006). Dong et al. (2006) reported oxidative refolding of denatured lysozyme assisted by artificial chaperones in RMs formed by nonionic surfactant of sorbitan trioleate modified with Cibacron blue F-3GA (CBF) in *n*-hexane.

14.9 AFFINITY-BASED REVERSE MICELLAR EXTRACTION AND SEPARATION

High selectivity for protein extraction in RME has been achieved using affinity partitioning. This technique involves the affinity interaction between a protein and its ligand coupled with incorporation and stripping of these complexes into and out of RMs, respectively. Three steps are commonly involved in any affinity-based separation process (Senstad and Mattiasson 1989): (i) formation of a reversible ligand–ligate complex, (ii) selective extraction of the complex into RMs, and (iii) dissociation of the complex, resulting in the isolation of pure ligate. The affinity ligand thus recovered could be further used in subsequent extractions. The salient features of affinity-based reverse micellar extraction and separation (ARMES) are the following: (1) affinity interaction is intraphasic, and thus, the mass transfer limitations in ligand–ligate interactions are absent; hence, the ligand utilization is very high, resulting in high productivity; (2) no chemical modification of the ligand is needed; (3) easy to operate and has inherent scalability of the process compared with other affinity-based separations such as affinity chromatography/affinity electrophoresis/affinity precipitation; (4) with dual selectivity (in binding and extraction stages) due to combination of biospecific interaction between ligand and ligate along with the driving force of RME (Paradkar and Dordick 1993). All the studies reported so far are on the application of ARMES technique for model systems using anionic surfactant (AOT).

Hatton and coworkers (1989) demonstrated the practical application of ARMES for the first time wherein selective extraction of concanavalin A (con A) into AOT-isooctane RMs was achieved. Choe et al. (1998) resolved structurally similar glyco-proteins (soybean peroxidase and acid glycoprotein) from each other using ARMES and con A as affinity ligand. Sun et al. (1999) have developed a novel RMS with crude soybean lecithin as a weak ionic surfactant. A relatively good separation between lysozyme and cytochrome c was achieved using affinity interaction with the CBF, attached to the inner layer of the reversed micelles of crude lecithin (Sun et al. 2000). Adachi et al. (2000) reported bioaffinity separation of chymotrypsinogen using antigen–antibody reaction in RMS composed of new nonionic surfactant tetra-oxyethylenemonodecylether ($C_{10}E_4$).

Liu et al. (2006) studied kinetics of protein transfer (lysozyme and BSA) to and from affinity-based RMs of Span 85 modified with CBF. The solubilization capacity of the RMS for lysozyme increased linearly with increasing the CBF concentration. Recently, Dong et al. (2010) reported the Ni(II)-chelated RMs of equimolar Triton X-45 and Span 80 for the purification of recombinant hexahistidine-tagged enhanced green fluorescent protein (EGFP) expressed in *E. coli*. The high binding affinity/specificity of the chelated Ni^{2+} ions toward the histidine tag produced electrophoretically pure EGFP, which was similar to that purified by immobilized metal affinity chromatography.

14.10   TECHNOLOGY DEVELOPMENT ASPECTS OF REVERSE MICELLAR EXTRACTION

RME combines the steps of concentration and purification of the desired enzyme in a single process and has several interesting advantages over other DSP processes. The various steps that need to be considered for the development of RME processes for the extraction and purification of biomolecules are: selection of the RMS, optimization of FE and BE conditions, selection of extraction equipment, recovery and recycling of the system components such as surfactants and solvents (Harikrisna et al. 2002). Selection of a suitable RMS is mainly based on the nature and charge of the biomolecule to be extracted. The optimization of FE and BE is carried out by studying the effect of various parameters on the extraction/stripping of proteins experimentally using full or fractional factorial designs to develop a meaningful system description. A literature survey suggests that the knowledge available on the recovery and reuse of surfactants is very little. However, the removal of surfactants from the stripping aqueous solution can be achieved by filtration and can be recycled. Use of ultrafiltration was also shown to be a successful technique for the separation of surfactants from reverse micellar solution (Hebbar et al. 2010).

The RME processes can be performed using four basic types of extractors, namely mixer-settler, agitated column extractor, centrifugal, and membrane extractors. The mixer-settler and centrifugal extractors seem to be more appropriate for solvent extraction using RMs (Cabral and Aires-Barros 1993). For some cases, mixer-settler and agitated column extractors cannot be used due to the formation of stable emulsions between the aqueous and reverse micellar phase as the presence of surfactants stabilizes these emulsions. In such cases, centrifugal extractors may be applied to reduce the settling time. Additionally, membrane extraction techniques may be adapted for use with these types of systems, the membrane serving to stabilize the reverse micellar/aqueous phase interface. For systems where short residence times are required, centrifugal extractors might be very useful. Process development has centered exclusively on the use of mixer-settlers (Dekker et al. 1986) and membrane extractors. Dekker et al. (1986) reported promising results for the continuous extraction of α-amylase using mixer-settler units. To maintain the system balance and to restore the extraction efficiency, surfactant should be replenished to the reverse micellar phase.

Efficient extraction of proteins has been reported with reverse micellar liquid membrane systems, where the pores of the membrane are filled with the reverse micellar phase and the enzyme is extracted from the aqueous phase on one side of the membrane, while the BE into a second aqueous phase takes place at the other side. By this, both the FE and BE can be performed using one membrane module. Armstrong and Li (1988) confirmed the general trends observed in a phase transfer using a glass diffusion cell with a reverse micellar liquid membrane. No significant denaturation of proteins (cytochrome C, myoglobin, lysozyme, and BSA) was observed on either side of the membrane. Further, the development of compatible synthetic membranes will have significant impact in minimizing the adsorption effects and in improving the feasibility of the process. Asenjo and coworkers (1994) studied the kinetics of protein extraction using RMs in a well-mixed system and a liquid–liquid spray column. The authors have shown that pH and ionic strength control the equilibrium distribution of proteins, whereas the partitioning kinetics was influenced by the operating conditions of the contacting device (rate of mixing, dispersed phase flow rate, volume ratio, etc.). In both systems studied, it was reported that the forward transfer is limited by diffusion in the aqueous boundary layer film while back transfer is limited by an interfacially controlled mechanism. It was suggested that for the design of RME equipment, kinetic considerations must be evaluated. Han et al. (1994) demonstrated the utility of a simple spray column for the RME of intracellular proteins from *Candida utilis*. The optimal forward transfer was observed after two circulations at a flow rate of 0.2 ml/s, while the optimal back transfer occurred after seven to nine circulations. Extraction of proteins using RME in hollow fibers was found to be substantially faster than the protein extractions with conventional equipment. Dahuron and Cussler (1988) performed continuous liquid membrane extraction of protein solutions employing RMs in hollow fibers. In particular, these systems routinely have surface areas of 30 cm^2/cm^3. While the mass transfer coefficient, K, is not unusually high in fibers, the product K_a is often 10–50 times larger than in conventional extraction towers. Moreover, in hollow fibers the two fluid flows are almost completely independent. As a result, there is no constraint due to flooding, loading, or channeling. Hollow fibers appear to be a superior way to achieve rapid mass transfer.

Lazarova and Tonova (1999) reported an integrated process for extraction and stripping of α-amylase using RMs in a stirred cell with separated compartments for each process. A comparison between the classical process and the integrated process resulted in 1.27-fold enhancement in the enzyme purification by the latter. Poppenborg et al. (2000) reported kinetics of simultaneous separation of lysozyme, cytochrome c, and ribonuclease in a stirred cell (Lewis cell) and Graesser contactor using RMS of AOT/isooctane, wherein all the proteins chosen were of similar molecular weight and isoelectric points. Technology development aspects of RME with respect to process scale-up and continuous operation are yet to develop for commercial application.

14.11 INTEGRATION OF REVERSE MICELLAR EXTRACTION WITH OTHER DSP TECHNIQUES

Integration of DSP methods is used in biotechnology to achieve higher selectivity/ purification of biomolecules. In this part, RME integrated with other DSP techniques to increases the overall productivity of the process is discussed. Chitosanases, which represent a class of hydrolytic enzymes from the fermentation broth of *Bacillus cereus* NTU-FC-4 was recovered of 70% total activity and purified by 30-fold using acetone precipitation followed by RME. Extraction of chitosanase directly from fermentation broth was reported to be more difficult as it contains nontarget contaminants that might interfere with the partition behavior of target protein. Besides, the cell debris, being larger molecules, would possibly be precipitated with surfactant in the interface layer during RME (Chen et al. 2006). Duarte-Vázquez et al. (2007) reported purification of peroxidase by acetone precipitation followed by chromatographic steps, which resulted in purification of 466-fold with activity recovery of 2.7%, whereas direct RME resulted in 60% activity recovery and 7-fold purification. This is higher than that obtained with acetone precipitation alone (1.6-fold). When a purification scheme similar to that followed initially was conducted for reverse micellar prepurified sample purification of 480-fold and an activity recovery of 3% was obtained with the advantage of recycling of the reverse micellar solution up to three times.

Recently, Hebbar et al. (2010) reported an integrated approach of coupling RME with ultrafiltration to improve the overall efficiency of extraction and purification of bromelain from aqueous extract of pineapple core. The purification of bromelain increased from 5.9- to 8.9-fold after ultrafiltration. Further, comparison of RME results with ATPE (activity recovery of 93.1% and purification of 3.2-fold) and the conventional ammonium sulfate precipitation (activity recovery of 82.1% and purification of 2.5-fold) indicated the improved performance of RME.

14.12   CONCLUDING REMARKS

One of the critical factors for DSP of biomolecules using RMs is the selection of appropriate RMS. Most of the systems reported in the literature used either anionic surfactant AOT or cationic surfactants TOMAC (trioctylmethyl ammonium chloride), BDBAC (*N*-benzyl-*N*-dodecyl-*N*-bis(2-hydroxy ethyl) ammonium chloride), or CTAB. These systems have some desirable characteristics such as suitable physical properties, ease of formation of RMs, etc. However, as the solubilization of proteins by these systems involves strong electrostatic interactions between protein and surfactant head groups, there is a possibility of loss in enzyme activity, and further, BE process is quite difficult to achieve, resulting in lower recovery of proteins. Despite the fact that the cost of the organic solvents used in RME is high, they can be easily recovered and reused several times, which automatically brings down the overall cost of the process. Studies with recovered phase components are rarely reported (Hemavathi et al. 2007). Recently, the trend of using the attractive "green solvents" instead of pollutant hydrocarbons has appeared. Ionic liquids have already proved their advantages as solvents in nonaqueous biocatalysis (Maruyama et al. 2002; Ulbert et al. 2005). The advantage of reusability of enzyme and ionic liquid has been recently reported (Tonova and Lazarova 2008). It has also been shown that AOT forms reversed micelles in hydrophobic ionic liquid where catalytic activity of lipase can be improved compared with RMS of AOT/*i*-octane (Moniruzzaman et al. 2008).

On the laboratory scale, RME for DSP of proteins is well established. To the best of our knowledge, there are no reports available on the larger-scale extraction of proteins using this technique, at levels above the 2-l scale (Keri et al. 1995; Pessoa and Vitolo 1997). The same is true with regard to continuous operation of RME, which was reported using mixer-settler units by Dekker et al. (1986) wherein the operating volumes of the two settlers were 900 and 650 ml for FE and BE, respectively. Hasmann et al. (2007) used packed column for continuous prepurification of glucose-6-phosphate dehydrogenase. Therefore, there is a need to investigate large scale extractive recovery of proteins from the fermentative media of real systems using industrial equipment to prove the process feasibility. In particular, there is a need for the applicability of RME for successful recovery of genetically engineered proteins to be investigated in detail, as very few reports are available in this area at the present time. The rate of phase separation after extraction in AOT-based RMS is slow. Keeping this in view, there is a need to study in detail the phase separation kinetics of this RMS to evolve means that would enhance phase separation rate. This is a very important aspect as far as industrial adaptability of RME is concerned. One possible approach to enhance phase separation is the application of external fields such as electric, acoustic, and microwave to RMSs. Employing RMSs that phase separate quickly without the need for any external effort could also be a possible solution. Some examples of such systems are DTDPA RMS (Goto et al. 1995), sugar esters DK-F-110 RMs (Naoe et al. 1999; Noritomi et al. 2006a,b; 2008), and NaDEHP (sodium di-2-ethyl hexyl phosphate)-RMs (Hu and Gulari 1996). Regarding the surfactant, more and more applications (biopurification, bioconversion, or refolding) are already created over biocompatible phospholipid-based surfactants (Sun et al. 2000; Cinelli et al. 2006; Wu et al. 2006; Hasmann et al. 2007) or sugar esters (Naoe et al. 1999, 2001; Noritomi et al. 2006a,b; 2008). Nonionic RMS sensitive to the temperature can be profitable to perform a temperature-controlled protein desolubilization (Noritomi et al. 2006b). Thus, nonionic RMSs could be used for rapid and efficient protein release (Rairkar et al. 2007). The searching of effective surfactants for solubilization of polypeptides in water-in-$SCCO_2$ reversed micelles is still underway. Using high-pressure CO_2, the increased transfer of lysozyme in conventional AOT/i-octane RMs was achieved under reduced extraction time (Chun et al. 2005). Moreover, pressurized CO_2 stripped lysozyme from the micelles avoids the need for conventional desolubilization step (Zhang et al. 2001).

Scale-up and engineering aspects appear to have received scant attention and mathematical modeling is to be given its due importance. Finally, it is important to continue fundamental work on the physicochemical state of the solubilized proteins, as this is a key factor to a rational selection of the reversed micelle composition. A special attention should be paid to the development of environment-friendly/biodegradable, biocompatible, and at the same time powerful surfactant systems that could extend the field of application over the pharmaceutical, cosmetic, and food industries. In conclusion, in the context of DSP accounting for a large share of the final product cost in many biotechnological processes, RME appears to be an efficient and promising bioseparation technique.

ACKNOWLEDGMENTS

The authors wish to thank Dr. V. Prakash, Director, CFTRI, for the keen interest in downstream processing. A. B. Hemavathi gratefully acknowledges CSIR, India, for the award of Senior Research Fellowship.

REFERENCES

Adachi, M., Shibata, K., Shioi, A., Harada, M., Katoh, S. 1998. Selective separation of trypsin from pancreation using bioaffinity in reverse micellar system composed of a nonionic surfactant. *Biotechnol Bioeng* 58: 649–653.

Adachi, M., Harada, M., Katoh, S. 2000. Bio-affinity separation of chymotrypsinogen using antigen–antibody reaction in reverse micellar system composed of a nonionic surfactant. *Biochem Eng J* 4: 149–151.

Aires-Barros, M.R., Cabral, J.M.S. 1991. Selective separation and purification of two lipases form *Chromobacterium viscosum* using AOT reversed micelles. *Biotechnol Bioeng* 38: 1302–1307.

Anarbaev, R.O., Khodyreva, S.N., Zakharenko, A.L., Rechkunova, N.I., Lavrik, O.I. 2005. DNA polymerase activity in water-structured and confined environment of reverse micelles. *J Mol Catal B: Enzymatic* 33: 29–34.

Aragon, M.J., Lavela, P., León, B., Pérez-Vicente, C., Tirado, J.L., Vidal-Abarca, C. 2010. On the use of the RMs synthesis of nanomaterials for lithium-ion batteries. *J Solid State Electrochem* 14: 1749–1753.

Armstrong, D.W., Li, W. 1988. Highly selective protein separations with reversed micellar liquid membranes. *Anal Chem* 60: 86–88.

Ayala, G., Kamat, S.V., Komives, C., Beckman, E.J., Russell, A.J. 1992. Solubilization and activity of proteins in compressible fluids based microemulsions. *Biotechnol* 10: 1584–1588.

Aydogan, O., Bayraktar, E., Parlaktuna, M., Mehmetoglu, M.T.U. 2007. Production of L-aspartic acid by biotransformation and recovery using reverse micelle and gas hydrate methods. *Biocatal Biotransform* 25: 365–372.

Bera, M.B., Panesar, P.S., Panesar, R., Singh, B. 2008. Application of reverse micellar extraction process for amylase recovery using response surface methodology. *Bioprocess Biosyst Eng* 31: 379–384.

Brandani, V., Giacomo, G.D., Spera, L. 1996. Recovery of α-amylase extracted by reverse micelles. *Process Biochem* 31: 125–128.

Bratko, D., Luzar, A., Chen, S.H. 1988. Electrostatic model for protein/reverse micelle complexation. *J Phys Chem* 89: 545–550.

Cabral, J.M.S., Aires-Barros, M.R. 1993. Reversed micelles in liquid–liquid extraction, in *Recovery Processes for Biological Materials,* J.F. Kennedy and J.M.S. Cabral (eds.), pp. 247–271. New York: Wiley.

Carlson, A., Nagarajan, R. 1992. Release and recovery of porcine pepsin and bovine chymosin from reverse micelles: A new technique based on isopropyl alcohol addition. *Biotechnol Prog* 8: 85–90.

Carvalho, C.M.L., Cabral, J.M.S. 2000. Reverse micelles as reaction media for lipases. *Biochimie* 82: 1063–1085.

Caselli, M., Luisi, P.L., Maestro, M., Roselli, R. 1988. Thermodynamics of the uptake of proteins by reverse micelles: First approximation model. *J Phys Chem* 92: 3899–3905.

Chen, Y.X., Zhang, Z.X., Chen, S.M., You, D.L., Wu, X.X., Yang, X.C., Guan, W.Z. 1999. Kinetically controlled syntheses catalyzed by proteases in revere micelles and separation of precursor dipeptides of RGD. *Enzyme Microb Technol* 25: 310–315.

Chen, S.X., Wei, D.Z., Hu, Z.H. 2001. Synthesis of galacto-oligosaccharides in AOT/isooctane reverse micellar system by β-galactosidase. *J Molecul Cata B: Enzy* 16: 109–114.

Chen, C.W., Ou-Yang, C.C., Yeh, C.W. 2003. Synthesis of galactooligosaccharides and transgalactosylation modeling in reverse micelles. *Enzyme Microb Technol* 33: 497–507.

Chen, C.T., Cheng, Y.C. 2005. Fe_3O_4/TiO_2 core/shell nanoparticles as affinity probes for the analysis of phosphopeptides using TiO_2 surface-assisted laser desorption/ionization mass spectrometry. *Anal Chem* 77: 5912–5919.

Chen, Y.L., Su, C.K., Chiang, B.H. 2006. Optimization of reversed micellar extraction of chitosanases produced by *Bacillus cereus*. *Process Biochem* 41: 752–758.

Chen, L.J., Cao, X.J. 2007. Avoiding inactivation of proteins in the process of AOT/isooctane reverse micellar extraction. *J Chem Eng Jpn* 40: 511–515.

Chen, Y., Dong, C.M. 2010. pH-sensitive supramolecular polypeptide-based micelles and reverse micelles mediated by hydrogen-bonding interactions or host–guest chemistry: Characterization and in vitro controlled drug release. *J Phys Chem B* 114: 7461–7468.

Choe, J. VanderNoot, V.A., Linhardt, R.J., Dordick, J.S. 1998. Resolution of glycoproteins by affinity based reversed micellar extraction and separation. *AIChE J* 44: 2542–2548.

Choe, J., Zhang, F., Wolf, M.W., Murhammer, D.W., Linhardt, R.J., Dordick, J.S. 2000. Separation of α-acid glycoprotein glycoforms using affinity-based reversed micellar extraction and separation. *Biotechnol Bioeng* 70: 486–490.

Choi, S.J., Decker, E.A., Henson, L., Popplewell, L.M., McClements, D.J. 2010. Inhibition of citral degradation in model beverage emulsions using micelles and reverse micelles. *Food Chem* 122: 111–118.

Chun, B.S., Park, S.Y., Kang, K.Y., Wilkinson, G.T. 2005. Extraction of lysozyme using reverse micelles and pressurized carbon dioxide. *Sep Sci Technol* 40: 2497–2508.

Cinelli, G., Cuomo, F., Hochkoeppler, A., Ceglie, A., Lopez, F. 2006. Use of *Rhodotorula minuta* live cells hosted in water-in-oil macroemulsion for biotransformation reaction. *Biotechnol Prog* 22: 689–695.

Cortez, E.V., Pessoa Jr, A., Felipe, M.G.A., Roberto, I.C., Vitolo, M. 2004. Optimized extraction by cetyltrimethyl ammonium bromide reversed micelles of xylose reductase and xylitol dehydrogenase from *Candida guilliermondii* homogenate. *J Chromatogr B* 807: 47–54.

Dahuron, L., Cussler, E.L. 1988. Protein extractions with hollow fibers. *AIChE J* 34: 130–136.

Das, D., Roy, S., Mitra, R.N., Dasgupta, A., Das, P.K. 2005. Head-group size or hydrophilicity of surfactants: The major regulator of lipase activity in cationic water-in-oil microemulsions. *Chem-A Eur J* 11: 4881–4889.

Dasgupta, A., Das, D., Mitra, R.N., Das, P.K. 2005. Surfactant tail length-dependent lipase activity profile in cationic water-in-oil microemulsions. *J Colloid Interface Sci* 289: 566–573.

De, T.K., Maitra, A. 1995. Solution behavior of Aerosol OT in non-polar solvents. *Adv Colloid Interface Sci* 59: 95–193.

Debnath, S., Das, D., Das, P.K. 2007. Unsaturation at the surfactant head: Influence on the activity of lipase and horseradish peroxidase in reverse micelles. *Biochem Biophys Res Commun* 356: 163–168.

Dekker, M., Riet, K.V., Weijers, S.R., Baltussen, J.W.A., Lanne, C., Birjsterbosch, B.H. 1986. Enzyme recovery by liquid–liquid extraction using reversed micelles. *Chem Eng J* 33: B27–B33.

Dekker, M., Riet, K.V., Bijsterbosch, B.H., Wolbert, R.B.G., Hilhorst, R. 1989. Modeling and optimization of the reversed micellar extraction of α-amylase. *AIChE J* 35: 321–324.

Dekker, M., Riet, K., Van-Der-Pol, J.J., Baltussen, J.W.A., Hilhorst, R., Bijsterbosch, B.H. 1991. Effect of temperature on the reversed micellar extraction of enzymes. *Chem Eng J* 46: 869–874.

Dong, X.Y., Wu, X.Y., Sun, Y. 2006. Refolding of denatured lysozyme assisted by artificial chaperones in reverse micelles. *Biochem Eng J* 31: 92–95.

Dong, X.Y., Meng, Y., Feng, X.D., Sun, Y. 2010. A metal chelate affinity reverse micellar system for protein extraction. *Biotech Prog* 26: 150–158.

Dovyap, Z., Bayraktar, E., Mehmetoglu, U. 2006. Amino acid extraction and mass transfer rate in the reverse micelle system. *Enzyme Microb Technol* 38: 557–562.

Duarte-Vázquez, M.A., García-Padilla, S., García-Almendárez, B.E., Whitaker, J.R., Regalado, C. 2007. Broccoli processing wastes as a source of peroxidase. *J Agric Food Chem* 55: 10396–10404.

Dungan, S.R., Bausch, T., Hatton, T.A., Plucinski, P., Nitsch, W. 1991. Interfacial transport processes in the reversed micellar extraction of proteins. *J Colloid Interface Sci* 145: 33–50.

Fan, K.K., Ouyang, P.,Wu, X.J., Lu, Z.H. 2001. Enhancement of the activity of papin in mixed reverse micellar systems in the presence of Tween 80. *J Chem Technol Biotechnol* 76: 27–34.

Feng, X., Zhang, J., Chen, J., Han, B., Shen, D. 2006. Enhanced solubilization of bovine serum albumin in reverse micelles by compressed CO_2. *Chem Eur J* 12: 2087–2093.

Ferreira, L.F.P., Taqueda, M.E., Vitolo, M., Converti, A., Pessoa Jr., A. 2005. Liquid–liquid extraction of commercial glucose oxidase by reversed micelles. *J Biotechnol* 116: 411–416.

Forney, C.E., Glatz, C.E. 1994. Reversed micellar extraction of charged fusion proteins. *Biotechnol Prog* 10: 499–502.

Forney, C.E., Glatz, C.E. 1995. Extraction of charged fusion proteins in reversed micelles. Comparison between different surfactant systems. *Biotechnol Prog* 11: 260–264.

Fraaije, J.G., Rijnierse, E.M., Hilhorst, E.J., Lyklema, J. 1990. Protein partition and ion co-partition in aqueous–apolar two-liquid-phase systems. *Colloid Polym Sci* 268: 855–863.

Gaikar, V.G., Kulkarni, M.S. 2001. Selective reverse micellar extraction of penicillin acylase from *E coli*. *J Chem Technol Biotechnol* 76: 729–736.

Goklen, K.E., Hatton, T.A. 1985. Protein extraction using reverse micelles. *Biotechnol Prog* 1: 69–74.

Goklen, K.E., Hatton, T.A. 1987. Liquid–liquid extraction of low molecular weight proteins by selective solubilization in reversed micelles. *Sepa Sci Technol* 22: 831–841.

Goncalves, A.M., Serro, A.P., Aires-Barros, M.R., Cabral, J.M.S. 2000. Effects of ionic surfactants used in reversed micelles on cutinase activity and stability. *Biochim Biophys Acta* 1480: 92–106.

Goto, M., Kuroki, M., Ono, T., Nakashio, F. 1995. Protein extraction by new reversed micelles with Di (Tridecyl) phosphoric acid. *Sepa Sci Technol* 30: 89–99.

Goto, M., Ono, T., Horiuchi, A., Furusaki, S. 1999. Extraction of DNA by reversed micelles. *J Chem Eng Jpn* 32: 123–125.

Goto, M., Hashimoto, Y., Fujita, T.A., Ono, T., Furusaki, S. 2000. Important parameters affecting efficiency of protein refolding by reversed micelles. *Biotechnol Prog* 16: 1079–1085.

Goto, M., Momota, A., Ono, T. 2004. DNA extraction by cationic reverse micelles. *J Chem Eng Jpn* 37: 662–668.

Hagen, A.J., Hatton, T.A., Wang, D.I. 1990. Protein refolding in reversed micelles: Interactions of the protein with micelle components. *Biotechnol Bioeng* 35: 966–975.

Haghtalab, A., Osfouri, S. 2003. Vacancy solution theory for partitioning of protein in reverse micellar systems. *Sepa Sci Technol* 38: 553–569.

Han, D.H., Lee, S.Y., Hong, W.H. 1994. Separation of intracellular proteins from Candida utilis using reversed micelles in a spray column. *Biotechnol Tech* 8: 105–110.

Harikrishna, S., Srinivas, N.D., Raghavarao, K.S.M.S., Karanth, N.G. 2002. Reverse micellar extraction for downstream processing of proteins/enzymes. *Adv Biochem Eng Biotechnol* 75: 119–183.

Hasmann, F.A., Cortez, D.V., Gurpilhares, D.B., Santos, V.C., Roberto, I.C., Pessoa Jr., A. 2007. Continuous counter-current purification of glucose-6-phosphate dehydrogenase using liquid–liquid extraction by reverse micelles. *Biochem Eng J* 34: 236–241.

He, Z.M., Wu, J.C., Yao, C.Y., Yu, K.T. 2001. Lipase-catalyzed hydrolysis of olive oil in chemically modified AOT/isooctane reverse micelles in a hollow fiber membrane reactor. *Biotechnol Lett* 23: 1257–1262.

Hebbar, H.U., Raghavarao, K.S.M.S. 2007. Extraction of bovine serum albumin using nanoparticulate RMs. *Process Biochem* 42: 1602–1608.

Hebbar, H.U., Sumana, B., Raghavarao, K.S.M.S. 2008. Use of reverse micellar systems for the extraction and purification of bromelain from pineapple wastes. *Bioresour Technol* 99: 4896–4902.

Hebbar, H.U., Sumana, B., Hemavathi, A.B., Raghavarao, K.S.M.S. 2010. Separation and purification of bromelain by reverse micellar extraction coupled ultrafiltration and comparative studies with other methods. *Food Bioprocess Technol* DOI 10.1007/s11947-010-0395-4.

Hecht, H.G., Peled, H.B. 2006. Structure modifications of AOT reverse micelles due to protein incorporation. *J Colloid Interface Sci* 297: 276–280.

Hemavathi, A.B., Hebbar, H.U., Raghavarao, K.S.M.S. 2007. Reverse micellar extraction of bromelain from *Ananas comosus* L. Merryl. *J Chem Technol Biotechnol* 82: 985–992.

Hemavathi, A.B., Hebbar, H.U., Raghavarao, K.S.M.S. 2008. Reverse micellar extraction of β-galactosidase from barley *(Hordeum vulgare). Appl Biochem Biotechnol* 151: 522–531.

Hemavathi, A.B., Hebbar, H.U., Raghavarao, K.S.M.S. 2010. Mixed reverse micellar systems for extraction and purification of β-glucosidase. *Sepa Purif Technol* 71: 263–268.

Hossain, M.J., Takeyama, T., Hayashi, Y., Kawanishi, T., Shimizu, N., Nakamura, R. 1999. Enzymatic activity of *Chromobacterium viscosum* lipase in an AOT/Tween 85 mixed reverse micellar system. *J Chem Technol Biotechnol* 74: 423–428.

Hu, Z., Gulari, E. 1996. Protein extraction using the sodium bis (2-ethylhexyl) phosphate (NaDEHP) reverse micellar system. *Biotechnol Bioeng* 50: 203–206.

Jarudilokkul, S., Poppenborg, L.H., Valetti, F., Gilardi, G., Stuckey, D.C. 1999. Separation and purification of periplasmic cytochrome c553 using reversed micelles. *Biotechnol Tech* 13: 159–163.

Jimenez-Carmona, M.M., Luque de Castro, M.D. 1998. Reverse micelle formation for acceleration of the supercritical fluid extraction of cholesterol from food samples. *Anal Chem* 70: 2100–2103.

Jolivalt, C., Minier, M., Renon, H. 1993. Extraction of cytochrome c in sodium dodecylbenzenesulfonate microemulsions. *Biotechnol Prog* 9: 456–461.

Kamyshny, A., Trofimova, D., Magdassi, S., Levashov, A. 2002. Native and modified glucose oxidase in reversed micelles. *Colloid Surf B: Biointerfaces* 24: 177–183.

Kim, J., Cho, Y.H., Park, W., Han, D., Chai, C.H., Imm, J.Y. 2003. Solubilization of water soluble anthocyanins in apolar medium using reverse micelle *J Agri Food Chem* 51: 7805–7809.

Kinugasa, T., Watanable, K., Tukeuchi, H. 1992. Activity and conformation of lysozyme in reversed micellar extraction. *Ind Eng Chem Res* 31: 1827–1829.

Kinugasa, T., Kondo, A., Mouri, E., Ichikawa, S., Nakagawa, S., Nishii, Y., Watanabe, K., Takeuchi, H. 2003. Effects of ion species in aqueous phase on protein extraction into reverse micellar solution. *Sepa Purif Technol* 31: 251–259.

Klyachko, N.L., Levashov, A.V. 2003. Bioorganic synthesis in reverse micelles and related systems. *Current Opinion Colloid Interface Sci* 8: 179–186.

Kouptsova, O.S., Klyachko, N.L., Levashov, A.V. 2001. Synthesis of alkyl glycosides catalysed by *b*-glycosidases in a system of reverse micelles. *Russian J Bioorganic chem* 27: 380–384.

Krei, G.A., Hustedt, H. 1992. Extraction of enzymes by reverse micelles. *Chem Eng Sci* 47: 99–111.

Krei, G., Meyer, U., Borner, B., Hustedt, H. 1995. Extraction of α-amylase using BDBAC-reversed micelles. *Bioseparation* 5: 175–183.

Lakshmi, M.C., Raghavarao, K.S.M.S. 2010. Downstream processing of soy hull peroxidase employing reverse micellar extraction. *Biotechnol Bioprocess Eng* 15: 937–945.

Lan, J., Zhang, Y., Huang, X., Hu, M., Liu, W., Li, Y. 2008. Improvement of the catalytic performance of lignin peroxidase in reversed micelles. *J Chem Technol Biotechnol* 83: 64–70.

Lazarova, Z., Tonova, K. 1999. Integrated reversed micellar extraction and stripping of α-amylase. *Biotechnol Bioeng* 63: 583–592.

Lee, S.S., Kim, B.G., Sung, N.C., Lee, J.P. 2004a. Back extraction processes of C.C. lipase with mediated AOT reverse micellar system. *Bull Korean Chem Soc* 25: 873–877.

Lee, B.K., Hong, D.P., Lee, S.S., Kuboi, R. 2004b. Analysis of protein back-extraction processes in alcohol and carboxylic acid-mediated AOT reverse micellar systems. *Biochem Eng J* 22: 71–79.

Leodidis, E.B., Hatton, T.A. 1989. Specific ion effects in electrical double layers: Selective solubilization of cations in aerosol-OT reversed micelles. *Langmuir* 5: 741–753.

Leodidis, E.B., Hatton, T.A. 1990. Amino acids in AOT reversed micelles. 2. The hydrophobic effect and hydrogen bonding as driving forces for interfacial solubilization. *J Phys Chem* 94: 6411–6420.

Leser, M.E., Luisi, P.L., Paimieri, S. 1989. The use of reverse micelles for the simultaneous extraction of oil and proteins from vegetable meal. *Biotechnol Bioeng* 34:1140–1146.

Levashov, A.V., Khmelnitsky, Y.L., Klyachko, N., Chernyak, V.Y., Martinek, K. 1982. Enzymes entrapped into reversed micelles in organic solvents. *J Colloid Interface Sci* 88: 444–457.

Levinger, N.E., Swafford, L.A. 2009. Ultrafast dynamics in RMs. *Annu Rev Phys Chem* 60: 385–406.

Liu, J.G., Xing, J.M., Shen, R., Yang, C.L., Liu, H.Z. 2004. Reverse micellar extraction of nattokinase from fermentation broth. *Biochem Eng J* 21: 273–278.

Liu, Y., Dong, X.Y., Sun, Y. 2006. Equilibria and kinetics of protein transfer to and from affinity based reverse micelles of Span 85 modified with Cibracron Blue F-3GA. *Biochem Eng J* 28: 281–288.

Liu, Y., Dong, X.Y., Sun, Y. 2007. Protein separation by affinity extraction with reversed micelles of Span 85 modified with Cibracron blue F3G-A. *Sep Purif Technol* 53: 289–295.

López-Jiménez, F.J., Rubio S., Pérez-Bendito, D. 2010. Supramolecular solvent-based microextraction of Sudan dyes in chilli-containing foodstuffs prior to their liquid chromatography–photodiode array determination. *Food Chem* 121: 763–769.

Manocha, B., Gaikar, V.G. 2006. Permeabilization of *Aspergillus niger* by reverse micellar solutions and simultaneous purification of catalase. *Sep Sci Technol* 41: 3279–3296.

Martinek, K., Levashov, V., Klyachko, N., Khmelnitski, L., Berezin, H. 1986. Micellar enzymology. *Eur J Biochem* 55: 453–468.

Maruyama, T., Nagasawa, S., Goto, M. 2002. Poly(ethylene glycol)-lipase complex that is catalytically active for alcoholysis reactions in ionic liquids. *Biotechnol Lett* 24: 1341–1345.

Mathew, D.S., Juang, R.S. 2005. Improved back extraction of papain from AOT reverse micelles using alcohols and a counter-ionic surfactant. *Biochem Eng J* 25: 219–225.

Mathew, D.S., Juang, R.S. 2007. Role of alcohols in the formation of inverse microemulsions and back extraction of proteins/enzymes in a reverse micellar system. *Sep Purif Technol* 53: 199–205.

Matzke, S.F., Creagh, A.L., Haynes, C.A., Prausnitz, J.M., Blanch, H.W. 1992. Mechanisms of protein solubilization in reverse micelles. *Biotechnol Bioeng* 40: 91–102.

Melo, E.P., Aires-Barros, M.R., Cabral, J.M.S. 2001. Reverse micelles and protein biotechnology. *Biotechnol Annu Rev* 7: 87–129.

Michizoe, J., Goto, M., Furusaki, S. 2001. Catalytic activity of laccase hosted in reversed micelles. *J Biosci Bioeng* 92: 67–71.

Michizoe, J., Uchimura, Y., Maruyama, T., Kamiya, N., Goto, M. 2003. Control of water content by reverse micellar solutions for peroxidase catalysis in a water-immiscible organic solvent. *J Biosci Bioeng* 95: 425–427.

Moniruzzaman, M., Hayashi, Y., Talukder, M.R., Saito, E., Kawanishi, T. 2006. Effect of aprotic solvents on the enzymatic activity of lipase in AOT reverse micelles. *Biochem Eng J* 30: 237–244.

Moniruzzaman, M., Hayashi, Y., Talukder, M.R., Kawanishi, T. 2007. Lipase-catalyzed esterification of fatty acid in DMSO (dimethyl sulfoxide) modified AOT reverse micellar systems. *Biocatal Biotransform* 25: 51–58.

Moniruzzaman, M., Kamiya, N., Nakashima, K., Goto, M. 2008. Water-in-ionic liquid microemulsion as a new medium for enzymatic reactions. *Green Chem* 10: 497–500.

Monterio, T.I., Porto, R.C., Carneiro-Leao, T.S., Silva, A.M.A., Carneiro-da-Cunha, M.P.C. 2005. Reversed micellar extraction of an extracellular protease from *Nacardiopsis sp.* fermentation broth. *Biochem Eng J* 24: 87–90.

Morales, M.D., Gonzales, M.C., Serra, B., Zhang, J., Reviejo, A.J., Pingarron, J.M. 2005. Biosensing of aromatic amines in reversed micelles with self-generation of hydrogen peroxide at glucose oxidase-peroxidase bienzyme electrodes. *Electroanal* 17: 1780–1788.

Morrison, S.A., Carpenter, E.E., Harris, V.G., Cahill, C.A. 2005. Atomic engineering of magnetic nanoparticles. *J Nanosci Nanotech* 5: 1323–1344.

Motlekar, N.A., Bhagwat, S.S. 2001. Activity of horseradish peroxides in aqueous and reverse micelles and back extraction from reverse micellar phases. *J Chem Technol Biotechnol* 76: 643–649.

Mutalik, R.B., Gaikar, V.G. 2003. Cell permeabilization for extraction of penicillin acylase from *Escherichia coli* by reverse micellar solutions. *Enzyme Microb Technol* 32: 14–26.

Mutalik, R.B., Gaikar, V.G. 2006. Reverse micellar solutions aided permeabilization of baker's yeast. *Process Biochem* 41: 133–141.

Nagayama, K., Katakura, R., Hata, T., Naoe, K., Imai, M. 2008. Reactivity of *Candida rugosa* lipase in cetyltrimethylammonium bromide microemulsion–gelatin complex organogels. *Biochem Eng J* 38: 274–276.

Nandini, K.E., Rastogi, N.K. 2009. Reverse micellar extraction for downstream processing of lipase: Effect of various parameters on extraction. *Process Biochem* 44: 1172–1178.

Nandini, K.E., Rastogi, N.K. 2010a. Separation and purification of lipase using reverse micellar extraction: Optimization of conditions by response surface methodology. *Biotechnol Bioprocess Eng* 15: 349–358.

Nandini, K.E., Rastogi, N.K. 2010b. Single step purification of lactoperoxidase from whey involving reverse micellar system assisted extraction and its comparison with reverse micellar extraction. *Biotechnol Prog* 26: 763–771.

Naoe, K., Nishino, M., Ohsa, T., Kawagoe, M., Imai, M. 1999. Protein extraction using sugar ester reverse micelles. *J Chem Technol Biotechnol* 74: 221–226.

Naoe, K., Ohsa, T., Kawagoe, M., Imai, M. 2001. Esterification by *Rhizopus delemar* lipase in organic solvent using sugar ester reverse micelles. *Biochem Eng J* 9: 67–72.

Naoe, K., Noda, K., Kawagoe, M., Imai, M. 2004a. Higher order structure of proteins solubilized in AOT reverse micelles. *Colloid Surf B: Biointerfaces* 38: 179–185.

Naoe, K., Awatsu, S., Yamada, Y., Kawagoe, M., Nagayama, K., Imai, M. 2004b. Solvent condition in triolein hydrolysis by *Rhizopus delemar* lipase using an AOT reverse micellar system. *Biochem Eng J* 18: 49–55.

Nishii, Y., Kinugasa, T., Nii, S., Takahashi, K. 2002. Transport behavior of protein in bulk liquid membrane using reversed micelles. *J Membr Sci* 195: 11–21.

Nishiki, T., Nakamura, K., Kato, D. 2000. Forward and backward extraction rates of amino acid in reversed micellar extraction. *Biochem Eng J* 4: 189–195.

Noh, K. H., Imm, J.Y. 2005. One-step separation of lysozyme by reverse micelles formed by the cationic surfactant, cetyldimethylammonium bromide. *Food Chem* 93: 95–101.

Noh, K.H., Rhee, M.S., Imm, J.Y. 2005. Separation of lactoferrin from model whey protein mixture by reverse micelles formed by cationic surfactant. *Food Sci Biotechnol* 14: 131–136.

Noritomi, H., Ito, S., Kojima, N., Kato, S., Nagahama, K. 2006a. Forward and backward extractions of cytochrome c using reverse micellar system of sucrose fatty acid ester. *Colloid Polym Sci* 284: 604–610.

Noritomi, H., Kojima, N., Kato, S., Nagahama, K. 2006b. How can temperature affect reverse micellar extraction using sucrose fatty acid ester? *Colloid Polym Sci* 284: 683–687.

Noritomi, H., Takasugi, T., Kato, S. 2008. Refolding of denatured lysozyme by water-in-oil microemulsions of sucrose fatty acid esters. *Biotechnol Lett* 30: 689–693.

Nowak, J.R., Maslakiewicz, P., Haber, J. 1996. The effect of linoleic acid on pH inside sodium bis(2-ethylhexyl)sulfosuccinate reverse micelles in isooctane and on the enzymatic activity of soybean lipoxygenase. *Eur J Biochem* 238: 549–553.

Okada, K., Nishii, Y., Nii, S., Kinugasa, T., Takahashi, K. 2001. Interfacial properties between aqueous and organic phases in AOT reversed micellar system for lysozyme extraction. *J Chem Eng Jpn* 34: 501–505.

Padalkar, K.V., Gaikar, V.G. 2003. Reverse micellar extraction of proteins: A molecular dynamic study. *Sep Sci Technol* 38: 2565–2578.

Pandey, A., Pandey, A. 2008. Reverse micelles as suitable microreactor for increased biohydrogen production. *Int J Hydrogen Energy* 33: 273–278.

Pandit, P., Basu, S. 2004. Removal of ionic dyes from water by solvent extraction using reverse micelles. *Environ Sci Technol* 38: 2435–2442.

Paradkar, V.M., Dordick, J.S. 1993. Affinity based reverse micellar extraction and separation: A facile technique for the purification of peroxidase from soybean hulls. *Biotechnol Prog* 9: 199–203.

Pessoa Jr., A., Vitolo, M. 1997. Separation of inulinase from *Kluyveromyces marxianus* using reversed micellar extraction. *Biotechnol Tech* 11: 421–422.

Pires, M.J., Aires-Barros, M.R., Cabral, J.M.S. 1996. Liquid–liquid extraction of proteins with reversed micelles. *Biotechnol Prog* 12: 290–301.

Poppenborg, L.H., Brillis, A.A., Stuckey, D.C. 2000. The kinetic separation of protein mixtures using reverse micelles. *Sep Sci Technol* 35: 843–858.

Rabie, H.R., Vera, J.H. 1998. A simple model for reverse micellar extraction of proteins. *Sep Sci Technol* 33: 1181–1193.

Raghavendra, T., Sayania, D., Madamwar, D. 2010. Synthesis of the 'green apple ester' ethyl valerate in organic solvents by *Candida rugosa* lipase immobilized in MBGs in organic solvents: Effects of immobilization and reaction parameters. *J Mol Catal B: Enzymatic* 63: 31–38.

Rairkar, M.E., Hayes, D., Harris, J.M. 2007. Solubilization of enzymes in water-in-oil microemulsions and their rapid and efficient release through use of a pH-degradable surfactant. *Biotechnol Lett* 29: 767–771.

Reddy, K.S.N., Salvati, L.M., Dutta, P.K., Abel, P.B., Suh, K.I., Ansari, R.R. 1996. Reverse micelle based growth of zincophosphate sodalite: Examination of crystal growth. *J Phys Chem* 100: 70–98.

Regalado, C., Asenjo, J.A., Pyle, D.L. 1996. Studies on the purification of peroxidase from horseradish roots using reverse micelles. *Enzyme Microb Technol* 18: 332–339.

Reta, M., Waymas, O., Silber, J.J. 2006. Partition of polyhydroxy compounds of biological and pharmacological significance between AOT reverse microemulsions and aqueous salt solutions. *J Phys Organic Chem* 19: 219–227.

Rojo, M., Gomez, M., Estrada, P. 2001. Polyphenol oxidase in reverse micelles of AOT/cyclohexane: A thermostability study. *J Chem Technol Biotechnol* 76: 69–77.

Sakono, M., Goto, M., Furusaki, S. 2000. Refolding of cytochrome c using reversed micelles. *J Biosci Bioeng* 89: 458–462.

Senstad, C., Mattiasson, B. 1989. Precipitation of soluble affinity complexes by a second affinity interaction: A model study. *Biotechnol Appl Biochem* 11: 41–48.

Shah, C., Sellappan, S., Madamwar, D. 2000. Entrapment of enzyme in water-restricted microenvironment-amyloglucosidase in reverse micelles. *Process Biochem* 35:971–975.

Shen, R., Liu, J.G., Xing, J.M., Liu, H.Z. 2005. Extraction of industrial lipase with CTAB-based mixed reverse micelles. *Chin J Process Eng* 5: 255–259.

Sheu, E., Goklen, K.E., Hatton, T.A., Chen, S.H. 1986. Small angle neutron scattering studies of protein-reversed micelle complexes. *Biotechnol Prog* 2: 175–185.

Shin, Y.O., Rodil, E., Vera, J.H. 2003a. Selective precipitation of lysozyme from egg white using AOT. *J Food Sci* 68: 595–599.

Shin, Y.O., Weber, M.E., Vera, J.H. 2003b. Effect of salt and volume ratio on the reverse micellar extraction of lysozyme using DODMAC. *Fluid Phase Equilib* 207: 155–165.

Shin, Y.O., Wahnon, D., Weber, M.E., Vera, J.H. 2004. Selective precipitation and recovery of xylanase using surfactant and organic solvent. *Biotechnol Bioeng* 86: 698–705.

Shkarina, T.N., Kuhn, H., Schewe, T. 1992. Specificity of soybean lipoxygenase-1 in hydrated reverse micelles of sodium bis(2-ethylhexyl)sulfosuccinate (aerosol OT). *Lipids* 27: 690–693.

Shiomori, K., Kawano, Y., Kuboi, R., Komasawa, I. 2000. Effect of electrostatic interaction on reverse micellar extraction of large molecular weight proteins. *J Chem Eng Jpn* 33: 800–804.

Shipovskov, S., Daria, T., Saprykin, E., Christenson, A., Ruzgas, T., Levashov, A.V., Ferapontova, E.E. 2005. Spraying enzymes in microemulsions of AOT in nonpolar organic solvents for fabrication of enzyme electrodes. *Anal Chem* 77: 7074–7079.

Shome, A., Roy, S., Das, P.K. 2007. Nonionic surfactants: A key to enhance the enzyme activity at cationic reverse micellar interface. *Langmuir* 23: 4130–4136.

Singh, A., Pandey, K.D., Dubey, R.S. 1999. Reverse micelles: A novel tool for H_2 production. *World J Microbiol Biotechnol* 15: 243–247.

Sivasamy, A., Rasoanto, P., Ramabrahmam, B., Swaminathan, G. 2005. Effects of nonpolar solvents on the solubilization of pepsin into bis(2-ethylhexyl)sodium sulfosuccinate and cetyltrimethylammonium bromide reverse micelles. *J Solution Chem* 34: 33–42.

Streitner, N., Voß, C., Flaschel, E. 2007. Reverse micellar extraction systems for the purification of pharmaceutical grade plasmid DNA. *J Biotechnol* 131: 188–196.

Streitner, N., Voß, C., Flaschel, E., Von, I. 2008. Plasmid–DNA durch in versmizellare Zweiphasensysteme—Optimierung der Rückextraktion. *Chem Ing Tech* 80: 831–837.

Subramani, S., Dittrich, N., Hirche, F., Hofmann, R.U. 1996. Characteristics of phospholipase D in reverse micelles of triton X-100 and phosphatidylcholine in diethyl ether. *Biotechnol Lett* 18: 815–820.

Su, C.K., Chiang, B.H. 2003. Extraction of immunoglobulin-G from colostral whey by reverse micelles. *J Dairy Sci* 86: 1639–1645.

Sun, Y., Bai, S., Gu, L., Tong, X.D., Ichikawa, S., Furusaki, S. 1999. Effect of hexanol as a cosolvent on partitioning and mass transfer rate of protein extraction using reverse micelles of CB-modified lecithin. *Biochem Eng J* 3: 9–15.

Sun, Y., Bai, S., Gu, L., Furusaki, S. 2000. Purification of lysozyme by affinity-based reversed micellar two-phase extraction. *Bioprocess Eng* 22: 19–22.

Talukder, M.R., Susanto, D., Feng, G., Wu, J.C., Choi, W.J., Chow, Y. 2007. Improvement in extraction and catalytic activity of *Mucor javanicus* lipase by modification of AOT reverse micelle. *Biotechnol J* 2: 1369–1374.

Tonova, K., Lazarova, Z. 2005. Influence of enzyme aqueous source on RME-based purification of α-amylase. *Sep Purif Technol* 47: 43–51.

Tonova, K., Lazarova, Z., Nemestothy, N., Gubicza, L., Belafi-Bako, K. 2006. Lipase-catalyzed esterification in a reversed micellar reaction system. *Chem Ind Chem Eng Quart* 12: 168–174.

Tonova, K., Lazarova, Z. 2008. Reversed micelle solvents as tools of enzyme purification and enzyme-catalyzed conversion. *Biotechnol Adv* 26: 516–532.

Treybal, R.E. 1980. *Mass-Transfer Operations*. 3rd edition, New York: McGraw-Hill.

Tsagkogeorgas, F., Petropoulou, O.M., Niessner, R., Knopp, D. 2006. Encapsulation of biomolecules for bioanalytical purposes: Preparation of diclofenac antibodydoped nanometer sized silica particles by reverse micelle and sol-gel processing. *Anal Chim Acta* 573–574: 133–137.

Tschapalda, K., Streitner, N., Voß, C., Flaschel, E. 2009. Generation of chromosomal DNA during alkaline lysis and removal by reverse micellar extraction. *Appl Microbiol Biotechnol* 84: 199–204.

Tzeng, Y.M., Tsun, H.Y., Chang, Y.N. 1999. Recovery of *thuringiensin* with cetylpyridinium chloride using micellar enhanced ultrafiltration process. *Biotechnol Prog* 15: 580–586.

Ugolini, L., De Nicola, G., Palmieri, S. 2008. Use of reverse micelles for the simultaneous extraction of oil, proteins, and glucosinolates from cruciferous oilseeds. *J Agric Food Chem* 56: 1595–1601.

Ulbert, O., Bélafi-Bakó, K., Tonova, K., Gubicza, L. 2005. Thermal stability enhancement of *Candida rugosa* lipase using ionic liquids. *Biocatal Biotransform* 23: 177–183.

Uskokovi, V.C., Drofenik, M. 2005. Synthesis of materials within RMs. *Surface Review Lett* 12: 239–277.

Vaidya, S., Patra, A., Ganguli, A.K. 2010. CdS@TiO$_2$ and ZnS@TiO$_2$ core–shell nanocomposites: Synthesis and optical properties. *Colloid Surfaces A: Physicochem Eng Aspects* 363: 130–138.

Vasudevan, M., Wiencek, J.M. 1996. Mechanism of the extraction of proteins into Tween 85 nonionic microemulsions. *Ind Eng Chem Res* 35: 1085–1089.

Wolbert, B.G., Hilhorst, R., Voskuilen, G., Nachtegaal, H., Dekker, M., Vantriet, K., Bijsterbosch, B.H. 1989. Protein transfer from an aqueous into reversed micelles, the effect of protein size and charge distribution. *Eur J Biochem* 184: 627–633.

Wu, J.C., Song, B.D., Xing, A.H., Hayashi, Y., Talukder, M.M.R., Wang, S.C. 2002. Esterification reaction catalyzed by surfactant-coated *Candida rugosa* lipase in organic solvents. *Process Biochem* 37: 1229–1233.

Wu, X.Y., Liu, Y., Dong, X.Y., Sun, Y. 2006. Protein refolding mediated by reverse micelles of Cibacron blue F-3GA modified nonionic surfactant. *Biotechnol Prog* 22: 499–504.

Yoksan, R., Chirachanchai, S. 2010. Silver nanoparticle-loaded chitosan-starch based films: Fabrication and evaluation of tensile, barrier and antimicrobial properties. *Materials Sci Eng C* 30: 891–897.

Yonker, C.R., Fulton, J.L., Phelps, M.R., Bowman, L.E. 2003. Membrane separations using reverse micelles in near-critical and supercritical fluid solvents. *J Supercrit Fluids* 25: 225–231.

Yu, Y.C., Chu, Y., Ji, J.Y. 2003. Study of the factors affecting the forward and back extraction of yeast-lipase and its activity by reverse micelles. *J Colloid Interface Sci* 267: 60–64.

Zamarro, M.T., Domingo, M.J., Ortega, F., Estrada, P. 1996. Recovery of cellulolytic enzymes from wheat straw hydrolysis broth with dioctyl sulphosuccinate/isooctane reverse micelles. *Biotechnol Appl Biochem* 23: 117–125.

Zhang, T.X., Liu, H.Z., Chen, J.Y. 2000. Extraction of yeast alcohol dehydrogenase using reversed micelles formed with CTAB. *J Chem Technol Biotechnol* 75: 798–802.

Zhang, H., Lu, J., Han, B. 2001. Precipitation of lysozyme solubilized in reverse micelles by dissolved CO$_2$. *J Supercrit Fluids* 20: 65–71.

Zhang, W., Liu, H., Chen, J. 2002. Forward and backward extraction of BSA using mixed reverse micellar system of CTAB and alkyl halides. *Biochem Eng J* 12: 1–5.

Zhao, X., Chen, F., Gai, G., Chen, J., Xue, W., Lee, L. 2008. Effects of extraction temperature, ionic strength and contact time on efficiency of bis(2-ethylhexyl) sodium sulfosuccinate (AOT) reverse micellar backward extraction of soy protein and isoflavones from soy flour. *J Sci Food Agric* 88: 590–596.

Zhu, K.X., Sun, X.H., Chen, Z.C., Peng, W., Qian, H.F., Zhou, H.M. 2010. Comparison of functional properties and secondary structures of defatted wheat germ proteins separated by reverse micelles and alkaline extraction and isoelectric precipitation. *Food Chem* 123: 1163–1169.

15 Aqueous Two-Phase Extraction of Enzymes for Food Processing

M. C. Madhusudhan, M. C. Lakshmi, and Karumanchi S. M. S. Raghavarao

CONTENTS

15.1 INTRODUCTION

In recent years, application of biotechnology for the production of biomolecules of importance for research, pharmaceutical/clinical, and industrial usage by research and industrial communities has increased. Enzymes used in food processing are derived from animal, plant, and microbial sources (Fernandes, 2010). Enzymes are broadly used in food industries for a wide range of applications, e.g., production of various types of syrups from starch or sucrose (α- and β-amylases, glucamylase, pullulanase, invertase, and glucose isomerase), meat/protein processing (proteases), removal of glucose and/or molecular oxygen (O_2) (glucose oxidase and catalase), dairy industry (lactase), and fruit juice and brewing industries (pectinases). Bioseparation/downstream processing of enzymes occupies a complex and important position in biotechnology and biochemical engineering in view of the recognized fact that product recovery costs become critical in the overall economics of modern biotechnological processes (Walter et al. 1985; Hatti-Kaul 2000). Improved downstream processing techniques are increasingly important for biotechnology because recovery is often the limiting factor for the commercial success of biological processes. The conventional filtration process employed for solid–liquid separation is not suitable for the bioseparation, where the size of the microorganisms to be separated is small, especially when the cells are disintegrated to release the intracellular biomolecules resulting in a system of increased viscosity (Kula et al. 1981). In case of conventional methods, like centrifugation, and even modern methods, such as electrophoresis or column chromatography, scale-up problems are considerable, making them uneconomical unless the product is of high value. The separation of many biomolecules is still performed by batch mode in small-scale processes such as column chromatography, salt- and solvent-induced precipitation, and electrophoresis. These unit operations have scale-up problems and are expensive at larger scale besides resulting in low product recovery (Diamond and Hsu 1992; Raghavarao et al. 1998). The downstream processing of biological materials requires purification techniques that are delicate enough to preserve the biological activity. There is, therefore, a need in the industry for efficient and economical methods of separation, purification, and concentration of the biomolecules.

Liquid–liquid extraction using aqueous two-phase systems (ATPSs) popularly known as aqueous two-phase extraction (ATPE) is one such method. Liquid–liquid extraction using organic–aqueous phase systems is extensively used in the chemical industry. Despite all the advantages, this method has not gained wide industrial recognition in the field of biotechnology. In fact, the commonly used organic solvent systems are unsuitable for the intended purpose due to the low solubility and denaturation of biomolecules in organic solvents.

ATPE has been successful to a large extent in overcoming these drawbacks of conventional extraction processes, since both the phases are aqueous. Liquid–liquid extraction employing ATPSs has proved to be a promising separation strategy for many biological products such as proteins, enzymes, viruses, cells, and other biological materials (Kula et al. 1982; Raghavarao et al. 1991). ATPSs were developed in Sweden during the mid-1950s for the separation of macromolecules, cells, and organelles. Since then, attention has been directed toward widening its application in biotechnology (Walter et al. 1985; Albertsson 1986; Zaslavsky 1995).

A wealth of information was reported in the literature on various aspects of ATPE and its applications (Kula et al. 1982; Walter et al. 1985; Albertsson 1986, Diamond and Hsu 1992; Zaslavsky 1995; Raghavarao et al. 1998; 2003). A few other liquid–liquid extraction methods employed for the extraction and purification of various biomolecules are reverse micellar extractions, cloud point extractions, and micellar extractions. However, ATPE is a better alternative to these existing methods of primary purification due to its high capacity, better yield/purity of product, low space time yield, biocompatible environment, lower process time, low energy, and ease of scale-up. Furthermore, application of ATPE permits easy adaptation of the equipment and the methods of conventional organic–aqueous extractions used in the chemical industry (Walter 1985; Raghavarao et al. 1991). This technique is effective in removal of by-products such as undesirable enzymes/proteins, unidentified polysaccharides, and pigments (Kula 1985). A broth can be directly subjected to ATPE by the addition of desired quantities of phase-forming polymers and salts. ATPE can be designed such that the desired biomolecule selectively partitions to one of the phases in a concentrated form, with considerable reduction in the volume of the stream to be handled during the subsequent purification steps.

In ATPE, the fermentation broth containing the product can be added to the polymer to make one of the phases for contacting the other phase during extraction, where the product selectively partitions to the other phase. Hence, ATPE can be considered complementary to more selective biomolecule purification methods. In some instances, ATPE has the potential to produce a concentrated and purified product in one step when compared to the number of steps involved in conventional downstream processing. An important feature of ATPE is that partitioning of enzymes/proteins, in general, does not depend on their concentration and the volume of the system over a wide range (Albertsson 1971). Therefore, it is relatively easier to scale-up these partitioning steps with greater precision when compared with conventional steps in enzyme/protein isolation and purification.

Although a number of research and some review papers have been published in recent years relating to various aspects of ATPE, no exhaustive review encompassing all aspects has been available. For example, mathematical modeling, which is very important for the prediction of the performance of the equipment, has not been given its due importance. In this chapter, an attempt has been made to provide comprehensive coverage of all the important aspects including industrial/food enzymes.

15.2   AQUEOUS TWO-PHASE SYSTEMS

15.2.1   Formation of ATPSs

ATPSs are of two types, polymer/polymer and polymer/salt. Some of the commonly used phase systems are listed in Table 15.1. Both components of these systems are

TABLE 15.1
Components for the Formation of ATPSs

Polymer/Polymer Phase Systems

Polyethylene glycol (PEG)	Dextran, Ficoll, pullulan, polyvinyl alcohol, Reppal PES-100 (hydroxypropyl starch), maltodexrin, xanthan, sodium polyacrylate (NaPA), Gemini/SDS, cashew-nut tree gum, mixture of 2-(dimethylamino) ethyl methacrylate, *t*-butyl methacrylate, and methyl methacrylate
Polypropylene glycol	Dextran, hydroxypropyl dextran, polyvinyl pyrrolidone, polyvinyl alcohol, methoxy PEG
Polyvinyl alcohol	Dextran, hydroxypropyl dextran, methyl cellulose
Polyvinylpyrrolidone	Dextran, hydroxypropyl dextran, methyl cellulose, Reppal PES-200 (hydroxypropyl starch)
Methyl cellulose	Dextran, hydroxypropyl dextran
Ethylhydroxyethyl cellulose	Dextran
Hydroxypropyl dextran	Dextran
Ficoll	Dextran
Ficoll	Agarose
Vinyl-2-pyrrolidone-guar gum	Dextran
Guar gum	Hydroxypropyl starch

Thermoseparating Polymer-Based Phase Systems

Breox (ethylene oxide–propylene oxide)	Dextran, Reppal PES-100, potassium phosphate
Ucon 50-HB-5100	Ammonium sulfate, polyvinyl alcohol, Reppal PES-100 (hydroxypropyl starch)
Poly(ethylene oxide-co-maleic anhydride)	Dextran

Detergent-Based Phase Systems

Triton X114	Water
Agrimul NRE 1205 (C12–18E5)	Water
Cetyltrimethylammonium bromide (CTAB)	Sodium dodecyl sulfonate (AS)
1,3-Propanediyl bis(dodecyl dimethylammonium bromide)	Sodium dodecyl sulfonate (AS)
Agrimul NRE 1205	Water
Triton X-114 + zwittergent + SDS	Polyethelene glycol
Dodecyltrimethylammonium bromide(DTAB)	Sodium dodecyl sulfonate (AS)

(continued)

TABLE 15.1 (Continued)
Components for the Formation of ATPSs

Polymer/Salt Phase Systems

PEG	Potassium phosphate, sodium sulfate, sodium formate, sodium potassium tartarate, magnesium sulfate, sodium citrate, ammonium sulfate, ammonium carbamate
Polypropylene glycol	Potassium phosphate
Methoxy-PEG	Potassium phosphate
Polyvinylpyrrolidone	Potassium phosphate
	Potassium citrate
Derivatives of PEG (PEG-benzoate (PEG-Bz), PEG-phosphate (PEG-PO$_4$), PEG-trimethylamine (PEG-TMA), PEG-palmitate (PEG-pal), and PEG-phenylacetamide (PEG-paa))	Sodium sulfate

Alcohol-Based Phase Systems

Ethanol	Dipotassium hydrogen phosphate, sodium thiosulfate, magnesium sulfate, ammonium sulfate, sodium dihydrogen phosphate, cesium carbonate, sodium chloride, tripotassium phosphate, calcium chloride, sodium carbonate
2-Propanol (propyl-alcohol)	Calcium chloride, magnesium chloride, ammonium sulfate, magnesium sulfate, lithium sulfate, sodium sulfate, ammonium sulfate, tripotassium phosphate, dipotassium hydrogen phosphate
1-Propanol	Magnesium sulfate, ammonium sulfate, tripotassium phosphate, dipotassium hydrogen phosphate, sodium carbonate, sodium chloride, tripotassium citrate, triammonium citrate
Methanol	Dipotassium hydrogen phosphate
Isopropanal	Ammonium sulfate, tripotassium phosphate, dipotassium hydrogen phosphate, sodium carbonate, sodium chloride
Acetone	Ammonium sulfate, tripotassium phosphate, dipotassium hydrogen phosphate, sodium carbonate, sodium chloride, calcium carbonate
Methanol	Dipotassium hydrogen phosphate, tripotassium phosphate

(*continued*)

TABLE 15.1 (Continued)
Components for the Formation of ATPSs

Ionic Liquid-Based ATPSs

1-Ethyl-3-methylimidazolium chloride, [C$_2$mim]Cl;	Tripotassium phosphate
1-butyl-3-methylimidazolium chloride, [C$_4$mim]Cl;	
1-hexyl-3-methylimidazolium chloride, [C$_6$mim]Cl;	
1-heptyl-3-methylimidazolium chloride, [C$_7$mim]Cl;	
1-decyl-3-methylimidazolium chloride, [C$_{10}$mim]Cl;	
1-allyl-3-methylimidazolium chloride, [amim]Cl;	
1-benzyl-3-methylimidazolium chloride, [C$_7$H$_7$mim]Cl;	
1-hydro xyethyl-3-methylimidazolium chloride, OHC$_2$mim]Cl;	
1-butyl-3-methylimidazolium bromide, [C$_4$mim]Br;	
1-butyl-3-methylimidazolium methanesulfonate, [C$_4$mim] [CH$_3$SO$_3$]	
1-butyl-3-methylimidazolium acetate, [C$_4$mim] [CH$_3$CO$_2$];	
1-butyl-3-methylimidazolium methylsulfate [C$_4$mim] [CH$_3$SO$_4$];	
1-butyl-3-methyl-imidazoliumtriflu oromethanesulfonate, [C$_4$mim][CF$_3$SO$_3$]	
1-butyl-3-methylimidazolium dicyanamide, [C$_4$mim] [N(CN)$_2$]	
1-Butyl-3-methylimidazolium chloride ([C$_4$mim]Cl)	Dipotassium hydrogen phosphate

separately miscible in water in all proportions and also with each other at low concentrations. As the concentration of these phase components in a common solvent (water) increases above a certain critical value, phase separation occurs. Each ATPS is characterized by an exclusive phase diagram that indicates the equilibrium composition for that particular system and constitutes the most fundamental data for any type of extraction. Bamberger et al. (1984) discussed in detail the methods for the construction of these phase diagrams. Albertsson (1971), Diamond and Hsu (1989), and Zaslavasky (1982) have compiled phase diagrams for a number of systems. Among these, ATPSs formed by PEG–dextran–water and PEG–salt–water systems are widely used for separation and purification of biomolecules. However, PEG–salt–water two-phase systems have certain advantages over PEG–dextran–water systems such as lower viscosity and cost.

15.2.2 Factors Affecting the ATPSs

15.2.2.1 Nature and Concentration of the Polymer

The hydrophobicity, molecular weight, and concentration of the phase-forming polymers in the case of polymer/polymer phase systems and the type of salt and its concentration in the case of polymer/salt phase systems influence the formation of

ATPS. The lower the molecular weight, the higher is the concentration of the polymer required for phase formation, and vice versa. Polymer molecular weight will influence the polymer concentrations required for phase separation and the phase diagram symmetry. As the difference in molecular size between the two polymers increases, the phase diagram becomes more asymmetric (Diamond and Hsu 1992).

15.2.2.2 Temperature

Temperature has considerable influence on the phase diagram of both types of phase systems. Decreasing the temperature of the polymer/polymer-type phase system will lead to lower polymer concentration required for phase separation; however, this reduction is marginal. The polymer/salt systems behaving in an opposite manner require a higher polymer concentration for separation at lower temperatures. Also, a decrease in temperature causes an increase in phase viscosity, which is not desirable for industrial scale extraction (Zaslavsky 1995).

15.2.2.3 Type and Concentration of Neutral Salts

Phase systems are significantly affected by the addition of neutral salts (univalent or multivalent) and their concentration. Zaslavsky et al. (1985) demonstrated that increasing the concentration of the univalent salts (up to 0.1 M) in a PEG/dextran system will alter the composition of the phases without significant effect upon the binodal position. However, multivalent salts, such as phosphate, sulfate, and tartrate, in the same system show an increasing tendency to partition to the bottom (dextran-rich) phase with an increase in salt concentration and distance from the critical point (Zaslavsky et al. 1995). These salts will significantly alter both phase composition and the binodal position of the PEG/dextran system. Generally, the binodal position is shifted to lower polymer concentration.

15.2.3 PHYSICOCHEMICAL PROPERTIES

The phase separation rate and also biomolecule mass transfer are largely determined by the physical properties of ATPS such as density, viscosity, and interfacial tension. Methods of measuring some of these properties were given by Albertsson (1986), Walter et al. (1985), and Zaslavsky (1995). Researchers measured these properties for the systems they used but most of them have not reported them. However, in the case of polymer/polymer-type ATPSs such compilations of data still appears to be scarce. Recently, Zuniga et al. (2006) reported physical properties of PEG/maltodextrin. There is a need for the systematic measurement and reporting of the physical properties of the other ATPSs.

15.3 TWO-PHASE EXTRACTION

15.3.1 ORGANIC/ATPE

The aim of extraction is to concentrate the product in a relatively small volume of solvent. The solvents to be used should be capable of dissolving the product of interest. Mostly polar solvents, such as acetone, methanol, and ethanol, are used as organic phases. These solvents should be checked over a wide range of pH for solubility behavior for determining the distribution coefficient. The distribution coefficient (m') is defined as a ratio of the equilibrium concentration of solute in extract to the concentration of solute in raffinate. A high value of the distribution coefficient would signify better separation in a single step, whereas a low value of m' will necessitate the use of multistage extraction procedures. The organic–aqueous extraction is well developed in the chemical/pharmaceutical industry and well documented in the literature. It is considered to have special advantages compared with other separation techniques for handling labile substances when distillation is impossible due to material properties or economic reasons. In recovery processes for natural products such as antibiotics (penicillin or erythromycin) from filtered or whole fermentation broth, organic–aqueous extraction was employed (Brunner et al. 1981). However, only very few proteins are soluble in commonly used organic solvents such as butanol (Craig and Craig 1956). Other solvents, such as phenol, lead to extensive denaturation of proteins, a fact commonly exploited for the purification of nucleic acids, which has been performed up to pilot plant scale. Hence, other ways and means to establish two immiscible liquid phases that provide a biocompatible environment for the extraction of biomolecules have to be considered. One such method that has gained significant importance is extraction using two aqueous phases.

15.3.2 Aqueous/ATPE

Unlike conventional liquid–liquid extraction involving organic–aqueous, ATPE employs two aqueous phases. ATPE has been successful to a large extent in overcoming the drawbacks of conventional extraction processes such as low solubility and denaturation of biomolecules in organic solvents. The important step in ATPE is the selection of suitable ATPS that gives the desired partitioning of the biomolecules (cells, bacteria, protein/enzymes, etc.) under consideration. After identification, the appropriate conditions must be fixed depending on the objective of the partition step. If ATPE is used as the primary purification step for the removal of cell debris from the fermentation broth containing the desired product, the aim is to partition the debris and the product into the opposite phases. Then, in subsequent partition steps, the desired/required degree of purity of the product is achieved. In all of these extraction steps, while fixing the system conditions, attention should be given to factors such as partition coefficient of the target protein/enzyme, contaminating materials, and volume ratio. It should be noted that the cell debris, itself a biopolymer, contributes to the formation of phases and decreases the phase volume ratio as its concentration increases. For example, if the desired protein/enzyme is to be partitioned to the top phase, system conditions must be adjusted in such a way that its partition coefficient is relatively much higher than that of cell debris as well as contaminating materials. To achieve this, the knowledge of the factors that affect the partitioning should be exploited.

Column contactors in which the phase separation occurs by gravity, which are commonly used in the chemical industry for organic–aqueous phase extraction, can be conveniently adapted in the case of ATPS and thereby eliminate the use of expensive centrifuges. Examples for the wide range of application of ATPE are listed in the Table 15.2, indicating its versatility.

TABLE 15.2
Application of ATPE for Purification of Biomolecules

Sl. No	Source/Biomolecule	System Employed	Significant Findings	Reference
1.	Plant esterase from wheat flour	27.0% PEG-1000/13.0%NaH$_2$PO$_4$, and 27.0% PEG-1000/13.0%NaH$_2$PO$_4$/6.0% (NH$_4$)$_2$SO$_4$	Purification factor (PF) of 18.46-fold and a yield of 83.16%	Yanga et al. 2010
2.	Lysozyme from hen egg white over four successive extractions	EO50PO50 [50% (w/w) ethylene oxide and 50% (w/w) propylene oxide] and K$_2$HPO$_4$	The specific activity of lysozyme increased from 38,438 to 42,907 U/mg	Dembczynski et al. 2010a
3.	Lysozyme from hen egg white using thermoseparating random copolymers	40% (w/w) EO50PO50, 10% (w/w) potassium phosphate, and 0.85 M sodium chloride at pH 9.0	85% yield and specific activity of 32,300 U/mg with a PF of 16.9	Dembczynski et al. 2010b
4.	2,3-Butanediol from fermentation broth	32% (w/w) ethanol and 16% (w/w) ammonium sulfate.	91.58% recovery of 2,3-BD with 7.10% partition coefficient	Li et al. 2010
5.	Acidic recombinant protein, glucuronidase (rGUS) from transgenic tobacco	13.4% (w/w) PEG 3400/18% (w/w) potassium phosphate system	74% activity recovery with PF of 20 and 90% of the native tobacco proteins were removed in the interphase	Srinivas et al. 2010
6.	Recombinant PheDH from *Bacillus badius*	ATPS (9% (w/w) PEG, 16% (w/w) K$_2$HPO$_4$, 16% (w/w) KCl) and chromatographic steps for large-scale downstream processing	Yield of 96.7% and specific activity 4231.4 U/mg	Ominidia et al. 2010
7.	Protease from the latex of *Calotropis procera*	System comprised of 12% PEG-4000/17% MgSO$_4$ with 6% (w/w) NaCl	Recovery of 74.6%	Chaiwut et al. 2010
8.	Vanillin	Ionic liquid based ATPS, 15% (w/w) of K$_3$PO$_4$, 60% (w/w) of an aqueous solution of vanillin and 25 wt% of ILs composed by halogenated anions, such as Cl⁻ or Br⁻	Vanillin preferentially partitioned to IL-rich phase presenting $K_{Van} > 1$ and vanillin concentration were more dependent on the IL cation nature	Filipa et al. 2010
9.	Lipoxygenase from soybeans	Precipitation, using poly ethylene glycol, followed by ATPE (PEG-6000 and ammonium sulfate)	125% activity recovery with the overall purification factor of 4.38 compared to crude enzyme extract	Lakshmi et al. 2009

(continued)

TABLE 15.2
Application of ATPE for Purification of Biomolecules

Sl. No	Source/Biomolecule	System Employed	Significant Findings	Reference
10.	Lipase from *Burkolderia pseudomallei* fermentation broth	PEG-6000/potassium phosphate	The purification factor of lipase was enhanced to 12.42-fold, with a high yield of 93%	Ooi et al. 2009
11.	Aspergillopepsin I, an acid protease, employing RSM	17.3% of PEG-4000 (w/w), 15% NaH₂PO₄ (w/w), and 8.75% NaCl (w/w)	Yield over 99% with purification factor of 5	Draginja et al. 2009
12.	Carmine, a natural dye	Polymer or copolymer with aqueous salt solutions (Na₂SO₄ and Li₂SO₄)	ATPSs have proven to be a powerful, attractive alternative to carmine purification	Magestea et al. 2009
13.	Aloe polysaccharide	18% PEG-2000, 25% ammonium sulfate, 0.3 M NaCl, and copolymer membrane of poly(acrylonitrile-acrylamide-styrene)	ATPE concentrated aloe polysaccharide in the salt-rich bottom phase. After ultrafiltration, retentate polysaccharide was found to consist of 95.28% of a high-mannose polymer	Xing et al. 2009a
14.	BSA and aloe polysaccharides, aloe polysaccharides concentrated by ultrafiltration	PEG-6000/(NH₄)₂SO₄ ATPS	The results show that mannose is mainly monosaccharide, and it only contains a few glucose	Xing et al. 2009b
15.	Pepsin, an acidic aspartic protease, from bovine abomasum	Combination of partition in ATPS (PEG-1450/phosphate) and chitosan precipitation	The combination resulted in a pepsin recovery of 48.5% with a purification factor of 9.0	Boeris et al. 2009
16.	Five model proteins (bovine serum albumin, Cytochrome C, lysozyme, myoglobin, and trypsin)	PEG/phosphate system	Five model proteins showed the different partition tendency in two phases	Qu et al. 2009

17.	Glycomacropeptide from whey proteins	ATPS composed of 15.0% (w/w) PEG-1500 + 18.9% (w/w) sodium citrate, at pH 8.0 and without NaCl	Glycomacropeptide partition coefficient was 33.4, and its recovery was 95% in the top phase	Da-Silva et al. 2009
18.	Alcohol dehydrogenase (ADH) from baker's yeast (*Saccharomyces cerevisiae*)	Precipitation followed by ATPE composed of PEG/potassium phosphate	90% enzyme activity recovery with 6.6-fold purification	Madhusudhan et al. 2008
19.	Proteases from *Clostridium perfringens* fermentation broth	8.0% sodium citrate, 22% PEG-10000, pH 8.5	Activity yield of 131% and purification factor of 4.2 was obtained	Porto et al. 2008
20.	Thermostable xylanase from the thermophilic fungus *Paecilomyces thermophila J18* in solid-state fermentation (SSF)	ATPS of 12.5% PEG-4000, 25% (NH4)$_2$SO$_4$, and 50% enzyme solution at pH 7.2	Purification factor (PF) of 5.54 and a 98.7% yield of enzyme activity in the top phase	Yang et al. 2008
21.	Pepsin from bovine stomach	PEG-600/potassium phosphate	Pepsin partitioned toward the top phase with a high partition coefficient	Imelio et al. 2008
22.	Mixture of enzymes (bromelain and polyphenol oxidase) from the pineapple (*Ananas comosus* L. Merr.)	18% PEG-1500 and 14% potassium phosphate	228% activity recovery and 4.0-fold purity of bromelain and 90% activity recovery and 2.7-fold purity of polyphenol oxidase	Babu et al. 2008
23.	C- and A-phycocyanin from *Spirulina platensis*	PEG-4000/potassium phosphate system	C-phycocyanin and allophycocyanin with a purity of 3.23 and 0.74, respectively. Multiple extractions (two) improved the purity of C-phycocyanin from 3.23 to 4.02.	Patil et al. 2008

(continued)

TABLE 15.2 (Continued)
Application of ATPE for Purification of Biomolecules

Sl. No	Source/Biomolecule	System Employed	Significant Findings	Reference
24.	Penicillin G from *Penicillium chrysogenum*	Solvent alone (physical extraction) and carrier–solvent combinations (reactive extraction)	An extraction of approximately 98% was achieved in an hour using Amberlite LA-2 with any of the solvents (Shellsol TK/ tributyl phosphate/butyl acetate) and at a recirculation flow rate of 4.17 L/s in a hollow-fiber membrane contactor	Hossain and Dean 2008
25.	Alpha galactosidase from *Aspergillus oryzae*	ATPS of PEG and potassium phosphate	Recovery of 89.6% and a 3.6-fold purity	Gautam and Simon 2008
26.	Hemoglobin, lysozyme and glucose-6-phospate dehydrogenase (G6PDH)	ATPS of PEG and sodium polyacrylate (NaPA)	Lysozyme partitioned to the PEG-phase, while G6PDH is partitioned to the NaPA phase	Rawdkuen et al. 2008
27.	Human antibodies using pure protein systems and mixture of proteins containing human immunoglobulin G (IgG), serum albumin, and myoglobin	ATPS of PEG and phosphate	A recovery yield of $101 \pm 7\%$, a purity of $99 \pm 0\%$, and a yield of native IgG of $97 \pm 4\%$ were obtained. The total extraction yield was 76% and the purity 100%	Rosa et al. 2007a
28.	Human immunoglobulin (IgG)	Polymer–polymer and polymer–salt ATPS along with several functionalised PEGs was studied	The best purification of IgG from the CHO cells supernatant was achieved in a PEG/ dextran ATPS in the presence of PEG-150–COOH with a recovery yield of 93%, a purification factor of 1.9, and a selectivity to IgG of 11	Rosa et al. 2007b

29.	Proteins such as lysozyme, ribonuclease A, and cytochrome c from corn	PEG-1450/Na$_2$SO$_4$ with 8.5% NaCl	Selectivity for lysozyme in the top phase relative to corn fraction proteins was 134 and 616 for endosperm and germ, respectively, in the top phase	Gu and Glatz 2007
30.	BSA, a model protein	PEG-8000/phosphate and PEG-600/phosphate systems	The results obtained showed that the VR and K for BSA were similar at different scales (10 g, 2 g, and 300 µl) for the same ATPS, which shows the reliability of the miniaturization of ATPS	Negrete et al. 2007
31.	Recombinant *Bacillus sphaericus* phenylalanine dehydrogenase (PheDH)	PEG and ammonium sulfate ATPSs	PEG-6000/(NH$_4$)$_2$SO$_4$ (pH 8.0) could be employed as an efficient and attractive tool for the large-scale recovery	Mohamadi et al. 2007
32.	Recombinant chymosin from inclusion bodies	Polyethylene-polypropylene oxide and PEG–phosphate	32% recovery in the polyethylene–polypropylene oxide, while in the PEG–phosphate the recovery was 50–59% higher than the standard method (43%)	Reh et al. 2007
33.	Caseinomacropetide (CMP)	PEG-1500/sodium citrate system	93.95% of CMP recovered	da Silva et al. 2007
34.	Model proteins bovine serum albumin, lysozyme, α-lactalbumin, β-lactoglobulin, and γ-globulin	PEO–PPO–PEO block copolymers F38 or F68 and ammonium carbamate	Partition coefficients were found to vary in the range of 0.1 to 0.8 for BSA, 0.5 to 2.0 for lysozyme, 1.0 to 2.5 for α-lactalbumin, 0.1 to 1.0 for α-lactoglobulin, and 0.3 to 1.0 for γ-globulin, depending on the polymer chain size and temperature	Oliveira et al. 2007
35.	Phenylalanine ammonia–lyase (PAL) from *Rhodotorula glutinis*	11.0% PEG-1000/14.0% Na$_2$SO$_4$, and 11.0% PEG-1000/14.0%Na$_2$SO$_4$/5.3% Na$_2$CO$_3$	PAL with purification factor of 9.3-fold and recovery yield of 80.6% was obtained	Yue et al. 2007

(continued)

TABLE 15.2 (Continued)
Application of ATPE for Purification of Biomolecules

Sl. No	Source/Biomolecule	System Employed	Significant Findings	Reference
36.	Recombinant *Bacillus badius* phenylalanine dehydrogenase (PheDH)	PEG-6000 and ammonium sulfate	The partition coefficient, recovery, yield, purification factor, and specific activity values were of 92.57, 141%, 95.85%, 474.3, and 10,424.97 U/mg, respectively	Mohamadia and Omidinia 2007
37.	C-phycocyanin from *Spirulina platensis*	PEG (4000)/potassium phosphate system	Purity of 3.52 from initial purity of 1.18 from single step and purity of 4.05 achieved in third extraction	Patil and Raghavarao 2007
38.	Betalains from beet root	PEG-6000/ammonium sulfate	Better differential partitioning of betalains and sugars at higher tie line length (34%), wherein betalains (about 70–75%) and sugars (about >90%) where in opposite phases and concentrated by 3.4-fold	Chethana et al. 2007
39.	*Cellulomonas fimi* β-mannanase containing a mannan-binding module	Affinity separations using galactomannan/hydroxypropyl starch ATPS	The new system includes only renewable polysaccharides and is therefore environment-friendly	Antov et al. 2006a
40.	Xylanase from *Polyporus squamosus* produced by SSF	ATPS of PEG/ammonium sulfate	Activity recovery of 97.37%, and purification factor 4.8 was obtained	Antov et al. 2006b
41.	Penicillin acylase (PA) from recombinant strain of *E. coli*	A direct comparison of a chromatography and ATPS (PEG-1450/potassium phosphate)	The proposed ATPS process recovered 97% of PA at the top phase (PEG rich phase) with a purity of 3.5	Aguilar et al. 2006
42.	Lysozyme from chicken egg white	PEG/salt ATPS	70% of lysozyme extracted from the diluted chicken egg white. The specific activity of was 39,500 unit/mg	Su and Chiang 2006

No.	Enzyme/Source	System	Results	Reference
43.	β-Glucosidase and total protein	(PEG) 4000 (8%, w/w) and potassium phosphate salt (13%, w/w)	β-Glucosidase yield greater than 92% was obtained in the bottom phase for pH values less than 7.0 and temperatures below 35°C	Gautam and Simon 2006
44.	Papain from wet *Carica papaya* latex	8% (w/w) PEG and 15% (w/w) ammonium sulfate	Highly pure papain was obtained in a much shorter processing time directly from unclarified latex	Nitsawang et al. 2006
45.	Polyphenol oxidase (PPO) from the potato tuber (*Solanum tuberosum*)	5% PEG-8000 and 28.5% phosphate	Purification factor of 15.7 and a yield of 97.0%	Vaidya et al. 2006
46.	C-Phycocyanin from *Spirulina platensis*	ATPE (PEG 12.28%, potassium phosphate 11.63%) and ion-exchange chromatography	The purity of 5.22 C-phycocyanin obtained after ATPS and ion-exchange chromatography resulted in 6.69-fold purity	Patil et al. 2006
47.	α-Amylase from *Bacillus subtilis*	PEG–citrate ATPS with RSM	Two-fold purification and over 90% yield of α-amylase is achieved at optimized condition in a single purification step	Zhi et al. 2005
48.	Alkaline protease from cell free fermentation broth of *Bacillus subtilis* TISTR25	PEG-1000/potassium phosphate	Purification fold of 6.1 and yield of 62.2%	Chouyyok et al. 2005
49.	Spleen proteinase from yellow fin tuna (*Thunnus albacores*)	PEG-1000 (15%, w/w) and magnesium sulfate (20%, w/w)	The specific activity of 47.0 units/mg protein and purification fold of 6.61 and a yield of 69.0% was obtained	Klomklao et al. 2005
50.	1,3-1,4-Glucanase, α-amylase, and neutral proteases from clarified and whole fermentation broths of *Bacillus subtilis* ZJF-1A5	A two-step process comprising of (PEG/MgSO$_4$ system)	13-Glucanase recovery of 65.3% and specific activity of 14027 U/mg, 6.6 times purity	Guo-qing et al. 2005

(continued)

TABLE 15.2 (Continued)
Application of ATPE for Purification of Biomolecules

Sl. No	Source/Biomolecule	System Employed	Significant Findings	Reference
51.	Proteins and fat from cheese whey	PEG-6000/potassium phosphate system of 23.9% w/w	Recovery of the proteins in top phase	Anandharamakrishnan et al. 2005
52.	B-Phycoerythrin from *Porphyridium cruentum*	PEG-1450 24.9% (w/w), phosphate 12.6% (w/w), and system pH of 8.0	Protein purity of 2.9 ± 0.2 and a yield of 77.0% (w/w)	Benavides and Palomares 2004
53.	Bromelain from *Ananas comosus*	poly(ethylene oxide) (PEO)– poly(propylene oxide) (PPO)– poly(ethylene oxide) (PEO) block copolymers	Enzyme activity recovery around 79.5%, PF ~1.25, and activity partition coefficient around 1.4	Rabelo et al. 2004
54.	Fungal cellulase endoglucanase I (EGI) and amphiphilic protein hydrophobin I (HFBI) from *Trichoderma reesei*	Nonionic detergent Agrimul NRE 1205 based ATPSs	Using the technical nonionic detergent Agrimul NRE 1205 the separation was successfully scaled up to 1200 L	Colléna et al. 2004
55.	Cysteine, phenylalanine, methionine, and lysine	PEG/phosphate	Partitioning and purification of amino acids	Shang et al. 2004
56.	Endopectinase and exopectinase, from *Polyporus squamosus*	System composed of PEG-4000 and crude dextran	Maximal value of partition coefficient of endopectinase was 2.45, accompanied by a maximal top phase yield of 80.22%	Antov 2004
57.	Chitinase from *Neurospora crassa*, cabbage, and puffballs	Affinity precipitation and ATPE comprising of chitosan in PEG–salt ATPS	86, 80, and 88% activity recoveries and purification fold of 34, 20, and 38 was obtained respectively for *Neurospora crassa*, cabbage, and puffballs	Teotia et al. 2004
58.	Recombinant, thermostable α-amylase (MJA1) from the hyperthermophile	Ethylene oxide–propylene oxide random copolymer (PEO–PPO)/(NH$_4$)$_2$SO$_4$, and PEG/ (NH$_4$)$_2$SO$_4$ ATPSs	Enzyme recovery of up to 90% with a purification factor of 3.31 was achieved using a single extraction	Li et al. 2004

59.	Phospholipase D (PLD) from peanuts and Carrots	Affinity partitioning (AP) using alginate as a macroaffinity ligand (PEG–salt)	Peanuts and carrots PLD could be purified 78- and 17-fold with 82 and 85% activity recovery, respectively	Teotia and Gupta 2004
60.	Xylose reductase (XR) from *Candida mogi*	PEG and phosphate salt	Yield 103.5% and PF of 1.89	Zea et al. 2004
61.	Lutein from the green microalga *Chlorella protothecoides*	PEG-8000 22.9% (w/w) and phosphate 10.3% (w/w)	Overall product yield of 81.0 ± 2.8%	Cisneros et al. 2004
62.	Endopolygalacturonase from *Kluyveromyces marxianus* strains	Thermoseparating polymer Ucon 50-HB-5100 (a random copolymer of 50% ethylene oxide and 50% propylene oxide) as one of the phase-forming compounds	10-fold enzyme concentration and a purification factor close to the expected maximum while maintaining more than 95% of the initial enzyme activity	Pereira et al. 2003
63.	Peroxidase from leaves of *Ipomoea palmetta*	(PEG-1550)/KH$_2$PO$_4$ system having 2% NaCl	By coupling ultrafiltration with ATPE, the enzyme IPP was purified to 6-fold, with 9.7-fold activity enrichment and 76% recovery	Srinivas et al. 2002
64.	Glucose-6-phosphate dehydrogenase and hexokinase	PEG/hydroxypropyl starch and PEG/phosphate ATPSs with free triazine dyes, Cibacron Blue F3GA, and Procion Red HE3B	In the PEG–PES system, Cibacron Blue F3GA changed the partition coefficient of G6PDH from 0.73 to 1.59 with phosphate buffer solution	Xua et al. 2002
65.	Endoglucanase Icore-hydrophobin I (EGIcore-HFBI) with HFBI as a fusion tag from *Trichoderma reesei* fermentation broth	Detergent/polymer ATPS	Total recovery of EGIcore-HFBI after the two separation steps was 90% with a volume reduction of six times; detergent removed in the second step by addition of thermoseparating polymer	Colléna et al. 2002

(continued)

TABLE 15.2 (Continued)
Application of ATPE for Purification of Biomolecules

Sl. No	Source/Biomolecule	System Employed	Significant Findings	Reference
66.	Geniposide from gardenia fruit	PE62 (ethylene oxide propylene oxide, 20:80) copolymer and phosphate system	39.0-g geniposide extracted as a final product (in powder form) with 77% purity from 500-g fruit	Pana et al. 2002
67.	Lactate dehydrogenase (LDH) from porcine muscle, *B. stearothermophilus* LDH with a fusion tag of six histidine residues (His6–LDH) from recombinant *Escherichia coli* homogenate	Affinity extraction of dye- and metal ion-binding proteins (polyvinylpyrrolidone [PVP40]– Reppal PES-100 two-phase system)	The partitioning of porcine LDH to the PVP phase was increased 100-fold, and a maximal recovery of 89% was obtained in the system loaded with 0.2% (w/w) Cibacron Blue. A more than 10-fold increase in the partition coefficient of His6–LDH was achieved in the two-phase system loaded with 0.4%	Fernandes et al. 2002
68.	Recombinant, thermostable α-amylase (MJA1) from the hyperthermophile, *Methanococcus jannaschii*	Thermo separating polymer PEO–PPO/salt two-phase systems	Resulted in a higher maltose yield from 19.0 to 15.5 mg/ml compared with control run. A 22% gain in maltose yield was obtained as a result of the increased productivity.	Li et al. 2002
69.	Penicillin acylase from *Escherichia coli*	PEG-3350/sodium citrate ATPS	A PF more than 5.5 and a yield about 80% in first step purification from crude extract	Marcos et al. 2002
70.	Trypsin	PEG/cashew-nut tree gum ATPSs	Maximum recovery of trypsin activity in the cashew-nut tree gum phase was obtained with PEG (molecular weight 8000) at pH 7.0 and 1.0 M NaCl	Oliveira et al. 2002
71.	Proteolytic enzymes from a commercial pectinase preparation (Pectinex-3XL)	PEG/potassium phosphate system	PF of 5.49 for exo-galacturonase, 16.28 for endopolygalacturonase, 16.64 for pectinase, and 14.27 for pectinase lyase	Lima et al. 2002

72.	Intracellular glyceraldehyde 3-phosphate dehydrogenase (G3PDH) and other proteins from baker's yeast	Cell disruption and ATPE (PEG 12% (w/w), phosphate 28% (w/w)) were operated simultaneously	Simultaneous disruption and ATPE can achieve the primary recovery of intracellular proteins from yeast	Palomares and Lyddiatt 2002
73.	Glycyrrhizin from an extract of *Glycyrrhiza uralensis* Fisch	ATPS (60% (v/v) ethanol and 15% (w/v) phosphate)	92% recovery of glycyrrhizin with 2.6-fold purification	Tianwei et al. 2002
74.	Endoglucanase from culture filtrate of *Trichoderma reesei*	PEG-4000/sodium/potassium phosphate ATPS	Partition coefficient of 54, a yield of 98%, and an almost complete purification by a single step with a fusion of (WP)4 to the EGI catalytic module after a pentaproline linker [EGI core PI-(WP)4]	Collena et al. 2001
75.	Endopolygalacturonase (endo-PG) from *Kluyveromyces marxianus*	PEG-8000/polyvinyl alcohol (PVA10 000) and PEG–hydroxypropyl starch (Reppal PES-100)	Enzyme recovery of 91% purification factor of 1.9 and a concentration factor of 5	Wu et al. 2001
76.	Amyloglucosidase from solid-state fermentation of *Aspergillus niger*	ATPE (PEG-6000/KH$_2$PO$_4$) followed by ultrafiltration	Resulted in 3.1-fold purification, 12-fold concentration, and 78% recovery of the enzyme	Tanuja et al. 2000
77.	*Fusarium solani pisi* recombinant cutinase	Thermoseparating random copolymer of ethylene oxide/propylene oxide 50/50 (%w/w), Breox, and hydroxypropyl starch–Reppal PES-100	Partitioning coefficient of 9 and a recovery yield of 60%	Cunha et al. 2000
78.	Endopolygalacturonase from *Kluyveromyces marxianus* fermentation broth	PEG-8000-(NH$_4$)$_2$SO$_4$ ATPE	A high enzyme recovery up to 95% and a concentration factor of 5 to 8 with a purification factor of about 1.25 in single ATPE step	Wu et al. 2000

(continued)

TABLE 15.2 (Continued)
Application of ATPE for Purification of Biomolecules

Sl. No	Source/Biomolecule	System Employed	Significant Findings	Reference
79.	Plant peroxidase from leaves of *Ipomoea palmetta*	Partial purification with ATPE (PEG/ ammonium sulfate/NaCl (24:7.5:2.0%, w/v) system) followed by gel filtration on a Sephadex G-100 column	ATPS resulted with PF 2.18 and recovery of 91.5%. The recovery of 75.3% and PF of 49 obtained by gel filtration	Srinivas et al. 1999
80.	Xanthine oxidase from milk	7% dextran and 5% PEG in the range of 5–25 mA sodium phosphate	The extraction of milk xanthine oxidase was substantially free from other contaminant proteins	Ortin et al. 1991
81.	Superoxide dismutase from comminuted bovine liver	20% biomass, 15% PEG, and 8% phosphate	The enzyme was purified fourfold with recovery of 83%	Boland et al. 1990

15.3.3   Factors Affecting the Biomolecule Partitioning in ATPSs

15.3.3.1   Phase-Forming Polymer Molecular Weight and Concentration

The molecular weight has significant effect on partitioning of biomolecules because it affects the phase composition. Generally, increasing the molecular weight of a phase-forming polymer will cause biomolecule to partition more toward the opposite phase. Similarly, when a phase-forming polymer's molecular weight is decreased, a biomolecule will tend to partition into that polymer rich top phase. The extent of this effect depends also on the molecular weight of the biomolecules. Albertsson et al. (1986) reported that the effect of polymer molecular weight was more prominent for the biomolecules of higher molecular weight (up to 250 kDa). Near the plait point of the binodal curve, the partition coefficient of biomolecule is one. As polymer concentration is increased (moving away from the plait point) protein partition coefficient becomes exceedingly greater or lower than one.

15.3.3.2   System Temperature

The effect of system temperature on protein partitioning has not yet been thoroughly investigated. The change in temperature causes a sharp change of the binodal curve, which in turn affects the partition behavior of the biomolecule during ATPE (Albertsson 1986). Hence, the change in system temperature has an indirect effect on partitioning of biomolecules.

15.3.3.3   Biomolecule Size

The size of the biomolecule has a significant role on its partitioning behavior during ATPE. Generally, the small molecules tend to partition themselves evenly between two phases, where as large molecules tend to distribute in an uneven manner, while very large biomolecules partition themselves to one of the phases (Albertsson 1986).

15.3.3.4   Solution pH

The change in system pH has an indirect effect on partitioning of biomolecules. The partitioning of proteins/enzymes in ATPS is affected by net charge on the biomaterial, which in turn depends on the pH of the solution. The net charge on the biomolecules can be varied by changing the pH of the solution. This is due to increased surface area of the biomolecules that causes more hydrophobic interactions.

15.3.3.5   Protein Concentration

In general, partitioning is not dependent on the concentration of the protein/enzymes. However, at very high concentration of the protein there could be a possibility of the formation of a third (protein) phase by itself.

15.3.3.6   Chemical Modification of Phase Polymers

The chemical modifications to PEG, such as covalent bonding of fatty acid chains, charged groups, hydrophobic derivatives, and biospecific affinity ligands, have considerable effect on partition behavior of proteins (Raghavarao et al. 1995). Charged PEG derivatives such as trimethylamino-PEG (TMA-PEG) and sulfonate-PEG

(S-PEG) gives the information about the net charges and isoelectric point of proteins as well as particles. Johansson and Shanbhag (1984) observed that an increase in concentration of the derivatized PEG resulted in a 10 times increase in partition coefficient compared to that of normal PEG. Many researchers employed hydrophobic polymer derivatives during the partitioning study of the biomolecules in ATPS. In recent years, metal ion and dye binding affinity partitioning (AP) was widely used to enhance the partitioning of various biomolecules such as human hemoglobin, bovine hemoglobin, whale and horse myoglobins, lactate dehydrogenase, glucose-6-phosphate dehydrogenase, and hexokinase (Wuenschell et al. 1990; Chung and Arnold 1990; Sheryl Fernandes et al. 2002; Xu et al. 2002).

15.4 EQUIPMENT FOR EXTRACTION OF BIOMOLECULES

Transport phenomena (mass, momentum, and heat transfer), along with their interaction with one another, is having considerable importance in ATPE. Further, hydrodynamics play a major role in the efficient design of extraction equipment. The design aspects of these extraction units have been discussed in detail elsewhere (Joshi et al. 1990; Raghavarao et al. 1995). Some hydrodynamics of mechanical agitated contactors and column contactors as well as separators are presented in the following sections.

15.4.1 SPRAY COLUMN

It is easy to carry out ATPE of biomolecules in spray columns. In spray columns one phase is made a continuous phase, while the other phase is dispersed in the form of droplets with the help of sieve plate or nozzle. During operation, the heavy continuous phase and light dispersed phase will flow in countercurrent direction. The major drawbacks of spray column are flooding and back mixing. Use of spray columns for the extraction of various biomolecules such as bovine serum albumin, horse radish peroxidase have been demonstrated well in the literature (Joshi et al. 1990; Raghavarao et al. 1991; Jafarabad 1992; Srinivas et al. 2002).

15.4.2 PACKED COLUMN

The packed column consists of a stack of packings (such as glass spheres, raschig rings, polystyrene rings) arranged regularly or irregularly in a column over a perforated support. The selection of the packing material is markedly important. The packing should be wetted preferentially by the continuous phase and its diameter should be less than one-eighth of the column diameter in random packing. Packed columns are more efficient than spray columns for the extraction of biomolecules by ATPE. The packing material used in the columns reduces back mixing and provides tortuous pathways for the two liquids and can also cause distortions and breakup of the drops. However, one of the drawbacks of packed columns is lower throughput compared to that of spray columns. Patil et al. (1991) and Igarashi et al. (2004) employed packed columns for the extraction of biomolecules using ATPSs.

15.4.3 PERFORATED ROTATING DISC CONTACTOR

Perforated rotating disc contactor (PRDC) is well suited for aqueous systems with low interfacial tension such as ATPS formed by PEG and phosphate salts. The PRDC contains a column equipped with perforated discs mounted on a central rotating shaft. There are a few research articles published on the use of PRDC for the extraction of various biomolecules such as cutinase (Cunha 2003), bovine serum albumin (Porto et al. 2000; Sarubbo et al. 2003), trypsin (Oliveira 2002), and α-toxin (Cavalcanti et al. 2006). The main advantage of PRDC is greater efficiency of contacting and better operational flexibility.

15.4.4 COUNTERCURRENT DISTRIBUTION

Countercurrent distribution (CCD) is adapted from aqueous/organic and organic/organic two-phase extraction methods. The principle of operation is very similar to that of column chromatography. Although CCD has been employed for effective fractionation of proteins, enzymes, cells, and cell organelles on a laboratory scale (Albertsson 1986; Walter et al. 1985), the unavailability of CCD units commercially is limiting the widespread use of the technique (Diamond and Hsu 1992).

15.4.5 GRAESSER RAINING BUCKET CONTACTOR

The graesser contactor is composed of a horizontal cylindrical shell containing an internal rotor assembly consisting of a series of circular discs mounted on a central shaft. Between each pair of discs, a series of buckets are supported by tie rods extending the length of the apparatus. Application of graesser raining bucket differential contactor in ATPS was reported in the literature (Coimbra et al. 1994; Zuniga et al. 2005, 2006).

15.4.6 YORK–SHEIBEL COLUMN

In York–Sheibel column dispersion is achieved in the stirring zone by the impeller and the coalescence is achieved in the wire-mesh region. The main advantage of this column is higher throughput compared to that of packed columns. Extraction of biomolecules using ATPS in York–Sheibel columns was reported by Jafarabad et al. (1992). The application of ATPE for the purification of various biomolecules has been demonstrated well in the literature.

15.5 THEORETICAL ASPECTS OF PARTITIONING

The prediction of the partitioning behavior of a given biomolecule can be possible without the measurement of an inordinately large number of parameters by mathematical models. Mathematical models associated with ATPSs are broadly categorized into three types: (1) lattice models, which are based on a lattice representation of the polymer solutions within each of the coexisting phases; (2) virial models, which use a virial type expansion in the concentration of the components of the system to describe thermodynamic properties; (3) a scaling thermodynamic approach, which utilizes recently developed polymer physics concepts aimed at describing polymer solution behavior. The quantitative modeling of biomolecule partitioning in ATPS is an extremely complex problem. Excellent reviews by Baskir et al. (1989a, b) and Diamond and Hsu (1992) are available, who analyzed the existing models while comparing them with their own models. Brooks et al. (1985) reported in detail the theoretical aspects of partitioning. The model for the phase behavior of ATPSs was explained by modified Flory–Huggins equation accounting for the solvation of polymer molecules by water molecules (Filho and Mohamed 2004). Models that examine the influence of protein surface properties, such as surface charge and hydrophobicity are explored (Madeira et al. 2005). Furthermore, investigation of whether the interaction of water with phase polymers, buffering salts, and proteins plays a key role in protein partitioning was reported (Shang 2006).

The mass transfer in a column contactor is influenced by drop dynamics (drop size and drop velocity) as studied in detail by Barhate et al. (2004). The study of drop formation deals with the fundamental understanding of the behavior of liquid drops under the influence of various external bodies as well as surface forces. These studies provide a basis for designing the extractions in column contactors (Barhate et al. 2004).

15.6 ADVANCES IN ATPSS

15.6.1 AFFINITY PARTITIONING

Affinity-based separations include precipitation, membrane-based purification, and two-/three-phase extractions. AP in ATPS is based on the preferential/biospecific interaction between the molecule and affinity polymer derivative. The interaction results in a biomolecule–polymer derivative complex, which selectively partitions to one of the phases leaving the contaminating substances or proteins in the other phase. Most of the reported investigations regarding AP pertain to polymer/polymer-type ATPSs (Diamond and Tsu 1992). However, only a few reports are available on polymer/salt-type ATPSs (Xu et al. 2002) mainly due to the interference of high salt concentrations with the biospecific interactions (Raghavarao et al. 1995). AP is influenced by many factors such as the ligand concentration and its binding characteristics, concentration and molecular weight of polymers, pH, temperature, salt type and concentration, number of thionyl chlorides, covalently linked to iminodiacetic acid (IDA), and the specific metal ligand Cu^{2+} attached to the PEG ligands per molecule (Johansson and Kopperschlager 1987).

AP as currently practiced requires the ligand to be covalently attached to one of the phase-forming polymer components, thereby causing the ligand–polymer to partition predominantly into one of the phases. Although triazine dyes covalently coupled to PEG can be produced on a large scale, the process is complicated and requires a chromatographic step and several organic solvent extractions (Johansson and Joelsson 1985). To simplify the AP, Giuliano (1991) used the free dyes, uncoupled to the phase-forming polymers, as affinity ligands for the partitioning of lysozyme in the polyvinylpyrrolidone–maltodextrin ATPS. In the PEG–phosphate system, it was reported that free triazine dyes, partitioned predominantly into polymer phase, showed the affinity effect on some dehydrogenases and kinases (Wang et al. 1992; Bhide 1995). Lin et al. (1996) also used free triazine dyes as ligands to affinity extraction of lactate dehydrogenase in PEG–hydroxypropyl starch (PES) ATPS. Cibacron Blue F3GA was used as the ligand for extraction of lactate dehydrogenase (LDH) from porcine muscle, while copper ions were used for extraction of *B. stearothermophilus* LDH with a fusion tag of six histidine residues (His6–LDH) from recombinant *Escherichia coli* homogenate in polyvinylpyrrolidone (PVP40)–Reppal PES-100 two-phase system (Fernandes 2002). Xu et al. (2002) studied the AP of glucose-6-phosphate dehydrogenase (G6PDH) and hexokinase in PEG–hydroxypropyl starch (PES). Further, PEG–phosphate ATPSs was investigated with free triazine dyes, Cibacron Blue F3GA and Procion Red HE3B, as their affinity ligands. Purification of PLD from peanuts and carrots was carried out in PEG–salt system with alginate, a known macroaffinity ligand for PLD (Teotia and Gupta 2004). The partitioning of a crude soybean peroxidase (*Glycine max*) was carried out in an ATPS by metal affinity. PEG-4000 was activated using 4000–IDA achieving a recovery of 64% of the purified enzyme (Silva and Franco 2000).

The chitosan was used as affinity ligand for the purification of chitinases from unclarified extracts/homogenates in poly (ethylene glycol) (PEG)–salt ATPS (Teotia et al. 2004). Alginate and Eudragit S-100 were also used as macro affinity ligands in ATPS for the purification of β-amylases and xylanses (Teotia et al. 2001a, b). *Cellulomonas fimi* β-mannanase was selectively partitioned to galactomannan top phase in galactomannan/hydroxypropyl starch ATPS (Antov et al. 2006). AP reported using derivatives of polymer (PEG) for the purification of enzymes such as glucose isomerase, penicillin acylase (Dolia and Gaikar 2006; Gavasane and Gaikar 2003).

Affinity ligands can also be linked to membranes. Ultrafiltration membranes (retaining particles in the range of 0.01–0.1 mm sizes) are used to separate proteins. There are a variety of formats/separation modules in which affinity membranes can be used. For example affinity cross flow filtration, stacked affinity membranes, membrane composites, etc. To minimize "fouling," various techniques such as rotating or vortex systems, high-frequency back-pulsing and high-performance tangential flow filtration have been used. Affinity membrane-based separations combine the virtue of high throughput with high selectivity of the separation process.

15.6.2 Extractive Bioconversion

"Extractive fermentation" is an emerging technique that involves the use of ATPS-based in situ fermentation processes. Integration of bioconversion and downstream processing steps not only increases the productivity of the bioprocesses but also provides the possibility of running the bioconversion in a continuous mode (Mattiasson 1988). Extractive bioconversion employing ATPSs can improve certain existing bioprocesses to make them economically viable (Banik et al. 2003). Simultaneous production and purification of a bioproduct obtained through the use of enzymes or microorganisms is the interesting feature of this technique. The advantages of such a system include rapid mass transfer due to low-interfacial tension, ease of operation under continuous mode, rapid and selective separation, biocompatibility, separation at room temperature, easy and reliable scale-up of bench scale results to production scale, eco-friendliness, suitability for systems with product inhibition, and high yield of biomolecules (Sinha et al. 2000).

Wennestern et al. (1983) studied the conversion of starch to glucose (90% conversion) using the substrate starch as the affinity ligand to keep the enzymes in the bottom phase, in a system composed of PEG/DX. A number of extractive bioconversions relating to the conversion of complex compounds such as cellulose to simpler compounds such as alcohol, using ATPS were reported (Su and Feng 1999).

Simultaneous extraction and purification of some biomolecules/enzymes such as β-galactosidases, penicillin G, α-amylase and serine proteinase, endoglucanase, cellulose, glyceraldehyde 3-phosphate dehydrogenase from various sources were reported with better purity and yield (Chen and Wang et al. 1991; Liao 1999; Ivanova et al. 2001; Ülger 2001; Palomares 2002).

An extractive fermentation using ATPS was studied for the simultaneous cell cultivation and downstream processing of extracellular lipase derived from *Burkholderia pseudomallei* (Ooi et al. 2010). Enzymatic hydrolysis of casein to produce free amino acids by papain was performed in a novel two-phase system composed of *n*-propanol, NaCl, and water (Zhang et al. 2010). The transformation yield was 99.5%, which was higher than those of *n*-propanol/water single phase, aqueous single-phase, *n*-hexane/water two-phase system, and PEG/phosphate two-phase system. Some other reports on extractive bioconversion are listed in Table 15.3.

15.6.3 New Phase Systems

Up to now, most ATPSs used for purification were based either on PEG/salt systems or polymer/polymer (e.g., PEG/dextran) systems. Due to the high cost of the polymers and the difficulty in isolating the extracted molecules from the polymer phase by back-extraction, these systems can be hardly used in a large scale production of bulk chemicals. Recently, short-chain alcohol/inorganic salt systems have been used as a novel ATPS to purify natural compounds because of its advantages such as low cost, low interfacial tension, good resolution, high yield, high capacity, and simple scale-up (Jiang et al. 2009). Because of its hydrophilic structure, ATPSs are suitable for hydrophilic compounds. Short-chain alcohols (ethanol, 2-propanol) can form stable and adjustable ATPS with inorganic salts (e.g., phosphate, sulfate), which might be due to a salting-out effect and the low solubility of inorganic salt in alcohols. In the existence of excessive inorganic salt, water molecules are attracted by the salt ions. Thus, salting-out leads to phase separation in ATPS (Albertsson 1986). When an ATPS is formed, the top phase is rich in alcohol and the bottom phase is rich in inorganic salt (Zhi and Deng 2006). Li et al. (2009) investigated the ethanol/ammonium sulfate ATP system in detail, including the effects of phase composition on the partition of 1,3-propanediol (1,3-PD) and removal of cells and biomacromolecules from the broth.

To make the process economical, gum-based systems were developed by various researchers. A new ATPS composed of derivatized guar and dextran (Chethana et al. 2006; Kumar et al. 2009) and PEG and xanthan were reported. Detailed parti-

TABLE 15.3
Extractive Bioconversion of Enzymes

Biomolecule	Organism/Source	Phase System	Reference
Alkaline protease	*Bacillus thuringiensis*	PEG-6000/phosphate	Hotha and Banik 1997
Galacto-oligosaccharides (GalOS)	*Pectinex Ultra-SP from Aspergillus aculeatus*	PEG-6000/phosphate	del-Val and Otero 2003
Amylase	*Methanococcus jannaschii*	PEO–PPO-2500/ magnesium sulfate	Li et al. 2002
α-Amylase	*Bacillus licheniformis, Aspergillus niger*	PEG/dextran	Karakatsanis and Kyriakides 2007
α-Chymotrypsin	α-Chymotrypsin immobilized on a water soluble polymer Eudragit S-100	PEG–dextran two-phase system	Sharma et al. 2003
Amyloglucosidase	*Aspergillus niger*	PEG/salt	Ramadas et al. 1996
L-Asparaginase	*Escherichia coli*	PEG/phosphate	Jiang and Zhao 1999
Nisin	*Lactococcus lactis*	PEG/magnesium sulfate	Li et al. 2000
Pectinases	*Polyporus squamosus*	PEG/crude dextran	Antov and Pericin 2000
Glyceraldehyde 3-phosphate dehydrogenase	Baker's yeast	PEG and potassium phosphate	Palomares and Lyddiatt 2002
Nisin	*Lactobacillus lactis*	PEG/magnesium sulfate	Li et al. 2000
Cephalexin	Penicillin G acylase	PEG and magnesium sulfate	Wei et al. 2001
β-Xylosidase	*Trichoderma koningii G-39*	PEG/sodium sulfate	Pan et al. 2001
α-Amylase	*Methanococcus jannaschii*	PEO–PPO/ammonium sulfate, PEO–PPO/ magnesium sulfate, PEG/ammonium sulfate, and PEG/ magnesium sulfate	Mian et al. 2002
α-Amylase	*Bacillus amyloliquifaciens*	PEG/dextran	Ulger et al. 2001
Lactic acid	*Lactococcus lactis*	Poly(ethyleneimine)/ (hydroxyethyl) cellulose	Kwon et al. 1996
Chitinase	*Serratia marcescens*	PEG-20000/dextran T500	Chen and Lee 1995
β-Galactosidase	*E. coli*	PEG/phosphate	Kuboi et al. 1995

tioning studies of some standard proteins are also attempted. A phase diagram was constructed based on the phase separation exhibited by the system at 25°C.

Gutowski et al. (2003) showed that aqueous solutions of imidazolium-based ionic liquids (ILs) can form ATPS under the addition of appropriate inorganic salts such as K_3PO_4. ATPS formed by K_3PO_4 and ILs are extremely advantageous for the partition of several biomolecules yielding larger partition coefficients than those conventionally obtained with polymer/inorganic salts or polymer/polymer ATPS (Ventura et al. 2009). The extractive potential of biomolecules using IL-based ATPS was previously studied for distinct compounds such as testosterone and epitestosterone, alkaloids, antibiotics, bovine serum albumin, penicillin G, tryptophan, and food colorants (Filipa et al. 2010).

The initial interest on ILs as alternative extractive fluids results from their physical and chemical advantages, namely, negligible flammability and vapor pressure, high solvation ability, high chemical stability, high selectivity, and ease in recovering and recycling them (Dreyer et al. 2008). Moreover, it was already shown that adequate ILs do not significantly inactivate enzymes, thus ensuring their structural integrity and enzymatic activity (Martínez-Aragón 2010). Filipa (2010) studied the partitioning of vanillin, a phenolic aldehyde, using ternary systems composed by imidazolium-based ILs, water, and K_3PO_4, for the extraction and the concentration of vanillin. For all the studied systems and at all the conditions analyzed, vanillin preferentially partitioned to IL-rich phase presenting $K_{Van} > 1$.

A new thermosensitive copolymer, poly (NIPA-co-BA) (abbreviated P_{NB}), was synthesized by Shu et al. (2009) using N-isopropylacrylamide and N-butyl acrylate as monomers. It was utilized for the formation of ATPS with another novel pH-sensitive copolymer poly (AA-co-DMAEMA-co-BMA) (abbreviated P_{ADB}). The new system was employed for the bioconversion of penicillin G with a yield of 95.4%.

15.7 SOME RECENT APPLICATIONS OF ATPE

The success of ATPSs in the efficient generation of bench scale prototype processes with commercial application has been reported for the recovery of a large number of biopharmaceutical products. A key feature of these systems is that they can be easily scaled up from laboratory scale to industrial scale with performance data retained. Biotechnology industry has recently been demanding nanoparticulate products (20–200 nm) such as viruses, plasmids, virus-like particles, and drug delivery assemblies. These products are mainly used as gene delivery systems in gene therapy protocols. The solvent extraction method with ATPS has been used to successfully recover these bioproducts on a large scale. The partition behavior of pure bacteriophage T4 ATPSs were reported and shown applicability of ATPE (Negrete et al. 2007; Brass et al. 2000). Technical feasibility has been reported up to 100,000-L scale for the purification of proteins. ATPE is adopted by the biomanufacturing industries as an alternative platform. One is related to the maximum capacity of these systems, particularly whether they can process very high-titer cells supernatants or whether the throughput will be limited by solubility problems (Rosa et al. 2010).

Synthesis and recovery of nanoparticles (radio active gold [198] (Au III)) using ATPSs is reported by Roy and Lahiri (2006). A few studies have shown the nanoparticle conjugations to improve the protein partitioning in ATPSs (Long and Keating 2006). Extractive fermentation/bioconversion in ATPSs is a meaningful approach to overcome low product yield in a conventional fermentation process and by proper design of the two-phase system it is possible to obtain the product in a cell free stream.

Electroextractions of the proteins were mainly carried out in the polymer–polymer phase systems (Levine and Bier 1990; Theos and Clark 1995). Application of electric field for electroextraction in a polymer salt system was also studied despite the high conductivity of the salt phase (Nagaraj et al. 2005). Recently, electroextraction of model protein in aqueous two-phase polymer/salt systems was reported (Madhusudhan et al. 2010).

A combination of free electrophoresis and multistage extraction called electrophoretic counter current extractor was explored. A simple mathematical model was developed to describe the mass and heat transfer during the electrophoretic separation and validated by the extraction of fixed human red blood cells and latex particles of different sizes suspended in phosphate buffer (Chethana et al. 2008).

15.8 SCALE-UP AND ECONOMIC ASPECTS

Large-scale extraction using ATPS and scale-up are facilitated by the availability of equipment and machinery used for extraction technology in the chemical industry. There are many reports to show large scale productions of enzymes from various sources (Dunhill 1991). The scale-up studies of extraction and isolation of a recombinant *Fusarium solani pisi* cutinase was reported for an extracellular mutant enzyme expressed in *Saccharomyces cerevisiae*. The performance of ATPS was reproducible within the scale range of 0.010–30 liter, provided a standard deviation of the yield lower than 8% with a yield over 95% (Costa et al. 2000). Technical feasibility of scale-up up to 100,000 L for the purification of proteins is reported (Rosa et al. 2010).

When developing a downstream procedure for process scale, it is mandatory to consider processing time, energy, manpower, good manufacturing practices, recycling of chemicals, sterilization and cleaning of equipment apart from separation efficiency. Major hindrances for implementation of ATPE in industries are the high cost of phase-forming polymers and low demixing rates of the phases after extraction. The former aspect is addressed to a good extent by employing low-cost phase systems and recovery and recycling of phase components, at times even by employing thermo-separating polymers. The slow demixing rate is due to small differences in densities between the phases, high viscosity, and low interfacial tension. Application of external fields such as electric (Raghavarao et al. 1998b, Nagaraj et al. 2005), magnetic, microwave (Nagaraj et al. 2003), and acoustic fields (Srinivas et al. 2000; Nagaraj et al. 2002) has been successfully employed for enhancing the demixing rate.

Process integration is one of the most effective ways to increase the overall productivity of any given process. However, not many successful attempts are reported in this regard. The productivity of a given bioprocess can be considerably improved by a relatively new strategy of process integration. It could be by design of extraction in such a way to integrate in itself, various unit operations such as solid separation, purification, concentration, etc. It also could be integration of ATPE with other downstream processes for achieving desired selectivity and purity of the biomolecules.

15.9   CONCLUSIONS

Downstream processing in many fermentation processes accounts for a large share of the total cost. The ATPE appears to be a promising technique for efficient downstream processing and some successful applications of ATPE on large/industrial scale have been demonstrated. Unfortunately, information in the literature on the engineering aspects of ATPE involving mass transfer and hydrodynamics is scant or remains proprietary and only a few reports are available. For ATPS, in which the interfacial tension and the density differences are low and the phase viscosities are high, the mechanism of drop formation needs to be investigated. Rational correlations need to be developed for the drop size especially for newer phase systems. The flooding characteristics of the column contactors need to be investigated. The design of coalescence zone needs close attention. The use of fibrous beds for promoting the coalescence needs to be developed for these systems. As the cost of phase forming chemicals increases linearly with scale, studies exploring the feasibility of recycling polymers other than PEG must be undertaken.

To develop effective downstream processing methods employing ATPSs, an interdisciplinary effort involving a combination of microbiological, biochemical, and engineering aspects is very essential. Although enough attention has been paid to microbiological and biochemical aspects, the engineering aspects have been largely neglected. Although, in principle, ATPE offers the advantage of easy adaptation of the extraction equipment used in the chemical/pharmaceutical industry to achieve efficient extraction, the drop dynamics and mass transfer aspects of these contacting equipment should be studied in detail employing various ATPSs as well as real systems involving the actual fermentation broths. Even if some of these aspects are addressed in greater depth by future researchers, the objective of this chapter can be considered fulfilled in view of the recognized scientific and industrial potential of ATPE.

ACKNOWLEDGMENTS

The authors wish to acknowledge Dr. V. Prakash, Director, CFTRI, for the keen interest in downstream processing. M. C. Madhusudhan and M. C. Lakshmi gratefully acknowledge CSIR, India, for the award of Senior Research Fellowship.

REFERENCES

Aguilar, O., Albiter, V., Serrano-Carreón, L., Palomares, M.R. 2006. Direct comparison between ion-exchange chromatography and aqueous two-phase processes for the partial purification of penicillin acylase produced by *E. coli. J Chromatogr B*, 835: 77–83.
Albertsson, P.A. 1971. *Partition of Cell Particles and Macromolecules*, 2nd ed. New York: Wiley.
Albertsson, P.A. 1986. *Partition of cell particles and macromolecules*, 3rd ed. New York: Wiley.

Anandharamakrishnan, C., Raghavendra, S.N., Barhate, R.S., Hanumesh, U., Raghavarao, K.S.M.S. 2005. Aqueous two-phase extraction for recovery of proteins from cheese whey. *Food Bioprod Process* 83: 191–197.

Antov, M.G., Pericin, D.M., Dasic, M.G. 2006. Aqueous two-phase partitioning of xylanase produced by solid-state cultivation of *Polyporus squamosus*. *Process Biochem* 41: 232–235.

Antov, M., Anderson, L., Andersson, A., Tjerneld, F., Stalbrand, H. 2006. Affinity partitioning of a *Cellulomonas fimi*-mannanase with a mannan-binding module in galactomannan/starch aqueous two-phase system. *J Chromatogr A* 1123: 53–59.

Antov, M.G. 2004. Partitioning of pectinase produced by *Polyporus squamosus* in aqueous two-phase system polyethylene glycol 4000/crude dextran at different initial pH values. *Carbohyd Polym* 56: 295–300.

Bamberger, S., Seaman, G.V.F., Sharp, K.A., Brooks, D.E. 1984. The effects of salts on the interfacial tension of aqueous dextran poly(ethylene glycol) phase systems. *J Colloid Interface Sci* 99(1): 194–200.

Banik, R.M., Santhiagu, A., Kanari, B., Sabarinath, C., Upadhyay, S.N. 2003. Technological aspects of extractive fermentation using aqueous two-phase systems. *World J Microb Biot* 19: 337–348.

Barhate, R.S., Patil, G., Srinivas, N.D., Raghavarao, K.S.M.S. 2004. Drop formation in aqueous two-phase systems. *J Chromatogr A* 1023: 197–206.

Baskir, J.N., Hutton, T.A., Suter, U.W. 1989a. Protein partitioning in two phase aqueous polymer systems. *Biotechnol Bioeng* 34: 541–558.

Baskir, J.N., Hutton, T.A., Suter, U.W. 1989b. Affinity partitioning in two-phase aqueous polymer systems: A simple model for the distribution of the polymer–ligand tail segments near the surface of a particle. *J Phys Chem* 93: 969–976.

Bassani, G., Farruggia, B., Nerli, B., Romanini, D., Picó, G. 2007. Porcine pancreatic lipase partition in potassium phosphate–polyethylene glycol aqueous two-phase systems. *J Chromatogr B* 859: 222–228.

Benavides, J., Palomares, M.R. 2004. Bioprocess intensification: A potential aqueous two-phase process for the primary recovery of B-phycoerythrin from *Porphyridium cruentum*. *J Chromatogr B* 807: 33–38.

Bernaudat, F., Bulow, L. 2006. Combined hydrophobic-metal binding fusion tags for applications in aqueous two-phase partitioning. *Protein Expres Purif* 46: 438–445.

Bhide, A.A., Patel, R.M., Joshi, J.B., Pangarkar, V.G. 1995. Affinity partitioning of enzymes using unbound triazine dyes in PEG/phosphate system, *Separ Sci Technol* 30(15): 2989–3000.

Boeris, V., Spelzini, D., Farruggia, B., Pico, G. 2009. Aqueous two-phase extraction and polyelectrolyte precipitation combination: A simple and economically technologies for pepsin isolation from bovine abomasum homogenate. *Process Biochem* 44: 1260–1264.

Boland, M.J, Hesselink, P.G.M., Papamichael, N., Hustedt, H. 1991. Extractive purification of enzymes from animal tissue using aqueous two phase systems: Pilot scale studies. *J Biotechnol* 19: 19–34.

Braas, G.M.F., Walker, S.G., Lyddiatt, A. 2000. Recovery in aqueous two-phase systems of nanoparticulates applied as surrogate mimics for viral gene therapy vector. *J Chromatogr B* 743: 409–419.

Brescancini, A.P.R., Tambourgi, E.B., Pessoa Jr, A. 2004. Bromelain partitioning in two-phase aqueous systems containing PEO–PPO–PEO block copolymers. *J Chromatogr B* 807: 61–68.

Brooks, D.E., Sharp, K.A., Fisher, D. 1985. Theoretical aspects of partitioning, in *Partitioning in Aqueous Two Phase Systems*, H. Walter, D.E. Brooks, and D. Fisher (Eds.), pp. 11–84. New York: Academic Press.

Brunner, K.H., Scherfler, H., Stolting, M. 1981. Erfahrungen mit dem gebenstromextraktionsdekanter bei der gewinnung von antibotika aus fermentationsbruhen. *VT Verfahrenstechnik* 15: 619–622.

Cavalcanti, M.T.H., Porto, T.S., deBarros Neto, B., Lima-Filho, J.L., Porto, A.L.F., Pessoa Jr., A. 2006. Aqueous two phase systems extraction of α-toxin from *Clostridium perfringens* type A. *J Chromatogr B* 833: 135.

Chaiwut, P., Rawdkuen, S., Benjakul, S. 2010. Extraction of protease from *Calotropis procera* latex by polyethylene glycol–salts biphasic system, *Process Biochem* 45: 1148–1155.

Chen, J.P., Lee, M.S. 1995. Enhanced production of *Serratia marcescens* chitinase in PEG/dextran aqueous two phase systems. *Enzyme Microb Technol* 17: 1021–1027.

Chen, J.P., Wang, C.H. 1991. Lactose hydrolysis by β-galactosidase in aqueous two-phase systems *J Fermt Bioeng* 71(3): 168–175.

Chethana, S., Nayak, C.A., Raghavarao, K.S.M.S. 2007. Aqueous two phase extraction for purification and concentration of Betalains. *J Food Eng* 81: 679–687.

Chethana, S., Patil, G., Madhusudhan, M.C., Raghavarao, K.S.M.S. 2008. Electrophoretic extraction of cells/particles in a counter current extractor. *Separ Sci Technol* 43: 1–16.

Chethana, S., Rastogi, N.K., Raghavarao, K.S.M.S. 2006. New aqueous two phase system comprising polyethylene glycol and xanthan. *Biotechnol Lett* 28: 25–28.

Chouyyok, W., Wongmongkol, N., Siwarungson, N., Prichanont, S. 2005. Extraction of alkaline protease using an aqueous two-phase system from cell free *Bacillus subtilis* TISTR 25 fermentation broth. *Process Biochem* 40: 3514–3518.

Chung, B.H., Arnold, F.H. 1990. Metal affinity partitioning of phosphoproteins in PEG/dextran two phase systems. *Biotechnol Lett* 13: 615–620.

Cisneros, M., Benavides, J., Brenes, C.H., Palomares, M.R. 2004. Recovery in aqueous two-phase systems of lutein produced by the green microalga *Chlorella protothecoides*. *J Chromatogr B* 807: 105–110.

Coimbra, J.S.R., Thommes, J., Kula, M.R. 1994. Continuous separation of whey proteins with aqueous two-phase systems in a Graesser contactor. *J Chromatogr A* 668: 85–94.

Colléna, A., Penttila, M., lbranda, S.H., Tjernelda, F., Veidec, A. 2001. Extraction of endoglucanase I (Cel7B) fusion proteins from *Trichoderma reesei* culture filtrate in a poly(ethylene glycol)–phosphate aqueous two-phase system. *J Chromatogr A* 943: 55–62.

Colléna, A., Persson, J., Linder, M., Nakari-Seta, T., Penttila, M., Tjerneld, F., Sivars, U. 2002. A novel two-step extraction method with detergent/polymer systems for primary recovery of the fusion protein endoglucanase I-hydrophobin. *Biochim Biophys Acta* 1569: 139–150.

Costa, M.J.L., Cunha, M.T., Cabral, J.M.S., Aires-Barros, M.R. 2000. Scale-up of recombinant cutinase recovery by whole broth extraction with PEG-phosphate aqueous two-phase. *Bioseparation* 9: 231–238.

Craig, L.C., and Craig, D. 1956. Extraction and distribution, in *Technique of organic chemistry*, A. Weissberger (Ed.), vol. 3, 171–311. New York: Interscience Publishers.

Cunha, M.T., Cabral, J.M.S., Tjerneld, F., Aires-Barros, M.R. 2000. Effect of salts and surfactants on the partitioning of *Fusarium solani pisi* cutinase in aqueous two-phase systems of thermoseparating ethylene oxide/propylene oxide random copolymer and hydroxypropyl starch. *Bioseparation* 9: 203–209.

Cunha, M.T., Costa, M.J.L., Calado, C.R.C., Fonseca, L.P., Aires-Barros, M.R., Cabral, J.M.S. Integration of production and aqueous two-phase systems extraction of extracellular *Fusarium solani pisi* cutinase fusion proteins. *J Biotechnol* 100: 55–64.

Da Silva, C.A.S., Coimbra, J.S.R., Garcia Rojas, E.E., Minima, L.A., da Silva, L.H.M. 2007. Partitioning of caseinomacropeptide in aqueous two-phase systems. *J Chromatogr B* 858: 205–210.

Da Silva, C.A.S., Coimbra, J.S.R., Rojas, E.E.G., Teixeira, J.A.C. 2009. Partitioning of glycomacropeptide in aqueous two-phase systems. *Process Biochem* 44: 1213–1216.

Da Silva, M.E., Franco, T.T. 2000. Purification of soybean peroxidase (*Glycine max*) by metal affinity partitioning in aqueous two-phase systems. *J Chromatogr B* 743: 287–294.

Dembczynski, R., Bialas, W., Jankowski, T. 2010a. Recycling of phase components during lysozyme extraction from hen egg white in the EO50PO50/K_2HPO_4 aqueous two-phase system. *Biochem Eng J* 51: 24–31.

Dembczynski, R., Białas, W., Regulski, W., Jankowski, T. 2010b. Lysozyme extraction from hen egg white in an aqueous two-phase system composed of ethylene oxide–propylene oxide thermoseparating copolymer and potassium phosphate. *Process Biochem* 45: 369–374.

Diamond, A.D., Hsu, J.T. 1989. Fundamental studies of biomolecule partitioning in aqueous two-phase systems. *Biotechnol Bioeng* 34: 1000–1014.

Diamond, A.D., Hsu, J.T. 1992. Aqueous two phase systems for biomolecules, separation. *Adv Biochem Eng Biot* 47: 89–133.

Dolia, S., Gaikar, V. 2006. Aqueous two-phase partitioning of glucose isomerase from *Actinoplanes missouriensis* in the presence of PEG—Derivatives and its immobilization on chitosan beads. *Separ Sci Technol* 41: 2807–2823.

Dreyer, S., Kragl, U. 2008. Ionic liquids for aqueous two-phase extraction and stabilization of enzymes. *Biotechnol Bioeng* 99(6): 1416–1424.

Draginja, M.P., Senka, Z.M.-P., Ljiljana, M.R.-P. 2009. Optimization of conditions for acid protease partitioning and purification in aqueous two-phase systems using response surface methodology. *Biotechnol Lett* 31: 43–47.

Fernandes, S., Kim, H.S., Hatti-Kaul, R. 2002. Affinity extraction of dye- and metal on binding proteins in polyvinylpyrrolidone-based aqueous two-phase system. *Protein Expres Purif* 24: 460–469.

Fernandes, P. 2010. Enzymes in food processing: A condensed overview on strategies for better biocatalysts. *Enzyme Res* DOI: 10.4061/2010/862537.

Filho, P.A.P., Mohameda, R.S. 2004. Thermodynamic modeling of the partitioning of biomolecules in aqueous two-phase systems using a modified Flory–Huggins equation. *Process Biochem* 39: 2075–2083.

Filipa, M.C.A., Freire, M.G., Freire, C.S.R., Silvestre, A.J.D., Coutinho, J.A.P. 2010. Extraction of vanillin using ionic-liquid-based aqueous two-phase systems. *Separ Purif Technol* 75: 39–47.

Gautam, S., Simon, L. 2007. Prediction of equilibrium phase compositions and β-glucosidase partition coefficient in aqueous two-phase systems. *Chem Eng Commun* 194:117–128.

Gautam, S., Simon, L. 2006. Partitioning of β-glucosidase from *Trichoderma reesei* in poly(ethylene glycol) and potassium phosphate aqueous two-phase systems: Influence of pH and temperature. *Biochem Eng J* 30: 104–108.

Gavasane, M.R., Gaikar, V.G. 2003. Aqueous two-phase affinity partitioning of penicillin acylase from *E. coli* in presence of PEG-derivatives. *Enzyme Microb Technol* 32: 665–675.

Giuliano, K.A. 1991. Aqueous two-phase protein partitioning using textile dyes as affinity ligands. *Anal Biochem* 197(2): 333–339.

Giuliano, K.A. 1992. Chromatography of proteins on columns of polyvinylpolypyrrolidone using adsorbed textile dyes as affinity ligands. *Anal Biochem* 200(2): 370–375.

Gu, Z., Glatz, C.E. 2007. Aqueous two-phase extraction for protein recovery from corn extracts. *J Chromatogr B* 845: 38–50.

Gutowski, K.E., Broker, G.A., Willauer, H.D., Huddleston, J.G., Swatloski, R.P., Holbrey, J.D., Rogers, R.D. 2003. *J Am Chem Soc* 125: 6632–6633.

Hatti-Kaul, R. (Ed). 2000. *Aqueous Two Phase Systems—Methods and Protocols*. Totowa, NJ: Humana Press.

He, G.-Q., Zhang, X.-Y., Tang, X.J., Chen, Q., Ruan, H.Z. 2005. Partitioning and purification of extracellular 1,3–1,4-glucanase in aqueous two-phase systems. *Univ Sci* 8: 825–831.

Hossain, M., Dean, J. 2008. Extraction of penicillin G from aqueous solutions: Analysis of reaction equilibrium and mass transfer. *Separ Purif Technol* 62: 437–443.

Igarashi, L., Kieckbusch, T.G., Franco, T.T. 2004. Mass transfer in aqueous two-phases system packed column. *J Chromatogr B* 807: 75–80.

Imelio, N., Marini, A., Spelzini, D., Picó, G., Farruggia, B. 2008. Pepsin extraction from bovine stomach using aqueous two-phase systems: Molecular mechanism and influence of homogenate mass and phase volume ratio. *J Chromatogr B* 873: 133–138.

Ivanova, V., Yankov, D., Kabaivanova, L. Pashkoulov, D. 2001. Simultaneous biosynthesis and purification of two extracellular Bacillus hydrolases in aqueous two-phase systems. *Microbiol Res* 156(1): 19–30.

Jafarabad, K.R., Patil, T.A., Sawant, S.B., Joshi, J.B. 1992. Enzyme and protein mass transfer coefficient in aqueous two-phase systems. York–Scheibel extraction column. *Chem Eng Sci* 47: 69–73.

Jiang, B., Li, Z.G., Dai, J.Y., Zhang, D.J., Xiu, Z.L. 2009. Aqueous two-phase extraction of 2,3-butanediol from fermentation broths using an ethanol/phosphate system. *Process Biochem* 44: 112–117.

Johansson, G., Joelsson, M., Olde, B. 1985. Affinity partitioning of biopolymers and membranes in Ficoll–dextran aqueous two-phase systems. *J Chromatogr* 331: 11–21.

Johansson, G., Shanbags, V.P. 1984. Affinity partitioning of protein in aqueous two phase systems containing polymer bound fatty acid: Effect of polyethylene glycol-palmiate on the partition of human serum albumin and α-lactalbumin. *J Chromatogr A* 284: 63–72.

Johansson, H., Magaldi, F.M., Feitosa, E., Pessoa Jr, A. 2008. Protein partitioning in poly(ethylene glycol)/sodium polyacrylate aqueous two-phase systems. *J Chromatogr A* 1178: 145–153.

Johansson, G., Joelsson, M. 1987. Affinity partitioning of enzymes using dextran-bound procion yellow HE-3G: Influence of dye–ligand density. *J Chromatogr A* 393(2): 195–208.

Joshi, J.B., Sawant, S.B., Raghavarao, K.S.M.S., Patil, T.A., Rostami, K.M., Sikdar, S.K. 1990. Continuous counter-current two-phase aqueous extraction. *Bioseparation* 1: 311–324.

Klomklao, S., Benjakul, S., Visessanguan, W., Simpson, B.K., Kishimura, H. 2005. Partitioning and recovery of proteinase from tuna spleen by aqueous two-phase systems. *Process Biochem* 40: 3061–3067.

Kuboi, R., Umakoshi, H., Komasawa, I. 1995. Extractive cultivation of *E. coli* using two phase aqueous polymer systems to produce intracellular beta glucosidase. *Biotechnol Progr* 11: 202–207.

Kula, M.R. 1985. Liquid–liquid extraction of biopolymers, in *Comprehensive Biotechnology*, A. Humphrey and C.L. Cooney (Eds.), Vol. 2, 451–471. New York: Pergamon Press.

Kula, M.R., Kroner, K.H., Hustedt, H. 1982. Purification of enzymes by liquid–liquid extraction. *Adv Biochem Eng* 24: 73–118.

Kula, M.R., Kroner, K.H., Hustedt, H., Schütte, H. 1981. Technical aspects of extractive enzyme purification. *Ann N Y Acad Sci* 369: 341–354.

Kumar, K., Narahari, Pujari, S., Golegaonkar, S.B., Ponrathnam, S., Nene, S.N., Bhatnagar, D. 2009. Vinyl-2-pyrrolidone derivatized guar gum based aqueous two-phase system. *Separ Purif Technol* 65: 9–13.

Kwon, Y.J., Kaul, R., Mattiasson, B. 1996. Extractive lactic acid fermentation in poly-(ethyleneimine)-based aqueous two-phase systems. *Biotechnol Bioeng* 50: 280–290.

Lakshmi, M.C., Madhusudhan, M.C., Raghavarao, K.S.M.S. 2009. Extraction and purification of lipoxygenase from soybean using aqueous two-phase system. *Food Bioprocess Tech* DOI: 10.1007/s11947-009-0278.

Li, C., Ouyang, F., Bai, J. 2000. Extractive cultivation of *Lactococcus lactis* using a polyethylene glycol/MgSO$_4$–7H$_2$O aqueous two-phase system to produce nisin. *Biotechnol Lett* 22: 843–847.

Li, M., Kim, J.W., Peeples, T.L. 2002. Amylase partitioning and extractive bioconversion of starch using thermoseparating aqueous two-phase systems. *J Biotechnol* 93: 15–26.

Li, M., Peeples, T.L. 2004. Purification of hyperthermophilic archaeal amylolytic enzyme (MJA1) using thermoseparating aqueous two-phase systems. *J Chromatogr B* 807: 69–74.

Li, Z.G., Jiang, B., Zhang, D., Xiu, Z.-L. 2009. Aqueous two-phase extraction of 1,3-propane-diol from glycerol-based fermentation broths. *Separ Purif Technol* 66: 472–478.

Li, Z., Teng, H., Xiu, Z. 2010. Aqueous two-phase extraction of 2,3-butanediol from fermentation broths using an ethanol/ammonium sulfate system. *Process Biochem* 45: 731–737.

Liao, L.-C., Ho, C.S., Wu, W.T. 1999. Bioconversion with whole cell penicillin acylase in aqueous two-phase systems. *Process Biochem* 34(5): 417–420.

Lima, A.S., Alegre, R.M., Meirelles, A.J.A. 2002. Partitioning of pectinolytic enzymes in polyethylene glycol/potassium phosphate aqueous two-phase systems. *Carbohyd Polym* 50: 63–68.

Lin, D.Q., Zhu, Z.Q., Mei, L.-H. 1998. Process design for purification of muscle lactate dehydrogenase by affinity partitioning using free reactive dyes. *Separ Sci Technol* 33(13): 1937–1953.

Long, M.S., Keating, C.D. 2006. Nanoparticle conjugation increases protein partitioning in aqueous two phase systems. *Anal Chem* 78: 379–386.

Madeira, P.P., Xua, X., Teixeira, J.A., Macedo, E.A. 2005. Prediction of protein partition in polymer/salt aqueous two-phase systems using the modified Wilson model. *Biochem Eng J* 24: 147–155.

Madhusudhan, M.C., Chethana, S., Raghavarao, K.S.M.S. 2010. Demixing, electrokinetic demixing of polymer salt systems with biomolecules. *Separ Sci Technol* 46(5): 727–733.

Madhusudhan, M.C., Raghavarao, K.S.M.S., Nene, S. 2008. Integrated process for extraction and purification of alcohol dehydrogenase from baker's yeast involving precipitation and aqueous two phase extraction. *Biochem Eng J* 38: 414–420.

Magestea, A.B., de Lemosa, L.R., Ferreira, G.M.D., Da Silva, M.C.H., da Silva, L.M.H., Bonomob, R.C.F., Minim, L.A. 2009. Aqueous two-phase systems: An efficient, environmentally safe and economically viable method for purification of natural dye carmine. *J Chromatogr A* 1216: 7623–7629.

Marcos, J.C., Fonseca, L.P., Ramalhoa, M.T., Cabral, J.M.S. 2002. Application of surface response analysis to the optimization of penicillin acylase purification in aqueous two-phase systems. *Enzyme Microb Technol* 31: 1006–1014.

Mattiasson, B. 1988. Immobilized enzymes and cells. *Method Enzymol* Part D. 137: 657–667.

Mayerhoff, D.V.L., Roberto, I.C., Franco, T.T. 2004. Purification of xylose reductase from *Candida mogii* in aqueous two-phase systems. *Biochem Eng J* 18: 217–223.

Mohamadi, H.S., Omidinia, E., Dinarvand, R. 2007. Evaluation of recombinant phenylalanine dehydrogenase behavior in aqueous two-phase partitioning. *Process Biochem* 42: 1296–1301.

Mohamadia, H.S., Omidinia, E. 2007. Purification of recombinant phenylalanine dehydrogenase by partitioning in aqueous two-phase systems. *J Chromatogr B* 854: 273–278.

Naganagouda, K., Mulimani, V.H. 2008. Aqueous two-phase extraction (ATPE): An attractive and economically viable technology for downstream processing of *Aspergillus oryzae* α-galactosidase. *Process Biochem* 43: 1293–1299.

Nagaraj, N., Chethana, S., Raghavarao, K.S.M.S. 2005. Electrokinetic demixing of aqueous two-phase polymer/salt systems. *Electrophoresis* 26: 10–17.

Nagaraj, N., Srinivas, N.D., Raghavarao, K.S.M.S. 2002. Acoustic field assisted demixing of aqueous two phase systems. *J Chromatogr A* 977: 163–172.

Nagaraj, N., Narayan, A.V., Srinivas, N.D., Raghavarao, K.S.M.S. 2003. Microwave field assisted enhanced demixing of aqueous two-phase systems. *Anal Biochem* 312(2): 134–140.

Nandini, K.E., Rastogi, N.K. 2010. Integrated process for downstream processing of lactoperoxidase from milk whey involving ATPE and ultrasound assisted ultrafiltration. *Appl Biochem Biotechnol* 163(1): 173–185.

Negrete, A., Ling, T.C., Lyddiatt, A. 2007. Aqueous two-phase recovery of bio-nanoparticles: A miniaturization study for the recovery of bacteriophage T4. *J Chromatogr B* 854: 13–19.

Nitsawang, S., Hatti-Kaul, R., Kanasawuda, P. 2006. Purification of papain from Carica papaya latex: Aqueous two-phase extraction versus two-step salt precipitation. *Enzyme Microb Technol* 39: 1103–1107.

Oliveira, L.A., Sarubbo, L.A., Porto, A.L.F., Campos-Takaki, M., Tambourgi, E.B. 2002. Partition of trypsin in aqueous two-phase systems of poly(ethylene glycol) and cashew-nut tree gum. *Process Biochem* 38: 693–699.

Oliveira, M.C., Americo, M., Filho, N.A., Pessoa Filho, P.L. 2007. Phase equilibrium and protein partitioning in aqueous two-phase systems containing ammonium carbamate and block copolymers PEO–PPO–PEO. *Biochem Eng J* 37: 311–318.

Omidinia, E., Mohamadi, H.S., Dinarvand, H., Taherkhani, H.A. 2009. Investigation of chromatography and polymer/salt aqueous two-phase processes for downstream processing development of recombinant phenylalanine dehydrogenase. *Bioproc Biosyst Eng* 33: 317–329.

Ooi, C.W., Hii, S.L., Kamal, S.M.M., Ariff, A., Ling, T.C. 2010. Extractive fermentation using aqueous two-phase systems for integrated production and purification of extracellular lipase derived from *Burkholderia pseudomallei*. *Process Biochem* 46(1): 68–74.

Ooi, C.W., Tey, B.T., Hii, S.L., Ariff, A., Wu, H.S., Wei, J.C., Juang, R.S., Kamal, S.M.M. Ling, T.C. 2009. Direct purification of *Burkholderia pseudomallei* lipase from fermentation broth using aqueous two-phase systems. *Biotechnol Bioproc E* 14: 811–818.

Ortin, A., Lopez-Perez, M.J., Muiro Blanco, M.T., Cebrian Perez, J.A. 1991. Extraction of xanthine oxidase from milk by counter-current distribution in an aqueous two-phase system. *J Chromatogr* 558: 357–367.

Palomares, M.R., Lyddiatt, A. 2002. Process integration using aqueous two-phase partition for the recovery of intracellular proteins. *Chem Eng J* 87: 313–319.

Pan, I.-H., Yao, H.-J., Li, Y.-K. 2001. Effective extraction and purification of β-xylosidase from *Trichoderma koningii* fermentation culture by aqueous two-phase partitioning. *Enzyme Microb Technol* 28: 196–201.

Pan, I.-H., Chiu, H.-H., Lub, H.-H., Lee, L.-T., Li, Y.-K. 2002. Aqueous two-phase extraction as an effective tool for isolation of geniposide from gardenia fruit. *J Chromatogr A* 977: 239–246.

Patil, G. and Raghavarao, K.S.M.S. 2007. Aqueous two phase extraction for purification of C-phycocyanin. *Biochem Eng J* 34: 156–164.

Patil, G., Chethana, S., Madhusudhan, M.C., Raghavarao, K.S.M.S. 2008. Fractionation and purification of the phycobiliproteins from *Spirulina platensis*. *Bioresource Technol* 99: 7393–7396.

Patil, G., Chethana, S., Sridevi, A.S., Raghavarao, K.S.M.S. 2006. Method to obtain C-phycocyanin of high purity. *J Chromatogr A* 1127: 76–81.

Patil, T.A., Rostami, J.K., Sawant, S.B., Joshi, J.B. 1991. Enzyme mass transfer coefficient in aqueous two phase system using a packed extraction column. *Can J Chem Eng* 69: 548–556.

Paula, A.J.R., Azevedo, A.M., Aires-Barros, M.R. 2007. Application of central composite design to the optimisation of aqueous two-phase extraction of human antibodies. *J Chromatogr A* 1141: 50–60.

Pereira, M., Wu, Y.-T., Venancio, A., Teixeira, J. 2003. Aqueous two-phase extraction using thermoseparating polymer: A new system for the separation of endo-polygalacturonase. *Biochem Eng J* 15: 131–138.

Peričin, D.M., Madārev-Popovic, S.Z., Radulović-Popovic, L.M. 2009. Optimization of conditions for acid protease partitioning and purification in aqueous two-phase systems using response surface methodology. *Biotechnol Lett* 31: 43–47.

Porto, A.L.F., Sarubbo, L.A., Lima-Filho, J.L., Aires Barros, M.R., Cabral, J.M.S., Tambourgi, E.B. 2000. Hydrodynamics and mass transfer in aqueous two-phase protein extraction using a continuous perforated rotating disc contactor. *Bioprocess Eng* 22: 215–218.

Porto, T.S., Medeiros e Silva, G.M., Porto, C.S., Cavalcanti, M.T.H., Neto, B.B., Lima-Filho, J.L., Converti, A., Porto, A.L.F., Pessoa Jr., A. 2008. Liquid–liquid extraction of proteases from fermented broth by PEG/citrate aqueous two-phase system, *Chem Eng Process* 47: 716–721.

Qu, F., Qin, H., Dong, M., Zhao, D.X., Zhao, X.Y., Zhang, Z.H. 2009. Selective separation and enrichment of proteins in aqueous two-phase extraction system. *Chin Chem Lett* 20: 1100–1102.

Raghavarao, K.S.M.S., Guinn, M.R., Todd, P.W. 1998. Recent developments in aqueous two-phase extraction in bioprocessing. *Separ Purif Method* 27: 1–50.

Raghavarao, K.S.M.S., Ranganathan, T.V., Srinivas, N.D., Barhate, R.S. 2003. Aqueous two phase extraction: An environmentally benign technique. *Clean Technol Envir* 5: 136–141.

Raghavarao, K.S.M.S., Rastogi, N.K., Gowthaman, M.K., Karanth, N.G. 1995. Aqueous two phase extraction for downstream processing of enzymes, proteins. *Adv Appl Microbiol* 41: 97–171.

Raghavarao, K.S.M.S., Srinivas, N.D., Chethana, S., Todd, P.W. 2002. Field assisted extraction of cells, particles and macromolecules. *Trend Biotechnol* 20: 72–78.

Raghavarao, K.S.M.S., Stewart, R.M., Todd, P. 1991. Electrokinetic demixing of immiscible aqueous solutions. II. Separation rates of polyethylene glycol-maltodextrin mixtures. *Separ Sci Technol* 26: 257–267.

Raghavarao, K.S.M.S., Stewart, R.M., Rudge, S.R., Todd, P. 1998. Electrokinetic demixing of aqueous two-phase systems III: Drop electrophoretic motilities and demixing rates, *Biotechnol Prog* 14: 922–930.

Raghavarao, K.S.M.S., Szlag, D., Sikdar, S.K., Sawant, S.B., Joshi, J.B. 1991. Protein extraction in a column using a polyethylene glycol/maltodextrin two-phase polymer system. *Chem Eng J* 46: 75–81.

Ravindra Babu, B., Rastogi, N.K., Raghavarao, K.S.M.S. 2008. Liquid–liquid extraction of bromelain and polyphenol oxidase using aqueous two-phase system. *Chem Eng Process* 47: 83–89.

Rawdkuen, S., Pintathong, P., Chaiwut, P., Benjakul, S. 2010. The partitioning of protease from *Calotropis procera* latex by aqueous two-phase systems and its hydrolytic pattern on muscle proteins. *Food Bioproduct Process* 89(1): 73–80.

Reh, G., Spelzini, D., Tubío, G., Picó, G., Farruggia, B. 2007. Partition features and renaturation enhancement of chymosin in aqueous two-phase systems. *J Chromatogr B* 860: 98–105.

Palomares, M.R., Lyddiatt, A. 2002. Process integration using aqueous two-phase partition for the recovery of intracellular proteins. *Chem Eng J* 87: 313–319.

Rosa, P.A.J., Azevedo, A.M., Ferreira, I.F., de Vries, J., Korporaal, R., Verhoef, H.J., Visser, T.J., Aires-Barros, M.R. 2007. Affinity partitioning of human antibodies in aqueous two-phase systems. *J Chromatogr A* 1162: 103–113.

Rosa, P.A.J., Ferreira, I.F., Azevedo, A.M., Aires-Barros, M.R. 2010. Aqueous two-phase systems: A viable platform in the manufacturing of biopharmaceuticals. *J Chromatogr A* 121: 2296–2305.

Ross, K.C., Zhang, C. 2010. Separation of recombinant-glucuronidase from transgenic tobacco by aqueous two-phase extraction. *Biochem Eng J* 49: 343–350.

Roy, K., Lahiri, S. 2006. A green method for synthesis of radioactive gold nano particles. *Green Chem* 8: 1063–1066.

Sarubbo, L.A., Oliveira, L.A., Porto, A.L.F., Lima-Filho, J.L., Campos-Takaki, G.M., Tambourgi, E.B. 2003. Performance of a perforated rotating disc contactor in the continuous extraction of a protein using the PEG cashew-nut tree gum aqueous two-phase system. *Biochem Eng J* 16: 221–227.

Selber, K., Tjerneld, F., Collén, A., Hyytiä, T., Nakari-Setälä, T., Bailey, M., Fagerström, R., Kand, J., Van der Laan, J., Penttil, M., Kula, M.R. 2004. Large-scale separation and production of engineered proteins, designed for facilitated recovery in detergent-based aqueous two-phase extraction systems. *Process Biochem* 39: 889–896.

Shang, Q.K., Lia, W., Jiaa, Q., Li, D.Q. 2009. Partitioning of glycomacropeptide in aqueous two-phase systems in aqueous two-phase systems containing polyethylene glycol and phosphate buffer. *Fluid Phase Equilibr* 219: 195–203.

Shu, M., Jinpeng, C., Xuejun, C. 2010. Preparation of a novel thermo-sensitive copolymer forming recyclable aqueous two-phase systems and its application in bioconversion of penicillin G. *Separ Purif Technol* DOI: 10.1016/j.seppur.2010.08.003.

Sinha, J., Dey, P.K., Panda, T. 2000. Aqueous two-phase: The system of choice for extractive fermentation. *Appl Microb Biotechnol* 54: 476–486.

Spelzini, D., Farruggia, B., Picó, G. 2005. Features of the acid protease partition in aqueous two-phase systems of polyethylene glycol–phosphate: Chymosin and pepsin. *J Chromatogr B* 821: 60–66.

Srinivas, N.D., Barhate, R.S., Raghavarao, K.S.M.S. 2002. Aqueous two-phase extraction in combination with ultrafiltration for downstream processing of Ipomoea peroxidase. *J Food Eng* 54: 1–6.

Srinivas, N.D., Narayan, A.V., Raghavarao, K.S.M.S. 2002. Mass transfer in a spray column during two-phase extraction of horseradish peroxidase. *Process Biochem* 38: 387–391.

Srinivas, N.D., Rashmi, K.R., Raghavarao, K.S.M.S. 1999. Extraction and purification of a plant peroxidase by aqueous two-phase extraction coupled with gel filtration. *Process Biochem* 35: 43–48.

Su, Z.G., Feng, X.L. 1999. Process integration of cell disruption and aqueous two-phase extraction. *J Chem Technol Biotechnol* 74: 284–288.

Su, C.-K., Chiang, B.-H. 2006. Partitioning and purification of lysozyme from chicken egg white using aqueous two-phase system. *Process Biochem* 41: 257–263.

Tanuja, S., Srinivas, N.D., Gowthaman, M.K., Raghavarao, KSMS. 2000. Aqueous two-phase extraction coupled with ultrafiltration for purification of amyloglucosidase. *Bioprocess Eng* 23: 63–68.

Teotia, S., Gupta, M.N. 2004. Purification of phospholipase D by two-phase affinity extraction. *J Chromatogr A* 1025: 297–301.

Teotia, S., Lata, R., Gupta, M.N. 2004. Chitosan as a macroaffinity ligand purification of chitinases by affinity precipitation and aqueous two-phase extractions. *J Chromatogr A* 1052: 85–91.

Tianwei, T., Qing, H., Qiang, L. 2002. Purification of glycyrrhizin from *Glycyrrhiza uralensis Fisch* with ethanol/phosphate aqueous two phase system. *Biotechnol Lett* 24: 1417–1420.

Ülger, C., Sağlam, N. 2001. Partitioning of industrial cellulase in aqueous two-phase systems from *Trichoderma viride* QM9414. *Process Biochem* 36(11): 1075–1080.

Ulger, C., Cerakoglu, C. 2001. Alpha amylase production in aqueous two phase systems with *Bacillus amyloliquifaciens. World J Microb Biotechnol* 17: 553–555.

Vaidya, B.K., Suthar, H.K., Kasture, S., Nene, S. 2006. Purification of potato polyphenol oxidase (PPO) by partitioning in aqueous two-phase system. *Biochem Eng J* 28: 161–166.

Ventura, S.P.M., Neves, C.M.S.S., Freire, M.G., Marrucho, I.M., Oliveira, J., Coutinho, A.P. 2009. *J Phys Chem B* 113: 9304–9310.

Walter, H., Johansson, G. 1994. Aqueous two-phase systems, in *Methods in Enzymology*, Vol. 228, pp. xx–xx. San Diego: Academic Press.

Walter, H., Brooks, D.E., Fisher, D. 1985. *Partitioning in Aqueous Two Phase Systems*. New York: Academic Press.

Wang, W.-H., Kuboi, R., Komasaw, I. 1992. Aqueous two-phase extraction of dehydrogenases using triazine dyes in PEG/phosphate systems. *Chem Eng Sci* 47(1): 113–121.

Wei, D.-Z., Zhu, J.-H., and Cao, X.-J. 2002. Enzymatic synthesis of cephalexin in aqueous two-phase systems. *Biochem Eng J* 11: 95–99.

Wennersten, R., Tjerneld, F., Larsson, M., Mattiasson, B. 1983. In *Proc Int Solvent Extraction Conference*, 506. Denver, CO.

Wu, Y.T., Pereira, M., Venâncio, A., Teixeira, J. 2001. Separation of endo-polygalacturonase using aqueous two-phase partitioning. *J Chromatogr A* 929: 23–29.

Wu, Y.-T., Pereira, M., Venâncio, M., Teixeira, J. Recovery of endo-polygalacturonase using polyethylene glycol-salt aqueous two-phase extraction with polymer recycling 2000. *Bioseparation* 9: 247–254.

Wuenschell, G.E., Naranjo, E., Arnold, F.H. 1990. Aqueous two phase metal affinity extraction of heme proteins. *Bioprocess Eng* 5: 199–202.

Xing, J.M., Li, F.F. 2009. Purification of aloe polysaccharides by using aqueous two-phase extraction with desalination. *Nat Prod Res* 23: 1424–1430.

Xing, J.-M., Li, F.-F. 2009. Separation and purification of aloe polysaccharides by a combination of membrane ultrafiltration and aqueous two-phase extraction. *Appl Biochem Biotechnol* 158:11–19.

Xu, Y., Vitolo, M., Albuquerque, C.N., Pessoa Jr., A. 2002. Affinity partitioning of glucose-6-phosphate dehydrogenase and hexokinase in aqueous two-phase systems with free triazine dye ligands. *J Chromatogr B* 780: 53–60.

Yang, S., Huang, Z., Jiang, Z., Li, L. 2008. Partition and purification of a thermostable xylanase produced by *Paecilomyces thermophila* in solid-state fermentation using aqueous two-phase systems. *Process Biochem* 43: 56–61.

Yanga, L., Huoa, D., Houa, C., Hea, K., Lva, F., Fab, H., Luoa, X. 2010. Purification of plant-esterase in PEG1000/NaH2PO4 aqueous two-phase system by a two-step extraction. *Process Biochem* 45: 1664–1671.

Yue, H., Yuan, Q., Wang, W. 2007. Purification of phenylalanine ammonia-lyase in PEG1000/Na2SO4 aqueous two-phase system by a two-step extraction. *Biochem Eng J* 37: 231–237.

Zaslavsky, B.Y. 1995. Aqueous two phase partitioning: Physical chemistry and bioanalytical applications. New York: Marcel Dekker.

Zaslavsky, B.Y., Mesetechkina, N.M., Miheeva, L.M., Rogozhin, S.V. 1982. Measurement of relative hydrophobicity of amino-acid side chains by partition in an aqueous two phase system—Hydrophobicity scale for non-polar and ionogenic side-chains. *J Chromatogr A* 240: 21–28.

Zhang, Y., Shi, G., Zhao, F. 2010. Hydrolysis of casein catalyzed by papain in *n*-propanol/NaCl two-phase system. *Enzyme Microb Technol* 46(6): 438–443.

Zhi, W.B., Deng, Q.Y. 2006. Purification of salvianolic acid B from the crude extract of *Salvia miltiorrhiza* with hydrophilic organic/salt-containing aqueous two-phase system by counter-current chromatography. *J Chromatogr A* 1116: 149–152.

Zhi, W., Song, J., Ouyang, F., Bi, J. 2005. Application of response surface methodology to the modeling of alpha-amylase purification by aqueous two-phase systems. *J Biotechnol* 118: 157–165.

Zuniga, A.D.G., Coimbra, J.S.R., Minim, L.A., Roja, E.E.G. 2005. Axial mixing in a Graesser liquid–liquid contactor using aqueous two-phase systems. *Chem Eng Process* 44: 441–447.

Zuniga, A.D.G., Coimbra, J.S.R., Minim, L.A., Roja, E.E.G. 2006. Dispersed phase hold-up in a Graesser raining bucket contactor using aqueous two-phase systems. *J Food Eng* 72: 302–306.

16 Enzyme-Assisted Aqueous Extraction of Oilseeds

Stephanie Jung, Juliana Maria Leite Nobrega de Moura, Kerry Alan Campbell, and Lawrence A. Johnson

CONTENTS

16.1   INTRODUCTION

World vegetable oil consumption reached 129.5 million metric tons in 2009, and soybeans contributed 28% of the total production (Soystats Web site). The use of flammable hydrocarbon solvents, mainly hexane, is the current industrial process applied to oilseeds to satisfy the world demand for vegetable oil. The use of petroleum-derived hexane has been the most efficient procedure considering both extraction yield (>95% oil extracted from the seeds) and economic profitability. For the conventional soybean extraction process, pretreatment steps for optimal hexane extraction include cleaning, tempering, cracking, dehulling, conditioning, flaking, and sometimes expanding followed by solvent extraction, meal desolventizing and toasting, and miscella evaporation and oil stripping (Johnson 2008). The defatted meal (solid residue) is then cooled, ground, and used for animal feeding applications. The normal meal desolventizing/toasting procedure is replaced with flash desolventizing when making food-grade defatted soy flour (Johnson 2008). Worldwide, soybean meal (SBM) is the primary source of protein for poultry and livestock industries. In the United States, about 54% of all SBM is used to feed poultry, while 26% is consumed by swine (Stein et al. 2008).

These standard practices have several disadvantages including dependence on petroleum, high operating cost, and high capital investment (Owusu-Ansah 1997). Furthermore, consumer concern regarding the use of hexane has increased over the years, supported by data on residual hexane and studies reporting risks associated with exposure to n-hexane, which amounts to two thirds of the mixture of hexane isomers, and its potential neurotoxicity (Barregard et al. 1991; Armstrong 1995). In a recent report of the Cornucopia Institute, hexane residues were reported to be <10 ppm in soybean oil and range from 14 to 21 ppm in SBM. The U.S. Food and Drug Administration has not set a maximum residual level for hexane in food and potential harm of consuming hexane residues in edible oils and other processed foods that contain soy protein (infant formula, meat analogs) is unknown.

Some restrictive measures have been taken including the addition of n-hexane in the list of chemicals on the U.S. Toxic Release Inventory in 1994 and the U.S. Environmental Protection Agency issued regulations in 2001 to control hexane emissions because of its neurotoxic effects (Galvin 1997) and adverse environmental effects (Johnson 1997; Wakelyn 1997). In addition to n-hexane being classified as a hazardous air pollutant, there is evidence that n-hexane can react with nitrogen oxides to form ground level ozone. Hexane is also highly flammable, and a number of explosions in soybean crushing units have resulted in loss of life and injury to people. As a consequence, there has been increasing interest especially during the past decade in identifying environmentally friendly alternatives to hexane extraction. These alternatives include the use of alternative solvents such as acetone, ethanol, and isopropyl alcohol, alone or in combination with hexane (Hron 1982; Johnson and Lusas 1983; Hron 1997; Lusas and Hernandez 1997; Russin et al. 2011; Seth et al. 2007, 2010). Environmentally- and user-friendly alternatives, such as supercritical carbon dioxide (SC-CO_2), have also been investigated, achieving improved quality of the extracted oil (Russin et al. 2011; King 1997; Han et al. 2009).

Another alternative is to integrate enzymes into the extraction processes, either in combination with solvent or in an aqueous environment (enzyme-assisted aqueous extraction). The use of water as an extraction medium (without enzyme), known as aqueous extraction processing (AEP), has also been investigated for use with different oil-bearing materials (Rosenthal et al. 1996; Johnson 2008; Campbell et al. 2011). Adding enzymes during aqueous extraction usually increases oil extraction yields. The concept of aqueous extraction has evolved over the years from a marginal process with low oil extraction yields to a well-defined concept, achieving, in the case of soybean oil, extraction yields comparable to those obtained with flammable hydrocarbon solvents. For any alternative to current flammable hydrocarbon solvent extraction to be economically valuable, high oil extraction yields must be achieved as well as value needs to be added to all coproduct streams, particularly protein and insoluble carbohydrates. Rosenthal et al. (1996) reviewed the status of AEP and enzyme-assisted AEP (EAEP) of various oilseeds, and due to increasing interest in alternative technologies to solvent extraction, other recent review articles are now available (Campbell et al. 2011; Russin et al. 2011). The purpose of this chapter is to focus on the treatment of soybeans through enzyme-assisted aqueous processing (proteases or cellulases/pectinases) to identify pitfalls and to summarize strategies that can be undertaken to circumvent those pitfalls. The economics involved with the development of such a process will be discussed as well. Understanding oilseed structure is important in order to better understand the principles behind aqueous extraction, assisted or unassisted with enzymes. Articles focusing on plant structure or summarizing key plant structural compounds are widely available in the scientific literature, and therefore, this aspect will not be addressed in this chapter.

16.2 PRINCIPLES OF AEP

16.2.1 EXTRACTING THE OIL FROM THE MATRIX

As with all extraction processes, fundamental principles of mass transfer apply to aqueous extraction of oilseeds. However, specific phenomena occur that differentiate aqueous extraction from hydrophobic solvent extraction processes, such as hexane extraction and supercritical CO_2 extraction, the greatest difference being that aqueous extraction separates both water-soluble components (i.e., protein, sugars, etc.) as well as insoluble liquid components (i.e., oil) from the insoluble plant matrix. In an aqueous medium, oil droplets can be of the same length scale as the pores created by physical communition (i.e., flaking), and therefore, the geometry/configuration of the oilseed matrix is especially important. In general, extraction from plant tissues is a complex, multicomponent, multiphase process that incorporates several steps: (1) diffusion of solvent into the plant matrix; (2) solubilization of cellular components; (3) transport of the solutes to the exterior of the solid matrix; and (4) transport of the solutes from the matrix surface to the bulk medium (Aguilera and Stanley 1999).

In AEP and EAEP of oilseeds, the relative rates of each of these processes and, therefore, the rate limiting step, have not been well characterized, but we may speculate based on diffusion principles. Water, being the smallest molecule in the AEP system, diffuses relatively quickly (process 1) compared with the larger protein and sugar molecules of oilseed cells. Likewise, in a well-agitated environment often used in aqueous extraction, the diffusion from the matrix surface into the bulk (process 4) would also occur relatively quickly. Therefore, rate-limiting steps for mass transfer of the soluble components are likely solubilization of cellular components (protein and oligosaccharides) and their diffusion to the surface of the plant matrix (processes 2 and 3).

Based on diffusion rates, however, removal of the soluble components is not likely to be a rate-limiting step for the removal of oil. For oil, matrix diffusion rates would be much slower than solutes, not only because the oil droplets are much larger than solute molecules (extracted droplet sizes range from 0.3 to 100 µm) (Chabrand et al. 2008) but also because droplets are of the same length scale as solid matrix pores, which can range from <1 to >55 µm, depending on the extent of the cell wall rupture and the mode of comminution used. For soybean flour, diffusivity estimates of soybean oil bodies using the Stokes–Einstein equation were consistent with mass transfer rates observed at short extraction times (<5 min) (Campbell and Glatz 2009a). As with small solutes, agitation may increase the diffusion rates of oil droplets not only by increasing the rate of mixing, but agitation can also decrease droplet diameter, increasing mobility (and yield) as well as diffusivity (Campbell and Glatz 2009a). Since the starting material is dispersed into water, a minimum agitation speed is required to achieve adequate exposure of the matrix to the extraction medium. High agitation speeds may lead to increased oil extraction if some additional disruption of the cytoplasmic cells occurs. Excessive agitation, however, might affect oil droplet size distribution, which could increase the stability of the extracted oil toward creaming, thereby making it difficult to separate the extracted oil from the aqueous fraction (Rosenthal et al. 1998).

Cell wall disruption and particle size reduction are critical steps in oil and protein extraction efficiencies in aqueous processing for reasons discussed above. Increments in oil and protein extraction are thought to be due to increased cell wall disruption, with increased surface area-to-volume ratio decreasing oil droplet diffusion path lengths. Campbell and Glatz (2009a) proposed an extraction model based in part on soy flour particle size where oil is extracted from disrupted cells on the surface of the flour particle, in support of an earlier model proposed by Rosenthal et al. (1998). When processing soybeans, size reduction has been achieved by using mechanical treatments such as grinding, flaking, extruding, and combinations of these processes. The choice of the size reduction process should consider the mechanical properties of the material (brittle, soft, or strong), extent of size reduction (large, medium, or fine), and energy efficiency of the process (Chen and Rosenthal 2009) with specific needs depending on product application.

The relationships between diffusion rate, pore size, and tortuosity emphasize the importance of matrix geometry, and hence the mode of comminution used to disrupt cells. Three different modes of comminution (extrusion, grinding to flour, flaking) yielded very different geometries. Images of extruded soybean flakes reveal complete destruction of the cellular matrix (Campbell and Glatz 2009a; Jung et al. 2009a), but extrusion transforms the cotyledon from a matrix of disrupted cells to a matrix of insoluble protein cross-linked by noncovalent and disulfide bonds, which render roughly 30% of the oil unextractable (Lamsal et al. 2006; Jung et al. 2009a). Proteases have been successfully used to achieve >98% yield from extruded soy flakes (de Moura and Johnson 2009; de Moura et al. 2009) by disrupting this network (Campbell and Glatz 2009a; Jung et al. 2009a).

Extrusion combines unit operations of mixing, cooking, shaping, and forming to produce a wide range of products. Typically, raw materials are fed into the extruder barrel and the screw(s) convey the food along it. Further down the barrel, the volume is restricted and the material is compressed. The screw then kneads the material under pressure into a semisolid, plasticized mass (Fellows 2009). Extruding full-fat soybean flakes before aqueous extraction improved oil extraction yield from 46 to 71% (Lamsal et al. 2006).

Grinding soybeans into fine full-fat flour can achieve up to 95% cell disruption of cotyledon cells (Campbell and Glatz 2009a). Flaking can also achieve high degrees of cellular disruption, but microscopic observation indicates that pore length and tortuosity are greater in flakes than in flours, coinciding with lower oil extraction yields than observed in flours (Jung 2009). Rosenthal et al. (1998) reported that oil extraction increased from approximately 23 to 64%, while protein extraction increased from 44 to 75% when reducing flour particle size from 900 to 100 µm, and AEP was carried out at pH 8.0, 1:10 SLR (solids-to-liquid ratio), 50°C, and 200 rpm for 1 h. Cell wall rupture is necessary for any appreciable mass transfer of soluble or insoluble species to occur. Although increased extraction yields are related to particle size reduction, excessive size reduction (i.e., very fine grinding) may produce small oil globules, which favors stable emulsion formation (Lamsal et al. 2006).

Factors other than just opening the matrix through communion significantly impact oil extraction yields, and among them, pH, SLR, and temperature are key parameters that impact mobility of oil in the insoluble matrix and creaming rates. Soybean oil and protein extraction yields are strongly dependent on pH. At pH near the isoelectric point of soy proteins (4.5), soy protein solubility is <10%, which also coincides with minimal oil extraction (Rosenthal et al. 1998). Insoluble proteins inside disrupted cells likely impede oil diffusion, but they can also adsorb and entrap oil that has escaped the cellular matrix. Temperature and SLR ratio can affect oil droplet diffusivity by thermokinetic mechanisms as well as by altering fluid viscosity. For flakes and flours, unextracted oil was in the form of coalesced oil, indicating that faster diffusion favors high yields by minimizing droplet coalesce prior to oil release (Campbell and Glatz 2009a). Generally, higher temperatures are associated with enhanced extraction. Mass transfer rate is favored by increased solute solubility and diffusion into the bulk solvent; however, degradation of thermolabile components (i.e., proteins) should be considered when working at high temperatures (Takeuchi et al. 2009). Rosenthal et al. (1998) observed a slight decrease in oil and protein extraction yields when increasing extraction temperature up to 50°C, after which a sharper decrease was observed when increasing temperatures up to 70°C. Lusas et al. (1982) reported maximum oil recovery from soybean flour at a temperature range of 40 to 60°C, with temperature being critical to protein extraction, but not significant to oil extraction.

Thermal denaturation reduces protein solubility (Kinsella 1979), which may affect both oil and protein extractability, depending on the extent of protein denaturation. Protein/lipid interaction seems to be favored by heat treatment, where denatured proteins seem to sequester oil before its release (Lamsal et al. 2006).

Low SLR has been associated with higher extraction yields in AEP of soybeans. Rosenthal et al. (1998) reported that protein extraction decreased from approximately 87 to 77% when increasing SLR ratio from 1:25 to 1:6; however, oil extraction (~75%)

did not vary significantly when using soybean flour. When using extruded full-fat soybean flakes for enzyme-assisted aqueous extraction, protein extraction decreased from 88 to 80% and oil extraction decreased from 95 to 90% when increasing SLR from 1:10 to 1:5 (de Moura and Johnson 2009).

16.2.2 Recovering the Oil from the Aqueous Extract

Once released from the insoluble matrix, oil must next be separated from the continuous (aqueous) phase, which is generally accomplished by centrifugation resulting in an oil-rich oil-in-water emulsion (cream) and an oil-lean, protein-rich, oil-in-water emulsion (skim). Centrifugal separation at speeds normally encountered in aqueous extraction (3000–5000 × g) provides incomplete separation, and between 10 to 20% of the total oil remains emulsified in the skim complicating downstream protein purification and limiting applications of any resulting protein concentrates and isolates. The factors preventing the oil in the skim from creaming are not well understood. Considering the small droplet sizes (<0.5 µm) that are sometimes observed, it is likely that the smallest droplets are kinetically stable (i.e., have very slow creaming velocity at economically feasible centrifugation speeds). In addition, droplets are likely covered with adsorbed proteins and phospholipids, which provide both steric and electrostatic repulsion, contributing to stability toward creaming. Because the emulsified oil has adsorbed proteins at the oil–water interface, the degree of protein adsorption can affect droplet density, especially for small droplet sizes. Specific protein coverage has been measured to be in the range of 10 to 15 mg/m^2 (Iwanaga et al. 2007; Chabrand et al. 2008).

16.3 EXTRACTABILITY WITH ENZYME-ASSISTED AEP

Oil and protein aqueous extractions from soybeans have been significantly improved by using enzymes having either proteolytic or cellulosic/pectic activities (Lamsal et al. 2006; Rosenthal et al. 2001; de Moura et al. 2008; Kapchie et al. 2008). The probable mechanism of protease action for extruded soybeans is to disrupt proteins which releases oil. In flour, protein hydrolysis changed the interfacial conditions of coalesced oil in disrupted cells, favoring droplet break-up due to turbulent forces (Campbell and Glatz 2009a). Proteases also hydrolyze, partially or totally, the membrane of the oil body (i.e., oleosome) as the oleosins (major proteins located at the oil body membrane) are likely to be hydrolyzed. In this case, free oil instead of intact oil bodies is released. The purpose of using cellulases/pectinases in the process is to free the oil bodies from the cotyledon cells by disrupting cell walls, leaving oil bodies intact. To differentiate the two processes, the abbreviations CAEP and PAEP will be used for cellulase/pectinase- and protease-assisted AEP, respectively.

16.3.1 Protease-Assisted AEP

16.3.1.1 Extraction Principles of PAEP

Enzymes are proteins that exhibit catalytic action and selectivity toward substrates (Parkin 1993). Considering that proteins are linear polymers of amino acids and that amino acids differ from each other with respect to only the side chain, differences in the occurrence and distribution of the side chains account for the enormous differences in the biological properties of enzymes. Proteases can be classified according to their source (animal, plant, microbial), their catalytic action (endopeptidase or exopeptidase), and the nature of their catalytic sites. Endopeptidases cleave the polypeptide chain at particular susceptible peptide bonds distributed along the chain, whereas exopeptidases hydrolyze one amino acid at a time from either the N-terminus (aminopeptidases) or the C-terminus (carboxypeptidases) (Adler-Nissen 1993). To enable reaction to occur, enzyme active sites must interact with substrate via covalent and/or noncovalent interactions and stabilize the activated species. Both endoproteases and exoproteases have been used for aqueous extraction of oilseeds, and the enzyme preparations were often from *Bacillus* or *Aspergillus* species and contained both activities, but usually one predominates (Russin et al. 2011).

When extracting ground material, where cellular disruption is incomplete and little heat denaturation has occurred, the mechanisms through which proteases increase oil extractability are not always clear. One hypothesis has been that hydrolysis of proteins inside the cytoplasm makes the inner structure less compact, thereby favoring removal protein and oil from the cells (Rosenthal et al. 1996). Alternatively, protein hydrolysates may act as emulsifiers, increasing yield by decreasing average oil droplet size and, therefore, increasing droplet mobility (Campbell and Glatz 2009a). The efficacy of protease to extract oil from oilseeds is highly dependent on the parameters described above, such as extent of cell disruption, pH, and temperature.

The effects of proteases are most pronounced on substrates that have undergone heat abuse causing extensive protein denaturation. For example, when using heat-treated soybean flour (heated at 121°C for 8 min) with Alcalase 2.4L (pH 8.0, 15 min, 2000 rpm, 50°C), extraction yields increased from 28 to 66% for protein and from 42 to 59% for oil. No significant improvement upon protease addition on extraction yields, however, was observed when the soybean flour was not submitted to thermal treatment (Rosenthal et al. 2001). Variability in enzyme performance over heated or unheated substrates has been reported, which could be related not only to the thermal treatment itself but also to specific enzyme aggressiveness. Extruding soybean flakes clearly improved proteolytic enzyme action over the exposed substrate compared with soybean flakes, significantly improving oil extraction yields from 56 to 88% (Lamsal et al. 2006). Enhanced enzyme performance seems to be a consequence of protein denaturation, which increases susceptibility of proteins to enzyme attack. Thermoplastic extrusion enhances the accessibility of enzymes on cell components by disrupting cells, thereby increasing the surface area and protein denaturation and increasing susceptibility to proteolysis.

In addition to substrate characteristics, optimum extraction conditions, such as enzyme concentration, temperature, pH, or optimum SLR ratio, need to be determined. The impact of optimum pH for best enzyme activity and its effects on extraction yields from extruded full-fat soybean flakes is illustrated by comparing the oil extraction yields of 90–93% obtained with alkaline proteases (Protex 6L, Protex 7L, and Protex 51FP) versus 71% for acidic proteases (Protex 50FP), indicating that low protein solubility at acid pH was a barrier to extracting oil from the matrix (Wu et al. 2009). Among alkaline proteases, variations in oil and protein extractabilities also exist. For example, oil and protein extractions of 88 and 77%, respectively, were achieved when using 0.5% Multifect Neutral to extract extruded soybean flakes at 1:10 SLR, pH 7.0 for 1 h, and pH 8.0 for 15 min at 50°C (Lamsal et al. 2006). The use of a different protease (0.5 % Protex 6L) and slight modifications of extraction parameters (1:10 SLR, pH 9.0, 50°C for 1 h) improved oil and protein extraction yields from extruded full-fat soybean flakes to 96 and 85%, respectively (de Moura et al. 2008). This difference was attributed to the more aggressive nature of Protex 6L, which achieved a greater extent of hydrolysis. Increasing the enzyme concentration beyond 1.0% did not significantly improve extraction yields, indicating a limit to the effects of proteolysis in increasing oil extraction yield.

16.3.1.2 Single-Stage versus Multistage PAEP Extraction

For soybean aqueous extraction, most studies were performed by using single-stage extraction (Rosenthal et al. 1996, 1998; Lamsal et al. 2006; de Moura et al. 2008), but there are many advantages in using a countercurrent multistage extraction process (de Moura and Johnson 2009; de Moura et al. 2009). Schematic diagrams for single-stage and countercurrent multistage extraction are presented in Figure 16.1.

In single-stage extraction, solutes and extraction solvents flow in the same direction. They are fed to the extractor, mixed, allowed to react, and then separated into two outflows: the extract (which constitutes a large amount of solvent and dissolved solutes) and the residue (containing the insoluble solids with entrained solution) (Wankat 2007; Takeuchi et al. 2009). Single-stage extractions are disadvantaged due to the considerable use of water, thereby producing a significant volume of dilute effluent that has to be processed for product recovery. In countercurrent systems, solutes and extraction solvents flow in opposite directions, allowing reduced solvent use (Wankat 2007).

The countercurrent process is characterized by the enrichment of the extract solution. Both the entrance of the feed and the exit of the final extract take place in the first extraction stage, and both the entrance of fresh solvent and exit of final residue take place in the last extraction stage. The extracted slurry obtained in one stage is used as the extraction solvent in the next stage (Takeuchi et al. 2009). The number of stages required to achieve the desired extraction must be determined for each specific substrate and extraction conditions used. The key principle of the countercurrent system for soybean oil extraction is that the freshest (most oil-rich) flakes are contacted with the richest extraction media, where flakes progress through the process until near-depleted flakes contact fresh (oil lean) extraction media (Johnson 2000).

The development of a multistage countercurrent extraction process for soybeans proved to be successful, not only from the perspective of reducing water use but also maintaining high extraction yields compared with single-stage extraction (de Moura and Johnson 2009). Countercurrent two-stage PAEP achieved greater oil, protein, and solids extraction with approximately one-half of the normal water used in single-stage PAEP. Up to 98% of the oil, 92% of the protein, and 80% of the solids were extracted by countercurrent two-stage PAEP, whereas 96% oil, 87% protein, and 77% solids extraction was achieved by the single-stage PAEP of soybeans. The 40% water reduction achieved by countercurrent two-stage PAEP represents an important energy savings in the recovery of protein and carbohydrates present in the dilute skim fraction.

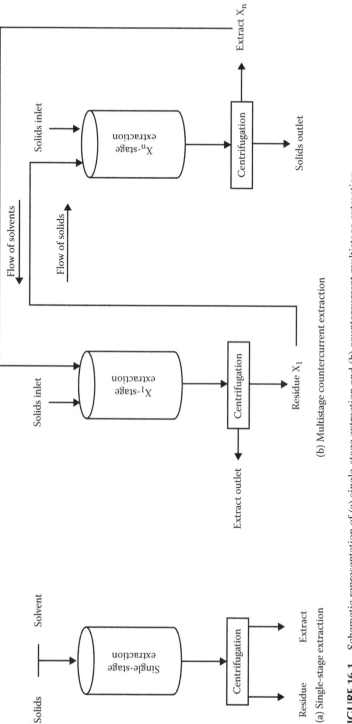

FIGURE 16.1   Schematic representation of (a) single-stage extraction and (b) countercurrent multistage extraction.

When scaling up countercurrent two-stage PAEP of soybeans from bench scale (80 g of extruded flakes, 1:6 SLR) to laboratory process simulation (2 kg of extruded flakes, 1:6 SLR), 99% oil, 94% protein, and 83% solids extraction were achieved; however, the distribution of the extracted oil among the liquid fractions (cream/free oil and skim) was altered. Modifying processing parameters improved the oil distribution among the fractions, increasing oil yield in the cream fraction (76 to 86%) and reducing oil yield in the skim fraction (23 to 12%) (de Moura et al. 2009). Low oil content in the skim fraction is critical to PAEP of soybeans since recovering the oil from this fraction as free oil remains a challenge. This important consideration is addressed in the section below on oil recovery.

16.3.1.3 Integrated Extraction and Demulsification Concept

The first attempt to integrate extraction and cream demulsification steps by recycling the enzyme from the demulsification into the extraction step in the PAEP of soybeans was recently reported by de Moura et al. (2011a). This recycling step reduces the total amount of enzyme used (extraction + cream demulsification) in PAEP by 25%, resulting in substantial economic savings. Countercurrent two-stage extraction and cream demulsification were fully integrated and demonstrated at laboratory scale (2 kg extruded full-fat soybean flakes) wherein the protease used for demulsifying the cream was recycled into the extraction stages. Oil, protein, and solids extraction yields of 96, 89, and 81%, respectively, were achieved using this integrated process.

Higher degrees of protein hydrolysis (DH) were obtained when using the integrated process compared to when the enzyme was not recycling (DH of 7.2 ± 1.2 and 16.4 ± 2.0% for the first and second extraction stages, respectively, versus 6.4 ± 0.1 and 10.1 ± 1.3%), which likely increased the emulsion stability toward creaming, thereby affecting oil distribution among the fractions as well as separation of extracted components. When separating the liquid fractions (skim and cream) obtained by the integrated process, more oil was partitioned into the skim fraction, thereby affecting free oil recovery from the cream fraction. Emulsification of oil by protein can be improved by increasing exposure of hydrophobic sites of proteins, which depends on both DH and the state of protein (native or denatured).

Jung et al. (2005) reported that low levels of hydrolysis of denatured soy protein increased surface hydrophobicity and emulsification capacities, whereas the same level of hydrolysis for less denatured protein decreased emulsification capacities. Qi et al. (1997) reported increased protein and oil absorption in emulsions prepared from soy protein isolate modified with pancreatin when increasing DH from 7 to 15%; however, they decreased when increasing DH of hydrolysis from 15 to 17%. In the integrated process (de Moura et al. 2011a), the amount of enzyme needed to break the cream emulsion is higher than the amount needed for the extraction step, and use of excessive amounts of enzyme in the extraction step can affect mechanisms involved in emulsion formation thereby imparting partitioning and stability of the formed emulsion. The integrated countercurrent two-stage PAEP of soybeans was successfully scaled-up in the pilot plant using 75 kg of extruded flakes and 480 kg of slurry (1:6 SLR) (de Moura et al. 2011b). Nearly 35% enzyme reduction was achieved through the recycling procedure at pilot-plant scale. Oil and protein extractions of 98% each were achieved and represent a significant achievement important to the feasibility each of industrial adoption of the PAEP concept. Flow diagrams of the integrated extraction and demulsification process are provided in Figures 16.2 and 16.3.

16.3.2 CELLULASE/PECTINASE-ASSISTED AEP

Work related to the extraction of oil from oilseeds with a protease-free process can be classified into two groups depending on the major objective of the process. If extracting intact oil bodies is the primary objective, oil extraction yields are usually not optimized and therefore are rather low. Extraction of oleosomes by conventional methods (hydrating intact soybeans, grinding, filtering, and centrifuging) results in <45% oil extraction yields, mainly due to incomplete cell wall rupture structure without disrupting oleosome membrane integrity (Kapchie et al. 2008). Oil yields of 36% were obtained by Iwanaga et al. (2007) when using an aqueous–flotation–centrifugation method, with the main focus being the characterization of the extracted oil bodies and not extraction yield as would be needed for a commercial process to recover oil. The use of cellulases/pectinases can potentially increase extraction yield and also allow simultaneous extraction of undenatured proteins, which could be an important advantage over the use of protease where peptides are formed.

16.3.2.1 Extraction Principles of CAEP

The use of cellulases, hemicellulases, and pectinases in EAEP of soybeans has been reported (Freitas et al. 1997; Rosenthal et al. 2001; Lamsal et al. 2006; Kapchie et al. 2008). Cellulases are carbohydrases that cleave the β-1,4 linkages of cellulose or its chemically modified forms, in addition to degrading cellodextrin or cellobiose. Generally, they are multienzyme complexes bearing endo-1,4-β-glucanase, cellobiohydrolase, and β-glucosidase activity. Hemicellulases are alkali-soluble polysaccharides of cellulosic or pectic substances found in plant cell walls and, therefore, able to break down these polysaccharides. Pectinases break down complex polysaccharides of plant tissues into simpler molecules like galacturonic acids. Kasai et al. (2004) reported that cellulase treatment of soybean fiber residue was effective in digesting the primary cell wall, whereas pectinase treatment was effective in digesting the secondary cell wall. In general, *Aspergillus niger* and *Trichoderma viride* are the main sources of cellulases and pectinases destined for food and nonfood applications, respectively, while fungi are the most common industrial source for hemicellulases (Bigelis 1993). For aqueous processing of oilseeds, the role of these hydrolytic enzymes is to break down the structure of the cotyledons cell walls, making the structure more permeable (Rosenthal et al. 1996).

16.3.2.2 Extraction Yields with CAEP

As with protease, the efficiency of cellulases/pectinases in improving extraction yields is closely related to the characteristics of the starting materials. Cellulase, hemicellulase, and pectinase do not improve oil and protein extraction yields from soybean flour, flakes, or extruded flakes (Rosenthal et al. 2001; Lamsal et al. 2006);

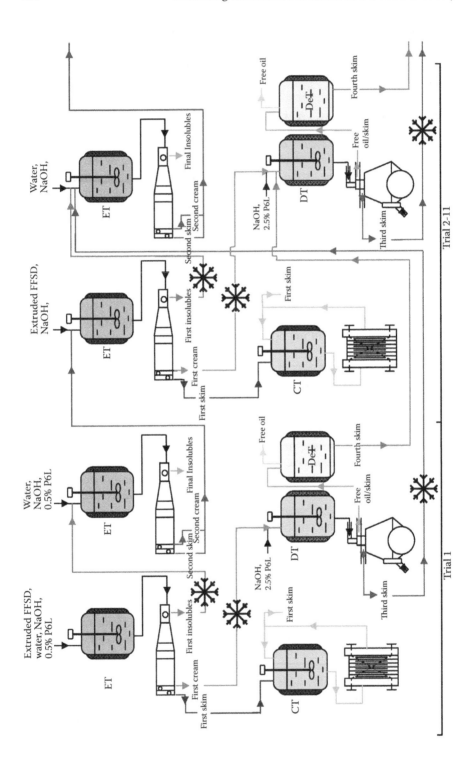

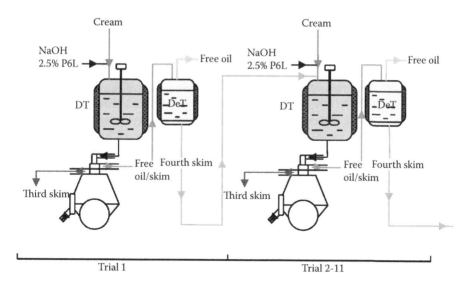

FIGURE 16.3 Flowchart for cream demulsification. DT: demulsification tank (65°C, 1.5 h, pH 9); DeT: decantation tank (4°C, 16 h). (From de Moura, J.M.L.N. et al., *J Am Oil Chem Soc,* 2011b. With permission.)

however, increased oil extraction yields from 35 to 60% was reported by Freitas et al. (1997) when using cellulases with extruded soybean flakes. The effects of pretreatments, such as high hydrostatic pressure and ultrasonication, prior to soybeans incubation were evaluated by Kapchie et al. (2008). Treatment of the slurry at 500 MPa pressure decreased oleosome extraction, which was likely related to reduced protein solubility due to denaturation/precipitation under pressure. Ultrasonication prior to enzyme treatment gave high oil recovery in the oleosome fraction (~80% with use of 3% of enzymes) when applied for 3 min. Oil recovery in the oleosome fraction as high as 85% (without pretreatment) was achieved using 3% (volume enzyme/weight of beans) of an enzyme cocktail (mixture of Multifect Pectinase FE: pectinase complex with cellulase, hemicellulase, and arabinose side activities; cellulase A: cellulase, β-glucanase hemicellulase, and xylanase activities; Multifect CX 13L: with significant activity toward cellulose, hemicelluloses, β-glucans, and arabinoxylans), in a four-stage extraction process (one main extraction with three successive extractions of the residue) (Kapchie et al. 2008). In this study, oil extraction yield of intact oleosomes of 63% was achieved in the first extraction, with the remaining 22% being achieved with three successive extractions.

FIGURE 16.2 Flowchart of the steps involved in the integrated protease-assisted aqueous extraction of extruded full-fat soybean flakes. ET: extraction tank (50°C, 1 h, pH 9); CT: cooling tank (10°C); DT: demulsification tank (65°C, 1.5 h, pH 9); DeT: decantation tank (4°C, 16 h); refrigerated storage (10°C, 16 h). (From de Moura, J.M.L.N. et al., *J Am Oil Chem Soc,* 2011b. With permission.)

A simplified procedure to extract intact oleosomes, in which less material and time are required, was recently developed (Kapchie et al. 2010a). Oil recoveries from oleosomes of 77.4 vs 60.7% were achieved when using the simplified procedure vs. the one previously described (Kapchie et al. 2008), respectively. In their more recent study (2010a) the authors used a different combination of enzymes (Multifect Pectinase FE, Multifect CX B – with cellulase complex standardized on β-glucanase, and Multifect GC – cellulase complex with hemicellulase, xylanase and glucanase side activities, at 1% each). The lower oil yield (60.7%) reported by Kapchie et al. (2010a) compared with 85% previously reported (Kapchie et al. 2008) was likely a consequence of differences in the enzyme mixture composition used in both studies. The much higher oleosome yield (85%) from their previous work was attributed to the combination of enzymes used, Multifect Pectinase FE, Cellulase A and Multifect CX 13L at 1% each which was very effective in digesting the soybean cell wall.

When using mixture of Multifect Pectinase FE, Multifect CX B, and Multifect GC (Kapchie et al. 2010a), higher yields in the modified procedure compared to the initial procedure (77.4 vs 60.7%) are likely a consequence of pre-treating soybean flour with an osmotic solution (0.4 M sucrose and 0.5 M NaCl) at 57°C for 16 h. Structural changes in plant tissues (contraction of cellular membranes, degradation of cell walls, and decrease of intercellular contact area) have been observed when contacting tissues with high osmotic solutions (Muntada et al. 1998), which could lead to cell disruption and higher surface area for enzyme action, both of which would improve extraction yields.

Towa et al. (2010a) scaled-up the simplified CAEP oleosome isolation procedure to pilot-plant scale (75 kg of soybean flour) and reported 93% oil recovery from oleosomes. In order to reduce the amount of aqueous sucrose- and sodium chloride-rich supernatant generated by CAEP of oleosomes, recycling the supernatant was evaluated (Kapchie et al. 2010b). The procedure comprised a complete primary extraction (four extraction stages) with recycling of the supernatant produced in further extractions with fresh full-fat flour, with or without adding enzyme. Recoveries of 82% oil and 36% protein in the oleosome fraction, and 2.5% oil and 52% protein in the supernatant fraction (skim) were achieved in the primary extraction. Recycling the supernatant (skim) to extract fresh soybean flour without adding enzyme (but containing residual enzyme active from primary extraction) for the first and second time yielded 73 and 70% oil recoveries, respectively. Protein recovery in the supernatant for both first and second recycling was approximately 60 and 50%, respectively, not being improved by the recycling procedure. Adding fresh enzyme in the supernatant from the primary extraction before the first recycling improved oil extraction from oleosomes (81 vs. 73%); however, protein recovery in the supernatant was reduced from ~ 60 to 54%.

16.3.3 Protease- versus Cellulase/Pectinase-Assisted AEP

Comparison of the current best oil extraction practices using PAEP and CAEP obtained with soybean-based starting material at a large scale (pilot plant) is shown in Figure 16.4.

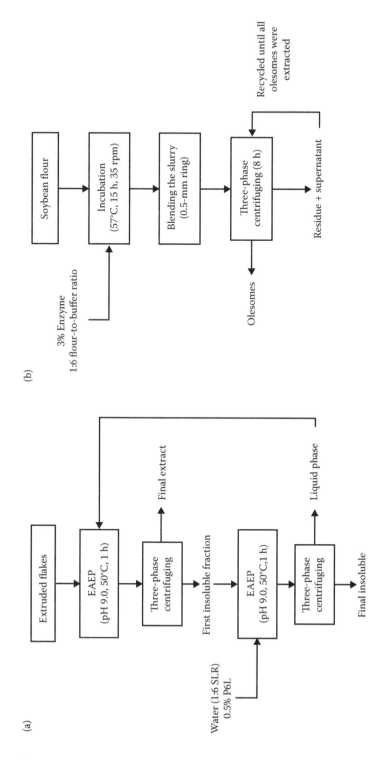

FIGURE 16.4 Best extraction practices for (a) protease and (b) cellulase/pectinase strategies in the EAEP of soybeans.

The use of protease (PAEP) (Figure 16.4a) in a countercurrent two-stage process enabled 98% of the oil and 96.5% of the protein to be extracted from extruded soybean flakes with the skim fraction mainly composed of hydrolyzed peptides with MWs < 25 kDa (de Moura et al. 2011b). With CAEP (Figure 16.4b), 93% of the oil was recovered from the starting material and 40–45% of the protein was recovered in the skim fraction. The extraction procedure described in Figure 16.4b is a simplification of the procedure described by Towa et al. (2010a), which was recently modified by Kapchie et al. (2011a).

In addition to economic considerations, the main advantages of using a proteolytic treatment are slightly higher oil yield (96 vs. 93%), much higher protein extractability (96 vs. 40–45%), and reduced reaction times when compared with the use of cellulases/pectinases. Although higher oil extraction was obtained with PAEP, the final amount of free oil recovered was similar for both PAEP and CAEP (~80%) because of the partitioning of more oil in the skim fraction (where it is difficult to be recovered) in PAEP versus CAEP. Reducing incubation and/or extraction times are essential to reducing microbial risks and costs in EAEP. Although longer incubation times were obtained when using cellulases/pectinases, the possibility of extracting intact oil bodies and minimally hydrolyzed proteins (due to presence of some protease side activity in the cellulase enzyme preparation) enables the utilization of these fractions in many different applications.

16.3.4 OTHER OIL-RICH AGRICULTURAL MATERIALS

In addition to soybeans, the use of water, with or without enzyme, has been used to extract edible oil from many other oil bearing materials including seeds, fruits, and other oil-rich plant materials and was recently reviewed by Moreau (2011). In the present section, only a few examples were selected to illustrate the extent to which the EAEP concept has been successfully applied to other oil-bearing materials in addition to soybeans.

For *Jatropha* kernels, oil and protein extraction yields up to 89 and 90%, respectively, were reported when performing extractions at 1:200 (enzyme/kernel, w/w, enzyme not identified), pH 7.5, 45°C, 2 h, 500 rpm shaking, and 1:8 SLR (Latif et al. 2010). In relation to the extracted oil, 73.8% free oil recovery was achieved, while the remaining 15.6% was unrecovered in the cream and skim fraction (aqueous phase). The effects of different enzymes (Protex 7L, Multifect Pectinase FE, Multifect CX 13L, Natuzyme) were evaluated in EAEP of canola (Latif et al. 2008). In this study, the use of enzymes improved oil extractability from 16 (control) to 22–26%, but extraction yields were less than those achieved by solvent extraction (43%). The use of Protex 7L yielded higher protein content in the aqueous phase compared with the control (5.9 vs. 4.4%).

For rapeseed, a different procedure was used in conjunction of PAEP (Zhang et al. 2007). In this study, the rapeseed slurry was initially treated with a mixture of carbohydrases followed by sequential alkaline extraction and protein hydrolysis. Oil and protein extractions of 73–76% and 80–83%, respectively, were achieved when treating the slurry with a 2.5% enzyme cocktail composed of pectinase, cellulase, and β-glucanase (4:1:1, v/v/v, 4 h, pH 5.0, 48°C); performing an alkaline extraction at pH 10 for 10 min and using 1.25–1.5% of Alcalase 2.4L for 150–180 min, 60°C, and pH 9 (Zhang et al. 2007).

Using 0.5% Protex 6L (g of enzyme/ g of extruded flakes) significantly improved oil (63 vs. 81%) and protein (61 vs. 77%) extraction yields when using extruded lupin flakes at 1:10 SLR, 600 rpm, 1 h, and pH 9.0 (Jung 2009).

Various commercial enzymes (cellulases, xylanases, pectinases, and proteases) have been evaluated in the extraction of oil from corn germ from a commercial corn wet mill (Moreau et al. 2004). Oven-dried corn germ from a commercial wet-milling process was extracted by churning with enzymes and buffer for 4 h at 50°C, and an additional 16 h at 65°C, leading to 91 and 93% extraction yields when using cellulases Multifect GC and Celluclast, respectively. A new enzyme-assisted extraction method was developed to extract corn oil from corn germ produced by dry milling corn or from a newly developed enzymatic wet-milling process (E-germ). The advantage of this new process is that no cooking or drying of the germ is necessary, thereby saving energy costs. In this new two-step process, combining both acidic cellulose and alkaline protease treatments achieved 50–65% and 80–90% oil extraction yields from dry-milled corn germ and E-germ, respectively, illustrating the importance of the procedure used to obtain the corn germ on oil extraction yields (Moreau et al., 2009). Treating condensed corn distillers solubles (CCDS), which contain up to 20% oil, with an enzyme cocktail (cellulase and protease) only slightly increased overall oil recovery compared to untreated material (80.7 vs. 78.4%) (Majoni et al. 2011).

Different enzymes (Protex 7L, Alcalase 2.4L, Viscozyme L, Natuzyme, and Kemzyme) were evaluated in EAEP from ground sunflower seed (Latif and Anwar 2009). Highest oil extraction yield of 39.7% was obtained when using Viscozyme L, whereas the lowest oil extraction yield was observed when using Alcalase 2.4L.

Different proteases (Protizyme, AS1398; Nutrase, Alcalase 2.4L, and Protamex) were evaluated individually or in combination with cellulases for enhancing peanut oil extraction (Jiang et al. 2010). The best oil (92%) and protein (88%) extraction yields were achieved by using two-stage extraction (first extraction: pH 9.5, 60°C, 1:5 SLR, alkaline extraction 1.5 h, 1.5% Alcalase 2.4L (w/w), 5 h hydrolysis time; second extraction: emulsion and residue were hydrolyzed with 1% AS1398 for 2 h).

A combination of protease, α-amylase and cellulase was also efficient in aqueous extraction of rice bran oil with 77% of the oil being recovered but required 17 h of processing time (Sharma et al. 2001).

Proteases have been assessed in the oil recovery from Kalahari melon seed (Nyam et al. 2009). Oil recovery of 69% was achieved when using Neutrase 0.8 L at 25 g/kg, pH 7, 58°C, and 31 h incubation, whereas 72% oil recovery was achieved when using Flavourzyme 1000 L at 21 g/kg, pH 6, 50°C, and 36 h incubation. EAEP has also been applied to fruit pulps such as coconuts (Chen and Diosady 2003). Approximately 84% oil and 44% of protein were extracted from desiccated coconut meat by the use of a 2% Gamanase™ 1.0L (w/w, endohemicellulase). Extractions were performed at 50–55°C and pH 4.5 for 5 h under gentle stirring, followed by a 15-h settling period without stirring and centrifuging at 9000 × g for 25 min.

Although EAEP has been applied to many different oil-bearing materials, the technology has achieved the highest degree of development with soybeans, in part due to its high protein content, which helps to maximize economic feasibility of the process, and also offers the opportunity to produce organic soy protein ingredients. Other rationale for the focus on soybeans includes the fact that soybeans are the major oil-bearing seed grown in the United States.

16.4   RECOVERY OF FREE OIL FROM EAEP

When alternatives to extraction with hydrocarbon solvents are sought, the main focus is usually on oil extraction yield (i.e., how much oil is removed from the starting material). However, the form in which this oil is recovered is as crucial as the extraction yield. Because of the different mechanistic principles of aqueous extraction versus hydrocarbon solvent extraction, in most of the cases and particularly with soybeans, only a very small percentage, if any, of the extracted oil is recovered as free oil. Depending on the form of the extracted oil, three categories can be made: (1) free oil that floats without need of any additional treatment, (2) intact oil bodies (oleosomes), or (3) oleosome-based emulsion (which includes what we call the cream fraction). This last category relates to the oil-rich extract containing all oleosome components but which did not maintain oleosome size due to potential rearrangements/modifications during extraction, but which can contain some intact oleosomes. The environment in which this extracted oil is present, i.e., oil-rich emulsion (cream fraction) or aqueous protein- and sugar-rich fraction (skim), is another important parameter to be considered. Additional steps to recover free oil are key as they can add processing costs. Therefore, the highest extraction yield might not be the best choice if the recovered oil is trapped in a stable emulsion and/or if some of the oil goes into the skim, from which oil is even more difficult to recover. The quality of the free oil will determine if additional refining steps are needed to remove minor impurities, such as phospholipids, free fatty acids (FFAs), or pigments.

Several procedures have been used to extract and purify oleosomes and the specific process conditions will affect the composition of the environment surrounding these structures and therefore oleosome stabilities. When extracted into water, the purified oleosomes are accompanied by other organelles and/or water-soluble compounds including soluble carbohydrates and extraneous proteins that reduce the purity of the extracted oleosomes. The extracted proteins can interact with the surfaces of the oleosomes and form a secondary layer that impacts the stability of the oleosomes (Nikiforidis and Kiosseoglou 2009). The importance of washing steps to remove extraneous proteins was illustrated by comparing the SDS-PAGE profile of sunflower oil bodies obtained before and after urea washing, resulting in 70% removal of proteins (White et al. 2008). In addition, some procedures add salt and acid to increase oleosome extraction yields. Without appropriate washing, these additional salts can interfere with oleosome stability. When protease is added during the extraction steps to improve oil extraction, the enzyme can hydrolyze proteins located at the oil body membrane and the proteins surrounding the oil bodies. Potentially, all components of natural oil bodies can be released, i.e., oil, phospholipids and intact or hydrolyzed proteins, and form a new emulsion along with oil bodies that remain intact, leading to complex colloidal systems.

Comparisons of the stabilities of oleosomes/emulsions (cream) between different studies must therefore be done with caution. Purity of the oleosomes, composition and concentration of the environment, as well as extraction procedures used are likely to impact the characteristics of the oil-rich fraction being extracted, thereby affecting its stability. The variation in cream fractions stabilities is illustrated by comparing the stabilities toward enzymatic demulsification of various soybean cream fractions obtained during aqueous extraction, assisted, or unassisted by protease addition, of different full-fat soybean substrates (flour, flakes, and extruded flakes) where free oil yield varied between 7.4 to 77.9% (Yao and Jung 2010).

Oleosomes and oleosome-based emulsions (cream) are often studied because of their potential use in various applications such as food applications (dressings, sauces, beverages, and desserts) as well as cosmetic and pharmaceutical applications (Nguyen et al. 2010; Guth and Ferguson 2009). For this reason, strategies to further improve their stabilities have also been investigated, including adding pectin coating and enzymatic cross-linking (Chen et al. 2010; Iwanaga et al. 2008). Artificial oil bodies have also been produced in order to understand the roles played by oleosome constituents (Chen et al. 2004). On the other hand, when aqueous extraction of oilseeds is applied, ways to totally destabilize these stable structures in order to recover free oil are sought. For free oil recovery, the cream layer fraction is usually submitted to a demulsification process without washing, and as a consequence, the fraction to be demulsified is rich in compounds dispersed in the aqueous phase of the emulsion. When oleosome-rich fractions have been studied for their potential as a naturally occurring emulsion for food or pharmaceutical applications, extensive washing is usually applied.

16.4.1 Stabilities of Oleosomes and Oleosome-Based Emulsions

Oleosomes are usually described as spherical organelles with particle sizes ranging from 0.5 to 2 μm (Tzenet al. 1993) and having a monolayer biological membrane comprised of phospholipids and proteins of low molecular weight (<30 kDa) that surrounds the triacylglycerol core forming a very stable organelle (Figure 16.5).

These oil bodies are composed of up to 98% neutral lipids (mainly triacylglycerols) and varied in their proportions of phospholipids (0.5–2%) and proteins (0.5–3.5%). Tzen et al. (1993) reported that 80% of the oleosome membrane is comprised of phospholipids, and the remainder is mostly oleosin proteins. There is much evidence from natural or artificial oil bodies showing that proteins play important roles on the stability of the oil bodies; however, the effects of phospholipid content and profile have received less attention. The proteins involved in these oil bodies are mainly oleosins, along with caleosins and steroleosin (Purkrtova et al. 2008). The role of the steroleosin, a sterol-binding protein, in the stability of oil bodies seems to be minor, failing to contribute to the formation of stable artificial oil bodies, unlike oleosins and caleosins (Chen et al. 2004). The oleosins have N- and C-terminal hydrophilic regions of various lengths and a central hydrophobic region of 11-nm length composed of 70 to 80 residues. Caleosins along with a 27-kDa protein represent up to 10% of the total oil body proteins and was also identified as a key player in imparting stability to oil bodies and having capacity to interact with calcium (Purkrtova et al. 2008). Natural oil bodies of various seeds including rape, mustard, cotton, and maize have isoelectric points varying between 5.7 and 6.6, indicating they have a net

(a)   (b)

(c)

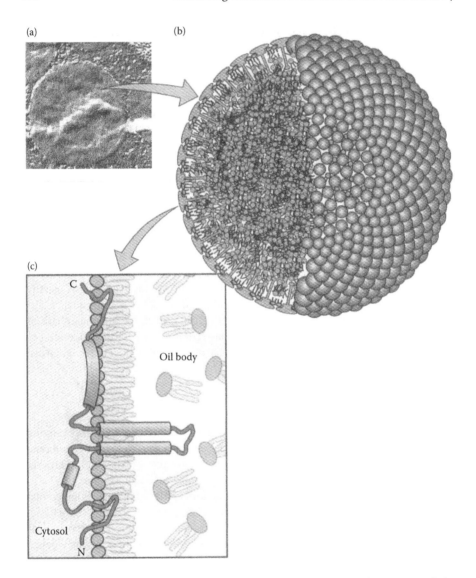

Oil body

Cytosol

FIGURE 16.5 Representation of oleosome and conformation of oleosin: (a) transmission electron microscopy of oleosome, (b) model of oleosome showing oleosins forming the outer surface of the oil body, and (c) model of the conformation of a maize oleosin. (From Buchanan, B.B. et al., *Biochemistry & Molecular Biology of Plants*. Oxford: Wiley-Blackwell. 2000. With permission.)

negative charge at neutral pH (Tzenet al. 1993). The presence of enzymes, including phospholipases and lipases, in the oleosome extract could modify the functional groups of the phospholipids at the oleosome interface and/or hydrolyze lipids producing FFAs and decreasing the isoelectric point of the oleosomes to a lower value (Iwanaga et al. 2007).

16.4.2 Stabilities of Purified Oleosomes

To understand oleosomes stability, a purification process is usually required in order to remove interferences of adsorbed proteins and carbohydrates present in the extracted solution. A soybean oleosome-rich fraction comprising 47% moisture and 40% fat, and having an approximated particle size of 0.3 μm at pH 8 was submitted to various treatments in order to correlate their other properties to stability. After dilution, oleosomes were subjected to pH changes (2–8), thermal treatment (30–90°C for 20 min), and salt addition (0–150 mM NaCl) (Iwanaga et al. 2007). These oleosomes were stable against thermal treatment with no modifications of the particle sizes of the droplets and no creaming, despite a decrease in ζ-potential from –20 to –5 mV at 30 to 90°C, respectively. Decreased creaming stability of these oleosomes was reported after adding salt to achieve ≥50 mM concentration or adjusting to pH 4 and 5. The instability observed due to pH adjustment was related to the isoelectric point of the oleosome (pH 4) and increased particle size. Similarly increased particle size of a sunflower seed oil body emulsion (with initial composition of 67.6% lipid, 5.4% protein, 25.2% moisture) was observed after adjusting the pH to 5–6 (isoelectric point of the oleosins), at which flocculation of the droplets was observed increasing the emulsion viscosities (White et al. 2008). Increasing $CaCl_2$ concentration to 5–150 mM modified the emulsion stability at pH 7.0 but had no impact when adjusted pH 3. In the study of Chen and Ono (2010), adding NaCl (5 to 500 mM) to diluted soybean oleosomes prevented the aggregation of the oil droplets that was observed at pH 5.7. These authors also observed the high stability of the oil bodies against thermal treatment with no effect on particle size, surface charge, or surface hydrophobicity.

16.4.3 Stabilities of Nonpurified Oil-Rich Fractions

Studies focusing on recovery of free oil do not involve extensive purification of the oil-rich fraction. These studies can be differentiated by the starting material from which the oil-rich fraction was obtained, which not only impacts the protein properties but also the type of compounds extracted and the phospholipid profile.

Chabrand et al. (2008) reported high stability of a nondiluted oil-rich fraction (60% oil, 35% water) recovered from an aqueous extract of soybean flour, with no trace of free oil after 4 weeks of refrigerated storage. This fraction, which was not submitted to washing steps, and therefore, more likely to contain extraneous material, was also very stable to thermal treatment (95°C for 30 min). A protease hydrolysis (1% w/w of cream, 3 h) followed by a freeze–thaw treatment led to 46% free oil recovery, which was higher than the application of each individual treatment (~25%). When protease was added during the extraction of full-fat soybean flour, the stability of the extracted cream fraction was severely reduced with an increase of free oil yield from 2 to 83% by adjusting to pH 4.5. Enzymatic demulsification yielded 72% free oil, which increased to 95% by applying a two-step protease demulsification process (Chabrand and Glatz 2009). The effect of protease addition on cream emulsion stability is also highly dependent on demulsification conditions, with no change in the stability of the cream obtained from flour, with or without protease during extraction, which was observed by Yao and Jung (2010), who used the same protease as Chabrand and Glatz (2009) but at different conditions (2.5%, 25°C, 10 min vs. 3%, 50°C, and 3 h). Soybean oleosomes were obtained from soybean flour with the aid of cellulases in an extraction medium containing 0.1 M potassium acetate, 0.5 M NaCl, 0.4 M sucrose, at pH 4.6 and 57°C for 15 h, and after submitting the oleosomes to a protease treatment (5% v/dw Protex 6L, 18 h), 55% free oil yield was obtained (Towa et al. 2010b). By further treating the hydrolyzed cream and residual fractions (skim) with a combination of treatments including heating followed by a process similar to butter churning, 88% total free oil yield was achieved. This procedure, however, has a total incubation time of 25 h and involves three centrifugation steps, which illustrates the challenges of recovering free oil in an aqueous-based, flammable hydrocarbon solvent-free process.

The stabilities of creams recovered from PAEP of extruded full-fat flakes have been studied in more detail than other creams because this process gives the greatest extraction yields. The presence, nature, concentration, and incubation time of the enzymes used during extraction of extruded flakes not only affect extraction yields but also affect the stability of the emulsion to demulsification (de Moura et al. 2008; Jung et al. 2009b; Wu et al. 2009). While adjusting the pH of the cream from single-stage PAEP to 4.5 gave similar free oil recovery as protease demulsification, lower demulsification yield was obtained after pH adjustment of cream recovered from countercurrent two-stage PAEP (76 vs. 91–95%) compared with protease demulsification (de Moura et al. 2008, 2010). The higher stability of the two-stage cream toward demulsification may be related to the type and size of peptides present at the interface, as well as phospholipids. Peptides generated by countercurrent two- and single-stage extraction had MWs < 25 and < 54 kDa, respectively. The use of enzyme has several advantages including avoiding the addition of undesired salt and the potential of enzyme recycling (Wu et al. 2009). As previously described (16.3.1.3), protease used in the demulsification step retains its activity and can be recycled into the extraction step; however, excess enzyme activity during extraction can increase emulsion stability toward creaming, decreasing the oil recovered as cream (de Moura et al. 2011a). Efficiency of the demulsification treatment is also dependent on the conditions in which the starting material is stored and processed. Stabilities of creams recovered from countercurrent two-stage extraction of extruded soybeans flaked at 12% moisture content and conditioned at 75°C was higher than when flaked at 10% moisture and conditioning at 65°C (~76 vs. 91–95%), for both enzymatic and pH adjustment procedures (de Moura et al. 2010).

Part of the variation observed in emulsion stability could be the result of endogenous phospholipase D activity. While the total phospholipid content is sometimes reported in stability studies involving oil bodies or cream fractions, the phospholipid profile is rarely determined. The phospholipid content of natural oil bodies isolated from rape, mustard, cotton, flax, maize, peanut, and sesame varies from 0.57 to 1.97%, with the major phospholipids being phosphatidylcholine (PC) with % v/v varying from 41 to 62% (Tzen et al. 1993). Not only the quantity but also the phospholipid profile may have a dramatic impact on the oil-rich fraction stability. In the case of soybeans, the endogenous phospholipases may modify the phospholipid composition and significantly affect membrane stability if not denatured during processing prior to bean conditioning or aqueous extraction (Yao and Jung 2010). The importance of phospholipid in the oil-rich fraction is also illustrated by the decreased emulsion stability after phospholipase addition (Lamsal and Johnson 2007; Chabrand and Glatz 2009; Wu et al. 2009; Jung et al. 2009b). Destabilization of the cream fraction recovered from soy flour was obtained with a lysophospholipase A_1 enzyme treatment (G-ZYME) performed at pH 4.5 (Chabrand and Glatz 2009). A similar decrease in soybean emulsion stability was obtained with cream recovered from extruded material and demulsified with phospholipase A_2 (LysoMax) at pH 8.0, with some variations in the efficiency of the phospholipase depending on the nature of the enzyme(s) used in the extraction step (Jung et al. 2009b). Because phospholipases are usually more expensive than proteases and because in some conditions, their use can increase emulsion stability (Wu et al. 2009), phospholipases need to be used with caution.

As reported above, the potential for recycling protease during countercurrent two-stage aqueous extraction has been successfully scaled-up from 2 kg of extruded flakes (laboratory scale) to 75 kg (pilot plant scale) (de Moura et al. 2011b, Figures 16.2 and 16.3). Although 91.6% enzyme-catalyzed (protease) cream demulsification efficiency was achieved, 93% free oil recovery was obtained by recycling the remaining cream from previous demulsification into the next cream demulsification. Despite achieving 98% oil extraction with the fully integrated process, the overall free oil recovery (relative to the initial amount of oil in the extruded full-fat soybean flakes) was about 79% due to partitioning of some extracted oil in the skim fraction (19%) and unextracted oil in the insolubles (2%). Oil in the skim remains the greatest challenge to protease-assisted aqueous extraction.

16.4.4 OIL QUALITY

Aqueous extraction of oilseeds, with or without enzyme, produces oil with similar or better qualities than conventional extraction methods such as direct hexane extraction or prepress solvent extraction. Oils from soybeans (Jung et al. 2009b), sunflower seed (Latif and Anwar 2009), canola (Latif et al. 2008), and corn germ (Bocevska et al. 1993) produced by aqueous extraction have low phosphorus (phosphatide) levels, making possible physical refining without degumming and neutralizing the extracted oil thereby reducing chemical use and effluent production. They also have low peroxide values and similar FFA contents compared with oils obtained by conventional processes.

For soybeans, the effects of enzyme extraction and type of demulsification procedure (pH adjustment or enzymatic) used to recover free oil from the cream were evaluated by Jung et al. (2009b). Oils obtained by chemical (pH adjustment) or enzyme-catalyzed demulsification had significantly lower phosphorus contents (14 and 54 ppm, respectively) compared with hexane-extracted oil (240 ppm) and similar peroxide values (<6.5) indicating similar initial oxidation rates. However, the use of chemical treatment to demulsify the cream produced oils with higher *p*-anisidine values (2.38 vs. 1.73), an indicator of secondary oxidation products. In summary, the use of a protease (Protex 6L) in both extraction and cream demulsification steps enabled the production of oil with superior quality.

Quality and stability of corn germ oil obtained by CAEP (Pectinex Ultra SLP, a commercial enzyme preparation containing β-galactosidase activity) were assessed and compared with degummed hexane-extracted oil and screw-pressed oil (Bocevska et al. 1993). Despite the low phospholipids level, the CAEP-extracted oil was relatively stable (14.6-h induction period), likely a consequence of its high tocopherol content. Other parameters, such as FFAs and primary and secondary oxidation products, were similar to those of oils obtained by conventional procedures.

The physicochemical properties and oxidative state of sunflower seed oil extracted using different enzymes (Protex 7L: protease; Kemzyme: containing mainly α-amylase, β-glucanase, cellulose complex, hemicellulase complex, protease, and xylanase activities; Alcalase 2.4L: protease; Viscozyme L: multienzyme complex containing carbohydrases including arabanase, cellulase, β-glucanase, hemicellulase, and xylanase; Natuzyme: mainly cellulase, zylanase, phytase, α-amylase, and pectinase activities) were evaluated by Latif and Anwar (2009). When comparing with hexane-extracted oil, EAEP oil had lower FFA content (0.64–0.69 vs. 0.94%), lower oxidative indices (peroxide value and p-anisidine), and longer oxidation induction periods. EAEP sunflower oil had higher amounts of total tocopherol compared with hexane-extracted oil (833–842 vs. 799 mg/kg), likely a consequence of cell wall rupture due to enzyme action, which could increase tocopherol release.

Similarly, better oil quality was observed for EAEP-extracted oil from canola (Latif et al. 2008). Oils extracted with Protex 7L (protease), Multifect CX13L (significant activity toward cellulose, hemicelluloses, β-glucans, and arabinoxylans), and Natuzyme (see activity description above) had lower FFA contents, better oxidative stability parameters, such as peroxide value, conjugated dienes and trienes, and similar FFA compositions compared with solvent-extracted oil. EAEP canola oil had higher total tocopherol concentration compared with solvent-extracted oil (15–17 vs. 13 mg/kg).

The amounts of biologically active components that improve antioxidative, health, and sensory qualities of oil (total phenols, o-diphenols, major free phenols) were significantly increased in olive oil through a mechanical percolation system while using an enzyme cocktail obtained from olive fruit (*Bioliva*, mixture of pectolytic, cellulolytic, hemicellulolytic enzymes) (Ranalli et al. 2003).

EAEP/AEP-extracted oils from various oilseeds are of much better quality and, therefore, less additional refining is necessary than is required for hexane-extracted oils. Hence, less post-extraction refining is needed, saving energy and reducing oil losses.

16.5   ADDING VALUE TO EAEP COPRODUCTS

The two main streams produced along with the oil-rich fraction during aqueous extraction of oilseeds are a protein- and sugar-rich aqueous extract (skim fraction) and a fiber-rich insoluble fraction (Figure 16.2), and economic feasibility of the process is highly dependent on the values and returns of these fractions. Several applications have been envisioned for these fractions. Soy proteins have unique functionalities that make them valuable for use as food ingredients. The recovery of the soluble proteins produced during AEP will add value to this fraction. Proteases produce peptides of low molecular weight with high solubility, which make the separation of the proteins from the sugars challenging. Because the skim is rich in both proteins and sugars (typically 55% protein and 35% sugars), it may be suitable for use as fermentation media. This application may require some concentrating of the skim by membrane filtration or evaporation. Finally, the fiber-rich fraction might be a valuable renewable resource of fermentable sugars for solid-state fermentation or saccharification/fermentation to produce bioethanol or other types of biomolecules such as biosurfactants.

16.5.1 CHALLENGES AND STRATEGIES IN RECOVERING AEP PROTEINS

With CAEP, the buffer used for oleosome isolation contains 0.1 M potassium acetate, 0.4 M sucrose, and 0.5 M sodium chloride at pH 4.6, and the resulting pH of the aqueous supernatant is 4.9, which is close to the isoelectric point of soybean protein (pH 4.5). Usually soy proteins have very low solubility at the isoelectric point; however, protein solubility at this pH during oleosome fractionation is high because of the high ionic strength and osmolarity of the extraction solution. Dilution of the aqueous supernatant with distilled water or ethanol weakens the ionic strength of the solution and increases attraction between protein molecules to allow protein precipitation. Precipitation with ethanol and ultrafiltration were the most effective methods to obtain significant amounts of protein from aqueous supernatants after harvesting oleosomes, with 34 and 39% overall recovery, respectively (Kapchie et al. 2011b).

Although high oil and protein extraction yields are desired with PAEP, small reductions in extraction yields might be justified to produce soy protein products with different degrees of hydrolysis and functionalities thereby diversifying applications for the extracts (de Moura et al. 2011c). Increased protein solubility due to the use of protease, which favors production of acid-soluble peptides, might be a drawback to protein recovery. Although using protease in single-stage aqueous extraction of extruded soybean flakes increased protein extraction from 45 to 75%, a

---

TABLE 16.1

Effects of Protease on Soybean Oil and Protein Extractabilities in Countercurrent Two-Stage and on Protein Recovery by Membrane Filtration and Isoelectric Precipitation

Extraction Conditions	Extraction Yields (%)		Protein Recovery from Skim (%)		Overall Protein Recovery from Extruded Flakes (%)		
	Oil	Protein	TSMF	IP	TSMF	IP	IP + WF
A (two uses)	99[a]	96[a]	91	27[a]	88	26	93
B (one use)	94[b]	89[b]	96	61[b]	86	54	87
C (no enzyme)	83[c]	66[c]	99	87[c]	65	57	65

Note: A: 0.5% protease (wt/g extruded flakes) in both extraction stages; B: 0.5% protease only in the second extraction stage; C: no enzyme in either stage. Extractions were performed at 1:6 SLR, 50°C, pH 9.0, and 120 rpm for 1 h. TSMF: two-stage membrane filtration, IP: isoelectric precipitation, WF: whey filtration (de Moura et al. 2011c, 2011d). Means within the same column followed by different letters are statistically different at $P < 0.05$.

10% decrease in protein recovery yield was observed due to higher protein solubility at pH 4.5, thereby reducing protein recovery by isoelectric precipitation (Jung and Mahfuz 2009). A comparison of different extraction treatments for countercurrent two-stage PAEP of soybeans is presented in Table 16.1 with their respective oil and protein extraction yields, as well as the protein recovery yields when using two-stage membrane filtration (TSMF, skim ultrafiltration followed by nanofiltration of permeate from the first ultrafiltration; the nanofiltration steps were added to dewater the protein extract) or a combination of isoelectric precipitation and acid whey nanofiltration (de Moura et al. 2011c, 2011d). Three different treatments were applied: (A) 0.5% protease (wt/g extruded flakes) in both extraction stages; (B) 0.5% protease only in the second extraction stage; and (C) no enzyme in either stage.

From an extractability perspective, increased use of enzyme (from C to B and A) leads to higher oil and protein extraction yields (from 66 to 89 and 96% for protein extractability). Protein recovery with two-stage membrane filtration from the skim fraction increased from 91 to 96% when decreasing enzyme use from treatment A to B; however, when combining extraction yields and protein recovery by membrane filtration, treatments A and B had similar overall protein recoveries (88 vs. 86%) due to the higher protein extractability of treatment A. Although the treatment with the protease in both stages (treatment A) had higher oil extraction yield, protein recoveries by two-stage membrane filtration were quite similar, and the choice of treatment should consider protein functionality, which will determine appropriate applications of those proteins.

When considering isoelectric precipitation as a process to recover protein, extent of hydrolysis significantly impacts protein precipitation. Higher extents of hydrolysis considerably reduced skim protein recovery by isoelectric precipitation (from 87% for the control to 61 and 27% as enzyme use increased), corresponding to 57, 54, and 26% overall protein recoveries, respectively. Adding a second recovery step of nanofiltering to the whey produced by isoelectric precipitation improved overall protein recovery to 93 and 87% for treatments A and B, respectively. Assessment of the functional, nutritional, and biological properties of the proteins recovered after PAEP needs to be performed in order to determine if there are advantages to adding processing steps.

16.5.2 Functionality of the Recovered Protein Fraction

The extent of hydrolysis significantly impacts protein functional properties such as solubility, emulsification properties, gel formation, and foaming properties (de Almeida et al. 2010). In the example described in the previous section (16.5.1), the use of enzyme during extraction increased the protein solubility to 93 and 66% at pH > 4.5 for skim proteins from treatment A and B, respectively, compared with 29% solubility for skim proteins from treatment C (without enzyme). In this study, a control (treatment D) was performed by extracting proteins from air-desolventized defatted soybean flakes without using enzyme in either extraction stage, having 16% solubility at pH 4.5. Emulsification properties of these skim fractions, however, were reduced as the amount of enzyme used during extraction increased. Reducing enzyme use from treatment A to C decreased emulsification capacity from 116 to 214 g oil/g sample at pH 7, compared with 274 g oil/g sample for the control. Rate of foaming and foaming stability were favored by higher extent of hydrolysis. Hydrolyzed proteins produced gels with lower strengths compared with unhydrolyzed proteins. In terms of nutritional quality, essential amino acid compositions and protein digestibilities in vitro were not adversely affected by either extrusion or extraction treatments applied (A, B, C, or D), indicating that PAEP of soybeans had no detrimental effect on the protein nutritional quality of soybeans (de Almeida et al. 2010).

Jung and Mahfuz (2009) reported on the effect of proteolysis on functionality of ISP (isolated soy proteins) obtained by isoelectric precipitation of skims obtained in single-stage extraction. When proteases were added during AEP, the surface hydrophobicity of isoelectric precipitated protein increased (18 vs. 24.8) and water solubility of these acid-precipitated proteins decreased (94 vs. 88%). Increased surface hydrophobicity was likely a consequence of protein unfolding due to enzyme attack, exposing hydrophobic sites. Decreased ability of the ISP to bind oil was also observed (oil-holding capacity of 23 vs. 28%) when adding enzyme. Care has to be taken when comparing results on protein functionality. In the study of de Almeida et al. (2010), all proteins present in the skim fractions obtained by countercurrent two-stage EAEP were characterized, whereas in the study of Jung and Mahfuz (2009), only the proteins recovered by isoelectric precipitation from the skim obtained by single-stage EAEP were analyzed, which are two totally different fractions. While extensive literature on soy protein functionality and effects of protease addition is available, these studies were performed on a protein extract obtained from defatted soybean flakes, while the protein fractions recovered from AEP of full-fat soybean substrates contain lipid, which likely impacts the functionality of the protein extract.

16.5.3 POTENTIAL APPLICATIONS OF THE SKIM

The protein- and sugar-rich, oil-lean skim fraction could also be used "as is" or concentrated for use in feed applications. This fraction is an excellent source of proteins and soluble carbohydrates, for example, the skim obtained from the countercurrent two-stage process contained 6.7% water-soluble protein, 6.36 mg/ml of stachyose, 0.8% residual oil, and 89% water (de Moura et al. 2011b). The presence of flatus-producing stachyose in these skim fractions constitutes a challenge regarding the use of this fraction for use in feed or food applications. A strategy to solve this problem is to hydrolyze the oligosaccharides. Stachyose, a tetrasaccharide, can be hydrolyzed by using α-galactosidase to yield galactose, glucose, and fructose (de Moura et al. 2008).

The skim fraction could be a source of proteins, carbohydrate, and water for the dry-grind corn ethanol fermentation process, thereby improving the feed quality of dried distiller's grains with solubles (DDGS) due to the addition of soy proteins. Yao et al. (2011) observed that the use of skim increased the ethanol production rate. Although including skim in the fermentation step did not improve the final ethanol yield, it did increase the solids, protein, and oil contents of the resulting whole stillage. Investigations are currently being performed on using the membrane-concentrated skim fractions (containing protein and sugars), as well as using the whey fraction (containing mainly hydrolyzed proteins and sugars) produced by isoelectric precipitation of skim proteins, as alternative substrates to slurry ground corn for fermentation.

16.5.4 Adding Value to the Fiber-Rich Coproduct

Removal of oil, protein, and soluble sugars from oilseeds during AEP leaves an insoluble carbohydrate-rich fraction, the composition of which varies depending on the initial composition of the starting material and extraction yields. In the case of soybeans, this fraction can be compared with the fiber fraction produced after solvent extraction followed by protein extraction and can be used as feed and fermentation media. For feed applications, the high fiber content limits the use of this fraction for feeding nonruminant animals, such as beef and dairy cattle. Converting this biomass to fermentable sugars for either producing value-added products, such as bioethanol or other industrial biochemicals, and/or improving their nutritional quality as feed resources, can dramatically impact its economic value. There is great interest among enzyme companies to commercialize appropriate enzyme cocktails that will convert lignocellulosic biomass into fermentable sugars. Another approach that is being investigated is to perform solid-state fermentation with filamentous fungi that are able to grow in a low-moisture content environment and produce enzyme mixtures including cellulases able to convert these complex carbohydrates to fermentable sugars (Yang et al. 2011).

16.5.4.1 Adding Value to PAEP Insoluble Fiber-Rich Fraction

Glucan and xylan contents of insoluble fiber-rich fractions recovered from AEP and EAEP of full-fat soybean flakes and extruded full-fat soybean flakes varied between 10–16% and 3–5%, respectively, and were lower than for other agricultural residues, such as corn stover or wheat straw, that are in the 30–50% and 9–22% range, respectively (Mosier et al. 2005) (Table 16.2).

Enzymatic saccharification yields in AEP and PEAP soy fiber fractions using Accellerase 1000 were not affected by variations in oil and protein contents (Karki et al. 2011a). As for the oil extraction step, extruding full-fat soybean flakes significantly

---

TABLE 16.2

Composition of Soybean-Insoluble Fractions Recovered after PAEP of Extruded Full-Fat Soybean Flakes and Hexane-Extracted Cakes from Various Oil-Bearing Materials

Material	Glucan (%)	Xylan (%)	Arabinan (%)	Galactan (%)	Mannan (%)	TL (%)
Insoluble from PAEP[a]	16.0	5.4	8.9	16.4	0.5	6.7
Soy[b]	15.2	2.9	3.8	9.1	2.2	3.3
Canola[b]	13.7	2.5	5.2	4.6	0.3	15.8
DDGS[b]	20.8	9.9	7	2.2	2.2	15.4

Note: Data are based on oven-dry weight of biomass. TL: total lignin; DDGS: dried distiller's grains with solubles.

[a] Data from Karki et al. (2011a).

[b] Data from Balan et al. (2009), cakes are the residues obtained after hexane extraction.

improved the saccharification yields of the insoluble fraction, increasing yields from 33 to 49% (Karki et al. 2011a). This increase was attributed to the disruption of the cotyledon cell wall during extrusion, increasing access of the saccharification enzymes to their substrates. While the increase of saccharification yield to about 50% is encouraging, it also illustrates that this lignocellulosic material requires pretreatment to maximize substrate availability to cellulosic enzymes.

Pretreatments for various biomasses often involve drastic mechanical, chemical, or biological interventions and result in the production of toxic waste. These pretreatments also increase the cost of converting the biomass and potentially reduce the nutritional value of the residual protein fraction for animal feed. Pretreatment methods, such as steam explosion (Corredor et al. 2008), acid treatment (Lloyd and Wyman 2005), alkali treatment (Kim and Holtzapple 2005; Zhao et al. 2008), ammonia fiber explosion (Teymouri et al. 2005), high-power ultrasound (Lomboy-Montalbo et al. 2010; Nitayavardhana et al. 2010), and extrusion (Dale et al. 1999) have been studied on a large variety of biomass. Pretreating fiber from PAEP of extruded soybean flakes with 15% (w/w, db of insoluble fraction) ammonium hydroxide, 15% (w/w, db of insoluble fraction) sodium hydroxide, and 1% (w/w, db of insoluble fraction) sulfuric acid achieved 63, 53, and 61% saccharification yields, respectively, vs. 37% for the fiber not chemically pretreated (Karki et al. 2011b). Saccharification yield was increased to 88% by soaking the PAEP fiber from extruded soybean flakes in aqueous ammonia (SAA) at 80°C for 12 h, indicating that SAA is a simple and technically feasible pretreatment method for converting the fiber-rich insoluble fraction to fermentable monomers via enzymatic hydrolysis (Karki et al. 2011). High-power ultrasound treatment, which has the advantage of not involving addition of any chemicals, applied for 30 and 60 s at 144 μm peak-to-peak ultrasonic amplitude, 20 kHz frequency, and 2.2 kW maximum power output to the fiber-rich fraction recovered from PAEP of extruded full-fat soybean flakes, did not increase saccharification yield (Karki et al. 2011b), possibly because of the high degree of cell disruption already obtained by extrusion. Current work is being conducted in order to convert this fraction into bioethanol using *Saccharomyces cerevisiae* in separate hydrolysis and fermentation, and simultaneous saccharification and fermentation processes.

Determination of cellulase production during solid-state fermentation of the PAEP fiber-rich fraction when growing *Aspergillusoryzae*, *Trichodermareesei*, or *Phanerochaetechrysosporium* on soybean fiber recovered from PAEP showed that soybean fiber was an effective feedstock for microbial production of cellulases and xylanases and at a concentration higher than was in obtained with other agricultural substrates (12.6 and 84.2 IU/g, respectively) (Yang et al. 2011).

16.5.4.2 Adding Value to Other Fiber-Rich Agricultural Material

For soybean hulls, extrusion pretreatment was less effective than grinding to reduce particle size in enhancing production of reducing sugars (Lamsal et al. 2010). Mielenz et al. (2009) reported 80% cellulose conversion of soybean hulls to ethanol by using *S. cerevisiae*, producing a protein-rich residue more suitable for the animal feed market.

Solvent-defatted meal from soybeans and other oilseeds, such as canola, sunflower seed, sesame, and peanuts, were compared with DDGS, as sources of fermentable sugars to produce bioethanol (Balan et al. 2009). These fiber-rich sources were pretreated with ammonia fiber expansion before being subjected to sequential and simultaneous saccharification and fermentation. While the compositions of these fractions differed from fractions obtained by AEP because of the presence of protein that was not extracted by the solvent process, this study confirmed the potential of the soybean glucan-rich fiber to produce soluble carbohydrate, with 75% glucan conversion versus 80–85% for DDGS.

16.6 ECONOMICS OF EAEP OF SOYBEANS

An inherent advantage of AEP of oilseeds, especially soybeans, over conventional hexane extraction is that oil and protein can be separated in a single extraction step. After hexane extraction, protein remains with the cellulose (fibers/cell walls) and seed sugars, while AEP separates the oil, fiber, and protein/sugar fractions simultaneously in one extraction step. This simultaneous separation may result in increased value for the EAEP fractions. Ironically, however, this additional separation imposes an important complication to EAEP economics, which is the matter of market size.

The meal resulting from hexane extraction is suitable for a low-cost feed for swine and poultry. The high protein fraction resulting from EAEP would be more suitable for food applications (in particular, a hexane-free organic soy protein isolate or concentrate), but the market for food-grade soy protein is only about 2–3% of the entire soy protein market (Goldsmith 2008). While this does not necessarily negatively impact the ability of EAEP technologies to compete with conventional technology, the relatively small market for soy protein ingredients does suggest that the development of EAEP will have limited impact on the use of hexane in the oilseed processing industry. One exception to this may be in rural areas in less developed countries, where access to a highly-skilled workforce as well as to the level of capital necessary for operation of a complex hexane extraction plant is limited or if a value-added application can be found for the skim.

At this time, whether EAEP can be competitive with conventional oil extraction technologies has not been yet been established. Literature data are now available in adequate quantity to investigate the economic feasibility of EAEP of soybeans at an order-of-magnitude level. The objective of this section is to use the available data and rule-of-thumb cost factors to estimate and compare the economic viability of various EAEP technologies. Three different extraction strategies were modeled using a SuperPro Designer® software package based on data from the following publications: AEP/EAEP of soybean flour (Campbell and Glatz 2009b), EAEP of soybean flour for the recovery of intact soybean oil bodies (Kapchie et al. 2010a, 2010b), and EAEP of extruded soybean flakes for the recovery of free oil (de Moura and Johnson 2009; de Moura et al. 2008, 2009, 2010, 2011a–c). Four variants of EAEP of extruded soybean flakes were modeled: single-stage enzyme-assisted extraction (de Moura et al. 2008, 2010), countercurrent two-stage EAEP with enzyme activity in both stages (de Moura and Johnson 2009; de Moura et al. 2010; de Moura et al. 2011c), countercurrent two-stage EAEP with enzyme in one stage (de Moura and Johnson 2009; de Moura et al. 2011c), and countercurrent two-stage EAEP with no enzyme (de Moura et al. 2011c) (Figures 16.6 and 16.7).

The design basis was 400 MT of soybeans processed per day, which is equivalent to a modern soy protein production facility, but smaller than would be typical of a hexane extraction process, which is consistent with the assumption that an AEP would be used to produce edible protein products as the major economic driver. All prices are based on 2008 values. Market values of oil and meal were based on historical commodity market data from the Chicago Board of Trade (2010). The selling prices of the resulting skim fractions were determined to achieve a target 12% internal rate of return (IRR). The skim price was then used as the feedstock cost for the downstream protein recovery processes. Three different protein recovery processes were investigated: isoelectric precipitate (IEP) to make a 90% soy protein isolate (SPI), ultrafiltration (UF) to make a 70% soy protein concentrate (SPC), or a combination of the two (IEP followed by UF) to make both a SPI and SPC products. Again, the selling prices of the products were determined to achieve a 12% IRR. For the combined IEP/UF process, the lowest SPI selling price achieved by any AEP/EAEP process was used, and the selling price for the SPC was the dependent variable.

Selling prices for skim and resulting soy protein for each of the AEP/EAEP processes are outlined in Table 16.3. The oil body extraction process is the most

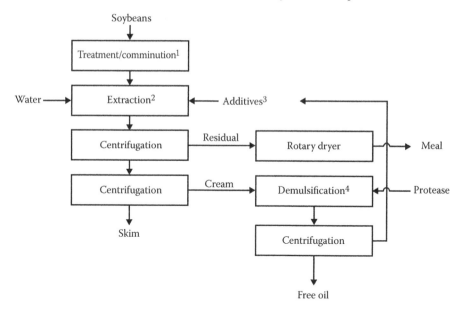

FIGURE 16.6 Extraction process flow diagram used for economic analysis of AEP/EAEP. [1]Pretreatment steps: flour process and oil body process, grinding; extrusion process, conditioning, flaking, then extrusion. [2]Extraction steps: flour process and extrusion process, agitation for 1 h at 50°C, pH 8 (flour) or 9 (extrudate); oil body process, incubation for 20 h at pH 4.5 with agitation. [3]Additives: for flour and extrusion process, sodium hydroxide; for oil body process, hydrochloric acid, 0.4 M sucrose, and 0.5 M sodium chloride. [4]Demulsification step was assumed to be 1 h agitation at pH 8 in the presence of protease Protex 6L at a concentration of 0.5% (wt protease/wt initial soybean mass). Aqueous fraction from demulsification is recycled to extraction step as the enzyme source. In the case of oil body extraction, no demulsification was conducted.

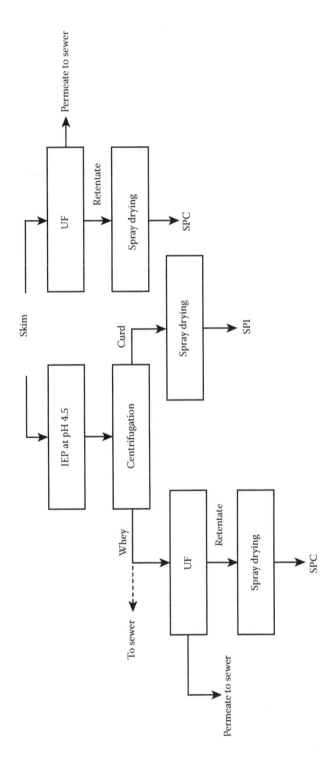

FIGURE 16.7 Process diagram of skim treatment options to make soy protein isolate (SPI) and/or soy protein concentrate (SPC). IEP: isoelectric precipitation; UF: ultrafiltration.

TABLE 16.3

Estimated Operating Costs and Product Selling Prices for Various Aqueous Extraction Processes for Soybeans

Extraction Process	Operating Cost ($/kg Soybean)	Skim Selling Price ($)		Protein Recovery Process	SPI Selling Price ($)		SPC Selling Price ($)	
		Per kg Skim	Per kg Protein		Per kg SPI	Per kg Protein	Per kg SPC	Per kg Protein
AEP flour[a]	0.573	0.038	1.249	IEP[b]	2.10	2.47	–	–
				UF[c]	–	–	1.33	2.03
				IEP + UF	ND	ND	ND	ND
EAEP flour[a]	0.567	0.035	1.131	ND	ND	ND	ND	ND
Oil body[d]	0.860	0.237	2.420	ND	ND	ND	ND	ND
One-stage extruded[e]	0.555	0.035	1.110	IEP[b]	6.54	7.27	1.61	2.60
				UF[b]	–	–	2.72	4.18
Two-stage extruded[f] Enzyme both stages	0.565	0.061	1.090	IEP + UF[f]	2.10	2.47	–	–
				IEP[f]	6.93	8.54	–	–
				UF (2.5 LMH)[f]	–	–	1.77	2.90
				UF (4 LMH)[f]	–	–	1.56	2.55
				IEP + UF (4 LMH)[f]	2.10	2.47	2.82	4.34
Two-stage extruded[f] Enzyme one stage	0.586	0.064	1.200	IEP[f]	3.00	3.53	–	–
Two-stage extruded[f] No enzyme	0.566	0.062	1.280	IEP[f]	2.08	2.42	–	–

Note: ND: no data; LMH: assumed permeate flux in $l/m^2 \cdot h$; IEP: isoelectric precipitation; UF: ultrafiltration; SPI: soy protein isolate; SPC: soy protein concentrate.

[a] Campbell and Glatz 2009a.
[b] Kapchie et al. 2010a,b.
[c] de Moura et al. 2008.
[d] de Moura and Johnson 2009, 2011a, 2011c.
[e] Campbell and Glatz 2009b.
[f] Lawhon et al. 1981.

expensive process because it is dependent on more expensive cellulase and pectinase enzymes and uses a large quantity of sucrose, resulting in higher operating costs. The sucrose might be recycled and this would decrease the overall processing cost but this approach was not considered as data were not yet available when this study was performed. For the other processes, which led to the recovery of free oil, differences in selling prices of the skim proteins are more strongly influenced by the extraction yields of oil and protein rather than operating costs.

Protein product prices are also strongly influenced by the yields of skim protein that can be recovered by each process; not only does reduced yield affect revenue flow, unrecovered protein incurs a large disposal cost, which was measured as biological oxygen demand and total Kjeldahl nitrogen loading in wastewater treatment. It may be more economical to evaporate unrecovered protein and sugars and mix it back with the fiber fraction to be sold as animal feed, similar to what is commonly done with the steepwater in the corn wet-milling industry. Protein hydrolysis makes protein recovery more difficult, as discussed above, which results in the high product prices (because of reduced yield) in the protease-assisted extruded soybean processes. Still, some of the numbers obtained are promising. In 2008, the selling price for a typical SPC was $1.43 per kg (personal communication), which after adjusting for protein content, would be around $2 per kg for SPI, suggesting some of these processes may be competitive.

Capital and annual operating costs are shown in Table 16.4. Although the countercurrent two-stage extruded soybean process is the most capital intensive, the extra capital does not have major impact on protein selling prices. For ultrafiltration, however, capital and operating costs are large enough to impact prices and are strongly dependent on the permeate flux, which determines the necessary membrane area. There are limited data for fluxes for hydrolyzed extruded soy protein, and therefore,

TABLE 16.4

Capital Expenditures and Annual Operating Costs for Different Extraction Processes

Extraction Process	Direct Fixed Capital Cost ($ million)				Annual Operating Costs ($ million)			
	Extraction	IEP	UF	IEP + UF	Extraction	IEP	UF	IEP + UF
AEP flour	24.6	3.3	10.2	—	75.7	77.9	73	—
EAEP flour	24.6	—	—	—	74.8	—	—	—
Oil body	21.6	—	—	—	113.6	—	—	—
One-stage extruded	26.2	3.3	27.1	51.2	73.3	97.3	83.9	100.8
Two-stage extruded	33.8	3.3	—	—	—	101.6	—	—
Flux = 4 LMH	—	—	52	51.2	74.6	—	87.7	100.8
Flux = 2.5 LMH	—	—	82.8	—	—	—	97.3	—

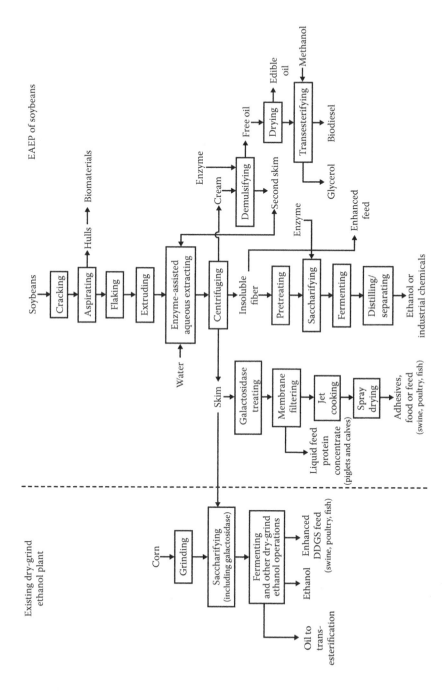

FIGURE 16.8 Integrated corn/soybean biorefinery concept.

two fluxes were studied. It is apparent that low permeate fluxes in protein recovery are potentially a major obstacle to commercialization of EAEP from extruded soybeans.

For all processes, the cost of soybeans was the largest single cost, which is common for commodity production. The 400 MT of soybean processed per day costs $51 million per year at $10 per bushel. Costs for protease enzymes used for extraction and demulsification, on the other hand, are relatively inexpensive, costing between $7 million and $9 million per year. For oil body extraction, the cellulase and pectinase enzymes necessary cost about $24 million per year, and the sucrose cost is $14 million, which pose more significant costs for this process. Although high water use may be a concern for this process in arid regions, the steam costs for drying the protein product after protein purification were relatively small, around $300,000 per year for isoelectric precipitation processes (assuming 50% moisture content after centrifugation). Steam costs for drying were considerably more for UF processes, which do not achieve the solids levels of isoelectric precipitation processes, ranging from $3.5 million to $5.5 million, although savings in steam cost for high solids UF feed was countered by an equal or greater increase in electricity usage for pumping, as well as an increase in capital cost and consumables for replacement membranes for the large area needed to accommodate reduced flux.

An order-of-magnitude economic analysis of various AEP and EAEP processes indicates that AEP/EAEP technologies may be commercially viable as a food-grade protein extraction process. However, costs associated with recovering hydrolyzed protein are greater than costs of recovering native protein and, in this case, outweigh the benefits gained by the use of enzymes needed to maximize free oil yields.

16.7   CONCLUSIONS

Future uses of EAEP of oilseeds seem very promising. Based on results obtained with soybeans, which has been the most extensively studied, strategies to overcome the major pitfalls of the protease- or cellulase/pectinase-based process have been identified and implemented. Feasibility of the process at a large scale (pilot plant) has been demonstrated and rendered the process more attractive to future commercial adoption. In addition, recent improvements in reducing water consumption, reducing enzyme use by recycling, and identifying means to add value to all downstream fractions are contributing to make the technology more attractive to commercial adoption. The concept of an integrated soybean/corn biorefinery, where each fraction can be valorized (Figure 16.8), will fit into the development of small biorefinery plants. With new local regulations such as in the state of California, where local laws prevent construction of new hydrocarbon solvent plants (Johnson 2008), and increasing environmental concerns regarding the use of organic solvents, alternative technologies to hexane extraction will likely become implemented by industry during the next decade. Life-cycle assessment of the process should be performed, which could be another important milestone in the development of EAEP. Processing technologies providing the right balance among environmental concerns, health/safety issues, and economic profits will constitute the next generation of vegetable oil extraction interventions.

ACKNOWLEDGMENTS

The authors would like to thank Dr. V. Kapchie for her review and useful comments. The authors acknowledge the financial support from USDA Cooperative State Research, Education, and Extension Service (USDA Special Research Grant 2008-34432-19325 and 2009-34423-20057); the Hatch Act; and the state of Iowa.

REFERENCES

Adler-Nissen, J. 1993. Proteases, in *Enzymes in Food Processing*, T. Nagodawithana and G. Reed (Eds.), pp. 159–199. New York: Academic Press.

Aguilera, J.M., Stanley D.W. 1999. *Microstructural principles of food processing and engineering*. Gaithersburg, MD: Aspen Publishers.

Armstrong, C. 1995. Longitudinal neuropsychological effects of *n*-hexane exposure: Neurotoxic effects versus depression. *Arch Clin Neuropsych* 10:1–19.

Balan, V., Rogers, C.A., Chundawat, S.P.S., Costa Sousa, L., Slininger P.A., Gupta, R., Dale, B.E. 2009. Conversion of extracted oil cake fibers into bioethanol including DDGS, canola, sunflower, sesame, soy, and peanut for integrated biodiesel processing. *J Am Oil Chem Soc* 86: 157–165.

Barregard, L., Sallsten, G., Nordborg, C., Gieth, W. 1991. Polyneuropathy possibly caused by 30 years of low exposure to *n*-hexane. *Scand J Work Environ Health* 17: 205–207.

Bigelis, R. 1993. Carbohydrases, in *Enzymes in Food Processing*, T. Nagodawithana and G. Reed (Eds.), pp. 121–147. New York: Academic Press.

Bocevska, M., Karlovic, D., Turkulov, J., Pericin, D. 1993. Quality of corn germ oil obtained by aqueous enzymatic extraction. *J Am Oil Chem Soc* 70:1273–1277.

Buchanan, B.B., Gruissem, W., Jones, R.L. (Eds.). 2000. *Biochemistry & Molecular Biology of Plants*. Oxford: Wiley-Blackwell.

Campbell, K.A., Glatz, C.E. 2009a. Mechanisms of aqueous extraction of soybean oil. *J Agr Food Chem* 57: 10904–10912.

Campbell, K.A., Glatz, C.E. 2009b. Protein recovery from enzyme-assisted aqueous extraction of soybean. *Biotechnol Progr* 26: 488–495.

Campbell, K.A., Glatz C.E., Johnson, L.A., Jung, S., de Moura, J.M.L.N., Kapchie, V., Murphy, P.A. 2011. Advances in aqueous extraction processing of soybeans. *J Am Oil Chem Soc* 88: 449–465.

Chabrand, R.M., Kim, H.J., Zhang, C., Glatz, C.E., Jung, S. 2008. Destabilization of the emulsion formed during aqueous extraction of soybean oil. *J Am Oil Chem Soc* 85: 383–390.

Chabrand, R.M., Glatz, C.E. 2009. Destabilization of the emulsion formed during the enzyme-assisted aqueous extraction of oil from soybean flour. *Enzyme Microb Technol* 45: 28–35.

Chen, B., McClements, D.J., Gray, D.A., Decker, E.A. 2010. Stabilization of soybean oil bodies by enzyme (Laccase) cross-linking of adsorbed beet pectin coatings. *J Agr Food Chem* 58: 9259–9265.

Chen, B.K., Diosady, L.L. 2003. Enzymatic aqueous processing of coconuts. *Int J Appl Sci Eng* 1: 55–61.

Chen, J., Rosenthal, A. 2009. Food processing, in *Food Science and Technology*, G. Campbell-Plat (Ed.), 207–246. West Sussex, UK: Wiley-Blackwell.

Chen, M.C.M., Chyan, C., Lee, T.T.T., Huang, S.-H., Tzen, J.T.C. 2004. Constitution of stable artificial oil bodies with triacylglycerol, phospholipid, and caleosin. *J Agr Food Chem* 52: 3982–3987.

Chen, Y., Ono, T. 2010. Simple extraction method of non-allergenic intact soybean oil bodies that are thermally stable in an aqueous medium. *J Agr Food Chem* 58: 7402–7407.

Chicago Board of Trade, http://www.cmegroup.com/trading/agricultural/index.html.

Corredor, D.Y., Sun, X.S., Salazar, J.M., Hohn, K.L., Wang, D. 2008. Enzymatic hydrolysis of soybean hulls using dilute acid and modified steam-explosion pretreatments. *J Biobased Mat Bioenergy* 2: 43–50.

Cornucopia Institute. 2009. *Behind the Bean: The Heroes and Charlatans of the Natural and Organic Soy Foods Industry*. Cornucopia, WI: Cornucopia Institute.

Dale, B.E., Weaver J., Byers, F.M. 1999. Extrusion processing for ammonia fiber explosion (AFEX). *Appl Biochem Biotechnol* 77–79, 35–45.

de Almeida, N.M., de Moura J.M.L.N., Johnson, L.A. 2010. Functional properties of protein produced by two-state aqueous countercurrent enzyme-assisted aqueous extraction. *American Oil Chemists' Society Annual Meeting Abstracts*, p. 130. Phoenix, AZ.

de Moura, J., Campbell, K., Mahfuz, A., Jung, S., Glatz, C.E., Johnson, L.A. 2008. Enzyme-assisted aqueous extraction of oil and protein from soybeans and cream de-emulsification. *J Am Oil Chem Soc* 85: 985–995.

de Moura, J.M.L.N., Johnson, L.A. 2009. Two-stage countercurrent enzyme-assisted aqueous extraction processing of oil and protein from soybeans. *J Am Oil Chem Soc* 86: 283–289.

de Moura, J.M.L.N., de Almeida, N.M., Johnson, L.A. 2009. Scale-up of enzyme-assisted aqueous extraction processing of soybeans. *J Am Oil Chem Soc* 86: 809–815.

de Moura, J.M.L.N., de Almeida, N.M., Jung, S., Johnson, L.A. 2010. Flaking as a pretreatment for enzyme-assisted aqueous extraction processing of soybeans. *J Am Oil Chem Soc* 87: 1507–1515.

de Moura, J.M.L.N., Maurer, D., Jung, S., Johnson, L.A. 2011a. Integrated two-stage countercurrent extraction and cream demulsification in the enzyme assisted aqueous extraction of soybeans. *J Am Oil Chem Soc* DOI 10.1007/S11746-011-J759-2.

de Moura, J.M.L.N., Maurer, D., Jung, S., Johnson, L.A. 2011b. Pilot-plant proof of concept for integrated, countercurrent, two-stage, enzyme-assisted aqueous extraction of soybeans. *J Am Oil Chem Soc* DOI 10.1007/S11746-011-1831-Y.

de Moura, J.M.L.N., Campbell, K., de Almeida, N.M., Glatz, C.E., Johnson, L.A. 2011c. Protein extraction and membrane recovery in enzyme-assisted aqueous extraction processing of soybeans. *J Am Oil Chem Soc* 88: 877–889.

de Moura, J.M.L.N., Campbell, K., de Almeida, N.M., Glatz, C.E., Johnson, L.A. 2011d. Protein recovery in aqueous extraction processing of soybeans using isoelectric precipitation and nanofiltration. *J Am Oil Chem Soc* DOI 10-1007/S11746-011-1803-2.

Fellows, P.J. 2009. Extrusion, in *Food Processing Technology: Principles and Practice*, 3rd edition, P.J. Fellow (ed.), pp. 294–308. Boca Raton: Woodhead.

Freitas, S.P., Hartman, L., Couri, S., Jablonka, F.H., de Carvalho, C.W.P. 1997. The combined application of extrusion and enzymatic technology for extraction of soybean oil. *Fett/Lipid* 99: 333–337.

Galvin J.B. 1997. Toxicity data for commercial hexane and hexane isomers, in *Technology and Solvents for Extracting Oilseeds and Nonpetroleum Oils*, P.J. Wan and P.J. Wakelyn (Eds.), pp. 75–85. Champaign, IL: AOCS Press.

Goldsmith, P.D. 2008. Economics of soybean production, marketing, and utilization, in *Soybeans: Chemistry, Production Processing, and Utilization*, L. Johnson, P.J. White and R. Galloway (Eds.), pp. 117–150. Urbana, IL: AOCS Press.

Guth, J., Ferguson, J. 2009. Controlled release of active agents from oleosomes, Patent WO 2009126301.

Han, X., Cheng, L., Zhang, R., Bi, J. 2009. Extraction of safflower seed oil by supercritical CO_2. *J Food Eng* 92: 370–376.

Hron, R.J. 1982. Renewable solvents for vegetable oil extraction. *J Am Oil Chem Soc* 59: 229–242.

Hron, R.J. 1997. Acetone, in *Technology and Solvents for Extracting Oilseeds and Nonpetroleum oils*, P.J. Wan and P.J. Wakelyn (Eds.), pp. 186–191. Champaign, IL: AOCS Press.

Iwanaga, D., Gray, D.A., Fisk, I.D., Decker, E.A., Weiss, J., McClements, D.J. 2007. Extraction and characterization of oil bodies from soy beans: A natural source of pre-emulsified soybean oil. *J Agr Food Chem* 55: 8711–8716.

Iwanaga, D., Gray, D.A., Decker, E.A., Weiss, J., McClements, D.J. 2008. Stabilization of soybean oil bodies using protective pectin coatings formed by electrostatic deposition. *J Agr Food Chem* 56: 2240–2245.

Jiang, L., Wang, Z., Xu, S. 2010. Aqueous enzymatic extraction of peanut oil and protein hydrolysates. *Food Bioprod Process* 88: 233–238.

Johnson, L.A., Lusas, E.W. 1983. Comparison of alternative solvents for oils extraction. *J Am Oil Chem Soc* 60: 229–242.

Johnson L.A. 1997. Theoretical, comparative and historical analyses of alternative technologies for oilseed extraction, in *Technology and Solvents for Extracting Oilseeds and Nonpetroleum Oils*, P.J. Wan and P.J. Wakelyn (Eds.), pp. 4–47. Champaign, IL: AOCS Press.

Johnson, L.A. 2008. Oil recovery from soybeans, in *Soybeans: Chemistry, Production, Processing and Utilization*, L.A. Johnson, P.J. White, and R. Galloway (Eds.), pp. 331–376. Champaign, IL: AOCS Press.

Johnson, L.A. 2000. Recovery of fats and oils from plant and animal sources, in *Introduction to Fats and Oils Technology*, R.D. O'Brien, W.E. Farr, P.J. Wan (Eds.), pp. 108–135. Champaign, IL: AOCS Press.

Jung, S. 2009. Aqueous extraction of oil and protein from soybean and lupin: A comparative study. *J Food Proc Pres* 33: 547–559.

Jung, S., Murphy, P.A., Johnson, L.A. 2005. Physicochemical and functional properties of soy protein substrates modified by low levels of protease hydrolysis. *J Food Sci* 70: C180–C187.

Jung, S., Mahfuz, A.A. 2009. Low temperature dry extrusion and high-pressure processing prior to enzyme-assisted aqueous extraction of full fat soybean flakes. *Food Chem* 114: 947–954.

Jung, S., Mahfuz, A., Maurer, D. 2009a. Structure, protein interactions and in vitro protease accessibility of extruded and pressurized full-fat soybean flakes. *J Am Oil Chem Soc* 86: 475–483.

Jung, S., Maurer, D., Johnson, L.A. 2009b. Factors affecting emulsion stability and quality of oil recovered from enzyme-assisted aqueous extraction of soybeans. *Bioresour Technol* 100: 5340–5347.

Kapchie, V.N., Wei, D., Hauck, C., Murphy, P.A. 2008. Enzyme-assisted aqueous extraction of oleosomes from soybeans (Glycine max). *J Agr Food Chem* 56: 1766–1771.

Kapchie, V.N., Towa, L.T., Hauck, C., Murphy, P.A. 2010a. Evaluation of enzyme efficiency for soy oleosome isolation and ultrastructural aspects. *Food Res Int* 43: 241–247.

Kapchie, V.N., Towa, L.T., Hauck, C., Murphy, P.A. 2010b. Recycling of aqueous supernatants in soybean oleosome isolation. *J Am Oil Chem Soc* 87: 223–231.

Kapchie, V.N., Hauck, C., Wang, H., Murphy, P.A. 2011a. Improvement in pilot-plant extraction of oleosome and isoflavone partitioning. *J Food Sci* (In press).

Kapchie, V.N., Towa, L.T., Hauck, C., Murphy, P.A. 2011b. Recovery and functional properties of soy storage proteins from lab- and pilot-plant scale oleosome production. *J Am Oil Chem Soc* (submitted).

Karki, B., Maurer, D., Kim, T.H., Jung, S. 2011a. Comparison and optimization of enzymatic saccharification of soybean fibers recovered from aqueous extractions. *Bioresour Technol* 102: 1228–1233.

Karki, B., Maurer, D., Jung, S. 2010b. Efficiency of pretreatments for optimal enzymatic saccharificationof soybean fiber. *Bioresour Technol* 102: 6522–6528.

Karki, B., Maurer, D., Kim, T.H., Jung, S. 2011c. Pretreatment of soybean fiber by soaking in aqueous ammonia prior to saccharification. Abstract # 42954, 102nd AOCS Annual meeting & Expo, May 1–4, Cincinnati, OH, USA.

Kasai, N., Murata, A., Inui, H., Sakamoto, T., Kahn, R.I. 2004. Enzymatic high digestion of soybean milk residue (Okara). *J Agr Food Chem* 52: 5709–5716.

Kim, S., Holtzapple, M.T. 2005. Lime pretreatment and enzymatic hydrolysis of corn stover. *Bioresour Technol* 96: 1994–2006.

King, J.W. 1997. Critical fluids for oil extraction, in *Technology and Solvents for Extracting Oilseeds and Nonpetroleum Oils*, P.J. Wan and P.J. Wakelyn (Eds.), pp. 283–310. Champaign, IL: AOCS Press.

Kinsella, J.E. 1979. Functional properties of soy proteins. *J Am Oil Chem Soc* 56: 242–258.

Lamsal, B.P., Murphy, P.A., Johnson, L.A. 2006. Flaking and extrusion as mechanical treatments for enzyme-assisted aqueous extraction of oil from soybeans. *J Am Oil Chem Soc* 83: 973–979.

Lamsal, B.P., Johnson, L.A. 2007. Separating oil from aqueous extraction fractions of soybean. *J Am Oil Chem Soc* 84: 785–792.

Lamsal, B., Yoo, J., Brijwani, K., Alavi, S. 2010. Extrusion as a thermo-mechanical pre-treatment for lignocellulosic ethanol. *Biomass Bioenerg* 34: 1703–1710.

Latif, S., Diosady, L.L., Anwar, F. 2008. Enzyme-assisted aqueous extraction of oil and protein from canola (*Brassica napus* L.) seeds. *Eur J Lipid Sci Tech* 110: 887–892.

Latif S., Anwar, F. 2009. Effect of aqueous enzymatic processes on sunflower oil quality. *J Am Oil Chem Soc* 86: 393–400.

Latif, S., Makkar, H.P.S., Becker, K. 2010. Aqueous enzyme-assisted oil and protein extraction from *Jatrophacurcas* kernels, in *American Oil Chemists' Society Annual Meeting Abstracts*, p. 121. Phoenix, AZ.

Lawhon, J.T., Rhee, K.C., Lusas, E.W. 1981. Soy protein ingredients prepared by new processes—Aqueous processing and industrial membrane isolation. *J Am Oil Chem Soc* 58: 377–384.

Lomboy-Montalbo, M., Johnson, L., Khanal, S.K., van Leeuwen, J., Grewell, D. 2010. Sonication of sugary-2 corn: A potential pretreatment to enhance sugar release. *Bioresour Technol* 101: 351–358.

Lloyd, T.A., Wyman, C.E. 2005. Combined sugar yields for dilute sulfuric acid pretreatment of corn stover followed by the enzymatic hydrolysis of remaining solids. *Bioresour Technol* 86: 1967–1977.

Lusas, E.W., Hernandez, E. 1997. Isopropyl alcohol, in *Technology and Solvents for Extracting Oilseeds and Nonpetroleum Oils*, P.J. Wan and P.J. Wakelyn (Eds.), pp. 199–266. Champaign, IL: AOCS Press.

Lusas, E.W., Lawhon, J.T., Rhee, K.C. 1982. Producing edible oil and protein from oilseeds by aqueous processing. *Oil Mill Gaz* 10: 28–34.

Majoni, S., Wang, T., Johnson L.A. 2011. Enzyme treatments to enhance oil recovery from condensed corn distillers solubles. *J Am Oil Chem Soc.* 88: 523–532.

Mielenz, J.R., Bradsley, J.S., Wyman, C.E. 2009. Fermentation of soybean hulls to ethanol while preserving protein value. *Bioresour Technol* 100: 3532–3539.

Moreau, R.A. 2011. Aqueous enzymatic oil extraction from seeds, fruits, and other oil-rich plant materials, in *Alternatives to Conventional Food Processing*, A. Proctor (Ed.), pp. 341–366. RSC Publishing, Cambridge, UK.

Moreau, R.A., Johnston, D.B., Powell, M.J., Hicks, K.B. 2004. A comparison of commercial enzymes for the aqueous enzymatic extraction of corn oil from corn germ. *J Am Oil Chem Soc* 81: 1071–1075.

Moreau, R.A., Dickey, L.C., Johnston, D.B., Hicks, K.B. 2009. A process for the aqueous enzymatic extraction of corn oil from dry milled corn germ and enzymatic wet milled corn germ (E-germ). *J Am Oil Chem Soc* 86: 469–474.

Mosier, N., Wyman, C., Dale, B., Elander, R., Lee, Y.Y., Holtzapple, M., Ladish, M. 2005. Features of promising technologies for pretreatment of lignocellulosic biomass. *Bioresour Technol* 96: 673–686.

Muntada, V., Gerschenson, L.N., Alzamora, S.M., Castro, M.A. 1998. Solute infusion effects on texture of minimally processed kiwifruit. *J Food Sci* 63: 616–620.

Nikiforidis, C.V., Kiosseoglou, V. 2009. Aqueous extraction of oil bodies from maize germ (Zea mays) and characterization of the resulting natural oil-in-water. *J Agr Food Chem* 57: 5591–5596.

Nitayavardhana, S., Shrestha, P., Rasmussen, M.L., Lamsal, B.P., van Leeuwen, J., Khanal, S.K., 2010. Ultrasound improved ethanol fermentation from cassava chips in cassava-based ethanol plants. *Bioresour Technol* 101: 2741–2747.

Nguyen, L., Barone, S.J., Macchio, R. 2010. Cosmetic compositions comprising a hydrosy-propyl methyl cellulose derivative, Patent WO 2010103008.

Nyam, K.L., Chin, P.T., Lai, O.M., Kamariah, L., Man, B.C. 2009. Enzyme-assisted aqueous extraction of Kalahari melon seed oil: Optimization using response surface methodology. *J Am Oil Chem Soc* 86: 1235–1240.

Owusu-Ansal, Y.J. 1997. Enzyme-assisted extraction, in *Technology and Solvents for Extracting Oilseeds and Nonpetroleum Oils*, P.J. Wan and P.J. Wakelyn (Eds.), pp. 323–332. Champaign, IL: AOCS Press.

Parkin, K.L. 1993. General Characteristics of Enzymes, in *Enzymes in Food Processing*. T. Nagodawithana and G. Reed (Eds.), pp. 7–37. New York: Academic Press.

Purkrtova, Z., Jolivet, P., Miquel, M., Chardot, T. 2008. Structure and function of seed lipid body-associated proteins. *CR Biologies* 331: 746–754.

Qi, M., Hettiarachchy, N.S., Kalapathy, U. 1997. Solubility and emulsifying properties of soy protein isolates modified by pancreatin. *J Food Sci* 62:1110–1115.

Ranalli, A., Pollastri L., Contento, S., Lucera, L., Del Re, P. 2003. Enhancing the quality of virgin olive oil by use of a new vegetable enzyme extract during processing. *Eur Food Res Technol* 216: 109–115.

Rosenthal, A., Pyle, D.L., Niranjan, K. 1998. Simultaneous aqueous extraction of oil and pro-tein from soybean: Mechanisms for process and design. *Trans Food Bioprod Process* 76: 224–230.

Rosenthal, A., Pyle, D.L., Niranjan, K., Gilmour, S., Trinca, L. 2001. Combined effect of operational variables and enzyme activity on aqueous enzymatic extraction of oil and protein from soybean. *Enzyme Microb Technol* 28: 499–509.

Rosenthal, A., Pyle, D.L., Niranjan, K. 1996. Aqueous and enzymatic processes for edible oil extraction. *Enzyme Microb Technol* 19: 402–420.

Russin, T.A., Boye, J.I., Arcand, Y., Rajamohamed, S.H. 2011. Alternative techniques for defatting soy: A practical review. *Food Bioprocess Technol* 4: 200–223.

Seth, S., Agrawal, Y.C., Ghosh, P.K., Jayas, D.S., Singh, B.P.N. 2007. Oil extraction rates of soya bean using isopropyl alcohol as solvent. *Biosyst Eng* 97: 209–217.

Seth, S., Agrawal, Y.C., Ghosh, P.K., Jayas, D.S. 2010. Effect of moisture content on the qual-ity of soybean oil and meal extracted by isopropyl alcohol and hexane. *Food Bioprocess Technol* 3: 121–127.

Sharma A., Khare, S.K., Gupta, M.N. 2001. Enzyme-assisted aqueous extraction of rice bran oil. *J Am Oil Chem Soc* 78: 949–951.

Soystats, http://soystats.com/2010/page_35.htm (accessed October 2010).

Stein, H.H., Berger L.L., Drackley, J.K., Fahey, G.C., Hernot, D.C., Parsons, C.M. 2008. Nutritional properties and feeding values of soybeans and their coproducts, in *Soybeans: Chemistry, Production, Processing and Utilization*, L.A. Johnson, P.J. White, and R. Galloway (Eds.), pp. 613–660. Champaign, IL: AOCS Press.

Takeuchi, T.M., Pereira, C.G., Braga, M.E.M., Jr., Marostica, M.R., Leal, P.F., Meireles, M.A.A. 2009. Low-pressure solvent extraction (solid–liquid extraction, microwave assisted, and ultrasound assisted) from condimentary plants, in *Extracting Bioactive Compounds for Food Products: Theory and Applications*, M. Angela and A. Meireles (Eds.), pp. 138–211. Boca Raton: CRC Press.

Teymouri, F., Laureano-Perez, L., Alizadeh, H., Dale, B.E. 2005. Optimization of the ammonia-fiber explosion (AFEX) treatment parameters for enzymatic hydrolysis of corn stover. *Bioresour Technol* 96: 2014–2018.

Towa, L.T., Kapchie, V.N., Hauck, C., Wang, H., Murphy, P.A. 2010a. Pilot plant recovery of soybean oleosome fractions by an enzyme-assisted aqueous process. *J Am Oil Chem Soc* 88: 733–741.

Towa, L.T., Kapchie, V.N., Hauck, C., Murphy, P.A. 2010b. Enzyme-assisted aqueous extraction of oil from isolated oleosomes of soybean flour. *J Am Oil Chem Soc* 87: 347–354.

Tzen, J.T.C., Cao, Y.-Z., Laurent, P., Ratnayake, C., Huang, A.H.C. 1993. Lipids, proteins, and structure of seed oil bodies from diverse species. *Plant Physiol* 101: 267–276.

Wakelyn, P.J. 1997. Regulatory considerations for extraction solvents for oilseed and other non-petroleum oils, in *Technology and Solvents for Extracting Oilseeds and Nonpetroleum Oils*, P.J. Wan and P.J. Wakelyn (Eds.), pp. 4–47. Champaign, IL: AOCS Press.

Wankat, P.C. 2007. *Separation Process Engineering*, 2nd edition, pp. 424–459. Boston: Prentice-Hall.

White, D.A., Fisk, I.D., Mitchell, J.R., Wolf, B., Hill, S.E., Gray, D.A. 2008. Sunflower-seed oil body emulsions: Rheology and stability assessment of a natural emulsion. *Food hydrocolloid* 22: 1224–1232.

Wu, J., Johnson, L.A., Jung, S. 2009. Demulsification of oil-rich emulsion from enzyme-assisted aqueous extraction of extruded soybean flakes. *Bioresour Technol* 100: 527–533.

Yang, S., Lio, J.Y., Wang, T. 2011. Solid state fermentation of soybean fiber and dried dis-tiller's grains. *Appl Biochem Biotech* (submitted).

Yao L., Jung, S. 2010. ^{31}P NMR phospholipid profiling of soybean emulsion recovered from aqueous extraction. *J Agr Food Chem* 58: 4866–4872.

Yao, L., Wang, T., Wang, H. 2011. Integration of soybean extraction with dry-grind corn fermentation for ethanol production. *Bioresour Technol* (submitted).

Zhang, S.B., Wang, Z., Xu, S.Y. 2007. Optimization of the aqueous enzymatic extraction of rapeseed oil and protein hydrolysates. *J Am Oil Chem Soc* 84: 97–105.

Zhao, Y., Wang, Y., Zhu, J.Y., Ragauskas, A., Deng, Y. 2008. Enhanced enzymatic hydrolysis of spruce by alkaline pretreatment at low temperature. *Biotechnol Bioeng* 99: 1320–1328.

Index